Y0-ARO-519

Texts and Monographs in Physics

Series Editors:

R. Balian, Gif-sur-Yvette, France
W. Beiglböck, Heidelberg, Germany
H. Grosse, Wien, Austria
E. H. Lieb, Princeton, NJ, USA
N. Reshetikhin, Berkeley, CA, USA
H. Spohn, München, Germany
W. Thirring, Wien, Austria

Kurt Bernardo Wolf

Geometric Optics on Phase Space

With 83 Figures

Professor Kurt Bernardo Wolf
Universidad Nacional Autónoma de Mexico
Centro de Ciencias Físicas
Apartado Postal 48-3
Cuernavaca, Morelos 62251, Mexico

ISBN 3-540-22039-9 Springer Berlin Heidelberg New York

Library of Congress Control Number: 2004105046

This work is subject to copyright. All rights are reserved, whether the whole or part of the material is concerned, specifically the rights of translation, reprinting, reuse of illustrations, recitation, broadcasting, reproduction on microfilm or in any other way, and storage in data banks. Duplication of this publication or parts thereof is permitted only under the provisions of the German Copyright Law of September 9, 1965, in its current version, and permission for use must always be obtained from Springer. Violations are liable to prosecution under the German Copyright Law.

Springer is a part of Springer Science+Business Media

springeronline.com

© Springer-Verlag Berlin Heidelberg 2004
Printed in Germany

The use of general descriptive names, registered names, trademarks, etc. in this publication does not imply, even in the absence of a specific statement, that such names are exempt from the relevant protective laws and regulations and therefore free for general use.

Typesetting by the author using LATEX
Final typesetting by Frank Herweg, Hirschberg-Leutershausen
Cover design: *design & production* GmbH, Heidelberg
Printed on acid-free paper 55/3141/rm 5 4 3 2 1 0

I dedicate this work to
Marcos Moshinsky and Yuval Ne'eman

who introduced me to symmetry

Preface

Optics is the physical science that best mirrors pure geometry. Light defines the metric of spacetime in the large, while in the infinitesimal it leads to the notions of Hamiltonian flow of phase space and of Lie evolution. Geometry is not mere land measurement, as its Greek root implies, nor only ray tracing with a good computer design program. In this volume we use the tactics of Lie algebras and groups to understand the geometric model of light. For readers with expertise in optics we indicate the path of symmetry methods to enter phase space; for theoreticians, optics will be presented as a particularly transparent realization of symplectic geometry, with models even richer than those of mechanics. Here we render geometric optics on the canvas of phase space with the tools, strategy and color of Lie algebras and groups.

It may have been a historical accident that the theory of Lie algebras and groups became occupied first with relativity and with quantum mechanics. Hermann Weyl and Eugene Wigner, among many others during the twentieth century, worked extensively with Lie methods to solve problems in the coupling of angular momenta for atomic and nuclear spectroscopy; they gradually refined these methods to the point of successfully tackling fundamental questions of Nature in elementary particle physics. Few physical systems are perfect though; among those that are (beside vacuum) we count the Kepler-Bohr model and the harmonic oscillator. The shape of their orbits (and the degeneracy of their quantum levels) point to a larger symmetry group behind the scenes; their scaling (and spacing between levels) implies still higher structures that encompass dynamics. The optical counterparts of these 'perfect' systems are the Maxwell fish-eye and the paraxial harmonic guide – both the geometric and the wave models.

The spacetime of special relativity became known to the learned community within a decade of its discovery; in contrast, phase space has taken more than a century to percolate our culture. Its symplectic metric is imbricate, but it is there that our senses operate (particularly hearing) and where Symmetry manifests herself. It should be emphasized that the Hamiltonian evolution paradigm is favored by Lie methods, while traditional optical methods, developed for image-forming devices, favor the Lagrangian construction which fixes the initial and final positions of rays.[1] The Hamilton-Lie formulation leads to a systematic and complete classification of aberrations of phase space (that is *distinct* from the scheme of Seidel – still the most common in the literature), and also allows the concatenation of aberrating subsystems through their ordered multiplication as elements of a group represented by matrices.

The text consists of four parts divided into 15 chapters, with the structure of a tree: in part I we present the basic concepts of optical phase space, Hamiltonian systems, and Lie algebras, rooted on two fundamental postulates. Parts II, III and IV cover the main branches of global, paraxial and metaxial optical models. One appendix gathers historical notes and two others establish contact with the homomorphic wave models of parts II and III. The plan of each chapter is laid out in a brief introduction so the reader can follow or skip one according to his interest. There is also a multitude of leaves (applications, ancillary notes, disgressions) that should not obstruct the view of a first reading; these are compacted into remarks and examples that end with a bullet. Cross-references, the subject index and the bibliography should be of assistance both to the browser and the methodical reader.

This book is dedicated to Professor Marcos Moshinsky (Instituto de Física UNAM) who introduced me to symmetries via nuclear collective models during my formative years, and to Professor Yuval Ne'eman (University of Tel-Aviv) who did so with elementary particle dynamics. I could not have guessed then that the same mathematical beings are manifest in the subtler world of geometric optics.

I thank my colleagues for their interest and collaboration in many of the topics that follow; and I must apologize beforehand for perhaps not citing all the important papers that nourished this book. I am indebted with Charles P. Boyer and Stanly Steinberg (University of New Mexico, Albuquerque) for company in exploring essential subjects of group theory, and who later pointed out to me the use of Lie series in optical models; Alex J. Dragt and Etienne Forest (University of Maryland, College Park) who eased me into the practice of phase space aberrations; Tetsundo Sekiguchi (University of Arkansas, Fayetteville) who shared my delight in the simple postulates of the geometric model of light; Natig M. Atakishiyev (Instituto de Matemáticas, UNAM, Cuernavaca) and Wolfgang Lassner (Naturwissenschaftlich-Theoretisches Zentrum, Leipzig) with whom I started work on relativistic aberrations and problems further afield; Sergey M. Chumakov, Andrei B. Klimov (Universidad de Guadalajara, México), S. Twareque Ali (Concordia University, Montréal) and again Natig M. Atakishiyev for work on the Wigner function on phase space for finite optical

[1] Many authors refer to the latter as *Hamiltonian*. Perhaps the best tribute to William Rowan Hamilton is to say that too many theories are associated with his name.

models; Vladimir I. Man'ko (P.N. Lebedev Physical Institute, Moscow) for driving company in the Euclidean and paraxial régimes; Darryl D. Holm (Los Alamos National Laboratory) who cut the hyperbolic onion with his Hamiltonian knife; Octavio Castaños, Alejandro Frank (Instituto de Ciencias Nucleares, UNAM, México DF) and François Leyvraz (Centro de Ciencias Físicas, UNAM, Cuernavaca) for translating the hydrogen atom into optics; David Mendlovic (University of Tel-Aviv) and Haldun M. Ozaktas (Bilkent University, Ankara) for kindling our interest in the fractional Fourier transform; R. Simon (The Institute of Mathematical Sciences, Chennai, India) for clarifying the structure of paraxial optical systems, and Sameen A. Khan (Middle East College of Information Technology, Muscat, Oman) for exploring their orbits; Miguel A. Alonso (University of Rochester, New York) and Gregory S. Forbes (Macquarie University, Sydney) for a further foray into Helmholtz phase space; and George S. Pogosyan (Joint Institute for Nuclear Research, Dubna, Russia and CCF-UNAM, Cuernavaca) for work yet to be translated into the language of light.

Special thanks are due to Guillermo Krötzsch (CCF-UNAM) because he provided the figures for this book, gave strong support with symbolic computation, and voiced the common sense indispensable for this venture. Students are the relish of research; I had the pleasure to work at various times with Miguel Navarro Saad, Enrique López Moreno, Ana Leonor Rivera, and presently with Luis Edgar Vicent. Work was done at the Instituto de Investigaciones en Matemáticas Aplicadas y en Sistemas, Unidad Cuernavaca, which in 1999 became part of the Centro de Ciencias Físicas, Universidad Nacional Autónoma de México. Activities connected with this book were funded by the Dirección General de Asuntos del Personal Académico, UNAM, through the *Óptica Matemática* serial PAPIIT projects, and by the Consejo Nacional de Ciencia y Tecnología in various ways.

Cuernavaca, January 2004 *Kurt Bernardo Wolf*

Contents

Part I Optical phase space, Hamiltonian systems and Lie algebras

Part I

Optical phase space, Hamiltonian systems and Lie algebras

Introduction

Chapter 1 begins with two basic postulates, one *geometrical* and the second *dynamical*, which define the Hamiltonian model of geometric light. It is a model because the physical phenomenon is abstracted, reduced in complexity, and treated with attention to the geometric properties of oriented lines that are subject to influence by generally inhomogeneous transparent media. The two Hamilton equations follow directly from these postulates. They are given in differential form in Chap. 2, and subjected to canonical transformations in Chap. 3 in a way that serves to introduce continuous Lie groups and their algebras. In Chap. 4 we return to the two postulates for the case of discontinuous media, where the Hamilton equations involve finite differences and the conservation laws for refraction and reflection.

The three subjects in the title of this part have been known by the scientific community in various fields for more than a century, and basic ideas can be traced back to the Greeks. In particular, the sine law of refraction commonly ascribed to Huygens, Descartes and/or Fermat, was found as a geometric construction – implying conservation laws – by Ibn Sahl in Baghdad, more than a thousand years ago. It is an impossible task to trace out all reference lines with justice, and at the same time guide amicably through shortcuts and anecdotes. In the appendix we present a brief but hopefully enjoyable historical essay on the manifold paths of geometric optics, on the Hamiltonian understanding of phase space, and on the life and work of Sophus Lie that closes the nineteenth century. The developments of the twentieth century are more difficult to summarize, so we shall be content to recall the realization that the dynamics of a system is part of its symmetry, when seen from a higher vantage point over phase space.

1 Two fundamental postulates

In geometric optics we model light by *rays*, i.e. by oriented lines in space, which obey two postulates that we deem evident. This formulation leads directly to the laws of refraction in the form of the two Hamilton equations. These equations generate the rays as trajectories in phase space, both in optics and in mechanics.

1.1 Geometric postulate

We denote a *ray* by a 3-vector of functions $\vec{q}(s) = \begin{pmatrix} q_x(s) \\ q_y(s) \\ q_z(s) \end{pmatrix}$ of the *arc length* parameter $s \in \mathsf{R}$, and require that it satisfy the following:

Geometric postulate: *Rays are continuous and piecewise differentiable.*

This means that the lines $\vec{q}(s)$ are connected and have tangent vectors $d\vec{q}(s)/ds$ *almost everywhere.*[1] Because s is arc length, $|d\vec{q}| = ds$.

We introduce now the *momentum* 3-vector $\vec{p}(s)$: it is a vector *parallel* to the tangent of the ray,

$$\vec{p}(s) = \begin{pmatrix} p_x(s) \\ p_y(s) \\ p_z(s) \end{pmatrix} \quad || \quad \frac{d\vec{q}(s)}{ds}, \tag{1.1}$$

whose norm will be determined below. As shown in Fig. 1.1, in an interval of smoothness of $\vec{q}(s)$ between s and $s + \Delta s$, the Mean Value theorem assures us that there exists a point s' where the increment vector[2] $\Delta\vec{q}(s) := \vec{q}(s + \Delta s) - \vec{q}(s)$ is parallel to $\vec{p}(s')$, and its magnitude is approximated[3] by the arc length increment Δs, as

$$\Delta\vec{q}(s) \approx \frac{\vec{p}(s')}{|\vec{p}(s')|}\Delta s. \tag{1.2}$$

---

[1] I.e., right- and left-derivatives exist at all $s \in \mathsf{R}$, and they are equal except possibly at isolated points, where the right- and left-derivatives may differ by a finite amount. There, the line *breaks* (i.e., *refracts*), but does not disconnect.

[2] We write ':=' for *is defined by*, and =: for *defines*.

[3] We write '≈' when the two sides differ at most by powers of the parameter higher than a certain order implied by the context.

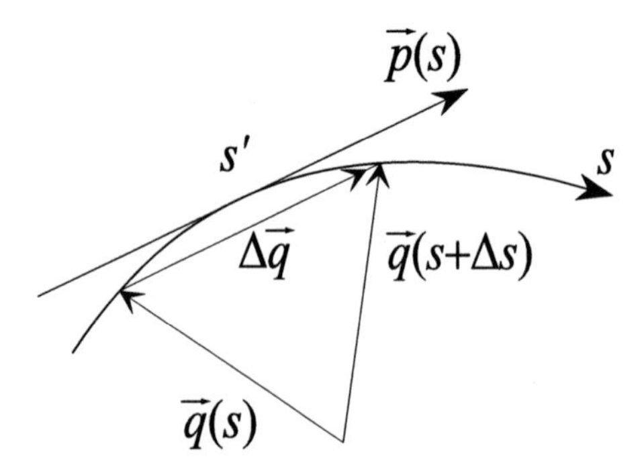

Fig. 1.1. A ray is a smooth line $\vec{q}(s)$, $s \in \mathrm{R}$. A secant vector $\Delta\vec{q} = \vec{q}(s+\Delta s) - \vec{q}(s)$ is parallel to the tangent vector of momentum $\vec{p}(s)$ at some point s' in the interval $[s, s+\Delta s]$.

Dividing by the arc length increment, the relation becomes an exact differential in the limit $\Delta s \to 0$, and yields the *first Hamilton* equation:

$$\frac{d\vec{q}(s)}{ds} = \frac{\vec{p}(s)}{|\vec{p}(s)|} = \frac{\partial H(\vec{p}, \vec{q})}{\partial \vec{p}}, \tag{1.3}$$

where the function $H(\vec{p}, \vec{q})$ that appears in the last term is called *the Hamiltonian*, and is determined by the middle term to have the form $H(\vec{p}, \vec{q}) = |\vec{p}| +$ any scalar function of $\vec{q}$.

1.2 Dynamic postulate

The second postulate is about the response of light rays to the inhomogeneity of an optical medium by changes in their momentum vectors. We assume that the complete characterization of the medium is given by a single scalar function, called the *refractive index* of the medium, $n(\vec{q})$, which depends only on the position $\vec{q} \in \mathrm{R}^3$.[4] For mathematical precision we must also assume that this index $n(\vec{q})$ is a *region-wise smooth* function, i.e., has a finite gradient 3-vector $\nabla n(\vec{q}) = \partial n(\vec{q})/\partial \vec{q}$ almost everywhere (except possibly at smooth interfaces between any two such regions).

In this context we propose the second,

Dynamic postulate: *The momentum of the ray changes along the arc length in proportional response to the local gradient of the refractive index.*

As suggested in Fig. 1.2, within a smooth region of the medium, the momentum increment vector $\Delta\vec{p}(s) := \vec{p}(s + \Delta s) - \vec{p}(s)$ is approximated by the gradient of the refractive index, $\nabla n(\vec{q}(s')) = \partial n(\vec{q})/\partial \vec{q}|_{s'}$, for some intervening point s' of the ray, times the arc length increment Δs,

[4] In Chap. 8 we shall relax this condition to study *anisotropic* media, where the refractive index may also depend on ray direction.

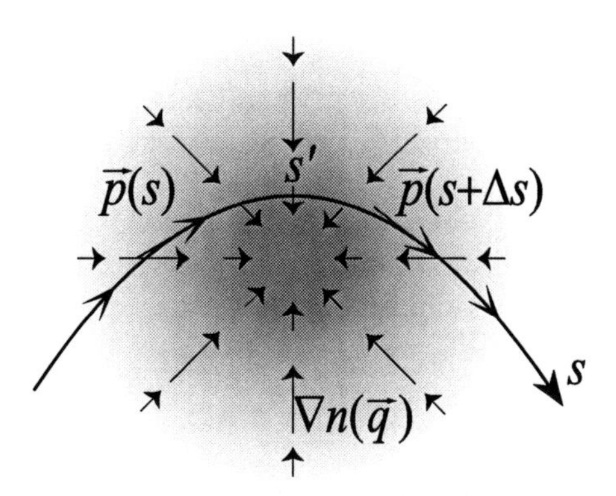

Fig. 1.2. The change of ray momentum is along the local gradient of the refractive index. Its increment $\Delta\vec{p} = \vec{p}(s+\Delta s) - \vec{p}(s)$ is parallel to the gradient vector $\nabla n(\vec{q}(s'))$ for some point s' of the ray in the interval $[s, s+\Delta s]$.

$$\Delta\vec{p}(s) \approx \nabla n(\vec{q}(s'))\,\Delta s. \tag{1.4}$$

We set the scale of n by choosing the proportion between the two members of the expression to be unity. Note that (1.4) implies $\Delta\vec{p}(s)\times\nabla n(\vec{q}(s')) \approx \vec{0}$.

Dividing (1.4) by the arc length increment, in the limit $\Delta s \to 0$ we obtain the *second Hamilton* equation:

$$\frac{d\vec{p}(s)}{ds} = \frac{\partial n(\vec{q})}{\partial \vec{q}} = -\frac{\partial H(\vec{p},\vec{q})}{\partial \vec{q}}. \tag{1.5}$$

The Hamiltonian function written in the last term must have the form $H(\vec{p},\vec{q}) = -n(\vec{q}) +$ scalar function of $\vec{p}$ [cf. (1.3)], and hence is

$$H(\vec{p},\vec{q}) := |\vec{p}(s)| - n(\vec{q}(s)), \tag{1.6}$$

up to a constant that serves to fix the base value $n = 1$ for vacuum.

Example: Homogeneous medium. An optical medium which appears the same after any translation is called homogeneous. Its refractive index is $n(\vec{q}) =$ constant. With no gradient to respond to, momentum vectors are constant along s, and the rays are straight lines:

$$\vec{q}(s) = \vec{q}(0) + s\frac{\vec{p}}{|\vec{p}|}, \quad \vec{p} = \vec{p}(0), \qquad s \in \mathsf{R}. \tag{1.7}$$

•

1.3 Conservation laws at discontinuities

Assume that the refractive index of a composite optical medium $n(\vec{q})$ has a surface S of finite discontinuity $S(\vec{q}) = 0$ between two regions of smoothness.

The Hamilton differential equations, found with the limit $\Delta s \to 0$ in the previous two sections, may not hold then. Here we show that the geometric and dynamic postulates can be turned into conservation laws for position and momentum, respectively.

Consider a ray $\vec{q}(s)$ which crosses the surface S with a value $\bar{s}$ that is between s_- in the first region and s_+ in the second. Then the limit $\Delta s = s_+ - s_- \to 0$ of (1.2) is valid and leads to the conservation of the position vectors $\vec{q}(s)$ at both sides $s_\pm \to \bar{s}_\pm$ of the impact point $\vec{q}(\bar{s})$,

$$\vec{q}(\bar{s}_-) = \vec{q}(\bar{s}_+). \tag{1.8}$$

This *first* conservation law follows from the geometric postulate alone, and states that rays are *connected* lines.

To derive a second conservation law at discontinuities of the medium from the dynamic postulate (1.4), we must explicitly assume that the interfaces between smooth R^3-regions of the medium are smooth R^2-surfaces $S(\vec{q}) = 0$, and so have a well-defined *normal* vector $\nabla S(\vec{q})|_S$. Although the gradient of the refractive index $\nabla n(\vec{q})$ becomes infinite at discontinuities of $n(\vec{q})$, its *direction* is well defined: it is parallel to $\nabla S(\vec{q})|_S$. As before, we set s_- and s_+ on opposite sides of the impact point $\bar{s}$; when $\Delta s \to 0$ and $s_\pm \to \bar{s}_\pm$, the vector increment $\Delta\vec{p} := \vec{p}(\bar{s}_+) - \vec{p}(\bar{s}_-)$ will now remain finite and satisfy $\Delta\vec{p}(\bar{s}) \times \nabla S(\vec{q})|_S = \vec{0}$. This leads to the conservation across the discontinuity of the vector

$$\vec{p}(\bar{s}_-) \times \nabla S(\vec{q})|_S = \vec{p}(\bar{s}_+) \times \nabla S(\vec{q})|_S. \tag{1.9}$$

Since $\nabla S(\vec{q})|_S$ appears on both sides of this equation, it implies the conservation of the components of optical momentum in the plane tangent to the surface S at all impact points.

The only component of momentum which suffers a finite change at the surface S is the normal one. If the angle between the vector $\vec{p}_\pm = \vec{p}(\bar{s})_\pm$ and $\nabla S(\vec{q})|_S$ is $\theta_\pm$, the norm of the vector equation (1.9) is $|\vec{p}_-| \sin\theta_- = |\vec{p}_+| \sin\theta_+$. This will yield the well-known sine law of refraction – once we find a consistent norm for the momentum 3-vectors $\vec{p}$.

1.4 Descartes' sphere and the Ibn Sahl law of refraction

The geometric and the dynamical postulates have led to the two Hamilton equations. A well-known property of these equations is that their Hamiltonian function (1.6) is constant along any ray and over all rays, because it is explicitly independent of the arc length parameter s, *viz.*,

$$\begin{aligned}\Delta H(\vec{p}, \vec{q}) &= \frac{\partial H}{\partial \vec{p}} \cdot \Delta\vec{p}(s) + \frac{\partial H}{\partial \vec{q}} \cdot \Delta\vec{q}(s) \\ &= \left(\nabla n \cdot \frac{\vec{p}}{|\vec{p}|} - \frac{\vec{p}}{|\vec{p}|} \cdot \nabla n\right) \Delta s = 0.\end{aligned} \tag{1.10}$$

All previous equations involving the arc length s, the refractive index n, and the value of the Hamiltonian function H, are insensitive to the addition of constants. We can set $H(\vec{p}, \vec{q}) = 0$ so the norm of the momentum vector be

$$|\vec{p}| = n(\vec{q}), \tag{1.11}$$

and $\vec{p}$ will be restricted by (1.6) to the surface of a sphere.

Definition: *At every point $\vec{q}$ of the optical medium $n(\vec{q})$, the locus of the momentum vectors $\vec{p}$ is the* **Descartes sphere**, *whose radius is $n(\vec{q})$.*

The attribution of a definite norm to the momentum vector in optical media was first made by *Abū Sa$^{\subset}$d al-$^{\subset}$Alā$^{\supset}$ Ibn Sahl* fully 650 years before Western natural philosophers (see the appendix to this part). When the angles between the surface normal $\nabla S(\vec{q})|_S$ and the incident and refracted rays are θ_- and θ_+ as before, and the norms (1.11) of the momenta in the two media are n_- and n_+, then from (1.9) now follows

$$n_- \sin\theta_- = n_+ \sin\theta_+, \tag{1.12}$$

and the three vectors are coplanar. The original Ibn Sahl diagram appears in Fig. 1.3. It has the virtue of explicitly showing the conserved quantity (1.12) on the common side of the two triangles.

Remark: Descartes' sphere guides the ray. The engravings adorning Descartes' *On Dioptrics* suggested the sphere that we called after him; see Fig. 1.4. But independently of his work, which pertains the *ray direction* sphere, our Descartes sphere of momentum has a variable radius $n(\vec{q})$. As

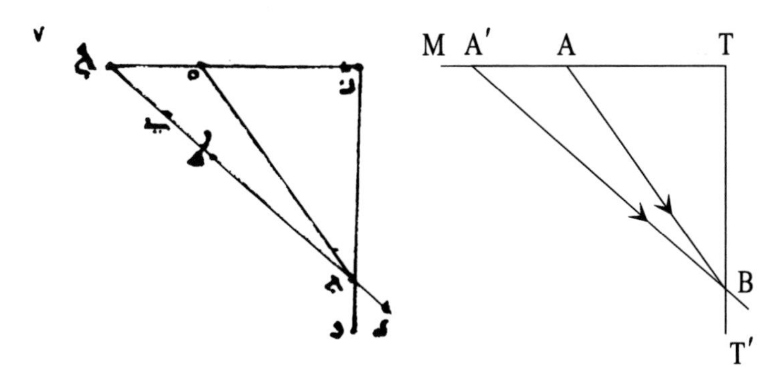

Fig. 1.3. Ibn Sahl diagram of refraction (*left*). The construction is performed (*right*) through the following steps: *(i)* draw a segment AB parallel to the incident ray, *(ii)* draw a plane TT′ tangent to the interface at the point of impact, containing B, whose perpendicular TM contains A, *(iii)*: draw a segment A′B with A′ on TM, with a length such that the ratio AB/A′B is constant (the ratio of the refractive indices, n_-/n_+). The segment A′B is then parallel to the refracted ray and TB is the quantity that is conserved under refraction.
Abū Sa$^{\subset}$d al-$^{\subset}$Alā$^{\supset}$ Ibn Sahl, On Burning Instruments (Baghdad, 984).

sketched in Fig. 1.5, each point of the ray $\vec{q}(s) \in \mathbb{R}^3$ is the center of a Descartes sphere. Progressing in s according to the Hamilton equations, the sphere expands or contracts in response to the local value of refractive index, while conserving the momentum component in the planes tangent to the surfaces $n(\vec{q}) =$ constant. •

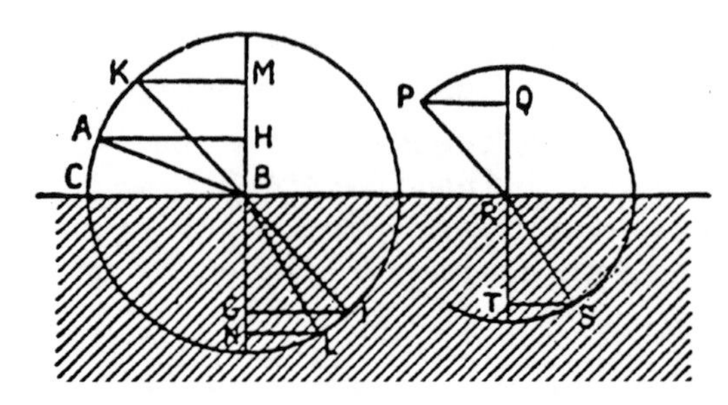

Fig. 1.4. Descartes' diagram for refraction. ... *As, for example, if there passes a ray in Air from* A *to* B *that finds at* B *the surface of Glass* CBR, *it detours to* I *in this Glass; and another from* K *to* B *detours to* L; *and another from* P *to* R *that detours to* S; *there must be the same proportion between the lines* KM *and* LN, *or* PQ *and* ST, *than between* AH *and* IG, *but not the same between the angles* KBM *and* LBN, *or* PRQ *and* SRT, *than between* ABH *and* IBG.
René Descartes, Discourse on the Method, On Dioptrics (Paris, 1633).

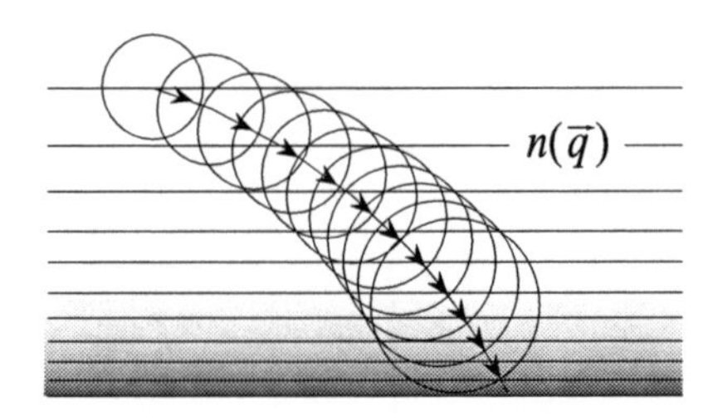

Fig. 1.5. Descartes' sphere 'guides' the ray through an inhomogeneous medium by changing its radius while accomodating the conserved components of momentum along $n(\vec{q}) =$ constant lines.

Remark: Range of the refractive index. The ratios n_-/n_+ of the refractive indices in Nature can be found from (1.12) by experiment. When vacuum is distinguished and used to set the base value $n_{\text{vac}} = 1$, all other indices are found to range up to about 2.4 (for diamond).[5] But in principle,

[5] Current wave theories describe properties of plasmas and of Bose-Einstein condensates through refractive indices with values much outside the optical range.

except for $n \geq 0$ as implied by (1.11), there is no further restriction on the range of n; but Nature is stricter than Mathematics. •

1.5 Geometric optics and classical mechanics

Most readers will be well acquainted with the Hamilton equations for mass-point particles in classical mechanics, where the evolution parameter is commonly time, the Hamiltonian has the prototypical form of kinetic plus potential energies,

$$H_{\rm mec}(\vec{p}, \vec{q}) := \frac{|\vec{p}|^2}{2m} + V(\vec{q}), \tag{1.13}$$

and where the potential function $V(\vec{q})$ characterizes the mechanical system fully. The Hamilton equations of mechanics in time are thus

$$\frac{d\vec{q}(t)}{dt} = \frac{\vec{p}(t)}{m} = \frac{\partial H_{\rm mec}(\vec{p}, \vec{q})}{\partial \vec{p}}, \qquad \frac{d\vec{p}(t)}{dt} = -\frac{\partial V(\vec{q})}{\partial \vec{q}} = -\frac{\partial H_{\rm mec}(\vec{p}, \vec{q})}{\partial \vec{p}}. \tag{1.14}$$

The Hamiltonian (1.13) is constant in time, but depends on the particular trajectory, and its value is the energy E.

The mechanical momentum of a point particle is its mass m times a velocity vector whose norm is ds/dt, the differential arc length ds advanced along the trajectory per differential time dt, i.e.,

$$\text{in mechanics:} \quad \vec{p} := m\frac{d\vec{q}}{dt} \quad \Rightarrow \quad \frac{ds}{dt} = \frac{|\vec{p}|}{m}. \tag{1.15}$$

Replacing the time derivative of the mechanical Hamilton equations (1.14) by the arc length, we find

$$\frac{d\vec{q}}{ds} = \frac{\vec{p}}{|\vec{p}|}, \qquad \frac{d\vec{p}}{ds} = -\frac{m}{|\vec{p}|}\frac{\partial V(\vec{q})}{\partial \vec{q}}. \tag{1.16}$$

Comparing the first equation with the corresponding optical one in (1.3), we note they are identical; comparing the second with (1.5) and recognizing the norm of momentum $n(\vec{q}) = |\vec{p}|$, we establish the following relation ($\leftrightarrow$) between their right-hand sides:

$$-\frac{\partial V(\vec{q})}{\partial \vec{q}} \leftrightarrow \frac{1}{2m}\frac{\partial n(\vec{q})^2}{\partial \vec{q}} \quad \Longrightarrow \quad 2m(E - V(\vec{q})) \leftrightarrow n(\vec{q})^2 - 1. \tag{1.17}$$

We have chosen to add constants $2mE$ and 1 (recalling the argument in Sect. 1.4) so that the classically allowed region of kinetic energy, $|\vec{p}|^2 = 2m(E - V(\vec{q})) > 0$, correspond to the region of a physical refractive index, $n \geq 1$. In this regard, mechanics in arc length is similar to optics. The subject of geometry and dynamics in mechanics is the subject of a recent, advanced book [93].

Example: Harmonic oscillator and elliptic refractive index profile. The mechanical harmonic oscillator is a well-known and studied system whose potential energy is $V(\vec{q}) = \kappa^2|\vec{q}|^2$ with a dimensional constant κ. The relation (1.17) establishes a corresponding geometric optical system whose refractive index $n(\vec{q})$ satisfies $n^2 + \kappa^2|\vec{q}|^2 = 2mE + 1 \geq 1$. This is the equation of an ellipse on the $|\vec{q}|$–n plane, i.e., the refractive index of the medium has a maximum value $n_0 := 2mE + 1$ at the origin $\vec{q} = \vec{0}$ and depends only on the distance $|\vec{q}|$. The radial index profile is the first quadrant of an ellipse, so we call it the *elliptic-index-profile* medium. The motion of the harmonic oscillator and the rays in this medium are thus related. •

Optical momentum (1.1) is very different from mechanical momentum (1.15) because it does not involve velocity. However, if independently we subscribe to the quasi-physical model of light points with velocity $c/n(\vec{q})$,

$$\text{in optics:} \quad \vec{p} := n(\vec{q})\,\frac{d\vec{q}}{ds} \quad \text{and} \quad \frac{ds}{dt} = \frac{c}{n(\vec{q})}, \tag{1.18}$$

then we can turn the optical Hamilton equations (1.3) and (1.5) into the form

$$\frac{d\vec{q}}{dt} = \frac{c\,\vec{p}}{n(\vec{q})^2} = \frac{\partial H'_{\text{opt}}}{\partial \vec{p}}, \qquad \frac{d\vec{p}}{dt} = \frac{c}{n(\vec{q})}\,\frac{\partial n}{\partial \vec{q}} = -\frac{\partial H'_{\text{opt}}}{\partial \vec{q}}, \tag{1.19}$$

with a new Hamiltonian given by

$$H'_{\text{opt}}(\vec{p}, \vec{q}) = c\,\frac{|\vec{p}|^2}{2n(\vec{q})^2}. \tag{1.20}$$

Remark: Newtonian *vs.* Huygensian light. If we believe the translation of the mechanical equations into optics, (1.13)–(1.17), then a light point would increase its velocity when falling from vacuum into a dense, high-n region; this was the conclusion of both Descartes and Newton. Hence, if a geometric optical 'time' is to be used at all, then the physically favored velocity is (1.18) because, when falling into denser regions, these light points slow down, as it actually happens in Huygens' wave optics. In Sect. 5.5–5.7 this 'time' will appear again. •

Remark: Optical path length. The differential $d\tau = c\,dt = n(\vec{q})\,ds$ is the *optical path* element. It comes from the Huygensian model (1.18) and appears with the Hamiltonian (1.20). This formulation is most useful in semi-classical theories, and will be used in Chap. 6 to study an optical system with the symmetry of the Kepler motion in mechanics: the Maxwell fish-eye. •

Remark: Newton and ray equations. The Newton equation of mechanics is found from the two Hamilton equations by applying $m\,d/dt$ twice to $\vec{q}$. In optics, when we apply $n(\vec{q})\, d/ds$ twice, we find the *ray equation*. The two equations are related to each other thus:

$$\left(m\,\frac{d}{dt}\right)^2 \vec{q} = -m\frac{\partial V(\vec{q})}{\partial \vec{q}} \quad \leftrightarrow \quad \left(n(\vec{q})\,\frac{d}{ds}\right)^2 \vec{q} = \tfrac{1}{2}\frac{\partial n(\vec{q})^2}{\partial \vec{q}}. \tag{1.21}$$

The two Hamilton equations constitute a *factorization* of the Newton and ray equations. The latter's second derivatives are thus resolved into the two first-derivative equations. •

2 Optical phase space

The manifold of straight oriented lines in three-dimensional (3-dim) space R^3 has less than the 6 coordinates of momentum and position $(\vec{p}, \vec{q})$ that we used in Chap. 1. As we saw, the momentum 3-vector is restricted to its Descartes sphere and the arc length origin $s = 0$ is arbitrary. This reduces the number of coordinates that define a ray by two, to the essential *four* dimensions of a manifold that we shall call *optical phase space*, $\wp$, associated to a 2-dim plane *screen*. We shall then rewrite the Hamilton equations on that screen to formulate the *evolution* of rays in an inhomogeneous optical medium as the screen moves normal to itself.

Most of our definitions are restricted in the sense that they are not as general as mathematical abstraction does allow. They are working definitions that serve for the Hamiltonian systems of geometric optics, where the distinction between the position and momentum coordinates originates from the two postulates. These are called Darboux coordinates for a general *symplectic* space that may have other coordinate systems which do not follow this separation. Distinct position and momentum coordinates are both useful and sufficient for our purposes.

2.1 Ray coordinates and their manifold

The rays of 3-dim geometric optics are commonly referred to Cartesian coordinate axes x, y, z which label the vector components. Using boldface for 2-vectors, we write

$$\begin{aligned} \vec{q} &= \begin{pmatrix} \mathbf{q} \\ q_z \end{pmatrix}, \quad \mathbf{q} := \begin{pmatrix} q_x \\ q_y \end{pmatrix}, \\ \vec{p} &= \begin{pmatrix} \mathbf{p} \\ p_z \end{pmatrix}, \quad \mathbf{p} := \begin{pmatrix} p_x \\ p_y \end{pmatrix}, \quad \begin{matrix} p_z = \sigma\sqrt{n(\mathbf{q},z)^2 - |\mathbf{p}|^2}, \\ \sigma = \operatorname{sign} p_z \in \{+,0,-\}. \end{matrix} \end{aligned} \tag{2.1}$$

We choose the 2-dim submanifold $q_z = 0$ and call it the *standard screen* with *optical center* at $\mathbf{q} = \mathbf{0}$ and *optical axis* q_z $(\mathbf{p} = \mathbf{0})$. See Fig. 2.1, where rays are characterized with reference to the standard screen by the coordinates $(\mathbf{p}, \sigma, \mathbf{q})$ as we detail below.

Definition: Optical phase space $\wp$ *(restricted definition) is a manifold with coordinates* $(\mathbf{p}, \sigma, \mathbf{q})$*, comprising two open regions of* R^4*, as follows:*

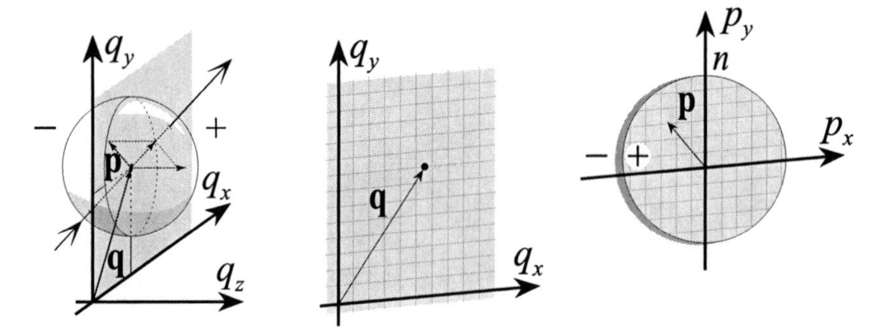

Fig. 2.1. Ray coordinates on the standard screen. *Left*: in 3-dim, a ray (bold arrow) is parametrized by the four coordinates of phase space $\wp$, namely $(\mathbf{p}, \mathbf{q})$ and the sign σ. This 4-dim manifold is resolved into the product of two planes. Middle: *positions* on the screen, $\mathbf{q} \in \mathsf{R}^2$. *Right*: for each point $\mathbf{q}$, two open disks of *momentum* $\mathbf{p} \in \mathcal{D}^2 \subset \mathsf{R}^2$ where $|\mathbf{p}| < n(\mathbf{q})$, distinguished by the *chart*-index $\sigma \in \{+, 0, -\}$ (forward + and backward − rays), sown at their common closure circle $\sigma = 0$. (Rays parallel to the screen cannot be thus parametrized.)

(i) *The two* **position** *coordinates* $\mathbf{q} \in \mathsf{R}^2$ *are the intersection of the ray with the standard screen (except for rays parallel to the screen, see below).*

(ii) *The two* **momentum** *coordinates* $\mathbf{p} \in \mathcal{D}^2_n \subset \mathsf{R}^2$ *of the 3-vector* $\vec{p}$ *tangent to the ray, projected on the plane of the screen. Their range is the disk* $\mathcal{D}^2_n$ *where* $|\mathbf{p}| < n(\mathbf{q})$.[1]

(iii) *The* **chart**-*index* $\sigma = \operatorname{sign} p_z \in \{+, 0, -\}$, *which distinguishes between forward* ($\sigma = +$), *and backward* ($\sigma = -$) *rays; these have the common boundary of rays that are parallel to the screen* ($|\mathbf{p}| = n$, $\sigma = 0$).[2]

Remark: Spherical coordinates on the Descartes sphere. Using spherical coordinates of colatitude and azimuth (θ, ϕ) for the 2-dim surface of the Descartes sphere, we can write

$$\vec{p} = \begin{pmatrix} \mathbf{p} \\ p_z \end{pmatrix} = n \begin{pmatrix} \sin\theta \sin\phi \\ \sin\theta\cos\phi \\ \cos\theta \end{pmatrix}, \qquad \begin{aligned} \mathbf{p} &= n \sin\theta \begin{pmatrix} \sin\phi \\ \cos\phi \end{pmatrix}, \\ p_z &= n\cos\theta, \end{aligned} \qquad \begin{aligned} 0 \le \theta &\le \pi, \\ -\pi < \phi &\le \pi, \\ \sigma &= \operatorname{sign}\cos\theta. \end{aligned} \tag{2.2}$$

The q_z-axis is the optical axis; it marks the north pole of the spherical coordinates and has zero colatitude. This axis is shown in optics always pointing to the right! •

[1] Note that, unless the medium is homogeneous (n = constant), phase space $\wp$ is **not** the direct product $\mathcal{D}^2 \otimes \mathsf{R}^2$.

[2] The ensemble of rays parallel to the screen is of dimension 3: orientation, and distance from and along the optical q_z-axis. This submanifold, having one dimension less than $\wp$, will be often ignored. Mostly we shall refer to the $\sigma = +$ chart only (supressing the index), and stay far from its boundary.

Remark: Two-dimensional optical media. Flat, 2-dim optical systems can be obtained from 3-dim ones through restricting q_y = constant and $p_y = 0$, and erasing all $\partial/\partial q_y$'s in the previous equations. In this plane world, the Descartes sphere is a circle and the standard screen is $q_z = 0$ (the q_x axis). As shown in Fig. 2.2, the phase space of rays in two dimensions, $\wp^{(2)} = \{p, \sigma; q\}$, has one position coordinate $q \in \mathsf{R}$ and one momentum coordinate $p = n \sin\theta \in (-n, n)$, and $\sigma = \text{sign}\cos\theta \in \{+, 0, -\}$. The two open intervals $|p| < n(q)$ are identified at their two endpoints (rays parallel to the screen). In our figures we illustrate mostly 2-dim optical media because the diagrams are then also two-dimensional, so we can draw them easily. •

In a world of $D + 1$ dimensions where rays are still lines, the standard screen $q_z = 0$ has the position coordinates $\mathbf{q} \in \mathsf{R}^D$ of a D-dim manifold. The Descartes sphere of momentum is the D-dim surface $\mathcal{S}^D$; its projection on the plane of the screen falls in the D-dim interior of two $\mathcal{S}^{D-1}$ spheres $|\mathbf{p}| < n$ (disks in $D = 2$), distinguished by their open hemisphere $\sigma = \pm$, and their common boundary $|\mathbf{p}| = n$ for $\sigma = 0$, which is the surface of $\mathcal{S}^{D-1}$ (a circle in $D = 2$). Optical phase space $\wp^{(D)}$ is then a $2D$-dim manifold on which all previous (and most subsequent) vector formulas hold.

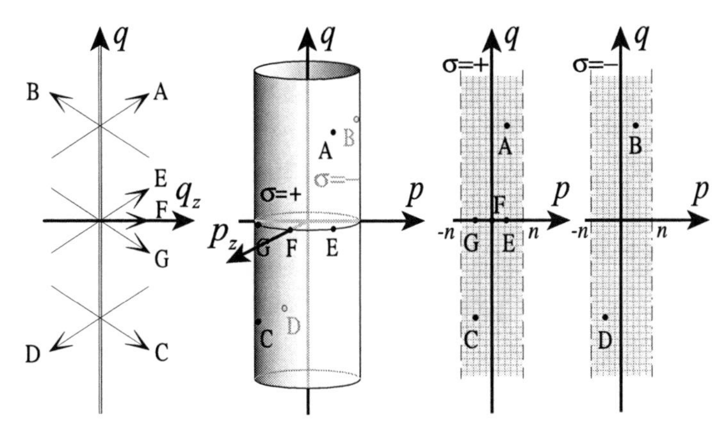

Fig. 2.2. Rays in optical and phase space in a flat world $D = 1$. *Left*: rays A, ..., G in the optical plane cross the standard screen line $q_z = 0$. Middle: phase space $\wp$ is a cylinder, where the rays are identified by points. *Right*: two strips (distinguished by $\sigma = \text{sign}\, p_z$) are the projection of the cylinder $\wp$ (whose radius n is generally a function of q) parametrized by the coordinate of position $q \in \mathsf{R}$ and of momentum $|p| < n(q)$. The two strips are identified at their edges $\sigma = 0$. (Rays that are parallel to the screen require a separate parametrization.)

2.2 Hamilton equations on the screen

The Hamilton equations on the screen relate the points of the 4-dim phase space $\wp$ for a screen at $q_z = z$, with their corresponding points for a screen at $z+\Delta z$, as $\Delta z \to 0$. These equations will thus describe the rays $(\mathbf{p}(z), \sigma; \mathbf{q}(z))$, as the screen is translated along the optical axis by drawing their *trajectories* on $\wp$, parametrized by the distance z from the standard screen $q_z = 0$.

The change of evolution parameter in the Hamilton equations (1.3)–(1.5) of Chap. 1, from arc length s along the ray to distance z from the standard screen, is summarized in Fig. 2.3. The triangle similarity between differential position and momentum is in fact the z-component of (1.3): $dz/ds = p_z/n = \cos\theta$ (for colatitude θ). Dividing the first two components of the vector equation (1.3) by the third, we obtain[3]

$$\frac{d\mathbf{q}}{dz} = \frac{d\mathbf{q}}{ds}\Big/\frac{dz}{ds} = \frac{\mathbf{p}}{|\vec{p}|}\Big/\frac{p_z}{n}, \tag{2.3}$$

$$\frac{d\mathbf{p}}{dz} = \frac{d\mathbf{p}}{ds}\Big/\frac{dz}{ds} = \frac{\partial n(\mathbf{q},z)}{\partial \mathbf{q}}\Big/\frac{p_z}{n}. \tag{2.4}$$

Since p_z is on the Descartes sphere $\sigma\sqrt{n(\mathbf{q},z)^2 - |\mathbf{p}|^2}$, we find the Hamilton equations on the phase space $\wp$ of the screen to be

$$\frac{d\mathbf{q}}{dz} = \frac{\mathbf{p}}{\sigma\sqrt{n(\mathbf{q},z)^2 - |\mathbf{p}|^2}} = \frac{\partial h(\mathbf{p},\sigma;\mathbf{q};z)}{\partial \mathbf{p}}, \tag{2.5}$$

$$\frac{d\mathbf{p}}{dz} = \frac{n(\mathbf{q},z)}{\sigma\sqrt{n(\mathbf{q},z)^2 - |\mathbf{p}|^2}}\frac{\partial n(\mathbf{q},z)}{\partial \mathbf{q}} = -\frac{\partial h(\mathbf{p},\sigma;\mathbf{q};z)}{\partial \mathbf{q}}, \tag{2.6}$$

where the *screen* Hamiltonian function is

$$h(\mathbf{p},\sigma;\mathbf{q};z) := -\sigma\sqrt{n(\mathbf{q},z)^2 - |\mathbf{p}|^2} = -n(\mathbf{q},z)\cos\theta = -p_z. \tag{2.7}$$

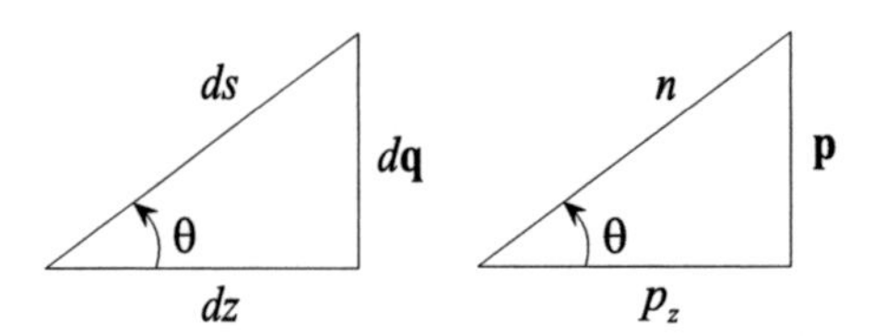

Fig. 2.3. The triangle of sides dz, $d\mathbf{q}$ and ds in *position* space is similar to the triangle of sides $p_z = -h$, $\mathbf{p}$, and $n = |\vec{p}|$ in *momentum* space; θ is the colatitude angle subtended between the ray and the z-axis in both spaces.

[3] There is trouble with the $\sigma = 0$ rays, where trajectories bend around in z, because the divisor equation becomes $0 = 0$ for that point. This problem is due to the coordinates we have chosen for $\wp$, and is not otherwise substantive.

Lastly, the z-component of the equation for momentum, dp_z/dz (see Fig. 2.3), is

$$\frac{dh}{dz} = \frac{\partial h}{\partial z} = \frac{n(\mathbf{q}, z)}{h(\mathbf{p}, \sigma, \mathbf{q}; z)} \frac{\partial n(\mathbf{q}, z)}{\partial z}. \tag{2.8}$$

Remark: The Hamiltonian in $\wp$ is not always constant. The value of the screen Hamiltonian function $h = h(\mathbf{p}, \sigma; \mathbf{q}; z)$ is generally not a constant. When the refractive index $n(\vec{q})$ depends on $q_z = z$, then (2.8) shows that the z-derivative of h is nonzero. When the index is independent of q_z, the medium is called a *guide*;[4] there, h is constant along every ray, but its value depends on the ray $(\mathbf{p}, \sigma, \mathbf{q})$. In a homogeneous medium $n =$ constant, (2.7) shows that $h = -n\cos\theta$ depends on the colatitude of the ray, being negative for forward rays and positive for backward ones. •

To formalize the structure of geometric optical systems that we have described up to this point, we introduce the following (restricted) definition (in position and momentum coordinates):

Definition: *A $2D$-dimensional* **Hamiltonian system** *is a model of D-dimensional Nature endowed with:*

(i) An essential set of D **position** *coordinates $\mathbf{q} = \{q_i\}_{i=1}^D$, and of D* **momentum** *coordinates $\mathbf{p} = \{p_i\}_{i=1}^D$ (called* **canonically conjugate** *sets of coordinates, with one or more charts σ).*
(ii) An **evolution** *parameter $z \in \mathsf{R}$, and a* **generating** *function (Hamiltonian) $h = h(\mathbf{p}, \mathbf{q}; z)$).*
(iii) A set of $2D$ **Hamilton equations** *with the structure (2.5)–(2.6),*

$$\frac{d\mathbf{q}}{dz} = \frac{\partial h}{\partial \mathbf{p}}, \qquad \frac{d\mathbf{p}}{dz} = -\frac{\partial h}{\partial \mathbf{q}}. \tag{2.9}$$

Remark: Canonical conjugacy. The physical meaning of the coordinates in a Hamiltonian system need not be position and momentum of a ray or of a point particle. *Any* two spaces with the same local R^D-geometry that can be woven into a pair of Hamilton equations will do, and will be called *canonically conjugate.* Also, in Chap. 3 and beyond, we shall exploit functions *other* than the optical Hamiltonians (1.6) or (2.7), to generate evolution along parameters other than distance along the optical axis. They also qualify as *canonically conjugate* generating function and evolution parameter. •

Example: Translation in a homogeneous medium. Rays in a homogeneous optical medium are governed by the Hamilton equations (2.5)–(2.6) with $\partial h/\partial\mathbf{q} = \mathbf{0}$. The solutions are

$$\mathbf{p} = \mathbf{p}(0), \quad \mathbf{q}(z) = \mathbf{q}(0) + \sigma\, z \frac{\mathbf{p}}{\sqrt{n^2 - |\mathbf{p}|^2}} \tag{2.10}$$

[4] The name '*wave*guide' would be unfortunate because it may conjure up wave theories of light.

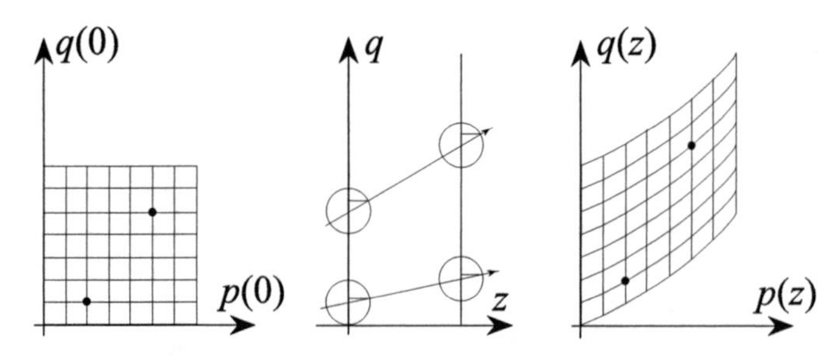

Fig. 2.4. Free propagation in optical and phase spaces. A patch of phase space $(p, q) \in \wp$ at the standard screen $z = 0$ (*left*), with two rays marked by bullets (•), under translation along the optical z-axis (middle), is deformed (*right*) to another patch $(p(z), q(z)) \in \wp$.

[cf. (1.7)]. In Fig. 2.4 we show how the translation of the screen (2.10) results in the 'motion' of the intersection point of the rays, within a deformation of a coordinate patch of optical phase space $\wp$ (for $D = 1$). This is a local patch of phase space involving no more than one chart; the global deformation of $\wp$ can be imagined in Fig. 2.2 noting that $q = 0$ rays (through the optical center) map to $(z/n)\tan\theta$ for colatitude θ. •

Remark: Expansions and reflections. The optical Hamiltonian in (2.7) and many expressions below can be expanded in Taylor series of $|\mathbf{p}|^2$, for which the following series are useful:[5]

$$\sqrt{n^2 - |\mathbf{p}|^2} = n - \frac{|\mathbf{p}|^2}{2n} - \frac{|\mathbf{p}|^4}{8n^3} - \frac{|\mathbf{p}|^6}{16n^5} - \frac{5|\mathbf{p}|^8}{128n^7} - \cdots$$
$$= -\sum_{m=0}^{\infty} \frac{(2m-3)!!}{(2m)!!} \frac{|\mathbf{p}|^{2m}}{n^{2m-1}}, \tag{2.11}$$

$$\frac{1}{\sqrt{n^2 - |\mathbf{p}|^2}} = \frac{1}{n} + \frac{|\mathbf{p}|^2}{2n^3} + \frac{3|\mathbf{p}|^4}{8n^5} + \frac{5|\mathbf{p}|^6}{16n^7} + \frac{35|\mathbf{p}|^8}{128n^9} + \cdots$$
$$= \sum_{m=0}^{\infty} \frac{(2m-1)!!}{(2m)!!} \frac{|\mathbf{p}|^{2m}}{n^{2m+1}}. \tag{2.12}$$

The series converge for $|\mathbf{p}| < n$; at $|\mathbf{p}| = n$, both functions have branch points and there are two Riemann sheets where the square roots are distinguished by the sign σ of $h = -p_z = -\sigma\,|h|$. The reflection $\sigma \leftrightarrow -\sigma$ of rays (forward $\leftrightarrow$ backward) produced by a mirror on the standard screen, is equivalent to $z \leftrightarrow -z$ in the evolution equations. (Or, since the series contain only odd powers of n, to a change of sign in the refractive index, $n \leftrightarrow -n$.) We shall study this further in Chap. 4. •

[5] The *double factorial* function on the integers is $N!! := N \times (N-2)!!$, with $1!! := 1$, $0!! := 1$; in particular $(-1)!! = 1$ and $(-3)!! = -1$. These series expand $\cos\theta$ and $\sec\theta$ in powers of $\sin\theta$.

2.3 Guides and their index profile

The Hamilton equations for an optical medium simplify considerably when the medium has *symmetry*, i.e., when its refractive index is invariant under some transformations. We consider here media which are invariant under translations of the screen in a fixed direction. We have called *guide* to an optical medium whose refractive index $n(\mathbf{q})$ is independent of the distance z along the optical axis. The function $n(\mathbf{q})$ is correspondingly called the refractive index *profile* of the guide. Note that homogeneous media are also guides; we are mainly interested in profiles $n(\mathbf{q})$ having a well-formed maximum near to the optical axis.

As we remarked above, the Hamiltonian (2.7) in guides has a constant value, independent of z,

$$h(\mathbf{p},\mathbf{q}) := h(\mathbf{p}(0);\mathbf{q}(0)) = h(\mathbf{p}(z);\mathbf{q}(z)) = -\sqrt{n(\mathbf{q})^2 - |\mathbf{p}|^2}, \tag{2.13}$$

which is determined by the initial conditions of the ray. The Hamilton evolution equations in guide media are obtained from (2.5)–(2.6) with this parameter h, as

$$\frac{d\mathbf{p}}{dz} = -\frac{1}{2h}\frac{\partial n(\mathbf{q})^2}{\partial \mathbf{q}}, \qquad \frac{d\mathbf{q}}{dz} = -\frac{1}{h}\mathbf{p}. \tag{2.14}$$

We can solve explicitly the Hamilton equations for the trajectories of rays in elliptic-index-profile guides (recall page 12), namely

$$n_e(\mathbf{q}) = +\sqrt{n_0^2 - \kappa^2|\mathbf{q}|^2}, \quad 0 \le |\mathbf{q}| \le n_0/\kappa. \tag{2.15}$$

This index profile is shown in Fig. 2.5; its maximum is on the optical z-axis, $n_e(\mathbf{0}) = n_0$, so the guide will keep rays spiraling along and around this axis. The Hamiltonian function (2.7)–(2.13) for this guide is

$$h_e(\mathbf{p},\mathbf{q}) = -\sqrt{n_0^2 - (|\mathbf{p}|^2 + \kappa^2|\mathbf{q}|^2)}, \tag{2.16}$$

and the Hamilton equations (2.5)–(2.5) are

$$h_e\frac{d\mathbf{q}}{dz} = \mathbf{p}, \qquad h_e\frac{d\mathbf{p}}{dz} = \tfrac{1}{2}\frac{\partial n_e(\mathbf{q})^2}{\partial \mathbf{q}} = -\kappa^2\mathbf{q}. \tag{2.17}$$

Their general solution yields the trajectories

$$\mathbf{p}(z) = \mathbf{p}(0)\cos\frac{z\kappa}{h_e} - \mathbf{q}(0)\kappa\sin\frac{z\kappa}{h_e}, \tag{2.18}$$

$$\mathbf{q}(z) = \mathbf{p}(0)\frac{1}{\kappa}\sin\frac{z\kappa}{h_e} + \mathbf{q}(0)\cos\frac{z\kappa}{h_e} \tag{2.19}$$

shown in Fig. 2.5.

Remark: Ellipses and the constant Hamiltonian. From (2.18)–(2.19) we see that $|\mathbf{p}|^2 + \kappa^2|\mathbf{q}|^2$ = constant; this is the equation of the concentric ellipses which contain the trajectories. The conserved quantity has the mechanical form of *energy* in a harmonic oscillator. Under evolution along z in the guide, the quantity of interest is the Hamiltonian, h_e in (2.16), which is by itself negative ($h_e = -n_0$ at the guide axis), grows with $|\mathbf{p}|^2 + \kappa^2|\mathbf{q}|^2$ up to the value zero; positive values correspond to backward rays. •

Remark: Dispersion of oscillation frequencies. As the screen advances along the z-axis of the elliptic-index-profile guide, phase space does *not* rotate rigidly (as it does in a mechanical harmonic oscillator) because the argument z of the trigonometric functions in (2.18)–(2.19) has the frequency κ/h_e (period $2\pi\, h_e/\kappa$) which depends on the size of the concentric ellipse, $|\mathbf{p}|^2 + \kappa^2|\mathbf{q}|^2$. The smallest ellipses (rays near to the axis) have the period $2\pi\, n_0/\kappa$, while the larger ellipses (rays at larger angles) have a decreasing oscillation period. Figure 2.5 shows the rays in optical space $(\mathbf{q}(z), z)$ and the z-evolution of trajectories in phase space $(\mathbf{p}(z), \mathbf{q}(z))$ along the guide. The excentricity of the set of ellipses is proportional to κ; it is drawn to be unity in the figure, so the trajectories appear as circles. Because trajectories of different frequencies do not oscillate in step, the elliptic-index guide is not harmonic, but *dispersive.* •

Remark: Constant angular momentum: meridional, saggital and skew rays. In 3-dim guides, from (2.18)–(2.19) we also see that $\mathbf{q} \times \mathbf{p}$ is independent of z; this is the (complete) analogue of mechanical *angular momentum*, which is related to the size and excentricity of the phase-space ellipses. Rays of zero angular momentum are *meridional* (i.e. $\mathbf{p} \parallel \mathbf{q}$), so their trajectories in space (and in phase space) lie in a plane with the optical axis. Rays with the maximum angular momentum (for a given 'energy') are *saggital* (i.e. $\mathbf{p} \perp \mathbf{q}$). In between, are all *skew* rays. •

Remark: Phase space of the *elliptic-index* guide is bounded. The phase space available to rays in the elliptic-index guide is bounded because the refractive index profile $n_e(\mathbf{q})$ in (2.15) falls to zero at the vertices of the ellipse, $|\mathbf{q}| = n_e/\kappa$ (see Fig. 2.5), – and is pure imaginary beyond. The Hamiltonian $h_e(\mathbf{p}, \mathbf{q})$ in (2.16) is the component of the ray direction vector on the surface of the Descartes sphere (of radius $n_e(\mathbf{q})$), and so it can be real only in the open ellipsoid of phase space $0 \leq |\mathbf{p}|^2 + \kappa^2|\mathbf{q}|^2 < n_0^2$, where n_0 is the refractive index at the axis of the guide. Nature further puts 1 for lower limit to exterior vacuum (Fig. 2.5), so in (2.15) we must further correct the bound to $1 \leq n_0^2 - \kappa^2|\mathbf{q}|^2$, and an upper limit $1 \leq n_0(\mathbf{q}) \lesssim 2.4$ (for diamond). Beyond this region there is nothing. •

Remark: Limit of wide guides. As we saw above, the width of the optical guide is proportional to κ^{-1}, so when κ decreases, the guide widens; in the limit $\kappa \to 0$, the guide becomes a homogeneous medium n_0. The elliptical trajectories in phase space degenerate into straight lines parallel to the vertical axis. See again Fig. 2.4. •

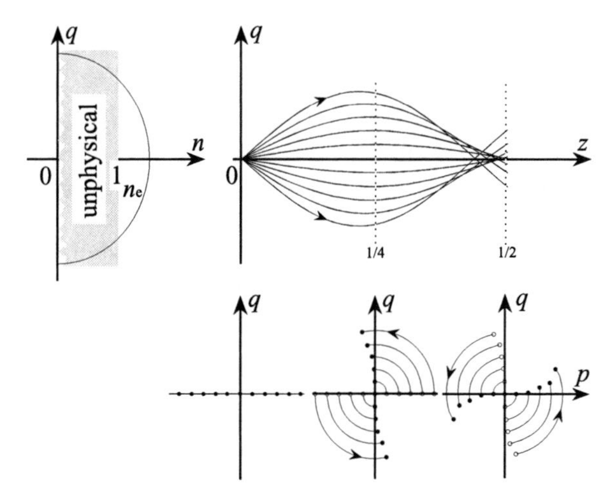

Fig. 2.5. Elliptic index-profile guide in a 2-dim ($D = 1$) optical medium. *Top left*: the refractive index profile $n_e(q)$ (the region $n < 1$ is unphysical). *Top right*: rays through the optical center $[q(0) = 0]$ suffer dispersion (they do not all meet again); dotted lines show the $\frac{1}{4}$- and $\frac{1}{2}$-periods of *paraxial* rays ($p \approx 0$). *Bottom*: trajectories of the rays in phase space between the $z = 0$ screen, and at quarter- and half-periods; phase space is deformed due to the differential rotation of the circles. (Generally the motion is along concentric ellipses.)

2.4 Paraxial optics and mechanics

Here we explore a relation between classical mechanics and a model of geometric optics which is the limit for rays which are *paraxial* (i.e., near to the axis) and media whose refractive is nearly constant. Basically, a neighborhood of rays on the Descartes sphere near the optical axis will be identified with a neighborhood of its tangent plane. We shall dedicate Chap. 9 to establish that this limit is by itself a consistent model of optics, to be called the paraxial régime, or simply *paraxial optics*.

We assume that medium is almost homogeneous, so its refractive index can be approximated by

$$n(\mathbf{q}, z) = n_0 - \nu(\mathbf{q}, z), \qquad \nu(\vec{0}) = 0, \quad \nu^2 \approx 0, \tag{2.20}$$

and that the colatitude angle of the rays [see (2.2)] is small, so we keep terms up to degree 2 in θ and degree 1 in ν. We use the series (2.11) to write

$$|\mathbf{p}| = n(\vec{q}) \sin\theta \approx n_0\theta, \quad h_0 = -n(\vec{q}) \cos\theta \approx -n_0(1 - \tfrac{1}{2}\theta^2) + \nu(\mathbf{q}, z). \tag{2.21}$$

The right-hand members of the approximation refer to a new *paraxial* momentum and Hamiltonian, $\mathbf{p}_{\text{prx}}$ and h_{prx}. The first simply replaces θ back to $|\mathbf{p}_{\text{prx}}|/n_0$ (with $\mathbf{p}_{\text{prx}} \parallel \mathbf{p}$) and (adding n_0) the second defines the *paraxial* optical Hamiltonian,

$$h_{\text{prx}}(\mathbf{p}, \mathbf{q}; z) := \frac{1}{2n_0}|\mathbf{p}_{\text{prx}}|^2 + \nu(\mathbf{q}, z). \tag{2.22}$$

This has a form analogous to the mechanical Hamiltonian (1.13) with the time parameter. The relation holds naturally between the momenta $\mathbf{p}_{\text{prx}} \leftrightarrow \mathbf{p}$, which can be freed from the bounded range $|\mathbf{p}| \le n$ of global optics, and extends to the Hamiltonians $h_{\text{prx}} \leftrightarrow H_{\text{mec}}$, evolution parameters $z \leftrightarrow t$, base refractive index and mass $n_0 \leftrightarrow m$, and refractive index anomaly with mechanical potential, $\nu(\mathbf{q}, z) \leftrightarrow V(\mathbf{q}, t)$. Guides in the paraxial régime thus correspond to time-independent potentials.

3 Canonical transformations

Light is neither created nor destroyed, only transformed, could be a quote pirated from Antoine Lavoisier's *dictum* on matter, to underline that it must neither appear nor vanish in passing through a passive optical apparatus. In this chapter we study those transformations which preserve light, and also maintain the Hamiltonian structure of geometric optics. These are the *canonical* transformations to be defined below.

It is a longer chapter because we also introduce the Poisson brackets on phase space; this will organize the structure of Hamiltonian systems by means of Lie algebras of operators that generate parametric groups of canonical transformations. We have the task of formally presenting the definitions of Lie algebras and groups in the context of geometric optics. Several important one-parameter Lie groups that we introduce as examples shall be much used later.

3.1 Beams and the conservation of light

We call 'light' to a *volume* of rays in optical phase space. This can be visualized for 2-dim optics ($D = 1$) in Figs. 2.4 and 2.5: the surface elements of a patch of phase space may change position, orientation, squeeze and/or shear, but they do preserve their area. On D-dim screens, phase space $(\mathbf{p}, \mathbf{q}) \in \wp$ is $2D$-dimensional, and its oriented volume element is the (totally antisymmetric) wedge product indicated $d\mathbf{p}{\wedge}d\mathbf{q}$. (In this chapter we omit the chart index σ; we shall mostly stay within the chart of 'forward' rays for convenience.)

We call *beam* any positive distribution of rays determined by a function $\rho(\mathbf{p}, \mathbf{q}) \geq 0$ on phase space. The total amount of light in the beam is the integral of the beam function over $\wp$. A single ray $(\mathbf{p}_\bullet, \mathbf{q}_\bullet)$ is thus described by a $2D$-dim Dirac δ on phase space, and associated ($\leftrightarrow$) to a space illumination (in optical space, see Fig. 3.1),

$$\delta^D(\mathbf{p} - \mathbf{p}_\bullet)\,\delta^D(\mathbf{q} - \mathbf{q}_\bullet) \quad \leftrightarrow \quad \delta^D(\mathbf{q} - \mathbf{q}_\bullet - z\mathbf{p}_\bullet/h_\bullet), \tag{3.1}$$

where $h_\bullet(\mathbf{p}, \mathbf{q}; z)$ is the Hamiltonian function on the ray (2.7), so $|\mathbf{p}_\bullet|/h_\bullet = \tan\theta_\bullet$ with the colatitude $\theta_\bullet$ of the ray at the screen. Accordingly, to a distribution function ρ on phase space is associated the following ray density, or

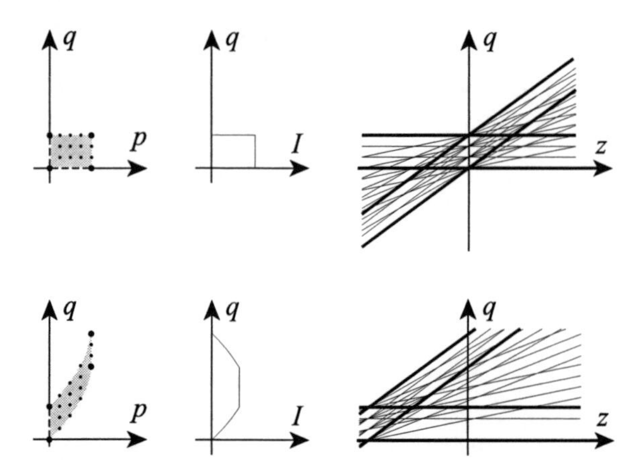

Fig. 3.1. Phase space and screen illumination of a 2-dim rectangular beam under translation. At the standard screen (*top row*) and at a z-displaced screen (*bottom row*), we show the phase space distribution (*left column*), the illumination of the screen (middle column) and the rays in optical space (*right column*). Heavy dots (*left*) correspond to heavy lines (*right*).

space illumination function:[1]

$$\rho(\mathbf{p},\mathbf{q}) = \int_{\wp} d\mathbf{p}_\bullet \wedge d\mathbf{q}_\bullet\, \rho(\mathbf{p}_\bullet,\mathbf{q}_\bullet)\, \delta(\mathbf{p}-\mathbf{p}_\bullet)\, \delta(\mathbf{q}-\mathbf{q}_\bullet), \tag{3.2}$$

$$\updownarrow \qquad\qquad\qquad\qquad \updownarrow$$

$$\begin{aligned} D(\mathbf{q},z) &:= \int_{\wp} d\mathbf{p}_\bullet \wedge d\mathbf{q}_\bullet\, \rho(\mathbf{p}_\bullet,\mathbf{q}_\bullet)\, \delta(\mathbf{q}-\mathbf{q}_\bullet - z\mathbf{p}_\bullet/h_\bullet) \\ &= \int_{\mathcal{D}} d\mathbf{p}\, \rho(\mathbf{p}, \mathbf{q} - z\mathbf{p}/\sqrt{n^2-\mathbf{p}^2}), \end{aligned} \tag{3.3}$$

where the last integral is over the momentum disk $\mathcal{D}$ of $\wp$.[2]

Remark: Screen illumination. We see a ray $(\mathbf{p}_\bullet,\mathbf{q}_\bullet)$ impinging on the standard screen as a dot at position $\mathbf{q}_\bullet$ – we do not see its momentum $\mathbf{p}_\bullet$. Rather, the *screen illumination* is an integral of the beam distribution $\rho(\mathbf{p},\mathbf{q})$ over all ray directions,[3]

$$I(\mathbf{q}) := D(\mathbf{q},0) = \int_{\mathcal{D}} d\mathbf{p}\, \rho(\mathbf{p},\mathbf{q}). \tag{3.4}$$

When a beam intersects a screen thus, the surface receives light from all directions of the Descartes sphere, integrated with the measure $d|\mathbf{p}| = n\cos\theta\, d\theta$

[1] The concepts in the field of radiometry are defined geometrically, yet there seem to be several conventions in the literature on names and units; we follow at least one of them.

[2] If also backward rays $\sigma = -$ are present, we should write (3.3) with a sum over σ, or consider separately the space illuminations on both sides of a the screen.

[3] Illumination is additive, as totally incoherent light in physical optics.

containing the geometric *obliquity factor* $\cos\theta$ of colatitude. The screen intensity can be read off the phase space distribution ρ (see the example below) by projecting it on the q-axis. Finally, the total light (*energy*) in the beam is

$$E := \int_{\mathsf{R}^D} d\mathbf{q}\, I(\mathbf{q}) = \int_{\wp} d\mathbf{p}{\wedge}d\mathbf{q}\, \rho(\mathbf{p},\mathbf{q}). \tag{3.5}$$

•

Example: Rectangular beams. Figures 3.1 show a 2-dim *rectangular* beam of forward rays,

$$\rho_{\mathrm{R}}(p,q) = \mathrm{Rect}_{[p_0,p_1]}(p)\,\mathrm{Rect}_{[q_0,q_1]}(q), \qquad \mathrm{Rect}_{[x_0,x_1]}(x) := \begin{cases} 1, & x_0 \le x \le x_1 \\ 0, & \text{otherwise,} \end{cases} \tag{3.6}$$

propagating in a homogeneous optical space. The space illumination (3.3) at the standard screen is

$$D_{\mathrm{R}}(q,0) = \int_{p_0}^{p_1} dp\, \mathrm{Rect}_{[q_0,q_1]}(q) = (p_1 - p_0)\,\mathrm{Rect}_{[q_0,q_1]}(q). \tag{3.7}$$

Displacement of the screen shears phase space, and with it, the distribution function of the beam. The lower and upper boundaries of the rectangle in Fig. 3.1, $q_{\mathrm{i}} = q_0$ and q_1, become the lines $q = q_{\mathrm{i}} + zp/\sqrt{n^2 - p^2}$; inverting for p, this defines the integration limits to find the screen illumination at z as

$$D_{\mathrm{R}}(q,z) = \int_{\max(p_0, P_1(q,z))}^{\min(P_0(q,z), p_1)} dp, \qquad P_{\mathrm{i}}(q,z) := \frac{n(q - q_{\mathrm{i}})}{\sqrt{z^2 + (q - q_{\mathrm{i}})^2}}. \tag{3.8}$$

•

A map of $2D$-dim phase space on itself given by $\mathbf{p} \mapsto \mathbf{P}(\mathbf{p},\mathbf{q})$, $\mathbf{q} \mapsto \mathbf{Q}(\mathbf{p},\mathbf{q})$ conserves the volume element, $d\mathbf{p}{\wedge}d\mathbf{q} = d\mathbf{P}{\wedge}d\mathbf{Q}$, when its Jacobian is unity,

$$J(\mathbf{p},\mathbf{q}) := \begin{vmatrix} \dfrac{\partial P_i}{\partial p_j} & \dfrac{\partial P_i}{\partial q_j} \\ \dfrac{\partial Q_i}{\partial p_j} & \dfrac{\partial Q_i}{\partial q_j} \end{vmatrix} = 1. \tag{3.9}$$

Then phase space behaves under the map as an incompressible fluid in $2D$ dimensions: no two points can coalesce and no point can divide.

Remark: Necessity and sufficiency of light conservation. In plane optics (with $D = 1$-dim screens), the transformations of phase space which conserve light, i.e. respect the incompressible-fluid condition (3.9), *also* conserve the Hamilton equations (2.5), (2.6) and (2.8). This can be verified longhand replacing p, q and $h(p,q;z)$ by P, Q and $H(P(p,q),Q(p,q);z) = h(p,q;z)$, writing out the partial derivatives for p, q, and using again the Hamilton equations with the Jacobian condition (3.9) – this will be performed in the next section. For screen dimensions $D \geq 2$ however, this Liouville-type incompressibility condition is necessary but not sufficient. As counter-example, consider maps that scale images only, i.e., $(p_x, p_y, q_x, q_y) \mapsto (p_x, p_y, aq_x, a^{-1}q_y)$, $a \neq 0$, which have unit Jacobian but do not respect the vector form of the Hamilton equations. •

Remark: On phase space distribution functions. Of great current interest for quantum and wave models is a concept of phase space which incorporates the Heisenberg uncertainty relation, and is therefore 'fuzzy'. There are several (*quasi*)-probability distribution functions, among them the Wigner function [144]; and it is worrisome that most of these have the unsavoury property of being *negative* in small regions of phase space. However, it turns out that the rules to describe the act of observation save the hurdle: measurements of probability will always be non-negative. The beam function $\rho(\mathbf{p}, \mathbf{q})$ used in this section has no limitation of mathematical origin. •

3.2 Conservation of the Hamiltonian structure

Admissible maps of optical phase space must satisfy a set of conditions *stronger* than Lavoisier's light conservation: they must conserve the two basic postulates of Chap. 1. As we said above, a map between rays in optical phase space $\wp$ must preserve the Hamilton equations on the screen (2.5) and (2.6).

The Hamilton equations establish relations between the phase space differentials dp_i, dq_i (in the *tangent* $2D$-dim vector space of $\wp$), and the partial derivatives $\partial/\partial p_i$, $\partial/\partial q_i$ (in the *cotangent* $2D$-dim vector space). While optical phase space $(p_i, q_j) \in \wp$ is not a *vector* space,[4] it is advantageous to use vector notation and to sum over repeated indices, to bring the Hamilton equations (2.5) and (2.6) to the following form:

$$\begin{pmatrix} \dfrac{dp_i}{dz} \\ \dfrac{dq_i}{dz} \end{pmatrix} = \begin{pmatrix} 0 & -\delta_{i,j} \\ \delta_{i,j} & 0 \end{pmatrix} \begin{pmatrix} \dfrac{\partial}{\partial p_j} \\ \dfrac{\partial}{\partial q_j} \end{pmatrix} h(\mathbf{p}, \mathbf{q}; z). \tag{3.10}$$

Under the (generally nonlinear) map $(\mathbf{p}, \mathbf{q}) \mapsto (\mathbf{P}(\mathbf{p}, \mathbf{q}), \mathbf{Q}(\mathbf{p}, \mathbf{q}))$, the vectors of differentials and partial derivatives transform linearly through the Jacobian and transpose Jacobian matrices, respectively

$$\begin{pmatrix} dP_i \\ dQ_i \end{pmatrix} = \begin{pmatrix} \dfrac{\partial P_i}{\partial p_j} & \dfrac{\partial P_i}{\partial q_j} \\ \dfrac{\partial Q_i}{\partial p_j} & \dfrac{\partial Q_i}{\partial q_j} \end{pmatrix} \begin{pmatrix} dp_j \\ dq_j \end{pmatrix}, \tag{3.11}$$

[4] Since the range of optical momentum is bounded, linear combinations of ray 'vectors' $(\mathbf{p}, \mathbf{q})$ can be outside $\wp$. In mechanics on the other hand, the range of momentum is unbounded, so $(\mathbf{p}, \mathbf{q}) \in \mathsf{R}^{2D}$ *is* a vector space that may be subject to linear transformations.

$$\begin{pmatrix} \dfrac{\partial}{\partial p_i} \\ \dfrac{\partial}{\partial q_i} \end{pmatrix} = \begin{pmatrix} \dfrac{\partial P_j}{\partial p_i} & \dfrac{\partial Q_j}{\partial p_i} \\ \dfrac{\partial P_j}{\partial q_i} & \dfrac{\partial Q_j}{\partial q_i} \end{pmatrix} \begin{pmatrix} \dfrac{\partial}{\partial P_j} \\ \dfrac{\partial}{\partial Q_j} \end{pmatrix}. \tag{3.12}$$

The Hamilton equations (3.10) in the new coordinates are

$$\begin{aligned} \begin{pmatrix} \dfrac{dP_i}{dz} \\ \dfrac{dQ_i}{dz} \end{pmatrix} &= \begin{pmatrix} \dfrac{\partial P_i}{\partial p_j} & \dfrac{\partial P_i}{\partial q_j} \\ \dfrac{\partial Q_i}{\partial p_j} & \dfrac{\partial Q_i}{\partial q_j} \end{pmatrix} \begin{pmatrix} \dfrac{dp_j}{dz} \\ \dfrac{dq_j}{dz} \end{pmatrix} \\ &= \begin{pmatrix} \dfrac{\partial P_i}{\partial p_j} & \dfrac{\partial P_i}{\partial q_j} \\ \dfrac{\partial Q_i}{\partial p_j} & \dfrac{\partial Q_i}{\partial q_j} \end{pmatrix} \begin{pmatrix} 0 & -\delta_{j,k} \\ \delta_{j,k} & 0 \end{pmatrix} \begin{pmatrix} \dfrac{\partial}{\partial p_k} \\ \dfrac{\partial}{\partial q_k} \end{pmatrix} h(\mathbf{p}, \mathbf{q}; z) \\ &= \begin{pmatrix} \dfrac{\partial P_i}{\partial p_j} & \dfrac{\partial P_i}{\partial q_j} \\ \dfrac{\partial Q_i}{\partial p_j} & \dfrac{\partial Q_i}{\partial q_j} \end{pmatrix} \begin{pmatrix} 0 & -\delta_{j,k} \\ \delta_{j,k} & 0 \end{pmatrix} \begin{pmatrix} \dfrac{\partial P_\ell}{\partial p_k} & \dfrac{\partial Q_\ell}{\partial p_k} \\ \dfrac{\partial P_\ell}{\partial q_k} & \dfrac{\partial Q_\ell}{\partial q_k} \end{pmatrix} \begin{pmatrix} \dfrac{\partial}{\partial P_\ell} \\ \dfrac{\partial}{\partial Q_\ell} \end{pmatrix} H, \end{aligned} \tag{3.13}$$

where $H = H(\mathbf{P}(\mathbf{p}, \mathbf{q}), \mathbf{Q}(\mathbf{p}, \mathbf{q}); z) = h(\mathbf{p}, \mathbf{q}; z)$.

We regain the Hamilton equations (3.10) in P_i and Q_i when the product of the three matrices in the last member of (3.13) is again the matrix in (3.10). When this happens, the product of their determinants will be that of $\begin{pmatrix} 0 & -1 \\ 1 & 0 \end{pmatrix}$, namely unity; the product of the Jacobian determinant and its transpose is therefore unity and, as seen in the previous section, rays will not be lost nor gained under the transformation.[5] Moreover, (3.13) contain a full set of further conditions; writing out the matrix product we find:

$$\begin{pmatrix} -\dfrac{\partial P_i}{\partial p_k}\dfrac{\partial P_j}{\partial q_k} + \dfrac{\partial P_i}{\partial q_k}\dfrac{\partial P_j}{\partial p_k} & -\dfrac{\partial P_i}{\partial p_k}\dfrac{\partial Q_j}{\partial q_k} + \dfrac{\partial P_i}{\partial q_k}\dfrac{\partial Q_j}{\partial p_k} \\ -\dfrac{\partial Q_i}{\partial p_k}\dfrac{\partial P_j}{\partial q_k} + \dfrac{\partial Q_i}{\partial q_k}\dfrac{\partial P_j}{\partial p_k} & -\dfrac{\partial Q_i}{\partial p_k}\dfrac{\partial Q_j}{\partial q_k} + \dfrac{\partial Q_i}{\partial q_k}\dfrac{\partial Q_j}{\partial p_k} \end{pmatrix} = \begin{pmatrix} 0 & -\delta_{i,j} \\ \delta_{i,j} & 0 \end{pmatrix}. \tag{3.14}$$

To write these restrictions compactly, we introduce the following operation between functions of phase space:

[5] It is allowed for the Jacobian determinant to be *minus* one; this is the case for reflection. But a continuous line of canonical transforms connected to the unit transformation must have Jacobian +1.

Definition: *The* **Poisson bracket** *of two functions* $f(\mathbf{p},\mathbf{q})$ *and* $g(\mathbf{p},\mathbf{q})$ *is the function on phase space indicated by*

$$\{f,g\}(\mathbf{p},\mathbf{q}) := \frac{\partial f(\mathbf{p},\mathbf{q})}{\partial \mathbf{q}} \cdot \frac{\partial g(\mathbf{p},\mathbf{q})}{\partial \mathbf{p}} - \frac{\partial f(\mathbf{p},\mathbf{q})}{\partial \mathbf{p}} \cdot \frac{\partial g(\mathbf{p},\mathbf{q})}{\partial \mathbf{q}}. \tag{3.15}$$

In particular, see that the coordinate functions satisfy

$$\{p_i,p_j\} = 0, \quad \{q_i,q_j\} = 0, \quad \{q_i,p_j\} = \delta_{i,j}. \tag{3.16}$$

This can be used to define the bracket as well (once its operational properties are provided – see below). Now, using Poisson brackets, the equations (3.14) become

$$\begin{pmatrix} \{P_i,P_j\} & \{P_i,Q_j\} \\ \{Q_i,P_j\} & \{Q_i,Q_j\} \end{pmatrix} = \begin{pmatrix} 0 & -\delta_{i,j} \\ \delta_{i,j} & 0 \end{pmatrix}. \tag{3.17}$$

We thus characterize the maps of our interest.

Definition: *A map of phase space* $\mathcal{M} : (\mathbf{p},\mathbf{q}) \mapsto (\mathbf{P}(\mathbf{p},\mathbf{q}),\mathbf{Q}(\mathbf{p},\mathbf{q}))$ *that preserves the Hamiltonian structure is a* **canonical** *transformation. This occurs when*

$$\{P_i,P_j\} = 0, \quad \{Q_i,Q_j\} = 0, \quad \{Q_i,P_j\} = \delta_{i,j}. \tag{3.18}$$

If moreover $\mathcal{M} : \wp \mapsto \wp$, *then the map is an* **optical** *transformation.*

Remark: Canonical transformations compose. Two canonical transformations applied consecutively compose to a canonical transformation. Their Jacobian matrices [see (3.11)–(3.12)] multiply following the chain rule and so the product Jacobian has ±-unit determinant, and since each factor transformation preserves the basic Poisson brackets from (3.16) to (3.18), so does their concatenation. The same assertion holds for optical transformations that are required to map optical phase space $\wp$ on itself. •

Remark: Canonical transformations invert. Given a canonical transformation, the inverse transformation exists and is also canonical. Since the Jacobian determinant [again see (3.11)–(3.12)] is nonzero, the inverse matrix exists and leads from (3.18) back to (3.16), and is therefore canonical. The inverse transformation exchanges upper- and lower-case coordinates in these formulas, showing that the partial derivatives of the Poisson bracket can be taken with respect to any set of canonically conjugate coordinates obtained from an 'original' $(\mathbf{p},\mathbf{q})$ by a canonical transformation. The product of a transformation with its inverse yields the identity transformation, which is evidently canonical. The same statements apply to optical transformations. •

Example: Fractional Fourier transformation. In plane optics ($D=1$), the maps

$$\mathcal{F}^\alpha : p \mapsto P = p\cos\tfrac{1}{2}\pi\alpha + q\sin\tfrac{1}{2}\pi\alpha, \tag{3.19}$$

$$\mathcal{F}^\alpha : q \mapsto Q = -p\sin\tfrac{1}{2}\pi\alpha + q\cos\tfrac{1}{2}\pi\alpha, \tag{3.20}$$

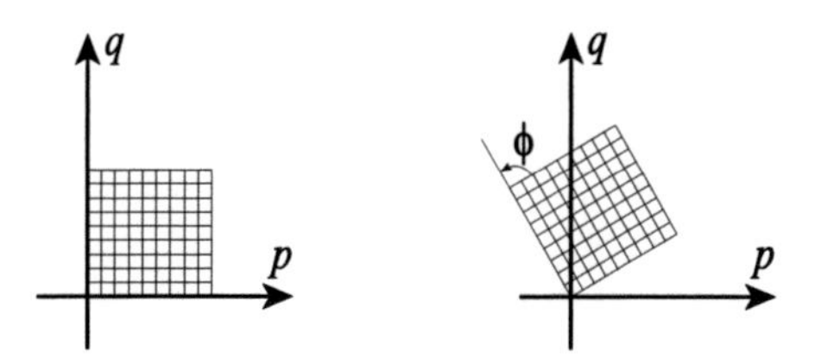

Fig. 3.2. A geometric fractional Fourier transformation $\mathcal{F}^\alpha$ rotates a patch phase space (*left*) by the angle $\phi = \frac{1}{2}\pi\alpha$ (*right*).

rotate the phase space plane $(p, q) \in \mathsf{R}^2$ by an angle $\phi = \frac{1}{2}\pi\alpha$, as shown in Fig. 3.2, and are called the (geometric) *fractional Fourier transformations* [112]. For $\alpha = 1$,

$$\mathcal{F} : p = q, \qquad \mathcal{F} : q = -p, \tag{3.21}$$

is the geometric counterpart of the Fourier *integral transform* between position and momentum representations of quantum mechanics [149], or frequency and time in wave optics.[6] In a fractional Fourier transformer, rays parallel to the optical axis, $(p{=}0, q)$ (the horizontal line of the figure) map to rays $(P, Q) = q(\sin\phi, \cos\phi)$ [cf. Fig. 2.5 for rays through the optical center $(p, q{=}0)$ of a guide], so the phase plane moves counterclockwise for $\phi > 0$, and[7] $\phi \equiv \phi + 2\pi$.[8] These transformations are canonical, but *cannot* be optical because they all map $\wp$ (where $|p| < n$ but $q \in \mathsf{R}$) out of itself. Fractional Fourier transformations compose as rotations do, i.e.,

$$\mathcal{F}^{\alpha_1}\mathcal{F}^{\alpha_2} = \mathcal{F}^{\alpha_1+\alpha_2}, \qquad \mathcal{F}^4 \equiv \mathcal{F}^0 = \mathit{1} \quad \text{(unit)}. \tag{3.22}$$

The subject of Sect. 10.3 is the generalization of the geometric fractional Fourier transform (3.19)–(3.20) to D dimensions. •

Example: Comatic maps. The family of maps $\mathcal{C}_\gamma : (\mathbf{p}, \mathbf{q}) \mapsto (\mathbf{P}, \mathbf{Q})$, $\gamma \in \mathsf{R}$, given by

$$\mathbf{P} = \frac{\mathbf{p}}{\sqrt{1 - 2\gamma|\mathbf{p}|^2}}, \quad \mathbf{Q} = \sqrt{1 - 2\gamma|\mathbf{p}|^2}\,(\mathbf{q} - 2\gamma|\mathbf{p}|^2\mathbf{p}\cdot\mathbf{q}\,\mathbf{p}), \tag{3.23}$$

are canonical because they satisfy the defining equations (3.18), as can be ascertained by working out the Poisson brackets. The maps are called *comatic* because, as we shall see in Sect. 3.7, $\mathbf{P}$ being a function only of $\mathbf{p}$, implies that the position coordinates $\mathbf{Q}$ will suffer the aberration of coma. Interestingly, $\mathcal{C}_{\gamma_1}\mathcal{C}_{\gamma_2} = \mathcal{C}_{\gamma_1+\gamma_2}$ and $(\mathcal{C}_\gamma)^{-1} = \mathcal{C}_{-\gamma}$. The comatic maps (3.23) are not *optical*

[6] Some people may prefer writing $\mathcal{F} : q = p$ and $\mathcal{F} : p = -q$; the signs we adopt here follow those of the Fourier integral transform for the 'Fourier–Schrödinger' operators of position and momentum, $q\cdot$ and $-i\frac{\lambda}{2\pi}d/dq$, detailed in appendix C.

[7] We write '$\equiv$' for *equivalent*, here meaning *modulo* 2π.

[8] In wave optics, motion along a harmonic waveguide results in equivalence of the angle ϕ modulo 4π, as we show in appendix C.

maps, because the limit circle of momentum, $|\mathbf{p}| = 1$, is sent to the circle $|\mathbf{P}| = 1/\sqrt{1-2\gamma}$, which for $\gamma \neq 0$ is different from unity. For $\gamma > 0$, all rays at angles greater than $|\sin\theta| = |\mathbf{p}| = 1/\sqrt{1+2\gamma}$ will thus be mapped out of $\wp$; for $\gamma < 0$ they have no inverse. •

Remark: The opening coma map. The comatic map (3.23) for $2\gamma = 1$ is very useful, since the circle $|\mathbf{p}| = 1$ is sent to infinity and $(\mathbf{P}, \mathbf{Q}) \in \mathsf{R}^{2D}$ *is* the phase space of mechanics. The transformation $\mathcal{C}_{1/2}$ has been called the *opening coma* map [92]. Its geometric interpretation is shown in Fig. 3.3: the map $\sin\theta \mapsto \tan\theta$ *opens* the rays of the $p_z > 0$ Descartes half-sphere onto the plane, projected from the center. The opening coma map provides another relation between optics and mechanics (which can be extended to the wave theories); it provides a 'half-global' transformation between forward rays and particle trajectories; backward rays $p_z < 0$ map onto *a second* mechanical momentum plane $\sigma = -$. •

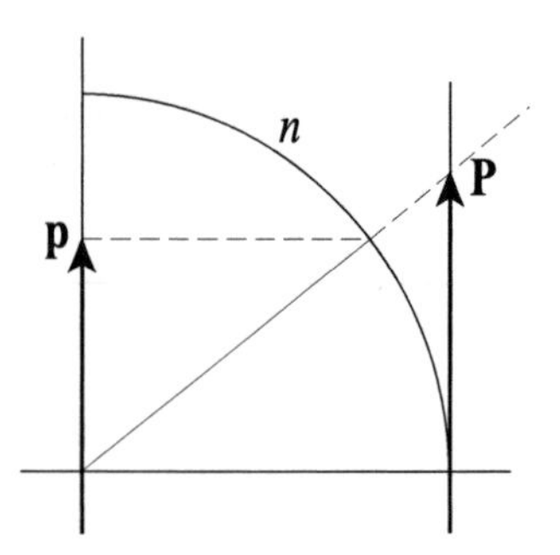

Fig. 3.3. The opening coma map $\mathcal{C}_{1/2}$ projects optical momentum $|\mathbf{p}| < n$ onto mechanical momentum $|\mathbf{P}| < \infty$. The Descartes hemisphere of forward rays (quarter-circle shown for $D = 1$) maps onto the plane (line in $D = 1$, to the right); backward rays map onto a second plane (not shown).

Remark: Where are the optical maps? The two previous examples were of canonical, but not *optical* transformations. Truly *optical* maps between $\wp$ and itself will be seen in Chap. 5: rigid Euclidean and relativistic Lorentz transformations of empty space, basically; and it would seem that few others can be found. But the opening coma map $\mathcal{C}_{1/2}$ remarked above allows us to produce, from every mechanical canonical transformation $\mathcal{M}^{\mathrm{M}} : \mathsf{R}^{2D} \mapsto \mathsf{R}^{2D}$, an optical transformation $\mathcal{M}^{\mathrm{O}} = \mathcal{C}_{1/2}^{-1}\mathcal{M}^{\mathrm{M}}\mathcal{C}_{1/2}$, for forward or for backward rays, or by properly stitching the two Descartes hemispheres together, for all of $\wp$. In Sect. 5.8 we find in this way an *optical* fractional Fourier transformation which *does* globally map $\wp$ onto itself. Optical transformations also occur in Chap. 6, in the guise of maps of the surface of a sphere $\mathcal{S}^D$. It must be admitted that the requirement of a map to be optical is a much too stringent condition for practical use in optical instruments. For this reason we will often neglect boundary conditions. •

We close this section with an enumeration of the properties of Poisson brackets, defined in (3.15) for functions $\rho_1, \rho_2, \ldots$ of phase space,

(i) **Linearity**: $\{\rho_1, c_1\rho_2 + c_2\rho_3\} = c_1\{\rho_1, \rho_2\} + c_2\{\rho_1, \rho_3\}, \quad c_i \in \mathsf{R}$.
(ii) **Skew-symmetry**: $\{\rho_1, \rho_2\} = -\{\rho_2, \rho_1\}$.
(iii) **Jacobi identity**: $\{\{\rho_1, \rho_2\}, \rho_3\} + \{\{\rho_2, \rho_3\}, \rho_1\} + \{\{\rho_3, \rho_1\}, \rho_2\} = 0$.
(iv) **Leibniz rule**: $\{\rho_1, \rho_2\,\rho_3\} = \{\rho_1, \rho_2\}\,\rho_3 + \rho_2\,\{\rho_1, \rho_3\}$.

These properties follow straightforwardly from the explicit form of the Poisson bracket as a first-order differential operator. Functions will be assumed to be as differentiable as necessary.

3.3 Hamiltonian evolution with Poisson brackets

The Poisson bracket defined in (3.15) invites elegance in writing and thinking about the evolution of phase space under the ægis of a Hamiltonian $h(z) = h(\mathbf{p}, \mathbf{q}; z)$. The Hamilton equations (2.5)–(2.6) for rays or beams are thus

$$\frac{d}{dz}\rho(\mathbf{p}, \mathbf{q}) = \frac{\partial \rho}{\partial \mathbf{p}} \cdot \frac{d\mathbf{p}}{dz} + \frac{\partial \rho}{\partial \mathbf{q}} \cdot \frac{d\mathbf{q}}{dz} = \frac{\partial \rho}{\partial \mathbf{p}} \cdot \frac{\partial h(z)}{\partial \mathbf{q}} - \frac{\partial \rho}{\partial \mathbf{q}} \cdot \frac{\partial h(z)}{\partial \mathbf{p}} = -\{h(z), \rho\}(\mathbf{p}, \mathbf{q}). \tag{3.24}$$

We can abstract this further by introducing the following:

Definition: *The* **Poisson operator** *of a function* $f(\mathbf{p}, \mathbf{q})$ *is*

$$\{f, \circ\} := \frac{\partial f}{\partial \mathbf{q}} \cdot \frac{\partial}{\partial \mathbf{p}} - \frac{\partial f}{\partial \mathbf{p}} \cdot \frac{\partial}{\partial \mathbf{q}}, \quad \text{i.e.} \quad \{f, \circ\}g(\mathbf{p}, \mathbf{q}) = \{f, g\}(\mathbf{p}, \mathbf{q}). \tag{3.25}$$

This operator is a vector of first phase-space derivatives. Using this notation, the Hamilton equations *and* their finite-difference approximations become

$$\frac{d}{dz} = -\{h(z), \circ\}, \qquad \Delta \approx -\{h(z), \circ\}\,\Delta z. \tag{3.26}$$

The difference form provides an algorithm for iterative numerical computation: To integrate the Hamilton equations (3.24) as we move the screen along the optical axis, we divide the journey from the standard screen $z = 0$ to the current screen at z in N small steps of length $\Delta z = z/N$. A ray or beam $\rho(z) = \rho(\mathbf{q}(z), \mathbf{p}(z))$, after the mth step between $z_m = m\Delta z$ and z_{m+1}, is

$$\rho(z_{m+1}) \approx \rho(z_m) + \left.\frac{d\rho(z)}{dz}\right|_{z_m} \Delta z = \Big(1 - \Delta z\{h(z_m), \circ\}\Big)\rho(z_m). \tag{3.27}$$

Thus we express $\rho(z)$ as an ordered product of N operators on $\rho(0)$,

$$\begin{aligned} \rho(z) &\approx (1 - \Delta z\{h(z_{N-1}), \circ\})(1 - \Delta z\{h(z_{N-2}), \circ\}) \times \cdots \\ &\qquad \cdots \times (1 - \Delta z\{h(z_1), \circ\})(1 - \Delta z\{h(z_0), \circ\})\rho(0) \\ &= \mathcal{T}_z \prod_{m=0}^{N-1} \Big(1 - \frac{z}{N}\{h(z_m), \circ\}\Big)\, \rho(0), \end{aligned} \tag{3.28}$$

where $\mathcal{T}_z$ is the *ordering symbol* that places earlier z's to the right and later ones to the left.

In guides where the Hamiltonian is explicitly independent of z (recall page 21), the ordering is unnecessary among the N equal factors in (3.28); for large N this yields the integration limit,

$$\rho(z) = \lim_{N\to\infty} \left(1 - \frac{z}{N}\{h, \circ\}\right)^N \rho(0). \tag{3.29}$$

3.4 One-parameter Lie groups

An optical guide medium exhibits symmetry under translations along the z-axis, and this leads to the important simplification in the evolution of its ray trajectories given by (3.29). We recall the product and sum limit definitions of the exponential function,[9]

$$\lim_{N\to\infty} \left(1 + \frac{x}{N}\right)^N = e^x = \lim_{N\to\infty} \sum_{m=0}^{N} \frac{x^m}{m!}, \tag{3.30}$$

which is formally valid also for first-order differential operator $x = -z\{h, \circ\}$.[10]

We use the limit series (3.30) to write the *evolution operator* that generates translation along the z-axis, (3.29), as the *Lie exponential* operator

$$\mathcal{G}(z) = \exp(-z\,\{h, \circ\}) := \sum_{m=0}^{\infty} \frac{(-z)^m}{m!}\{h, \circ\}^m \tag{3.31}$$

where $\{h, \circ\}^0 := 1$,

$$\{h, \circ\}^m := \{h, \circ\}\{h, \circ\}^{m-1} = \overbrace{\{h, \{h, \cdots \{h, \circ\} \cdots\}\}}^{m \text{ times}}. \tag{3.32}$$

This operator maps rays and beams from the standard screen to the screen at $z \in \mathsf{R}$, through

$$\mathcal{G}(z) : \mathbf{p} = \mathbf{p}(z) = \mathbf{P}(\mathbf{p}, \mathbf{q}; z), \quad \mathcal{G}(z) : \mathbf{q} = \mathbf{q}(z) = \mathbf{Q}(\mathbf{p}, \mathbf{q}; z). \tag{3.33}$$

When two or more z-translations are applied, they *compose* adding the parameters in their exponents:

$$\mathcal{G}(z_1)\,\mathcal{G}(z_2) = \mathcal{G}(z_1 + z_2), \quad \mathcal{G}(0) = 1, \quad (\mathcal{G}(z))^{-1} = \mathcal{G}(-z). \tag{3.34}$$

The properties of this family of evolution operators are the defining axioms of the following important concept:

---

[9] To check these formulas, see that the value of the product and series at $x = 0$ is 1 and that as $N \to \infty$, their first derivatives reproduce those of e^x.

[10] The numerical (*pointwise*) convergence of the series depends on the differentiability of the function space on which the operators act; we may assume *smooth* functions to be infinitely differentiable and of faster than exponential decrease.

Definition: *A* **one-parameter Lie group** *is a set G of elements $g(\alpha)$ parametrized by $\alpha \in \mathsf{R}$, with a (commutative) composition rule with the following properties,*

(i) **Closure***: $g(\alpha_1)\, g(\alpha_2) = g(\alpha_1 + \alpha_2)$ is in the set G.*
(ii) **Unit***: there exists one $g_o = g(0) \in G$ such that $g_o\, g(\alpha) = g(\alpha) = g(\alpha)\, g_o$ for all $g(\alpha) \in G$.*
(iii) **Inverse***: for every $g(\alpha) \in G$ there exists one inverse $g(\alpha)^{-1} = g(-\alpha) \in G$, such that $g(\alpha)\, g(-\alpha) = g_o$.*
(iv) **Associativity***: as implied by the sum, the composition of elements can be associated in any order, $(g(\alpha_1)\, g(\alpha_2))\, g(\alpha_3) = g(\alpha_1)\, (g(\alpha_2)\, g(\alpha_3))$.*

The (single) **generator**[11] *of the Lie group*[12] *is*

$$\widehat{f} := \left.\frac{dg(\alpha)}{d\alpha}\right|_{\alpha=0}, \qquad g(\alpha) = \exp(\,\alpha\, \widehat{f}\,). \tag{3.35}$$

When the equivalence relation $g(\alpha) \equiv g(\alpha + \mu)$ holds for a fixed μ, the 1-parameter Lie group is said to be cyclic and of length (or modulo) μ.[13]

Example: Free propagation in a homogeneous medium. This example was given before in (1.7) and (2.10). Here we find again the solution by using the evolution operator (3.31) in exponential series form. Since the free Hamiltonian $h^{\text{F}} := -\sqrt{n^2 - \mathbf{p}^2}$ depends only on $\mathbf{p}$, the series truncates: $\{h^{\text{F}}, \mathbf{p}\} = \mathbf{0}$, while $\{h^{\text{F}}, \mathbf{q}\} = \mathbf{p}/h^{\text{F}}$ and $\{h^{\text{F}}, \circ\}^2 \mathbf{q} = \mathbf{0}$, so

$$\begin{aligned} \exp(-z\{h^{\text{F}}, \circ\}) &: \mathbf{p} = \mathbf{p} - z\{h, \circ\}\mathbf{p} = \mathbf{p}, \\ \exp(-z\{h^{\text{F}}, \circ\}) &: \mathbf{q} = \mathbf{q} - z\{h, \circ\}\mathbf{q} = \mathbf{q} + z\mathbf{p}/\sqrt{n^2 - |\mathbf{p}|^2}. \end{aligned} \tag{3.36}$$

•

Example: Propagation in a paraxial oscillator guide. In Sect. 2.4 we introduced the small-angle limit of optical Hamiltonians and compared them with mechanical Hamiltonians. When the refractive index is $n(\mathbf{q}) \approx n_0 - \frac{1}{2} n_2 |\mathbf{q}|^2$, with $n_2 > 0$, the Hamiltonian is that of a harmonic guide, or mechanical oscillator,

$$h^{\text{H}}_{\text{prx}}(\mathbf{p}, \mathbf{q}) := \frac{1}{2n_0} |\mathbf{p}|^2 + \frac{n_2}{2} |\mathbf{q}|^2. \tag{3.37}$$

Using a vector-and-matrix arrangement, we write now the evolution as

[11] In (3.35) we use the hat notation for the generator f; this encompasses the Poisson operators $\{f, \circ\}$ that we use in most of this chapter, also the adjoint commutator $[f, \circ]$, and generally the Lie bracket $\{f, \circ\}$ defined in Sect. 3.7.

[12] When the parameter has a discrete range, for example $\alpha \in \mathcal{Z}$ (the integers), the group G is called a *discrete* Lie group with generator $g(1)$. We shall be interested in *continuous* one-parameter Lie groups, where $\alpha \in \mathsf{R}$.

[13] This marks the difference between the 1-parameter Lie groups of translations $\alpha \in \mathsf{R}$, and of rotations, where congruence is modulo $\mu = 2\pi$.

$$\exp(-z\{h^{\mathrm{H}}_{\mathrm{prx}},\circ\}) : \begin{pmatrix}\mathbf{p}\\ \mathbf{q}\end{pmatrix} = \begin{pmatrix}\cos\omega z & -\sqrt{n_0 n_2}\sin\omega z\\ \dfrac{1}{\sqrt{n_0 n_2}}\sin\omega z & \cos\omega z\end{pmatrix}\begin{pmatrix}\mathbf{p}\\ \mathbf{q}\end{pmatrix}. \tag{3.38}$$

where $\omega := \sqrt{n_2/n_0}$. To verify the equality it is sufficient to show that both sides have the same value and first z-derivative at the origin $z = 0$ of the 1-parameter Lie group. Note that to first order in z, the position $\mathbf{q}$ of the ray evolves similarly, $\mathbf{q}(z) \approx \mathbf{q} + z\mathbf{p}$ in the guide (3.38) as in free space (3.36). •

Remark: Fractional Fourier transformation. The fractional Fourier transform appears in (3.19)–(3.20). Note carefully that it differs by a *sign* from the previous harmonic guide evolution:

$$\mathcal{F}^\alpha : \begin{pmatrix}\mathbf{p}\\ \mathbf{q}\end{pmatrix} = \exp\left(\phi\{\tfrac{1}{2}(|\mathbf{p}|^2+|\mathbf{q}|^2),\circ\}\right)\begin{pmatrix}\mathbf{p}\\ \mathbf{q}\end{pmatrix} = \begin{pmatrix}\cos\phi & \sin\phi\\ -\sin\phi & \cos\phi\end{pmatrix}\begin{pmatrix}\mathbf{p}\\ \mathbf{q}\end{pmatrix}, \tag{3.39}$$

for $\alpha = 2\phi/\pi$. Thus, the generator of the 1-parameter cyclic Lie group of fractional Fourier transformations is *minus* the waveguide Hamiltonian for $n_0 = 1 = n_2$. •

As the previous examples indicate, Lie exponential operators and 1-parameter Lie groups of transformations of phase space can be generated by the Poisson operator (3.25) of any differentiable function $f = f(\mathbf{p},\mathbf{q})$. The exponential of a Poisson operator,

$$\mathcal{E}_f(\alpha) := \exp(\alpha\{f,\circ\}) = \sum_{m=0}^{\infty}\frac{\alpha^m}{m!}\{f,\circ\}^m, \tag{3.40}$$

applied to beam functions on phase space, $\rho = \rho(\mathbf{p},\mathbf{q})$ has the following properties [135]:

(i) **Linearity:** $\mathcal{E}_f(\alpha) : (c_1\rho_1 + c_2\rho_2) = c_1\mathcal{E}_f(\alpha) : \rho_1 + c_2\mathcal{E}_f(\alpha) : \rho_2$.
(ii) **Product distribution:**
$\mathcal{E}_f(\alpha) : (\rho_1\rho_2) = (\mathcal{E}_f(\alpha) : \rho_1)(\mathcal{E}_f(\alpha) : \rho_2)$, $\mathcal{E}_f(\alpha) : \rho^m = m\,\rho^{m-1}\,\mathcal{E}_f(\alpha) : \rho$.
(iii) **Poisson bracket preservation:**
$\mathcal{E}_f(\alpha) : \{\rho_1,\rho_2\} = \{\mathcal{E}_f(\alpha) : \rho_1,\ \mathcal{E}_f(\alpha) : \rho_2\}$.
(iv) **Action on function arguments:**
$\mathcal{E}_f(\alpha) : \rho(\mathbf{p},\mathbf{q}) = \rho(\mathcal{E}_f(\alpha) : \mathbf{p},\ \mathcal{E}_f(\alpha) : \mathbf{q})$.

Proof: The linearity of the transformation, *(i)*, is a consequence of the linearity of the Poisson operator $\{f,\circ\}$, its powers, and hence of each summand $\{f,\circ\}^m$ in the exponential series (3.40). The key property to examine is the distribution of products, *(ii)*, which derives from the Leibniz rule of the Poisson operators, $\{f,\circ\}(\rho_1\rho_2) = \{f,\rho_1\rho_2\} = \{f,\rho_1\}\rho_2 + \rho_1\{f,\rho_2\}$. We proceed by induction, stating now that its mth power distributes as[14]

[14] Here $\binom{m}{n} := m!/(m-n)!\,n!$ is the binomial coefficient. This computation involves the reordering of terms that leads to the Newton binomial theorem.

$$\{f,\circ\}^m(\rho_1\rho_2) = \sum_{n=0}^{m} \binom{m}{n} (\{f,\circ\}^n\rho_1)(\{f,\circ\}^{m-n}\rho_2). \tag{3.41}$$

and we prove it for $m+1$. Acting with $\{f,\circ\}$ on (3.41), where each summand is the product of two functions, $\{f,\circ\}^n\rho_1$ and $\{f,\circ\}^{m-n}\rho_2$, and so the Leibniz rule applies,

$$\begin{aligned}\{f,\circ\}^{m+1}(\rho_1\rho_2) &= \sum_{n=0}^{m} \binom{m}{n} (\{f,\circ\}^{n+1}\rho_1)(\{f,\circ\}^{m-n}\rho_2) \\ &\quad + \sum_{n=0}^{m} \binom{m}{n} (\{f,\circ\}^{n}\rho_1)(\{f,\circ\}^{m+1-n}\rho_2) \\ &= \sum_{n=0}^{m+1} \binom{m+1}{n} (\{f,\circ\}^{n}\rho_1)(\{f,\circ\}^{m+1-n}\rho_2). \end{aligned} \tag{3.42}$$

Now, replacing (3.41) in each summand of the Lie exponential series (3.40), and reordering the double summation, we find

$$\begin{aligned}\mathcal{E}_{f}(\alpha)\,(\rho_1\,\rho_2) &= \sum_{m=0}^{\infty} \frac{\alpha^m}{m!}\{f,\circ\}^m(\rho_1\,\rho_2) \\ &= \sum_{m=0}^{\infty}\sum_{n=0}^{m} \frac{\alpha^m}{(m-n)!\,n!}(\{f,\circ\}^n\rho_1)(\{f,\circ\}^{m-n}\rho_2) \\ &= \sum_{n=0}^{\infty}\sum_{\ell=0}^{\infty} \frac{\alpha^n}{n!}\frac{\alpha^\ell}{\ell!}(\{f,\circ\}^n\rho_1)(\{f,\circ\}^{\ell}\rho_2) \quad \text{is } (ii). \end{aligned} \tag{3.43}$$

This result extends evidently to any number of factors or powers. □

Proof: (Cont.) The proof of the preservation of Poisson brackets $\{\rho_1,\rho_2\}$ in *(iii)* follows from the previous one for the products $(\partial\rho_1/\partial q_i)(\partial\rho_2/\partial p_i)$ and $(\partial\rho_1/\partial p_i)(\partial\rho_2/\partial q_i)$. Most important is the ability of Lie transformations $\mathcal{M}_f$ to 'jump into' a function's arguments, *(iv)*, and thus act directly on the coordinates of phase space (cf. the remark on page 30). To prove this, one expands $\rho(\mathbf{p},\mathbf{q})$ in *its* Taylor series in powers of p_i, q_j – wherever convergent. Linearity and product distribution apply to the terms in this series, which is then summed back with the transformed coordinates $\mathcal{E}_f : p_i$ and $\mathcal{E}_f : q_j$. □

The literature on modern Hamiltonian techniques is very wide; we can claim neither complete rigor (see the texts [132], [135] and [62, Chap. 1] with application to the mathematical view of geometric optics), nor complete coverage of its applications.

3.5 Hamiltonian flow of phase space

The action of a 1-parameter Lie group of Lie exponential operators $\mathcal{E}_f(\alpha) = \exp\alpha\{f,\circ\}$, $\alpha \in \mathsf{R}$, on phase space $(\mathbf{p},\mathbf{q}) \in \wp$ (or on R^{2D}), is to generate a foliation of $\wp$ (or of R^{2D}) composed of all trajectories $(\mathbf{p}(\alpha), \mathbf{q}(\alpha))$. The Hamiltonian *flow* of $f = f(\mathbf{p}, \mathbf{q})$ is the $2D$-dim vector field which is tangent to the trajectories, i.e.

$$\begin{pmatrix} d\mathbf{p}(\alpha)/d\alpha \\ d\mathbf{q}(\alpha)/d\alpha \end{pmatrix} = \begin{pmatrix} \{f, \circ\}\mathbf{p} \\ \{f, \circ\}\mathbf{q} \end{pmatrix} = \begin{pmatrix} \partial f/\partial \mathbf{q} \\ -\partial f/\partial \mathbf{p} \end{pmatrix}. \tag{3.44}$$

[Cf. (1.3)–(1.5) and (2.5)–(2.6).] It can be easily checked that the $2D$-vector (3.44) is orthogonal to the $2D$-gradient of f:

$$\nabla_{\wp} f := \begin{pmatrix} \partial/\partial \mathbf{p} \\ \partial/\partial \mathbf{q} \end{pmatrix} f \quad \perp \quad \{f, \circ\} \begin{pmatrix} \mathbf{p} \\ \mathbf{q} \end{pmatrix}. \tag{3.45}$$

Finally, from $\{f, f\} = 0$ it follows that $\mathcal{E}_f(\alpha) f = f$, and therefore the function that generates the flow is invariant: $f(\mathbf{p}(\alpha), \mathbf{q}(\alpha)) = f(\mathbf{p}(0), \mathbf{q}(0))$; the flow is along the lines $f(\mathbf{p}, \mathbf{q}) =$ constant.

Remark: Lie exponential flows are canonical. A consequence of the orthogonality of the vectors in (3.44) is that the transformations of phase space generated by functions $f = f(\mathbf{p}, \mathbf{q})$, the Lie exponential operators $\mathcal{E}_f(\alpha) = \exp \alpha\{f, \circ\}$, are *canonical* maps of phase space. To prove this, consider a sequence of small increments $\Delta\alpha$ in the parameter [cf. the difference (3.26)]; the coordinate or beam function $\rho(\alpha) = \rho(\mathbf{p}(\alpha), \mathbf{q}(\alpha))$ will evolve according to

$$\rho(\alpha + \Delta\alpha) \approx \rho(\alpha) + \Delta\alpha \left. \frac{d\rho(\alpha')}{d\alpha'} \right|_{\alpha'=\alpha} = (1 + \Delta\alpha \{f, \circ\}) \rho(\alpha). \tag{3.46}$$

For two such functions ρ_1, ρ_2, and up to terms of first degree in $\Delta\alpha$, their Poisson bracket evolves through

$$\begin{aligned} &\{\rho_1(\alpha + \Delta\alpha), \rho_2(\alpha + \Delta\alpha)\} \\ &\approx \{\rho_1(\alpha), \rho_2(\alpha)\} \\ &\quad + (\{\rho_1(\alpha), \{f, \rho_2(\alpha)\}\} + \{\{f, \rho_1(\alpha)\}, \rho_2(\alpha)\})\, \Delta\alpha \\ &= \{\rho_1(\alpha), \rho_2(\alpha)\} + \{f, \{\rho_1(\alpha), \rho_2(\alpha)\}\}\, \Delta\alpha. \end{aligned} \tag{3.47}$$

When ρ_1 and ρ_2 are phase space coordinates [cf. (3.18)], then $\{\rho_1, \rho_2\} =$ constant and the $\Delta\alpha$ term in (3.47) is zero. The flow of phase space from α to $\alpha + \Delta\alpha$ is thus canonical *modulo* second-order terms. When $\Delta\alpha \to 0$, evolution becomes continuous in α, and the canonicity becomes exact. (This statement can be proven also by noting the property of Lie exponential transformations to preserve Poisson brackets, seen in page 36 above.) •

3.6 Some aberrations and their Fourier conjugates

We have seen that a function of phase space $f = f(\mathbf{p}, \mathbf{q})$ generates a 1-parameter Lie group, consisting of elements

$$\mathcal{E}_f(\alpha) = \exp(\alpha\{f, \circ\}), \qquad \alpha \in \mathsf{R}. \tag{3.48}$$

In this section, we consider two generic classes of functions for which $\{f, \circ\}$ can be exponentiated explicitly, giving closed-form canonical maps of phase space. These transformations are generally nonlinear, and thus in optics they deserve the name of *aberrations*. We do not impose the condition that they be *optical* transformations. A contextual account of aberrations is given in Chap. 4 and occupies Part IV.

3.6.1 Spherical aberration and pocus

Scalar functions $f(\mathbf{p})$ of momentum *only*, generate the action

$$\exp(\alpha\{f(\mathbf{p}), \circ\}) \begin{pmatrix} \mathbf{p} \\ \mathbf{q} \end{pmatrix} = \begin{pmatrix} \mathbf{p} \\ \mathbf{q} - \alpha \nabla_{\mathbf{p}} f(\mathbf{p}) \end{pmatrix}. \tag{3.49}$$

Strictly speaking, this is called *spherical aberration* only when when the generating function $f(\mathbf{p})$ is $-\sqrt{n^2 - |\mathbf{p}|^2}$, the Hamiltonian of a homogeneous medium. We extend this generic name to any map which leaves the momentum (ray direction) invariant and thus only moves the ray position on the screen by a (derivative) function of momentum.

Remark: Linear and quadratic functions. There are two cases of special interest: when $f(\mathbf{p})$ is a linear or a quadratic function in the components of momentum. Linear functions $f(\mathbf{p}) = \mathbf{t} \cdot \mathbf{p}$ generate *translations* within the screen,

$$\exp(\alpha\{\mathbf{t} \cdot \mathbf{p}, \circ\}) \begin{pmatrix} \mathbf{p} \\ \mathbf{q} \end{pmatrix} = \begin{pmatrix} \mathbf{p} \\ \mathbf{q} - \alpha \mathbf{t} \end{pmatrix}. \tag{3.50}$$

Quadratic functions generate linear transformations of ray positions on the screen; in particular,

$$\exp(\alpha\{|\mathbf{p}|^2, \circ\}) \begin{pmatrix} \mathbf{p} \\ \mathbf{q} \end{pmatrix} = \begin{pmatrix} \mathbf{p} \\ \mathbf{q} - 2\alpha \mathbf{p} \end{pmatrix} = \begin{pmatrix} \mathbf{1} & \mathbf{0} \\ -2\alpha\mathbf{1} & \mathbf{1} \end{pmatrix} \begin{pmatrix} \mathbf{p} \\ \mathbf{q} \end{pmatrix}, \tag{3.51}$$

This is free propagation in a homogeneous medium of the paraxial régime (Part III), where the Hamiltonian is (2.22). •

One-parameter Lie groups of maps can be subject to *Fourier transformation*, i.e.,

$$\mathcal{E}_{\widetilde{f}}(\alpha) = \mathcal{F}\,\mathcal{E}_f(\alpha)\,\mathcal{F}^{-1}, \qquad \widetilde{f}(\mathbf{p}, \mathbf{q}) = f(\mathbf{q}, -\mathbf{p}). \tag{3.52}$$

The Fourier transform of the functions of momentum $f(\mathbf{p})$ that generate spherical aberrations, are functions of position only, $\widetilde{f}(\mathbf{q})$. These have Poisson operators $\{\widetilde{f}, \circ\} = \nabla_{\mathbf{q}} \widetilde{f} \cdot \partial/\partial \mathbf{p}$ which act now only on the momentum coordinate. Again, the exponential series (3.48) truncates, yielding

$$\exp(\alpha\{\widetilde{f}(\mathbf{q}), \circ\}) \begin{pmatrix} \mathbf{p} \\ \mathbf{q} \end{pmatrix} = \begin{pmatrix} \mathbf{p} + \alpha \nabla_{\mathbf{q}} \widetilde{f}(\mathbf{q}) \\ \mathbf{q} \end{pmatrix}. \tag{3.53}$$

Under these aberrations, the positions of rays on the screen and the beam illumination are invariant; there is no change in the image. But ray momenta $\mathbf{p}$ translate to $\mathbf{p}+$ vector function of $\mathbf{q}$. These are $\mathbf{p}$-unfocusing *aberrations*, for which the cute name of *pocus* is suggested.[15]

Remark: Linear and quadratic functions. A linear function $\mathbf{u}\cdot\mathbf{q}$ generates translations of momentum by α in the direction of $\mathbf{u}$,

$$\exp(\alpha\{\mathbf{u}\cdot\mathbf{q},\circ\})\begin{pmatrix}\mathbf{p}\\ \mathbf{q}\end{pmatrix}=\begin{pmatrix}\mathbf{p}+\alpha\mathbf{u}\\ \mathbf{q}\end{pmatrix}. \tag{3.54}$$

For small $|\alpha|$, this is realized by a *thin prism* which inclines all rays by the same amount and direction. The next important case is that of quadratic functions $\mathbf{q}\cdot\mathbf{G}\mathbf{q}$, with $\mathbf{G}=||G_{i,j}||$ a $D\times D$ symmetric matrix, which generates *linear* transformations

$$\exp(\alpha\{g(\mathbf{q}),\circ\})\begin{pmatrix}\mathbf{p}\\ \mathbf{q}\end{pmatrix}=\begin{pmatrix}\mathbf{p}+2\alpha\mathbf{G}\mathbf{q}\\ \mathbf{q}\end{pmatrix}=\begin{pmatrix}\mathbf{1} & 2\alpha\mathbf{G}\\ \mathbf{0} & \mathbf{1}\end{pmatrix}\begin{pmatrix}\mathbf{p}\\ \mathbf{q}\end{pmatrix}. \tag{3.55}$$

This transformation has the effect of an astigmatic *thin lens.* •

3.6.2 Distorsion and coma

There is a class of functions that will generate *distorsion* of images on the screen, but without affecting their sharpness. This happens when $\mathbf{q}\mapsto\mathbf{q}'(\mathbf{q})$, i.e., when the image position depends only on the object position, and not on the directions of the rays joining the two. These functions are *linear* in momentum and generic in position. For simplicity, here we work out the $D=1$-dim case only.

Consider phase space functions of the form $f=f(p,q)=p\,g(q)$ with a positive generic function g of position only. Because $\{p\,g,\circ\}q=-g(q)$, the terms of the Lie exponential operator series (3.31) will produce functions of q only, and thus points q on the object screen are mapped on points of the distorted image $q(\alpha)=Q(q,\alpha)$. To find this function, we make the change of variable

$$\{f,\circ\}Q=-g(q)\,\frac{d}{dq}Q=\frac{d}{d\chi}Q, \tag{3.56}$$

$$\text{for}\quad \chi(q)=-\int^{q}\frac{dq'}{g(q')}, \tag{3.57}$$

[15] The nickname *pocus* passed the referees of [108]; H.A. Buchdahl, in [25, Sect. 121 and Sect. 125], recognizes this aberration when it appears in composition with free flight, and names it g_4. Except for *phase*, no other distinctive name seems to have been suggested in the literature. Ludwig Seidel [125] did not include pocus among his first list of aberrations because it leaves images invariant.

up to an additive constant. Now, the exponential series of $d/d\chi$ is easy to integrate: $\exp(\alpha\, d/d\chi)\, X(\chi) = X(\chi + \alpha)$ on any function $X(\chi)$. Denoting the function inverse to χ by χ^{-1}, we find that $q = \chi^{-1}(\chi(q)) \mapsto Q(q, \alpha) = \chi^{-1}(\chi(q) + \alpha)$ is the generic distorsion of the ray position on the image.

Next, the action of distorsion on the momentum, $p(\alpha) = P(p, q, \alpha)$, can be determined easily through recalling that the generating function of the flow is an invariant: $f(P, Q) = P\, g(Q) = f(p, q) = p\, g(q)$, and hence $P = p\, g(q)/q(Q)$. The full phase space map is thus

$$\exp(\alpha\{p\, g(q), \circ\}) \begin{pmatrix} p \\ q \end{pmatrix} = \begin{pmatrix} \dfrac{p\, g(q)}{g\big(\chi^{-1}(\chi(q) + \alpha)\big)} \\ \chi^{-1}(\chi(q) + \alpha) \end{pmatrix}, \tag{3.58}$$

and we are assured that it is canonical.

The Fourier transform of distorsion is called *coma* (for reasons that will become apparent on $D = 2$-dim screens in Chap. 5). It is generated by the functions $\widetilde{f} = q\, g(-p)$, found by replacing $p \mapsto q$ and $q \mapsto -p$ in (3.58). It follows that

$$\exp(\alpha\{q\, g(p), \circ\}) \begin{pmatrix} p \\ q \end{pmatrix} = \begin{pmatrix} \eta^{-1}(\eta(q) + \alpha) \\ \dfrac{q\, g(p)}{g\big(\eta^{-1}(\eta(p) - \alpha)\big)} \end{pmatrix}, \tag{3.59}$$

where $\eta(p) = -\int^p dp'/g(p')$ [cf. (3.57)].

Example: Scaling. One special case of both distorsion (3.58) and coma (3.59) is provided by the function $f(p, q) = pq$. In (3.57) this corresponds to $g(q) = q$ and $\chi(q) = -\int^q dq'/q' = -\ln q$. So we have a *scaling* map $\chi^{-1}(\chi(q) + \alpha) = \exp(-\chi(q) + \alpha) = e^{-\alpha} q$ or, in vector-and-matrix form,

$$\exp(\alpha\{\mathbf{p} \cdot \mathbf{q}, \circ\}) \begin{pmatrix} \mathbf{p} \\ \mathbf{q} \end{pmatrix} = \begin{pmatrix} e^{\alpha}\mathbf{p} \\ e^{-\alpha}\mathbf{q} \end{pmatrix} = \begin{pmatrix} e^{\alpha} & 0 \\ 0 & e^{-\alpha} \end{pmatrix} \begin{pmatrix} \mathbf{p} \\ \mathbf{q} \end{pmatrix}. \tag{3.60}$$

•

Remark: Spot diagrams. In real 3-dim optical media with $D = 2$-dim screens, phase space is four-dimensional. There, understanding canonical transformations by phase space diagrams such as Figs. 2.2, 2.4, or 2.5, can be a problem. To get a better view we can draw spot or spots diagrams. A **spot diagram** is the *image* of a point source at $\mathbf{q}_o$, i.e., the plot of $\mathbf{Q}(\mathbf{p}, \mathbf{q}_o)$ as a function of two-dimensional ray direction $\mathbf{p}$. Although Cartesian coordinates can be used for the momentum plane, it is best to use polar coordinates [which are spherical coordinates (2.2) on the Descartes sphere], drawing the coordinate grid for several values of θ and ϕ. In Fig. 3.4 we show the spot diagram for coma, generated by a $D = 2$ scalar function $\mathbf{q} \cdot \mathbf{p}\, g(\mathbf{p})/|\mathbf{p}|$: the ray along the optical axis $\theta = 0$ maps on a point while the ray cones $\theta =$ constant map on circles that open like a funnel, and the meridians $\phi =$ constant map on lines that issue from the apex $\theta = 0$. •

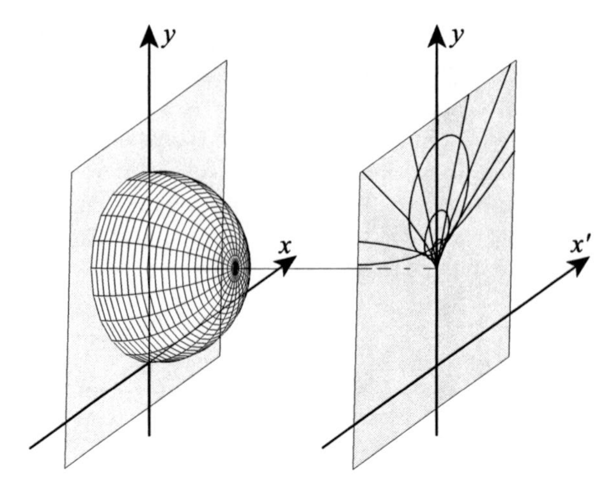

Fig. 3.4. Free rendering of the sphere of rays issuing from a point and its spot diagram on a screen under comatic aberration. Cones of fixed collatitude of the sphere map on circles, while meridian planes map on lines that issue from the apex of the spot.

Remark: Spots diagrams. In fact, coma is a *field dependent* aberration, which means that the *size* of the spot (i.e., the rays inside a cone of a given colatitude θ_o) depends on the *location* of the point source $\mathbf{q}_o$ on the screen. In Chap. 14 we shall show grids of spot diagrams corresponding to grids of point sources. We will refer to them as **spots diagram** (no confusion should arise due to the **s** since they are recognizably different). In spherical aberration (3.49), the circles of the spot diagram are concentric and the *size* of the spot is independent of its location on the screen, so the spots diagram repeats the same spot all over the screen. Thus, spherical aberration is said to be *field-independent.* •

3.7 Multiparameter Lie algebras and groups

We now consider two or more 1-parameter Lie groups of transformations of phase space, generated by distinct functions $f_1(\mathbf{p},\mathbf{q})$, $f_2(\mathbf{p},\mathbf{q})$, etc., of canonical maps

$$\mathcal{G}_1(\alpha) = \exp(\alpha\,\{f_1,\circ\}), \qquad \mathcal{G}_2(\beta) = \exp(\beta\,\{f_2,\circ\}). \tag{3.61}$$

[See (3.31) and (3.35).] The flows generally do not preserve each other.

When the two distinct concatenations of transformations

$$\mathcal{G}_{12}(\alpha) := \mathcal{G}_1(\alpha)\,\mathcal{G}_2(\alpha) \quad \text{and} \quad \mathcal{G}_{21}(\alpha) := \mathcal{G}_2(\alpha)\,\mathcal{G}_1(\alpha) \tag{3.62}$$

exist,[16] the information about *how* they differ is contained in the *group commutator*:

$$\mathcal{G}_{[1,2]}(\alpha) := \mathcal{G}_{12}(\alpha)\,(\mathcal{G}_{21}(\alpha))^{-1} = \mathcal{G}_1(\alpha)\,\mathcal{G}_2(\alpha)\,\mathcal{G}_1(-\alpha)\,\mathcal{G}_2(-\alpha). \tag{3.63}$$

If $\mathcal{G}_{[1,2]}(\alpha) = \mathit{1} = \mathcal{G}_{[2,1]}(\alpha)$ (the identity transformation), then the two 1-parameter Lie groups (3.61) *commute.* If not, the composition of the four sides of the square in (3.63) will not return to the starting point *1*, but to a point that we can picture as being outside the plane of the two 1-parameter group lines.

For small α let us expand the four exponential series in (3.63) keeping terms up to α^2, so that $\mathcal{G}(\pm\alpha) \approx \mathit{1} \pm \alpha\,\{f,\circ\} + \frac{1}{2!}\alpha^2\,\{f,\circ\}^2$. Replacing this into (3.63), a straightforward computation shows that all terms linear in α and all $\{f,\circ\}^2$'s *cancel.* What remains is

$$\mathcal{G}_{[1,2]}(\alpha) \approx \mathit{1} + \alpha^2\,(\{f_1,\circ\}\{f_2,\circ\} - \{f_2,\circ\}\{f_1,\circ\}). \tag{3.64}$$

The group commutator is thus on a *line* $\alpha \in \mathsf{R}$, with the value $\mathcal{G}_{[1,2]}(0) = \mathit{1}$ at the origin, and its generator can be found by writing $\mathcal{G}_{[1,2]}(\alpha) \approx 1 + \alpha^2\{f_{[1,2]},\circ\}$, with the *commutator* of the two Poisson operators, i.e.

$$\begin{aligned} \{f_{[1,2]},\circ\} &:= \{f_1,\circ\}\{f_2,\circ\} - \{f_2,\circ\}\{f_1,\circ\} \\ &= \{f_1,\{f_2,\circ\}\} - \{f_2,\{f_1,\circ\}\} \\ &= \{f_1,\{f_2,\circ\}\} + \{f_2,\{\circ,f_1,\}\} = \{\{f_1,f_2\},\circ\}. \end{aligned} \tag{3.65}$$

The intermediate steps can be verified applying the Poisson operators to the coordinates of a ray $(\mathbf{p},\mathbf{q})$ or to a beam function $g(\mathbf{p},\mathbf{q})$. Thus, given two functions f_1, f_2, the commutator of their Poisson operators is the Poisson operator of the *Poisson bracket* of the two functions. As a corollary, if the Poisson bracket of two functions is zero, then their Poisson operators will commute, and so will their two corresponding flows of phase space.

Example: Composition of distinct phase space maps. In Sect. 3.6 we introduced several families of functions and their maps. Any two spherical aberration maps (3.49) will commute since their Poisson bracket is zero. Comas (3.59) however, have nonzero commutators within the family and with the previous one, thus:

$$\begin{gathered} \{f_1(p), f_2(p)\} = 0, \quad \{f(p), q\,g(p)\} = -(f'\,g)(p), \\ \{q\,g_1(p), q\,g_2(p)\} = q\,(g_1\,g_2' - g_1'\,g_2)(p), \end{gathered} \tag{3.66}$$

where primes indicate derivatives. Similar relations hold, after Fourier transformation, for the families of pocus (3.53) and distorsion (3.58). Scaling (3.60), under Poisson brackets with any of the previous four families, yields a member of the same family. All these relations will be used systematically in Part IV. •

After having defined 1-parameter Lie groups of transformations, and seen how two of them, generated with different Poisson operators, compose into

[16] Note that $\mathcal{G}_1$ and $\mathcal{G}_2$ have the same parameter α, and that at $\alpha = 0$ the two products are the identity transformation $\mathcal{G}(0) = \mathit{1}$.

a third 1-parameter Lie group, we proceed with the formal definition of Lie algebras for the generators.

Definition: *A* **Lie algebra** a *is a linear vector space (whose elements we indicate by* $f, g, h, \ldots$*), subject to an extra operation called* **Lie bracket**, *which is a map* $\{\cdot,\cdot\} : \mathsf{a} \times \mathsf{a} \mapsto \mathsf{a}$, *satisfying*

(i) **Linearity***:* $\{f, c_1 g + c_2 h\} = c_1\{f, g\} + c_2\{f, h\}, \quad c_i \in \mathsf{R}.$
(ii) **Skew-symmetry***:* $\{f, g\} = -\{g, f\}.$
(iii) **Jacobi identity***:* $\{\{f, g\}, h\} + \{\{g, h\}, f\} + \{\{h, f\}, g\} = 0.$

The **dimension** *of the Lie algebra* a *is its dimension as a vector space; when this is infinite, we call* a *a* **functional** *Lie algebra.*

Definition: *A Lie algebra* a *is called a* **derivation**, *when among its elements there is defined one further (generally noncommutative) operation of associative multiplication,* $\diamond : \mathsf{a} \times \mathsf{a} \mapsto \bar{\mathsf{a}}$ *and* $\diamond : \mathsf{a} \times \bar{\mathsf{a}} \mapsto \bar{\mathsf{a}}$, *to the* **covering algebra** $\bar{\mathsf{a}}$, *which contains all formal products* $f \diamond g \diamond \cdots \diamond h$. *The operation* $\diamond$ *distributes with respect to the vector operations in* a*; with respect to the Lie bracket in* a *it satisfies a form of the Leibniz rule, called*

(iv) **Derivation** *property:* $\{f, g \diamond h\} = \{f, g\} \diamond h + g \diamond \{f, h\}.$

A denumerable basis $\{\widehat{f}_i\}$ of a (as a vector space) can be used to present its algebraic structure through the Lie brackets between the vectors of that basis, as

$$\{\widehat{f}_i, \widehat{f}_j\} = \textstyle\sum_k c^k_{i,j} \widehat{f}_k, \tag{3.67}$$

which is fully characterized by the *structure constants* $\{c^k_{i,j}\}$. Due to the properties of the Lie bracket, these constants must be skew-symmetric, i.e. $c^k_{i,j} = -c^k_{j,i}$, and satisfy the Jacobi identity in the form $\sum_l c^l_{i,j} c^m_{l,k}$ + (cyclic permutations of i, j, k) $= 0$.

Example: Poisson brackets are Lie brackets. The properties *(i)–(iii)* of a Lie algebra are satisfied by functions of phase space under Poisson brackets (see page 33). When the multiplication $\diamond$ is the ordinary commutative pointwise product of functions, $(f \diamond g)(\mathbf{p}, \mathbf{q}) = f(\mathbf{p}, \mathbf{q})\, g(\mathbf{p}, \mathbf{q}) = (fg)(\mathbf{p}, \mathbf{q})$, then the set of functions forms a functional derivation Lie algebra with the property *(iv)*, and the covering algebra coincides with the algebra itself. Alternatively, one can start with the basic Lie algebra of phase space coordinates (3.16) and, using the derivation property, build its covering algebra of all analytic functions of phase space. •

Example: Commutators are Lie brackets. Let now $\widehat{f}, \widehat{g}, \ldots$ be operators whose product $\diamond$ is ordered, such as $\widehat{f} = \{f, \circ\}$. When the Lie bracket is the commutator,

$$\{\widehat{f}, \widehat{g}\} = [\widehat{f}, \widehat{g}] := \widehat{f} \diamond \widehat{g} - \widehat{g} \diamond \widehat{f}, \tag{3.68}$$

then the derivation property *(iv)* is satisfied, as can be seen from $[\widehat{f}, \widehat{g} \diamond \widehat{h}] = \widehat{f} \diamond \widehat{g} \diamond \widehat{h} - \widehat{g} \diamond \widehat{h} \diamond \widehat{f}$, by adding and subtracting $\widehat{g} \diamond \widehat{f} \diamond \widehat{h}$, and regrouping terms.

In (3.66) we see that the commutator of two Poisson operators is a Poisson operator, but we note carefully that the single product of two operators, $\widehat{f} \diamond \widehat{g}$, is a differential operator of *second* order, and thus *not* a Poisson operator. In this case the covering algebra $\bar{\mathsf{a}}$ properly contains a. •

Warning: Algebras of Poisson brackets and of commutators may behave differently. When $\widehat{f} \diamond \widehat{g} \neq \widehat{g} \diamond \widehat{f}$, as is generally the case between operators, the derivation property holds but the derivation Lie algebras may be *different* from those of the commutative case. The most famous example of this non-correspondence is the 'quantization problem' of classical observables [148, Sect. IV]. Classical Poisson brackets and commutators of their Schrödinger operators will follow each other only for phase space functions of the form $q\,f(p) + g(p)$ (coma + spherical aberration), of the form $p\,f(q) + g(q)$ (distorsion + pocus), or of the form $p^m q^n$ for $m + n \leq 2$. When $m + n > 2$, there is *no* quantization scheme in which Poisson brackets and commutators can form the same algebra. •

Example: Commutators of matrices. Let the vector space m be the set of $N \times N$ matrices $\mathbf{m}$; by providing the Lie brackets given by the commutator of these matrices we obtain the Lie algebra (also denoted m). Here it is natural to introduce the $\diamond$ operation to be the usual (noncommutative) matrix multiplication, which maps $\mathsf{m} \times \mathsf{m} \mapsto \mathsf{m}$ and makes m a derivation Lie, and again the covering algebra coincides with the algebra itself, $\mathsf{a} = \bar{\mathsf{a}}$. •

In Sect. 3.4 we saw that 1-parameter Lie algebras generate 1-parameter *Lie groups*. Now, multiparameter Lie algebras a are expected to generate corresponding groups with the same number of parameters. But first we should publicly define an abstract *group* (without further qualification) by its four classical axioms:

Definition: *A* **group** *is a set G of elements $g_1, g_2, \ldots$ (finite or infinite), with a composition rule (omitted in the product notation) that has the following properties,*

(i) **Closure**: *for all $g_1, g_2 \in G$, also $g_1\,g_2 \in G$.*
(ii) **Unit**: *there exists one $g_o \in G$ such that $g_o\,g = g = g\,g_o$, for all $g \in G$.*
(iii) **Inverse**: *for every $g \in G$, there exists one element $g^{-1} \in G$ such that $g\,g^{-1} = g_o = g^{-1}g$.*
(iv) **Associativity**: *for all elements in G, $g_1(g_2 g_3) = (g_1 g_2) g_3$.*

Groups can be finite, discrete, continuous (or *wild*) according to whether the set of group elements is finite, countable, or is characterized by (local) coordinates from a parameter patch $\boldsymbol{\alpha} = \{\alpha_i\} \in \mathsf{R}^N$ [56]. In this case they are called topological groups [71],[17] and have elements $g(\boldsymbol{\alpha})$, where one may

[17] Mathematicians have studied abstract groups for more than a century. Axioms have been weakened to lack the inverse (*semigroups*, with or without unit), lack unit and inverse, etc. Additional strictures have been also introduced, one by one, on the *kind* of continuum of the group manifold and its differentiability. We

ask for connectivity, differentiability and integrability conditions, and find a Lie algebra as a vector space *tangent* to the group at its unit element.

Definition: *When a Lie algebra* a *generates a (large neighborhood of the identity*[18] *in the) group* G*, we call it a* **Lie group**. *For* $\widehat{f} \in \mathsf{a}$ *and* $g \in \mathsf{G}$,

$$\widehat{f}_i = \left. \frac{\partial g(\boldsymbol{\alpha})}{\partial \alpha_i} \right|_{\boldsymbol{\alpha} = \mathbf{0}}, \qquad g(\boldsymbol{\alpha}) = \exp\left(\textstyle\sum_i \alpha_i \widehat{f}_i \right). \tag{3.69}$$

One important theorem (see *e.g.* [71]) states that under composition $g(\boldsymbol{\alpha}_1)\, g(\boldsymbol{\alpha}_2) = g(\boldsymbol{\alpha}_3)$, in a neighborhood of the unit element $g(\mathbf{0})$, the function $\boldsymbol{\alpha}_3(\boldsymbol{\alpha}_1, \boldsymbol{\alpha}_2)$ is *analytic* in all its arguments. This is evident in groups of matrices, where the product parameters are bilinear in those of the factors. Also, rotations with Euler angles (although they are arduous to write out), have composition functions that are analytic in large neighborhoods.

Definition: *When a linear subspace* a' *of a Lie algebra* a *forms a Lie algebra by itself (under the same Lie bracket), it is a* **subalgebra** $\mathsf{a}' \subset \mathsf{a}$. *When a subset* G' *of a Lie group* G *forms a Lie group by itself (under the same product rule), it is a* **subgroup** $\mathsf{G}' \subset \mathsf{G}$.

Subalgebras generate subgroups of the group generated by the algebra. And of course there are the trivial cases of the subalgebras $\mathsf{a}' = \{0\}$ and $\mathsf{a}' = \mathsf{a}$, and of the subgroups $\mathsf{G}' = g_o$ and $\mathsf{G}' = \mathsf{G}$.

In this chapter we have presented the formalism of Hamiltonian evolution and some of the associated philosophy. We assumed that the optical media had smooth refractive indices; we were interested in guides, because our techniques were kept close to classical point-particle mechanics, where potentials are commonly analytic and time-independent. In the following chapter we leave this familiar ground to treat the maps which lie closest to the heart of geometric optics: 'sudden' transformations of phase space due to refracting and reflecting surfaces. We do this going back to the original Hamilton equations (1.2)–(1.4) in their finite-difference form.

---

are aiming our discussion at the continuous Lie groups that will be useful in this volume.

[18] It occurs (in our symplectic algebras!) that the *whole* group manifold cannot be covered by single exponentials of its Lie algebra elements, so it has a *non-exponential* region.

4 The roots of refraction and reflection

Figure 4.1 shows the phenomenon of refraction in phase space. A surface S between two optical media, $n \neq n'$, can be seen as an interface between two interpenetrating spaces, where rays in one medium are matched to rays in the other through conservation laws. The geometry of refraction has been known for over one millenium; restating it in the coordinates of phase space $(\mathbf{p}, \mathbf{q})$ will enrich our understanding of Hamiltonian systems, as we shall see in this chapter. Because in mechanics we do not have lenses, 'sudden' transformations of phase space have been seldom considered interesting before.[1] Refraction – and also reflection – will be decomposed into two factors, called *root* transformations, which are of purely geometric origin, and canonical. Root transformtions will be both useful for aberration expansions, and intriguing for their simplicity.

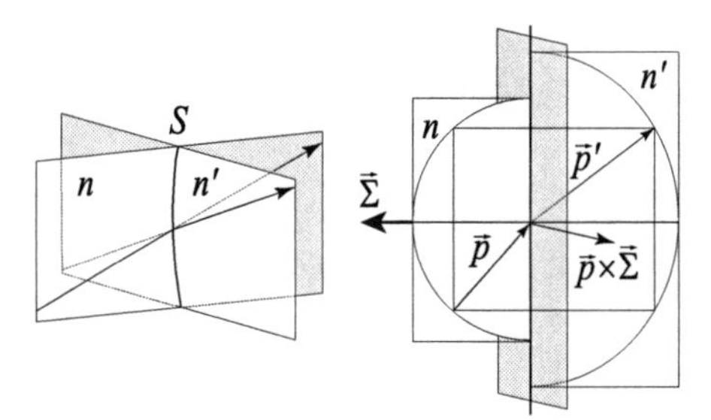

Fig. 4.1. *Left*: Two plane optical media n, n' (in $D = 1$) 'interpenetrate' at a refracting line S (shown here as if they were set at an angle); rays must match at S through the conservation of position (ray continuity). *Right*: At the point of impact, where $\vec{\Sigma}$ is the normal to S, there holds the conservation of tangential momentum in $\vec{p} \times \vec{\Sigma}$.

[1] In quantum mechanics, potential 'jolts' $\delta(t - t_o)\, V(\vec{q})$ produce finite transformations of the state of the system at the instant $t = t_o$. But these jolts occur all over space at the same time t_o, so the closest analogue we can find in (wave) optics are plane *Fresnel lenses*.

4.1 Refraction equations in screen coordinates

We saw in Sect. 1.4 that the passage of a ray through a refracting surface $S(\vec{q}) = 0$ conserves the position coordinate of the point of impact (1.8), and the component of momentum tangential to S at that point (1.9). We now write these equations in Cartesian coordinates x and y on the standard screen, and take as before z for the optical axis.

We assume that the surface $S(\vec{q}) = 0$ is single-valued in z so that we can define it as

$$S(q_x, q_y, z) = \zeta(q_x, q_y) - z = 0, \quad \text{or} \quad z = \zeta(\mathbf{q}). \tag{4.1}$$

The vector normal to this surface is

$$\vec{\Sigma} = \frac{\partial S}{\partial \vec{q}} = \begin{pmatrix} \mathbf{\Sigma} \\ -1 \end{pmatrix}, \qquad \mathbf{\Sigma} = \begin{pmatrix} \partial\zeta/\partial q_x \\ \partial\zeta/\partial q_y \end{pmatrix}. \tag{4.2}$$

Again, bars will be used for quantities at the surface S; primes for quantities in the 'second' optical medium n' and unprimed in the 'first', n.

Figure 4.2 shows the simple geometric relation between the ray coordinates $(\mathbf{p}, \mathbf{q})$ and $(\mathbf{p}', \mathbf{q}')$ at the standard screen $z = z' = 0$. The common point of impact is $\vec{\bar{q}} = (\bar{\mathbf{q}},\, z{=}\zeta(\bar{\mathbf{q}}){=}z')^\top$; with (2.10) this leads to the *first refraction equation*:

$$\mathbf{q} + \zeta(\bar{\mathbf{q}})\frac{\mathbf{p}}{p_z} = \bar{\mathbf{q}} = \mathbf{q}' + \zeta(\bar{\mathbf{q}})\frac{\mathbf{p}'}{p'_z}. \tag{4.3}$$

The point of impact of the ray on the surface, $\bar{\mathbf{q}}(\mathbf{p}, \mathbf{q}) = \bar{\mathbf{q}}(\mathbf{p}', \mathbf{q}')$, is thus a *conserved quantity*.

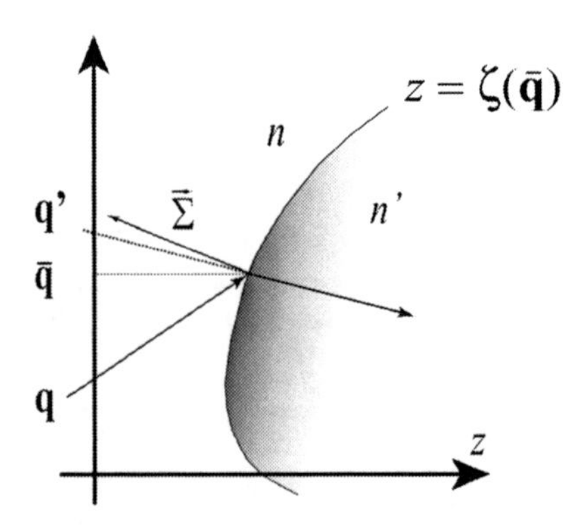

Fig. 4.2. A ray crossing the standard screen at $\mathbf{q}$ impacts the surface S ($z = \zeta(\mathbf{q})$) at $\bar{\mathbf{q}}$, where the surface normal is $\vec{\Sigma}(\bar{\mathbf{q}}, z)$. Refered to the standard screen, the refracting surface maps $\mathbf{q} \mapsto \bar{\mathbf{q}} \mapsto \mathbf{q}'$ (and $\mathbf{p} \mapsto \bar{\mathbf{p}} \mapsto \mathbf{p}'$, not shown), and thus factorizes into two root transformations – forth and back from the screen to the refracting surface.

The *second refraction equation* derives from the conservation of tangential momentum at S in (1.9). It reads

$$\mathbf{p} + \boldsymbol{\Sigma}(\bar{\mathbf{q}})p_z = \bar{\mathbf{p}} = \mathbf{p}' + \boldsymbol{\Sigma}(\bar{\mathbf{q}})p_z', \tag{4.4}$$

and thus defines the *conserved quantity* to be $\bar{\mathbf{p}}(\mathbf{p}, \mathbf{q}) = \bar{\mathbf{p}}(\mathbf{p}', \mathbf{q}')$. To prove this, we write the 3-vector $\vec{\Sigma}\times\vec{p}$ in components:

$$\begin{aligned} \begin{pmatrix} \boldsymbol{\Sigma}(\bar{\mathbf{q}}) \\ -1 \end{pmatrix} \times \begin{pmatrix} \mathbf{p} \\ p_z \end{pmatrix} &= \begin{pmatrix} \left(\begin{smallmatrix} 0 & 1 \\ -1 & 0 \end{smallmatrix}\right) (\mathbf{p} + \boldsymbol{\Sigma}(\bar{\mathbf{q}})\, p_z) \\ \boldsymbol{\Sigma}(\bar{\mathbf{q}})\times\mathbf{p} \end{pmatrix} \\ &= \begin{pmatrix} \mathbf{0} \\ -1 \end{pmatrix} \times \begin{pmatrix} \bar{\mathbf{p}} \\ 0 \end{pmatrix} + \begin{pmatrix} \mathbf{0} \\ \bar{p}_z \end{pmatrix}. \end{aligned} \tag{4.5}$$

The first two components contain the first equality in (4.4), and similarly for the primed quantities.

Remark: Coplanarity. The third component of (4.5) is

$$\boldsymbol{\Sigma}(\bar{\mathbf{q}})\times\mathbf{p} = \boldsymbol{\Sigma}(\bar{\mathbf{q}})\times\mathbf{p}', \tag{4.6}$$

which is also a consequence of the two refraction equations. In 3-vector form it is clear that the finite momentum difference between the incident and refracted ray is parallel to the surface normal at the point of impact $\bar{\mathbf{q}}$,

$$\vec{\Sigma}(\bar{\mathbf{q}})\times(\vec{p} - \vec{p}') = \vec{0}. \tag{4.7}$$

and that the three vectors are therefore *coplanar.* •

4.2 Factorization of refraction

The structure of the two refraction equations (4.3) and (4.4) separates the left-hand side in medium n, from the right-hand side in medium n'. This evident property led us to introduce in [108] the *root transformation* associated to the surface $z = \zeta(\mathbf{q})$, for *each* medium. It is

$$\mathcal{R}_{n;\zeta} : \mathbf{q} := \bar{\mathbf{q}} = \mathbf{q} + \zeta(\bar{\mathbf{q}}) \frac{\mathbf{p}}{\sqrt{n^2 - |\mathbf{p}|^2}}, \tag{4.8}$$

$$\mathcal{R}_{n;\zeta} : \mathbf{p} := \bar{\mathbf{p}} = \mathbf{p} + \boldsymbol{\Sigma}(\bar{\mathbf{q}})\sqrt{n^2 - |\mathbf{p}|^2}. \tag{4.9}$$

We have called these the *first* and *second root equations*, respectively. They have the same standing as the first and second Hamilton equations had for smooth inhomogeneous media, but for finite discontinuities in the refractive index.

The two refraction equations (4.3) and (4.4) now read

$$\mathcal{R}_{n;\zeta} : \begin{pmatrix} \mathbf{p} \\ \mathbf{q} \end{pmatrix} = \begin{pmatrix} \bar{\mathbf{p}} \\ \bar{\mathbf{q}} \end{pmatrix} = \mathcal{R}_{n';\zeta} : \begin{pmatrix} \mathbf{p}' \\ \mathbf{q}' \end{pmatrix}, \tag{4.10}$$

where the left-hand side contains the transformation from the (plane) standard screen to the (generally warped) surface S in medium n, and the right-hand side is the same transformation, but in the second medium n'.

To solve for the refracted ray $(\mathbf{p}', \mathbf{q}')$ in terms of the incident ray $(\mathbf{p}, \mathbf{q})$ we must *invert* the root transformation to:

$$\mathcal{R}_{n;\zeta}^{-1} : \bar{\mathbf{q}} = \mathbf{q} = \bar{\mathbf{q}} - \zeta(\bar{\mathbf{q}}) \frac{\mathbf{p}}{\sqrt{n^2 - |\mathbf{p}|^2}}, \tag{4.11}$$

$$\mathcal{R}_{n;\zeta}^{-1} : \bar{\mathbf{p}} = \mathbf{p} = \bar{\mathbf{p}} - \boldsymbol{\Sigma}(\bar{\mathbf{q}}) \sqrt{n^2 - |\mathbf{p}|^2}. \tag{4.12}$$

With this we have proven constructively the proposition of the *factorization of refraction*: The refracting surface transformation is the composition of a direct and an inverse root transformations,

$$\mathcal{S}_{n,n';\zeta} = \mathcal{R}_{n;\zeta} \, \mathcal{R}_{n';\zeta}^{-1}. \tag{4.13}$$

Remark: The order of the factors. To verify the consistency of the left-right order of action, we note that the action of this map on the arguments of the beam functions $\rho(\mathbf{p}, \mathbf{q})$, i.e., of 'jumping into' their arguments [cf. the property *(iv)* of Lie transformations on page 36]:

$$\begin{aligned} \mathcal{S}_{n,n';\zeta} : \rho(\mathbf{p}, \mathbf{q}) &= (\mathcal{R}_{n;\zeta} \, \mathcal{R}_{n';\zeta}^{-1}) : \rho(\mathbf{p}, \mathbf{q}) \\ &= \mathcal{R}_{n;\zeta} : (\mathcal{R}_{n';\zeta}^{-1} : \rho(\mathbf{p}, \mathbf{q})) \\ &= \mathcal{R}_{n;\zeta} : \rho(\mathcal{R}_{n';\zeta}^{-1} : \mathbf{q}, \mathcal{R}_{n';\zeta}^{-1} : \mathbf{p}) \\ &= \rho(\mathcal{R}_{n';\zeta}^{-1} : [\mathcal{R}_{n;\zeta} : \mathbf{q}], \mathcal{R}_{n';\zeta}^{-1} : [\mathcal{R}_{n;\zeta} : \mathbf{p}]) \\ &= \rho(\mathcal{R}_{n';\zeta}^{-1} : \bar{\mathbf{q}}, \mathcal{R}_{n';\zeta}^{-1} : \bar{\mathbf{p}}) \\ = \rho(\mathbf{q}', \mathbf{p}') &= \rho(\mathcal{S}_{n,n';\zeta} : \mathbf{q}, \mathcal{S}_{n,n';\zeta} : \mathbf{p}). \end{aligned} \tag{4.14}$$

Finally, when $n \to n'$, then $\mathcal{S}_{n,n';\zeta} \to \mathit{1}$. •

Example: The root of a flat surface is free displacement. When the surface is a plane, $z = \zeta_o =$ constant, the root transformation (4.8)–(4.9) becomes

$$\mathcal{R}_{n;\zeta_o} = \mathcal{D}_n(\zeta_o), \tag{4.15}$$

i.e., simple free propagation (3.36) along the optical axis by z in medium n.•

Remark: Translated surfaces. Note that if a surface $z = \zeta(\mathbf{q})$ is shifted by ζ_o to $\zeta_o + \zeta(\mathbf{q})$, the new root transformation is the composition of free propagation to the surface composed with the root transformation of the unshifted surface,

$$\mathcal{R}_{n;\zeta_o+\zeta} = \mathcal{D}_n(\zeta_o) \, \mathcal{R}_{n;\zeta}. \tag{4.16}$$

This can be verified in (4.8) with $\zeta_o + \zeta(\mathbf{q})$, introducing an intermediate $\mathbf{q}_{\text{int}} = \mathcal{R}_{n;\zeta} : \mathbf{q}$ and $\bar{\mathbf{q}} = \mathcal{D}_n(\zeta_o) : \mathbf{q}_{\text{int}}$. •

Remark: Implicit and explicit equations. The first root equation (4.8) is *implicit* in $\bar{\mathbf{q}}(\mathbf{p}, \mathbf{q}; \zeta)$, and is generally the more difficult one to solve. Once this equation has been solved, the second root equation (4.9) yields $\bar{\mathbf{p}}(\mathbf{p},\mathbf{q};\zeta)$ explicitly. Among the inverse root equations, (4.12) is implicit in $\mathbf{p}(\bar{\mathbf{p}},\bar{\mathbf{q}};\zeta)$ and must be solved first; the result is then replaced in (4.11), which solves for $\mathbf{q}(\bar{\mathbf{p}},\bar{\mathbf{q}};\zeta)$ explicitly. In Sect. 4.5 we use Taylor series (aberration expansion) to find recursively the generic transformation due to revolution-polynomial refracting surfaces. •

Remark: Domain and range of the root transformation. The *domain* of the root transformation $\mathcal{R}_{n;\zeta}$ and of its inverse are determined by the surface $z=\zeta(\mathbf{q})$. If the origin of phase space (ray along the optical axis) belongs to it, the domain is the neighborhood of rays that *intersect* S before tangency occurs, i.e., for the subset of $\wp$ such that $|\vec{\Sigma}(\bar{\mathbf{q}})\cdot\vec{p}|>0$. The *range* of $\mathcal{R}_{n;\zeta}$ is generally *not* completely contained in optical phase space $\wp$ however, because according to (4.9), $\bar{\mathbf{p}}$ is in the disk $|\mathbf{p}|=n\sin\theta$ translated by the unbounded vector $\mathbf{\Sigma}(\bar{\mathbf{q}})\,n\cos\theta$. In the products $\mathcal{R}_{n;\zeta}\,\mathcal{R}^{-1}_{n';\zeta}$ some rays in the unphysical range of the first factor may be brought back to $\wp$ by the second, but – as is obvious geometrically – some regions of phase space may be lost if the rays miss the surface S. •

4.3 Canonicity of the root transformation

We should expect the refracting surface transformation $\mathcal{S}_{n,n';\zeta}$ to be canonical because rays could be created or destroyed otherwise, but we remarked above that this cannot be true globally for the root transformation. Yet it is rather pleasant that the following result does hold:

$$\text{The root transformation } \mathcal{R}_{n;\zeta} \text{ is locally canonical.} \tag{4.17}$$

It is valid *locally* because we cannot easily go beyond the central domain of phase space around the optical center and axis of the surface.

For the pedestrian proof of (4.17) we can take $\bar{q}_i(\mathbf{p},\mathbf{q})$ and $\bar{p}_j(\mathbf{p},\mathbf{q})$ from (4.8)–(4.9) and verify that the basic Poisson brackets satisfy (3.18), i.e., $\{\bar{q}_i,\bar{p}_j\}=\delta_{i,j}$, etc. Yet it turns out surprisingly difficult to solve for the necessary partial derivatives. In $D=1$ dimensions the work may be performed with a steady hand, but for $D=2$ it already needs symbolic computation.[2] A different tactic to prove canonicity is fruitful. It relies on the *Hamilton characteristic function* [58, Chap. 7].

Various objects and structures characterized with Hamilton's name should *not* be confused: our main tool in Lie theory are the Hamilton *evolution* equations; yet in this section we shall use Hamilton's *characteristic* functions to define canonical transformations of $2(D+1)$-dim phase space $(\vec{p},\vec{q})\mapsto(\vec{P},\vec{Q})$. The characteristic function depends on $D+1$ *initial* phase space

[2] See [106, Appendix A], where the verification of canonicity for the $D=2$ case was done using symbolic computation in `REDUCE`.

coordinates, and $D+1$ *final* coordinates. Of the four basic choices,[3] we use a function of initial momentum and final position, $F(p_i, Q_j)$, to define

$$q_k = \frac{\partial F(p_i, Q_j)}{\partial p_k}, \qquad P_k = \frac{\partial F(p_i, Q_j)}{\partial Q_k}. \tag{4.18}$$

That the transformation thus defined is canonical follows from

$$\{q_i, p_j\} = \left\{\frac{\partial F}{\partial p_i}, p_j\right\} = \frac{\partial^2 F}{\partial q_j\, \partial p_i} = \frac{\partial q_i}{\partial q_j} = \delta_{i,j}, \tag{4.19}$$

$$\{Q_i, P_j\} = \left\{Q_i, \frac{\partial F}{\partial Q_j}\right\} = \frac{\partial^2 F}{\partial P_i\, \partial Q_j} = \frac{\partial P_i}{\partial P_j} = \delta_{i,j}, \tag{4.20}$$

and along similar lines we find $\{q_i, q_j\} = 0$ and $\{P_i, P_j\} = 0$.

We note that the particular characteristic function

$$F_{\rm id}(p_i, Q_j) := \vec{p} \cdot \vec{Q} = \sum_k p_k\, Q_k, \tag{4.21}$$

defines the *identity* transformation, because $q_k = \partial F_{\rm id}/\partial p_k = Q_k$ and $P_k = \partial F_{\rm id}/\partial Q_k = p_k$. But in optical phase space, momentum $\vec{p}$ is restricted to the Descartes sphere [(1.11) and (2.1)], so the essential momentum coordinates are those of $\mathbf{p}$ *on the disk* $\mathcal{D}$. Correspondingly, the final position vector $\vec{Q}$ is restricted to the points $z = \zeta(\bar{\mathbf{q}})$ *on the surface* S. Replacing capitals by bars for the final phase space, the restricted identity characteristic function (4.21) becomes

$$F_{\rm root}(\mathbf{p}, \bar{\mathbf{q}}) := F_{\rm id}(p_i, \bar{q}_j)|_{\mathcal{D},S} = \mathbf{p} \cdot \bar{\mathbf{q}} + p_z|_{\mathcal{D}}\, \bar{q}_z|_S = \mathbf{p} \cdot \bar{\mathbf{q}} + \sqrt{n^2 - |\mathbf{p}|^2}\, \zeta(\bar{\mathbf{q}}). \tag{4.22}$$

It is straightforward to verify that $F_{\rm root}(\mathbf{p}, \bar{\mathbf{q}})$ replaced in (4.18) yields the root equations (4.8)–(4.9).[4] Since the root transformation $\mathcal{R}_{n;\zeta}$ *has* a Hamilton characteristic function, it is *canonical*; and its inverse exists and is *also* canonical. Since the refracting surface transformation $\mathcal{S}_{n,n';\zeta}$ is the composition of two roots (4.13), it is thus a *canonical* transformation, as expected.

The root transformation is thus a change of coordinates between optical phase space $(\mathbf{p}, \mathbf{q}) \in \wp$ referred to the standard screen, and the phase space $(\bar{\mathbf{p}}, \bar{\mathbf{q}})$ which is referred to the warped surface S; see [9].

---

[3] The choices of Hamilton characteristic functions, in Goldstein's classic notation, [58, Chap. 8], are: point-point $F_1(q_i, Q_j)$, point-angle $F_2(q_i, P_j)$, angle-point $F_3(p_i, Q_j)$, and angle-angle $F_4(p_i, P_j)$.

[4] This insight was provided to the author by Prof. Joaquín Delgado (Departamento de Matemáticas, Universidad Autónoma Metropolitana–Iztapalapa).

4.4 Aberration series expansion for the root transformation

Solving the first root equation (4.8) is the toughest part of computing refraction and reflection. As we remarked before, it is an *implicit* equation in $\bar{\mathbf{q}}$ that depends on the surface S through the function $\zeta(\mathbf{q})$. Yet it is a *simple* kind of implicit equation, of the form $\bar{\mathbf{q}} = \mathbf{q} - \zeta(\bar{\mathbf{q}})/h(\mathbf{p})$, instead of the *coupled* refraction equations (4.3)–(4.4) that would have to be solved if we decided to ignore the middle terms. The strategy of *aberration expansions* expands all functions of phase space in Taylor series (without regard to convergence), so that an implicit equation between these functions becomes a set of recursion relations between successive Taylor coefficients. In this section we perform this expansion for the root transformation in plane optics, $D = 1$, to fifth aberration order. The results for the real 3-dim world $D = 2$ will be given in Chap. 14 to order seven.

Aberration expansions are particularly effective when $\zeta(q)$ is at least quadratic in $\bar{q}$. We consider the generic symmetric polynomial surface,

$$\zeta(q) = \zeta_2 q^2 + \zeta_4 q^4 + \zeta_6 q^6 + \cdots, \tag{4.23}$$

that *osculates* the standard screen at the optical center. In the root equations (4.8)–(4.9) we use (4.23) and the Taylor expansion of free propagation (2.11)–(2.12), keeping terms up to degree 5 in $(\mathbf{p}, \mathbf{q})$, to find

$$\begin{aligned} \bar{q} &= q + p\,(\zeta_2 \bar{q}^2 + \zeta_4 \bar{q}^4 + \cdots) \times \left(\frac{1}{n} + \frac{p^2}{2n^3} + \cdots\right) \\ &\approx q + p\left(\frac{\zeta_2 \bar{q}^2}{n} + \frac{\zeta_4 \bar{q}^4}{n} + \frac{\zeta_2 p^2 \bar{q}^2}{2n^3}\right). \end{aligned} \tag{4.24}$$

We start with the *first-order* approximation $\bar{q}^{[1]} := q$, replacing it in the right-hand side of (4.24) cut to *third*-degree terms. Thus we find the *third-order* approximant $\bar{q}^{[3]} := q + p\,\zeta_2 q^2/n$. Then we replace this again in the right-hand side of (4.24) cut now to *fifth*-degree terms. Since now a term in $\bar{q}^2$ exists, the *fifth-order* aberration will now include more cross terms. In this way we compute

$$\begin{aligned} \bar{q}^{[5]} := {} & q + \zeta_2 p\, q^2/n && 1^{\text{st}} \text{ and } 3^{\text{rd}} \\ & + \frac{\zeta_2}{2n^3} p^3 q^2 + \frac{2\zeta_2^2}{n^2}\, p^2\, q^3 + \frac{\zeta_4}{n} p\, q^4. && 5^{\text{th}} \text{ order} \end{aligned} \tag{4.25}$$

To find the approximants to $\bar{p}$, we use the explicit second root equation (4.9), with $\boldsymbol{\Sigma}(\bar{q}) = 2\zeta_2 \bar{q} + 4\zeta_4 \bar{q}^3 + \cdots$, times the Taylor series (2.12). To first order, for $\boldsymbol{\Sigma}(\bar{q})$ we use $\bar{q}^{[1]}$ to find $\bar{p}^{[1]} = p + 2n\zeta_2\, q$, and similarly for the higher approximants. To fifth aberration order, the root transformation thus yields

$$\begin{aligned}\bar{p}^{[5]} := {} & p + 2n\zeta_2\, q && 1^{\rm st} \text{ order}\\ & - \frac{\zeta_2}{n} p^2 q + 2\zeta_2^2 p\, q^2 + 4n\zeta_4\, q^3 && 3^{\rm rd} \text{ order}\\ & - \frac{\zeta_2}{4n^3} p^4 q + \Big(\frac{4\zeta_2^3}{n} - \frac{2\zeta_4}{n}\Big) p^2 q^3 && \\ & + 14\zeta_2\zeta_4\, p\, q^4 + 6n\zeta_6\, q^5. && 5^{\rm th} \text{ order}\end{aligned} \qquad (4.26)$$

To find the refracting surface transformation (4.13), we still have the task to find the inverse root transformation, $p'(\bar{p},\bar{q})$, $q'(\bar{p},\bar{q})$, and replace (4.25)–(4.26) in the former to obtain $p'(p,q)$, $q'(p,q)$. In Chap. 14, the results for aberration order seven and axis-symmetric surfaces in $D = 2$ dimensions are obtained using the symbolic computation algorithms in the program `mexLIE` [161] that we developed in Cuernavaca. Part IV is devoted to a structured analysis of aberrations of phase space.

4.5 Refraction between inhomogeneous media

We now discuss the general case when the refractive index has a discontinuity at a surface between two smooth inhomogeneous regions of the medium. To abstract the forth-and-back argument that led to the factorization of refraction in Sect. 4.1 for generic dimension D, we follow the rays $(\mathbf{p},\mathbf{q})$ in the medium $n(\mathbf{q},z)$ and $(\mathbf{p}',\mathbf{q}')$ in $n'(\mathbf{q}',z')$ that evolve along the common optical axis as $(\mathbf{p}(z),\mathbf{q}(z)) = (\mathbf{P}(\mathbf{p},\mathbf{q};z),\mathbf{Q}(\mathbf{p},\mathbf{q};z))$ and $(\mathbf{p}'(z'),\mathbf{q}'(z')) = (\mathbf{P}'(\mathbf{p},\mathbf{q};z'),\mathbf{Q}'(\mathbf{p},\mathbf{q};z'))$.

The rays meet at the surface S with $z = \zeta(\bar{\mathbf{q}}) = z'$ and their common impact point is

$$\bar{\mathbf{q}} = \mathbf{Q}(\mathbf{p},\mathbf{q};\zeta(\bar{\mathbf{q}})) = \mathbf{Q}'(\mathbf{p}',\mathbf{q}';\zeta(\bar{\mathbf{q}})). \qquad (4.27)$$

This yields the first refraction equation in generic form [cf. (1.8) and (4.3)], again factorized by the conserved quantity $\bar{\mathbf{q}}$, which is implicitly determined by the first equality. The second generic refraction equation is the conservation of the cross product between momentum and the common surface normal, and reads

$$\begin{aligned}\bar{\mathbf{p}} &= \mathbf{P}(\mathbf{p},\mathbf{q};\zeta(\bar{\mathbf{q}})) + \boldsymbol{\Sigma}(\bar{\mathbf{q}})P_z(\mathbf{p},\mathbf{q};\zeta(\bar{\mathbf{q}})) \qquad (4.28)\\ &= \mathbf{P}'(\mathbf{p}',\mathbf{q}';\zeta(\bar{\mathbf{q}})) + \boldsymbol{\Sigma}(\bar{\mathbf{q}})P_z'(\mathbf{p}',\mathbf{q}';\zeta(\bar{\mathbf{q}})),\end{aligned}$$

$$P_z(\mathbf{p},\mathbf{q};z) = \sqrt{n(\mathbf{q},z)^2 - |\mathbf{P}(\mathbf{p},\mathbf{q};z)|^2}. \qquad (4.29)$$

[Cf. (1.9) and (4.4).] The gist of the solution, again, is to compute refraction through the root transformation. The canonicity of the (generic) root transformation was proven in [43, Sect. 4.5] using the Lagrangian formalism.

Remark: Refraction between two guides. In the case of two guide media $n(|\mathbf{q}|)$ and $n'(|\mathbf{q}|)$, the refraction equations yield to the aberration expansion algorithm that we outlined in the previous section. Particularly, for guides with elliptic index profiles (2.18)–(2.18), the solution was found in

[152] and [157] to third and seventh aberration order, respectively. Refraction between two guides is included among the standard functions of `mexLIE` [161]. •

4.6 Factorization of reflection

Reflection has in common with refraction the conservation of the impact point and of the momentum tangential to the reflecting surface; the momentum component *normal* to the surface however, changes sign under reflection. As remarked on page 20, the reflected ray – we denote it $(\mathbf{p}^{\mathrm{R}}, \mathbf{q}^{\mathrm{R}})$ – propagates as the refracted ray would, but in the opposite chart index $\sigma \mapsto -\sigma$, with negative third component p_z. This in turn is equivalent to reflecting the surface $\zeta(\mathbf{q}) \mapsto -\zeta(\mathbf{q})$, and its normal $\mathbf{\Sigma}(\mathbf{q}) \to -\mathbf{\Sigma}(\mathbf{q})$, as can be seen from (4.9). With these replacements, the refraction equations (4.3)–(4.4) become the first and second *reflection* equations:

$$\mathbf{q} + \zeta(\bar{\mathbf{q}})\frac{\mathbf{p}}{p_z} =: \bar{\mathbf{q}} = \mathbf{q}^{\mathrm{R}} - \zeta(\bar{\mathbf{q}})\frac{\mathbf{p}^{\mathrm{R}}}{p_z^{\mathrm{R}}}, \tag{4.30}$$

$$\mathbf{p} + \mathbf{\Sigma}(\bar{\mathbf{q}})p_z =: \bar{\mathbf{p}} = \mathbf{p}^{\mathrm{R}} - \mathbf{\Sigma}(\bar{\mathbf{q}})p_z^{\mathrm{R}}. \tag{4.31}$$

The first of these equations is implicit in $\bar{\mathbf{q}}$, as was the case before, and then the equation for $\bar{\mathbf{p}}$ is explicit.

Equations (4.30)–(4.31) yield thus the *factorization of reflection*: The reflecting surface transformation is the composition of a direct and a reflected root transformation,

$$\mathcal{S}^{\mathrm{R}}_{n;\zeta} = \mathcal{R}_{n;\zeta}\,\overline{\mathcal{R}_{n;\zeta}}, \qquad \text{where} \quad \overline{\mathcal{R}_{n;\zeta}} := \mathcal{R}^{-1}_{n;-\zeta}. \tag{4.32}$$

Remark: Parallelism and coplanarity. Again, as in (4.5)–(4.7), we find that the difference between the momentum 3-vectors before and after reflection, $\vec{p}\,' - \vec{p} = (\mathbf{p}^{\mathrm{R}} - \mathbf{p}, -p_z^{\mathrm{R}} - p_z)^{\top} = (p_z^{\mathrm{R}} + p_z)\,\vec{\Sigma}(\bar{\mathbf{q}})$, is parallel to $\vec{\Sigma}(\bar{\mathbf{q}}) = (\mathbf{\Sigma}(\bar{\mathbf{q}}), -1)^{\top}$ [cf. (4.2)]. Since in the same medium $\vec{p}$ and $\vec{p}\,'$ have the same length, $\vec{\Sigma}\times\vec{p} = \vec{\Sigma}\times\vec{p}\,'$ and $\vec{\Sigma}\cdot\vec{p} = -\vec{\Sigma}\cdot\vec{p}\,'$, we can write explicitly

$$\vec{p}^{\,\mathrm{R}} = \vec{p}\left(1 - 2\frac{\cdot\,\vec{\Sigma}(\bar{\mathbf{q}})\vec{\Sigma}(\bar{\mathbf{q}})}{|\vec{\Sigma}(\bar{\mathbf{q}})|^2}\right). \tag{4.33}$$

•

The *reflected* root transformation, defined in (4.32) and indicated by the overline, plays a fundamental role in the definition of reflection of all optical systems $\mathcal{M}$ built out of free spaces and refracting surfaces. Our paradigm here is that a flat mirror duplicates the real world to its left (with all its optical systems) into a mirror world (equally ‘real’) to its right.

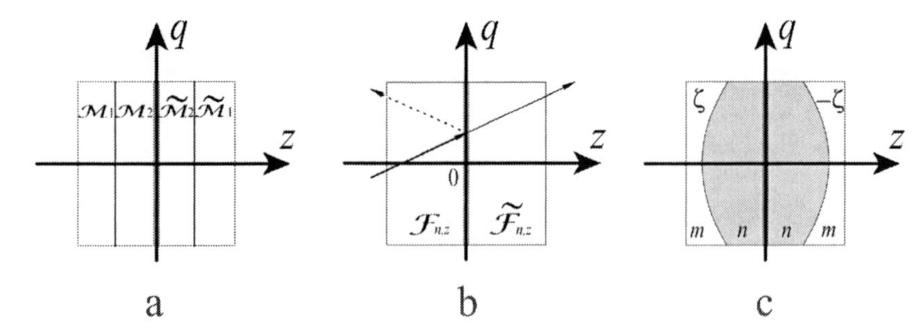

Fig. 4.3. The three conditions that define reflection by a mirror (thick vertical line). (**a**) Reflection is an antihomomorphism. (**b**) The reflection of free displacement is free displacement. (**c**) The reflection of a refracting surface ζ between media n and n', is a refracting surface $-\zeta$ between n' and n.

Consider two optical systems $\mathcal{M}_1$ and $\mathcal{M}_2$ which are reflected to $\overline{\mathcal{M}_1}$ and $\overline{\mathcal{M}_2}$ respectively; then, the operation of concatenation will be reflected as in Fig. 4.3a, i.e. inverting their order:

$$\mathcal{M} = \mathcal{M}_1\,\mathcal{M}_2 \mapsto \overline{\mathcal{M}} = \overline{\mathcal{M}_1\,\mathcal{M}_2} = \overline{\mathcal{M}_2}\,\overline{\mathcal{M}_1}, \quad \overline{\overline{\mathcal{M}}} = \mathcal{M}. \tag{4.34}$$

Reflection is thus an *antihomomorphism* of transformations, whose square is unity. Now, the two optical elements out of which we construct all systems, reflect as follows:

Free propagation by ζ_o = constant, shown in Fig. 4.3b, is invariant [see (4.15)],

$$\overline{\mathcal{D}_n(\zeta_o)} = \overline{\mathcal{R}_{n;\zeta_o}} = \mathcal{R}^{-1}_{n;-\zeta_o} = \mathcal{D}_n(-\zeta_o)^{-1} = \mathcal{D}_n(\zeta_o). \tag{4.35}$$

Refracting surfaces, as shown in Fig. 4.3c [see (4.13) and (4.34)], reflect as

$$\overline{\mathcal{S}_{n,n';\zeta}} = \overline{\mathcal{R}_{n;\zeta}\,\mathcal{R}^{-1}_{n';\zeta}} = \overline{\mathcal{R}^{-1}_{n';\zeta}\mathcal{R}_{n;\zeta}} = \mathcal{R}_{n';-\zeta}\,\mathcal{R}^{-1}_{n;-\zeta} = \mathcal{S}_{n',n;-\zeta}. \tag{4.36}$$

Remark: Reflection and/or inversion. Both reflection and inversion of optical systems are antihomomorphisms whose square is unity, and they commute. But they are clearly *distinct*, since from (4.32) we see that $(\overline{\mathcal{R}_{n,\zeta}})^{-1} = \mathcal{R}_{n,-\zeta} = (\overline{\mathcal{R}^{-1}_{n,\zeta}})$. Therefore, they *can not* related by a global homomorphism. Yet, as we shall see in Chap. 10, the *paraxial part* of these systems, represented by matrices, *can* be related by a simple similarity transformation. The *aberration part* of reflection will be studied in Part IV. •

Remark: Reflection and/or refraction? Experimentally, at the surface of all transparent objects, part of the light refracts and part reflects. The relative amount of each can be found knowing about light polarization and the conservation of electric and magnetic vectors at the surface. The postulates of geometric optics proposed in Chap. 1 allow the determination both rays, but leave their proportions undecided. •

A Some historical comments

Optics and Mathematics have been intertwined for too long for us to attempt a comprehensive panorama. But we do want to bring to the reader's attention the recent historical discovery that the geometric study of refraction, hitherto attributed to Snell, Descartes, and/or Fermat in its sine law form, was known and written upon by Ibn Sahl, working at the Abassid court, substantially more than 600 years before the European natural philosophers. This Appendix ends tentatively in the twentieth century, when the understanding of phase space, Hamiltonian systems and Lie algebras, was mature enough to recognize both geometry and dynamics as manifestations of symmetry.

A.1 Antiquity

The phenomenon of reflection was understood in the Hellenic world by the time Archimedes tried to burn the Roman fleet approaching Alexandria; the interplay of lines with planes and conics suffuses the *Elements* of Euclid, Ptolemy's *Optics*, and the *Conic Sections* of Apollonius. Applications of parabolic mirrors for burning instruments, first described by Anthemius of Tralles, were well known to the scientific circles of the late tenth century at the hub of Greco-Arabic civilization: Baghdad. ABŪ SA^cD AL-^cALĀ^ɔ IBN SAHL wrote around 984 his book *On Burning Instruments*, from which we reproduce Fig. 1.3 [119]. Ibn Sahl's book is both experimental (he provides mechanical means to draw the conic sections) and theoretical; in effect, he develops the geometry of refraction as the Greeks did for reflection, with the abstract lines of his diagrams. Suffice Fig. A.1 to show his mastery of the art: a biconvex hyperboloidal lens that concentrates the light of a near source at a point. Rock crystal was excavated at nearby Bassorah (Basra), so we may assume that such lenses were actually built.

Ibn Sahl was well known among his colleagues and students. During the early eleventh century, Ibn al-Hayatham wrote the *Book of Optics* acknowledging his mentor and giving notice of his name to the world. *On Burning Instruments* was never available to readers in Europe, however. One library in Damascus and one in Baghdad contained catalogued manuscripts of what were thought to be copies of the same book, and apparently they were never compared before 1990 when Professor Roshdi Rashed, a science historian

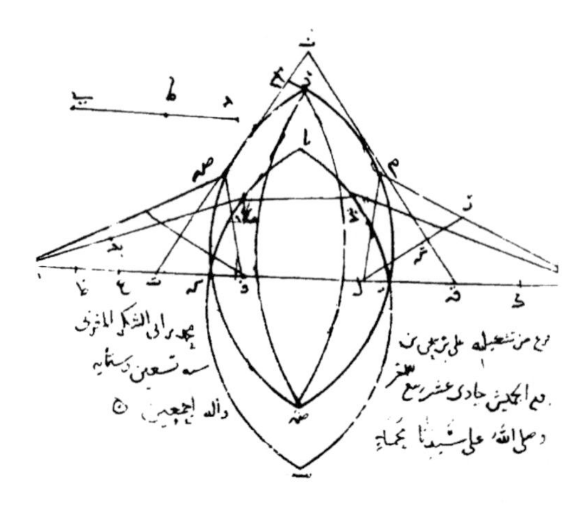

Fig. A.1. Ibn Sahl's diagram for a biconvex hyperbolic lens.

working in France, undertook the task to organize and study them.[1] The pages turned out to be parts of the same dishevelled manuscript, pages mixed and some lost, unnumbered, but providentially containing the key results and revealing the overall structure of the work.

A.2 Age of Reason

Whereas Ibn Sahl had read the translated Greek books on optics, recounting Archimedes' legend of high-tech defense, the seventeenth century philosophers in Flanders and France had little background information. Reading today RENÉ DESCARTES' *Discourse on the Method of Highly Conducting the Reason and Seeking the Truth for the Sciences* provides a quaint picture of the beginnings of rational thought; indeed, of the wooly frame of mind of his times. In one of its chapters, *On Dioptrics*, Descartes reflects on refraction comparing light with a cannonball fired at an angle into water. His conclusions are geometrically correct: the component of momentum vector that is tangential to the surface is preserved because the only vector of the interaction with the surface is its normal. Figure A.2 is Descartes' rendering of the cannonball theory. As to experiment, in his own words:

> *"... Ce qu'on a quelquefois expérimenté avec regret, lorsque, faisant tirer pour plaisir des pièces d'artillerie vers le fond d'une rivière, on a blessé ceux quie étaient sur l'autre côté sur le rivage."*[2]

[1] For his brilliant research, Professor Rashed was bestowed the 1991 Prize of the Third World Academy of Sciences.

[2] *"... Something we have tested with regret, having shot for pleasure artillery pieces towards the bottom of a river, wounding those that were on the other shore.*

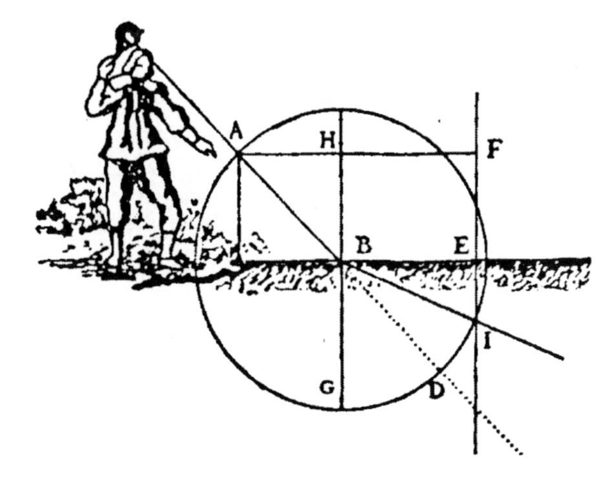

Fig. A.2. Descartes' diagram for cannon trajectory in water.

A.3 Fermat's principle and the Lagrangian

No dispute over the paternity of the sine law of refraction ever happened between Descartes and WILLEBRORD SNELLIUS (see the Introduction of [133]) who apparently did not know of each other's work during the fateful 1630s. It was Pierre de Fermat who tried to spite his elder rival, apparently out of chagrin after discovering that his new Principle of Least Action led, in the end, to Descartes' result. This very elegant principle states[3] that when the cost of a process between two configurations A and B is a function $L(\vec{q}, \dot{\vec{q}})$ of position $\vec{q}$ and velocity $\dot{\vec{q}} = d\vec{q}/ds$ with respect to a parameter s, the trajectory chosen by Nature is such that the integral

$$I := \int_A^B ds\, L(\vec{q}, \dot{\vec{q}}) \quad \text{is an extremal, i.e.,} \quad \delta I = 0. \tag{A.1}$$

Euler and Lagrange later transformed this to a formulation more amenable to actual solution, given by the *Euler-Lagrange* equations

$$\frac{d\vec{p}}{ds} = \frac{\partial L(\vec{q}, \dot{\vec{q}})}{\partial \vec{q}} \quad \text{where } \textit{momentum} \text{ is} \quad \vec{p} := \frac{\partial L(\vec{q}, \dot{\vec{q}})}{\partial \dot{\vec{q}}}. \tag{A.2}$$

The cost L is called the *Lagrangian* function of the system; in mechanics [58] it has the classical form $\frac{1}{2}m\,|\dot{\vec{q}}|^2 - V(\vec{q})$, i.e., kinetic minus potential energies.

Fermat's principle in optics minimizes the time of transit between the two points A and B of a 'light particle' postulated to move with velocity $ds/dt = c/n$, where n is the refractive index of the medium and c some universal

[3] *"Je reconnois premièrement... la vérité de ce principe, que la nature agit toujours par les voies les plus courtes."*

(constant) velocity, stating that $\delta \int_A^B dt = 0$. When we replace $dt = (n/c)\, ds$ into Fermat's integral, we come up with an 'optical Lagrangian' $L = n/c$. For the most common inhomogeneous but isotropic optical media $n(\vec{q})$, this Lagrangian will be independent of ray direction $\dot{\vec{q}}$ and the momentum will be identically zero, rendering (A.2) unfit to yield solutions; the case of isotropic media happens to be a singular limit of the theory. Actually, a Lagrangian formalism of optics that *is* successful chooses for parameter the distance along the optical axis z, and minimizes the *optical path* length $n\, ds = c\, dt$ between the points [41, 84],

$$\int_A^B dz\, n(\vec{q}, \dot{\vec{q}}) \sqrt{\left(\frac{dq_x}{dz}\right)^2 + \left(\frac{dq_y}{dz}\right)^2 + 1}. \tag{A.3}$$

Then, indicating by $\mathbf{q}' = d\mathbf{q}/dz$, the 2-dim momentum on the screen is

$$\mathbf{p} = \frac{\partial}{\partial \mathbf{q}'} n(\vec{q}, \dot{\vec{q}}) \sqrt{(\mathbf{q}')^2 + 1} = \frac{n(\vec{q}, \dot{\vec{q}})\, \mathbf{q}'}{\sqrt{(\mathbf{q}')^2 + 1}}. \tag{A.4}$$

The Legendre transformation $H = \mathbf{p} \cdot \mathbf{q}' - L$ returns the Hamilton equations, i.e., those we derived in Sect. 1.1, as a succint basis for the Hamiltonian *evolution* formulation of optics on the screen.

The *Lagrangian* formulation of optics that we outlined above is the standard basis for many modern works on Hamiltonian optics, such as those by Buchdahl [25, 26]. It was also the foundation of the Lie algebraic theory of geometric optics of Dragt *et al.* [42, 43, 44]. In this book, our point of view is that the *local ansätze* (difference–differential equations) in Sect. 1.1 are more economic than the *global* (variational integral) postulate of Fermat contained in (A.3). Before doing justice to the formulations of Hamilton himself, we must re-examine the prevailing paradigms of geometric optics during the nineteenth century.

A.4 Geometric optics in the nineteenth century

In his excellent doctoral thesis, Eisso Atzema [8] accounts for the mathematization undergone by the geometric study of light in nineteenth century Europe. Thomas Young's experiments had convinced the scientific community that light was a wave phenomenon; one that would have to wait for Maxwell and Hertz to be explained fully. So geometric optics became a science within geometry; one that would occupy the best minds of the century.

Atzema places one theorem in the 1810 book *Traité d'Optique* by ÉTIENNE MALUS, at the origin of one line of mathematical discourse that examined *systems* of rays $R(S)$ which are normal to a surface S. Clearly, random bunches of rays may not have such a common, generating surface. The theorem of Malus (completed by Gergonne and Hamilton [8]) states that $R(S)$, when subject to refraction or reflection by a surface S_o, will produce another system of rays $R(S')$ normal to another surface S'. The contrapositive of this theorem was proven in 1844 by Joseph Bertrand in the same vein as his classification of systems with closed orbits; it states that *only* the sine law of refraction at S_o will have the Malus property. In other words, no other 'refraction' laws can give rise to true, canonical optics.

A.5 Hamiltonian formulations

Abuse of language in mathematics refers to the designation of several related objects by one symbol that must be understood in context. By *Hamiltonian* geometric optics we want to refer to the manifold strategies that treat beams of light as infinitely thin pencils [8, Chap. 5], i.e., rays as points moving in phase space with their Cartesian neighborhood, or more precisely, to their determination with the two Hamilton equations of motion (1.3) and (1.5) from a ruling *Hamiltonian* function. The term Hamiltonian optics [26] has been also applied when the basic concept is the Hamiltonian *characteristic* function that enters (4.18), which we found most useful for the treatment of 'finite' refraction.

William Rowan Hamilton published in 1828 the *Theory of Systems of Rays*, based on his characteristic functions. Atzema [8] points out that Hamilton's first drafts of the book bear the title *Theory of the Infinitely Thin Pencil*, to be based on the ray paradigm and what we call the two Hamilton equations. His change of preference swayed both classical mechanics and geometric optics towards characteristic functions; Ludwig Seidel [125] based his 1853 classification of *bildfehler* (optical aberrations) on its Taylor expansion. The loom of mathematics involves many names: Fourier, Poisson, Liouville, Schultén, Sturm, Jacobi. Only at the turn of the twentieth century were the mathematical tools of geometric optics up to the task of designing complex optical systems. By then, Kirchhoff had shown geometric optics to be the short limit of the wave theory so – for the optical engineers – it had to work. Ernst Abbe, at the firm of Carl Zeiss at Jena during the last two decades of the century, first suffered and then confronted the limitations of the geometric theory: diffraction had not been taken into account. As one century before, physics and mathematics started a new dialogue on the nature of light.

A.6 The evolution of Sophus Lie

Marius Sophus Lie left his native Norway in 1869, after studying Galois theory with Sylow in Christiania, present Oslo. At the footloose age of 27, in quest for a wider education, he went to Berlin to hear seminars by Weierstrass and Kummer, then to Paris with Darboux and Jordan. Eventually, he reached Leipzig and stayed there for 20 years as Professor at the Leipziger Universität [63]. Belately, he accepted the Chair of *Professor für Transformationsgruppen* at the University of Christiania, but died in February 1899 of pernicious anemia, shortly after his prodigal return. Many mathematical objects are attributed to his name (and we add to the list that of a point of view of optics itself), because his basic work, the three volumes of his *Theorie der Transformationsgruppen*, published between 1888 and 1893, introduces fruitfully groups with continuous parameters.

The list of Lie's publications does not include the keywords of optics in any prominent way. Only one bears the name in the title: a $2\frac{1}{2}$-page report on his lecture of January 13, 1896, at the Royal Academy of Sciences of Saxony, in Leipzig [88]. In true academic style, he starts reminding his audience of his expertise on the subject:

> *"In meinem vieljärigen Vorlesungen (*Theorie der Trfgr. Bd. I, S. *58, Leipzig 1888; Bd. II, S. 250, 1890; Bd. IIIm Vorrede S. VII, 1893; Leipziger* Berichte *1889, S. 145 [hier* Abh. *VI, S. 237]) über die von mir begründeten Theorien an den Universitäten Christiania (1872 bis 1885) und Leipzig pflege ich regelmäßig, die Aufmerksamkeit meiner Zuhörer darauf zu lenken, daß verschiedene Gebiete der Mechanik und Physik (insbesondere der Optik) in schönster Weise die Begriffe eingliedrige Gruppe von Punkt- beziehungsweise Berürungstransformationen illustrieren und gleichzeitig durch explizite Einfürung dieser Begriffe geförderert werden."*[4]

He is emphatic in having given lectures to many mathematicians, astronomers, physicists and chemists, urging their interest in *... meine Theorien für die Naturerklärung zu verwerten.*[5] Lie seems to have been told by a colleague how some of his *Berührungstransformationen*[6] addressed a problem of optics. His statement is flat:

> *"Reflexionen sind Berührungstransformationen, die jene infinitesimale Berührungstransformation invariant lassen und mit ihr vertauschbar sind. Refraktionen bei Übergang zu einem anderen isotropen Medium sind Berührungstransformationen, die ebenfalls jene infinitesimale Berührungstransformation invariant lassen."*[7]

[4] *For many years in my lectures (···) on the theories founded by me at the universities of Christiania and Leipzig, I regularly call the audience's attention to the beauty with which the various fields of mechanics and physics (particularly optics) illustrate the concepts of one-parameter groups of point and contact transformations, and which are furthered by their explicit introduction.*

[5] ... to value my theories for the explanation of Nature.

[6] Contact transformations, used here in the sense of *canonical.*

[7] *Reflections are canonical transformations that commute and leave invariant any infinitesimal canonical transformation. Refractions to another isotropic medium*

The end of the note reminds the readers of a 1872 paper of his that applies, *selbstverständlich* – of course, to all isotropic and anisotropic wave movement. The renowned theorem of Malus has been generalized, Lie states, in the terms quoted above.

A.7 Symmetries and dynamics in the twentieth century

HERMANN WEYL's 1928 article and book *Gruppentheorie und Quantenmechanik* [140, 141] tamed the mathematical work of Élie Cartan on Lie algebras and groups [30] to place it squarely within the formalism of quantum mechanics (see also [142]). In quick succession, PAUL A.M. DIRAC [38] and EUGENE P. WIGNER [143] published their cornerstone books in 1930 and 1931. In particular, the Wigner-Eckart theorem [143, 46] allowed the recognition and calculation of rotational spectra of atoms and nuclei by separating their geometry, usually contained in the angular momentum quantum numbers, from the reduced radial matrix elements, where lurk the possibly unknown dynamics.

After this beginning, 'geometry' included not only the discrete groups of translations and rotations which had previously determined the possible crystalographic structures, nor only the *manifest* symmetries common to many central-force systems, but also the *hidden* symmetries denounced by degeneracies in the spectral lines of atoms [95]. The energy levels of the Bohr model do not depend on their angular momentum quantum number. Primacy in explaining this fact was claimed by VLADIMIR I. FOCK in early 1935. He writes [50]:

> *"Es ist längst bekannt, daß die Energieniveaus des Wasserstoffatoms in bezug auf die Azimutalquantenzahl ℓ entartet sind; mann spricht gelegentlich von einer „zufälligen" Entartung. Nun ist aber jede Entartung der Eigenwerte mit der Transformationsgruppe der betreffenden Gleichung verbunden: so z. B. die Entartung in bezug auf die magnetische Quantenzahl m mit der gewöhnlichen Drehgruppe. Die Gruppe aber, welche der „zufälligen" Entartung entspricht, war bis jetzt unbekannt."*[8]

---

are canonical transformations that also leave invariant any infinitesimal canonical transformation.

[8] *It is long known that the energy levels of the hydrogen atom are degenerate with respect to their azimuthal quantum number ℓ; one speaks occasionally of an "accidental" degeneracy. But every degeneracy in the eigenvalues is in correspondence with a transformation group: as for example, the magnetic quantum number m is in connection with the ordinary rotation group. However, the group which corresponds to the "accidental" degeneracy of the hydrogen atom was, up to now, unknown.*

In the compact article that follows, Fock derives the hydrogenic wavefunctions from the hyperspherical harmonics and provides sum rules for Na^+, Al^{+++}, Cu^+ and Zn^{++}. Six months later VALENTIN BARGMANN comments on Fock's article [14], presenting the $\mathsf{SO}(4)$ theory essentially in the form we use it in chapter 6.[9]

The dynamics of a quantum system determine the spacing between its discrete energy levels, E_n, $n \in \mathcal{Z}^+$. In the Bohr atom, the levels follow the inverse-quadratic rule $\sim 1/n^2$, and thus one still requires some manœuvres to uncover the Lie algebra which is responsible for the regularity [10, 13, 145, 19, 47]. Indeed, the clearest indicator of a dynamical group is an equally-spaced spectrum, $\sim n$, such as that of the quantum harmonic oscillator [100]. Widespread recognition that dynamics is a higher kind of geometry may be ascribed to the $\mathsf{SU}(3)$ model of elementary particles, where hadrons were accomodated into multiplets with specific 3-quark permutation symmetries, and mesons were the transition operators and generators of the Lie algebra $\mathsf{su}(3)$ (see *e.g.*, the papers contained in [23]); reaction rates could be broadly predicted from purely group-theoretical arguments. Quantum optics has also profited greatly from the simple oscillator model [80] and its inseparable reference to phase space [123].

A complete and even-handed list of relevant papers would turn out hundreds, if not thousands of names of those who during the 60s and 70s participated in the heyday of the search for noninvariance (actually, *co*variance) in quantum mechanics and in optics. That the same methods apply to geometric optics was not generally recognized, even though classical mechanics was seen early in a new perspective [76]. The advantage of refering to classical systems to understand hidden symmetry —as we shall see in parts II and III— is that one can explicitly integrate the Lie algebra of conserved quantities to a group of transformations. These transformations act on the trajectories of the system, on its phase space manifold, and on all objects based on Lie-theoretic foundations, including observables relevant to wave models, and to continuous or discrete signal and image analysis [86, 112]. The proceedings volumes of the biannual International Colloquia on Group Theoretical Methods in Physics may be the best repository of field notes from the trek to find the symmetries of Nature during the last quarter of that century. To analyze them is a task for future historians.

[9] Bargmann adds under a footnote in a 1933 article, that Hulthén has remarked that Klein commented to him about the commutation relations of the Pauli-Lubanski vector back in 1930. This is none other than the Runge-Lenz vector [122], whose true discoverer seems to have been Hamilton, who in 1847 published an article [66] where he solves the equations of Keplerian motion —at about the same time he worked on the fish-eye. This string of anecdotes underlines our conundrum on quoting all first and relevant references [57].

Part II

Symmetry and dynamics of optical systems

Introduction

This part contains transformation groups of optical phase space within the model that we should call *global*, i.e., where ray directions naturally range over the full sphere. The paraxial and metaxial models, where rays are assumed to be near (and not-so near) to the optical axis, will occupy the following two parts. Here, the chapters are essentially independent and illustrate complementary group-theoretical tactics.

In Chap. 5 we explore the symmetry of vacuum, namely the *Euclidean* group of translations and rotations, and the Lorentz group of rotations and boosts. In particular, boosts produce a global aberration that we call relativistic coma. In Chap. 6 we introduce the *conformal* model of optics, where space is curved into a sphere or hyperboloid, and where free trajectories are great circles or hyperbola arcs, rather than straight lines. A stereographic projection of phase space maps this sphere onto an optical system called the Maxwell fish-eye, which is the Kepler system of geometric optics. On the manifold of all fish-eyes we shall see the action of *dynamical* algebras and groups.

We study guides in Chap. 7. These are optical systems which are invariant under translations along, and rotations around, one axis; on the other hand, they can have arbitrary radial refractive index profile. There follows a reduction which is useful for visualizing the four-dimensional optical phase space in three, prescinding of the ignorable coordinate of the symmetry while respecting those of the Hamiltonian dynamic. In the manifold of the corresponding group, the two invariants (angular momentum and Hamiltonian) determine the ray trajectories. In Chap. 8 we generalize the optical model of Chap. 1 by relaxing the postulate that the refractive index depend only on the position, by allowing that it depend also on ray direction; this mimicks anisotropic optical media. Recalling Maxwellian anisotropy serves to highlight the restrictions which physics imposes on the freedoms of geometric optics.

In the appendix we present the recipe for *wavization* of the global geometric model of light by lines in space, to the model of monochromatic wave optics where the wavefields are solutions of the free Helmholtz equation. It is a process parallel to the quantization of classical mechanics, but based on the Euclidean algebra, rather than the Heisenberg-Weyl algebra (Appendix

C). We thus recognize Euclidean rotations and translations to be the *mother group* of all optical models.

Classical mechanics is an old and sometimes envious friend of geometric optics, because transformations of optical phase space are decidedly richer than those of its mechanical companion. Symmetries have a way of catching the eye and pleasing the human mind. The ancient Greeks were fascinated by regular polyhedra, and their interest has eventually led to study N-dimensional polytopes. In this part we shall see symmetries with meanings beyond those of mere invariance of a figure under discrete or continuous transformations. The concepts developed by physicists over much of the twentieth century, such as covariance, hidden symmetry and dynamical groups, probably have their most transparent manifestation in light.

5 Euclidean and Lorentzian maps

The transformations of Euclid and Lorentz are of particular importance because they are the symmetries of vacuum. Applications of symmetry often use the covariance of two structures under a group of transformations, one simple and/or well-known, and the other subject to scrutiny. The first are rigid maps of the plane and of the sphere, while the second will be here canonical transformations of $\wp$, optical phase space.

5.1 Presentations and realizations of symmetry

We first list some designations which are used loosely in the literature, but whose proper meanings serve to distinguish the many faces and rôles of symmetry.

Definition: *The* **presentation** *of a Lie* **group** G *is the product law* $\mathsf{G} \times \mathsf{G} \mapsto \mathsf{G}$ *given by analytic composition functions in some (complete, global and convenient) set of coordinates. The presentation of a Lie* **algebra** a *will specify the Lie brackets between the elements of a vector basis [see (3.67)].*

A group G acts on itself through the product rule $\mathsf{G} \times \mathsf{G} \mapsto \mathsf{G}$. In fact, every element $g' \in \mathsf{G}$ defines three distinct maps:

$$\textit{left}\text{ action}\quad g \mapsto g'g \tag{5.1}$$

$$\textit{right}\text{ action}\quad g \mapsto g\,g'^{-1}, \tag{5.2}$$

$$\textit{adjoint}\text{ action}\quad g \mapsto g'g\,g'^{-1}. \tag{5.3}$$

The action of a Lie group on a manifold derives into a corresponding action of its Lie algebra by differential *operators* on the manifold. We have a name for these manifolds:

Definition: *A* **homogeneous space** V *for a* **group** G *is a manifold whose points transform among themselves under the action of the group. The action is said to be* **effective** *when no point in* V *is invariant under all elements of* G*, and* **transitive** *when every point in* V *can be mapped onto any other point by some element of* G*. When* G *is a Lie group,* V *will also serve as a homogeneous space for its Lie* **algebra**.

Definition: *The* **realization** *of a Lie* **group** G *on a homogeneous space* V *is the map* $\mathsf{G} \times V \mapsto V$ *specified by a set of analytic composition functions in their coordinates. The realization of its Lie* **algebra** *entails specifying its elements (usually) by differential operators on* V.

In this chapter we realize several Lie groups by transformations of the homogeneous phase space $\wp$ of geometric optics. We also realize the groups with $N \times N$ matrices acting on the homogeneous space R^N of column vectors. In $\wp$, group elements are realized as Lie exponentials of first-order differential Poisson operators, while matrices are most amenable to express the composition rule of the group (and to be the group *presentation*). There is a correspondence between the two realizations:

Definition: *The points of two spaces* V *and* V' *(understood to be distinct), which are homogeneous spaces under the effective action of same group* G, *are said to be* G-**covariant**.

Definition: *The* **representations** *of a Lie group or algebra are their realization by matrices. The matrices must be square and, for groups, nonsingular (restricted definition).*

The above definition is restricted because representations exist not only by finite-dimensional matrices but, when unitarity is required in noncompact Lie groups (i.e., those of infinite volume, see Sect. B.3) 'half-infinite' and infinite matrices appear, as well as 'matrices' of *continuous* rows and columns (i.e., Hilbert-Schmidt operators), which are then called **integral transform** representations.[1] The analysis of group representations can be described as Fourier analysis generalized to noncommutative group manifolds, and is one of the major fields in the theory of Lie groups and algebras [56, 71]. There are many texts, both advanced and introductory published by Springer-Verlag, which treat the subject of Lie algebra and group representations *per se*, such as those by Baker [12], Curtis [36], Fulton and Harris [53], and Hall [65]. Nevertheless, classical systems – including the geometric optics of this volume – generally require only of a rather restricted set of (totally-symmetric or completely-degenerate) representations, which are *finite-dimensional*; unitarity is usually not an issue.

The classical (*Cartan*) Lie groups and algebras that we shall use in this and the following parts of this book, have two very useful representations: the *fundamental* and the *adjoint*. The **fundamental** representation of the groups of $D \times D$ orthogonal, unitary, and (when D even) symplectic matrices, are these matrices themselves. As we shall see in this chapter, when translations in R^D are included (as in the Euclidean group), the enlarged fundamental representation can be built by using $(D{+}1) \times (D{+}1)$ matrices with one dummy row. The fundamental representation of the group can be

[1] In this case, the realization of the Lie algebra is through *second*-order differential operators, as we shall see in Appendix C [149, Sect. 9.3]).

derived with respect to the parameters and valuated at the origin; this yields the fundamental representation of its Lie algebra.

The **adjoint** representation of an N-parameter Lie group G, and of its Lie algebra a, consists of $N \times N$ matrices. These matrices are built by recognizing that a natural homogeneous space for the linear action of G is the R^N vector space of a. The action on the algebra derives from the adjoint action on the group given in (5.3), when the adjoint map is applied to group elements $g(\boldsymbol{\alpha})$ near to the identity g_o at $\boldsymbol{\alpha} = \mathbf{0}$ (so $|\boldsymbol{\alpha}|^2 \approx 0$). Recalling the expression (3.40) for Lie exponential operators and (3.69) for their Lie algebra, the α_j-derivatives of (5.3) at the unit are

$$\frac{\partial}{\partial \alpha_j} g' g(\boldsymbol{\alpha})\, g'^{-1}\bigg|_{\boldsymbol{\alpha}=\mathbf{0}} = g' \, \frac{\partial g(\boldsymbol{\alpha})}{\partial \alpha_j}\bigg|_{\boldsymbol{\alpha}=\mathbf{0}} g'^{-1} = g' \, \widehat{f}_j \, g'^{-1} \in \mathsf{a}, \qquad (5.4)$$

and since every vector in a is expressible as a linear combination of the basis, the **adjoint map** $\mathsf{G} \times \mathsf{a} \mapsto \mathsf{a}$ of $g' \in \mathsf{G}$ on $\widehat{f}_j \in \mathsf{a}$ is defined by

$$\widehat{f}_j \mapsto g' \, \widehat{f}_j \, g'^{-1} = \sum_{k=1}^{N} D^{\mathrm{ad}}_{j,k}(g')\, \widehat{f}_k. \qquad (5.5)$$

The adjoint representation of G is the $N \times N$ matrix $\mathbf{D}^{\mathrm{ad}}(g') := \|D^{\mathrm{ad}}_{j,k}(g')\|$. The product rule $\mathbf{D}^{\mathrm{ad}}(g_1)\, \mathbf{D}^{\mathrm{ad}}(g_2) = \mathbf{D}^{\mathrm{ad}}(g_1\, g_1)$ follows because the linear group action distributes over the vector space sum.

The adjoint representation of G provides a representation of the action of the Lie algebra a on *it*self, $\mathsf{a} \times \mathsf{a} \mapsto \mathsf{a}$, which is also called *adjoint*. This is obtained from (5.5) for $g'(\boldsymbol{\alpha})$ near to the identity, so

$$\frac{\partial}{\partial \alpha_j} g'(\boldsymbol{\alpha})\, \widehat{f}_k \, g'^{-1}(\boldsymbol{\alpha})\bigg|_{\boldsymbol{\alpha}=\mathbf{0}} = \frac{\partial g'(\boldsymbol{\alpha})}{\partial \alpha_j}\bigg|_{\boldsymbol{\alpha}=\mathbf{0}} \widehat{f}_k + \widehat{f}_k \, \frac{\partial g'(-\boldsymbol{\alpha})}{\partial \alpha_j}\bigg|_{\boldsymbol{\alpha}=\mathbf{0}} = \widehat{f}_j \, \widehat{f}_k - \widehat{f}_k \, \widehat{f}_j$$

$$= \sum_{\ell=1}^{N} c^{\ell}_{j,k} \widehat{f}_\ell, \qquad c^{\ell}_{j,k} = \frac{\partial}{\partial \alpha_j} D^{\mathrm{ad}}_{k,l}(g'(\boldsymbol{\alpha}))\bigg|_{\boldsymbol{\alpha}=\mathbf{0}}. \qquad (5.6)$$

In the last line we have recognized the commutator between the generators of the algebra and introduced the structure constants $\{c^l_{j,k}\}^N_{j,k,l=1}$ from their basic definition in (3.67)–(3.68). The matrix $\mathbf{C}(\widehat{f}_j) := \|C_{k,l}(j)\|$, $C_{k,l}(j) := c^l_{j,k}$, is the adjoint representation of $\widehat{f}_j \in \mathsf{a}$ on the basis $\{\widehat{f}_k\}^N_{k=1}$ of a. Representations naturally extend to the whole Lie algebra through linearity, $\mathbf{C}\Big(\sum_j a_j \, \widehat{f}_j\Big) = \sum_j a_j \, \mathbf{C}(\widehat{f}_j)$.

Remark: Adjoint representation of translation groups. The adjoint representation becomes trivial for groups all of whose elements commute. Because then $g'\, g\, g'^{-1} = g$, the adjoint representation of such a group is $\mathbf{D}^{\mathrm{ad}}(g') = \mathbf{1}$, and representation of its algebra is $\mathbf{C}(\widehat{f}_j) = \mathbf{0}$ [see (5.5) and

(5.6)] – a trivial representation, but a representation nevertheless. The advantage of the adjoint representation of the algebra is that it always exists and allows us to isolate the commuting and non-commuting parts by bringing the matrices to block-triangular form and separating any translation part. What remains is a *semi-simple* algebra. The eigenvalues of the adjoint representation matrices were instrumental for Cartan [30] in his exhaustive classification of all such algebras. •

5.2 Translations in 3-space

Translations in a 3-dim homogeneous optical medium have been used to describe displacements of a screen along the optical axis [the z-direction in (3.36)], and for translations within its plane [the x- and y-translations in (3.50)]. These are three mutually commuting one-parameter Lie groups of transformations of optical phase space $(\mathbf{p}, \sigma, \mathbf{q}) \in \wp$,[2] whose generic element we denote by

$$\begin{aligned} \mathcal{T}(\alpha_x, \alpha_y, \alpha_z) &:= \exp(\alpha_x\{p_x, \circ\}) \exp(\alpha_y\{p_y, \circ\}) \exp(\alpha_z\{p_z(\mathbf{p}), \circ\}) \\ &= \exp\{\vec{\alpha} \cdot \vec{p}, \circ\}, \end{aligned} \tag{5.7}$$

where $\vec{\alpha} = (\alpha_x, \alpha_y, \alpha_z)^\top \in \mathsf{R}^3$ is a 3-vector and $p_z(\mathbf{p}) = \sigma\sqrt{1 - |\mathbf{p}|^2}$.

The set of translations (5.7) for $\vec{\alpha} \in \mathsf{R}^3$ provides the *presentation* of the three-dimensional translation Lie group, denoted T_3, as

$$\mathcal{T}(\vec{\alpha}_1)\, \mathcal{T}(\vec{\alpha}_2) = \mathcal{T}(\vec{\alpha}_1 + \vec{\alpha}_2), \qquad \mathcal{T}(\vec{0}\,) = \mathit{1}. \tag{5.8}$$

The generators of T_3 in (5.7) are here realized by the Poisson operators

$$\{p_x, \circ\} = -\frac{\partial}{\partial q_x}, \quad \{p_y, \circ\} = -\frac{\partial}{\partial q_y}, \quad \{p_z(\mathbf{p}), \circ\} = \frac{\sigma}{\sqrt{1 - |\mathbf{p}|^2}} \mathbf{p} \cdot \frac{\partial}{\partial \mathbf{q}}, \tag{5.9}$$

acting on the manifold of rays $\wp$, or of beam functions $\rho(\mathbf{p}, \sigma, \mathbf{q})$, and commuting among themselves. The Poisson bracket $\{p_i, p_j\} = 0$ (with structure constants all zero) presents the Lie algebra denoted $\mathfrak{t}_3$, which generates the Lie group of translations T_3.

Definition: *A Lie algebra all of whose Lie brackets are zero is called* **abelian.** *Such algebras generate groups all of whose elements commute, and are also called abelian. Prototypical of these are the translation groups, in space, on cylinders or on tori.*

[2] Recall Sect. 2.1, in particular our convention of indicating 3-vectors by arrows, as $(\vec{p}, \vec{q})$, and 2-vectors on the screen by boldface, as $(\mathbf{p}, \mathbf{q})$. The ranges of the latter are $|\mathbf{p}| < 1$ ($n = 1$, since we are in vacuum), $\sigma = \operatorname{sign} p_z \in \{-1, +1\}$, and $\mathbf{q} \in \mathsf{R}^2$, with the disk boundary points identified as specified in Sect. 2.1.

But translations T_3 have a realization even simpler than (5.9): On the homogeneous space of (smooth) functions $F(\vec{q})$ of 3-space, $\vec{q} = (q_x, q_y, q_z)^\top \in \mathsf{R}^3$, their group of transformations is

$$\mathcal{T}(\vec{\alpha}\,) : F(\vec{q}) = F(\vec{q} - \vec{\alpha}\,) =: \exp\{\vec{\alpha} \cdot \vec{p}, \circ\}_{(3)} \, F(\vec{q}). \tag{5.10}$$

In the last member we wrote the group element as the exponential of the Lie algebra operators, realized by

$$\{p_x, \circ\}_{(3)} := -\frac{\partial}{\partial q_x}, \quad \{p_y, \circ\}_{(3)} := -\frac{\partial}{\partial q_y}, \quad \{p_z, \circ\}_{(3)} := -\frac{\partial}{\partial q_z}. \tag{5.11}$$

We use the 3-dim Poisson-bracket notation,

$$\{f, \circ\}_{(3)} := \sum_{i=1}^{3} \left(\frac{\partial f(\vec{p}, \vec{q})}{\partial q_i} \frac{\partial}{\partial p_i} - \frac{\partial f(\vec{p}, \vec{q})}{\partial p_i} \frac{\partial}{\partial q_i} \right) \tag{5.12}$$

[cf. (3.15)] to introduce a canonically conjugate 3-dim space of *momentum* coordinates $\vec{p}$. But we do not have here any particular dynamics for this 6-dim space; there is no distinguished Hamiltonian function.

Remark: Passive *vs.* active viewpoints. We can interpret (5.10) in two complementary ways: either we are *passively* moving the reference coordinates to a new origin at $\vec{\alpha}$, so a value previously at $F(\vec{q})$ is now at $F(\vec{q} - \vec{\alpha}\,)$; or *actively* transporting the medium – rays and beams included, but not the reference screen – by $-\vec{\alpha}$. The two viewpoints are equivalent for rigid Euclidean motions (translations and rotations). When transformations are driven by a Hamiltonian, or are nonlinear in the coordinates so they deform phase space, the active picture is the more intuitive. •

The realization of the translation Lie algebra t_3 on 3-space (i.e., 6-dim phase space) in (5.11), *reduces* to the previous realization on $\wp$ [4-dim optical phase space, given by (5.9)], when we restrict $\vec{q}$ to the standard screen and $\vec{p}$ to the Descartes sphere. The reduction (see Sect. 2.1) consists of making the points $\vec{q}$ belong to *rays* that are straight lines along $\vec{p}$, and characterizing the rays by their intersection $\mathbf{q}$ and projection $\mathbf{p}$ on the side σ of the standard screen. The coordinate q_z becomes the evolution parameter z under the (canonically conjugate) Hamiltonian $h = -p_z(\mathbf{p})$. According to (2.8), (3.24) and (3.26), this implies that (see Fig. 2.3),

$$\mathbf{p}\, dq_z = p_z(\mathbf{p})\, d\mathbf{q} \quad \Rightarrow \quad \frac{d}{dz} = \frac{\partial}{\partial q_z} = \{p_z(\mathbf{p}), \circ\}, \tag{5.13}$$

and so (5.11) reduce to (5.9).

The concomitant restriction of momentum to the Descartes sphere, $|\vec{p}| = n = 1$ (in this chapter we are in vacuum), replaces the coordinate p_z by the function $p_z(\mathbf{p})$ with σ. The reduction of the 3-dim Poisson operator of

$F(\vec{p},\vec{q})$ to the 2-dim Poisson operator of a corresponding $f(\mathbf{p},\mathbf{q})$ on optical phase space $\wp$ (indicated $f = F|_\wp$) is

$$f(\mathbf{p},\sigma,\mathbf{q}) = F(\vec{p},\vec{q})|_\wp = F\Big(\mathbf{p},\ p_z{=}\sigma\sqrt{1-|\mathbf{p}|^2};\ \mathbf{q},\ q_z{=}0\Big), \tag{5.14}$$

$$\{f(\mathbf{p},\sigma,\mathbf{q}),\circ\} = \{F(\vec{p},\vec{q}),\circ\}_{(3)}\Big|_\wp \quad \begin{cases} \left.\dfrac{\partial}{\partial q_z}\right|_\wp = \dfrac{1}{p_z(\mathbf{p})}\mathbf{p}\cdot\dfrac{\partial}{\partial\mathbf{q}}, \\ p_z|_\wp = p_z(\mathbf{p}) = \sigma\sqrt{1-|\mathbf{p}|^2}. \end{cases} \tag{5.15}$$

Remark: Preservation of Lie algebras under reduction. The reader can verify directly that the reduction (5.15) extends to Poisson brackets between functions, and to commutators between Poisson operators, namely

$$\{F,G\}_{(3)}\Big|_\wp = \{f,g\}, \qquad \big[\{F,\circ\}_{(3)},\{G,\circ\}_{(3)}\big]\Big|_\wp = \big[\{f,\circ\},\{g,\circ\}\big]. \tag{5.16}$$

This means that if we have a multiparameter Lie algebra, realized on the phase space space $(\vec{p},\vec{q}) \in \mathsf{R}^6$, the restrictions (5.14)–(5.15) will lead to a 'reduced' realization of the *same* Lie algebra on the optical phase space $\wp$, referred to the standard screen. The two realizations will be covariant under the generated Lie group. •

5.3 Rotations of 3-space

The best-known nontrivial Lie group is the rotation group of 3-space, denoted SO(3).[3] It is inextricably woven with angular momentum theory, classical and quantum, from which we shall take the nomenclature freely. We first write the generators of SO(3) as 3-dim Poisson operators (5.11), and in the next section we reduce them to Poisson operators on optical phase space, referred

[3] Let this be a brief welcome to the arcane (but common and logical) nomenclature of classical Lie groups: 'S' stands for *special* linear transformations – by matrices of unit determinant. Next, there is a choice for orthogonal 'O', unitary 'U', symplectic 'Sp', or plain linear 'L' transformations. The dimension follows in parentheses, separated if need be into positive- and negative-metric components and the specification of the vector space field as real R, complex C, or quaternionic Q. Thus SO(3) is the special (real) orthogonal group in three dimensions, SO(3,1) is the Lorentz group of (3 + 1)-relativity, and Sp(2,R) = SL(2,R), the symplectic and its isomorphic group of 2×2 real matrices. The letter 'I' stands for *inhomogeneous* groups (in *semidirect* product – see the end of Sect. 5.4) with the translation groups T of appropriate dimension. This combines with the previous names to spell the Euclidean group as ISO(3) (translations and rotations) and the Poincaré group ISO(3,1) (including spacetime translations and boosts). The Lie *algebras* that generate those Lie groups are indicated by lower-case letters, such as so(3), iso(3), so(3,1), or isp(2,R). All groups that we use will be defined and denoted as they appear.

to the standard 2-dim screen (5.9) by means of the restrictions (5.15). A good text on the rotation and Lorentz groups is [118].

Rotations around an axis $k \in \{x, y, z\}$ act on $\vec{p} = (p_x, p_y, p_z)^\top \in \mathsf{R}^3$ through multiplication by 3×3 matrices, $\mathbf{R}_k(\beta)$, which are the fundamental (*and* the adjoint) representation of the 3-dim *orthogonal* group.[4] These rotations can be also realized by Lie exponential 3-dim Poisson operators, $\mathcal{R}_k(\beta) = \exp(\beta\{J_k, \circ\})$ [with $J_k(\vec{p}, \vec{q})$ to be found], acting on functions $F(\vec{p}, \vec{q})$. For rotations around the three Cartesian axes, their correspondence ($\leftrightarrow$) is as follows:

$$\mathbf{R}_x(\beta_x) = \begin{pmatrix} 1 & 0 & 0 \\ 0 & \cos\beta_x & -\sin\beta_x \\ 0 & \sin\beta_x & \cos\beta_x \end{pmatrix} \leftrightarrow \mathcal{R}_x(\beta_x) = \exp(\beta_x\{J_x, \circ\}_{(3)}), \quad (5.17)$$

$$\mathbf{R}_y(\beta_y) = \begin{pmatrix} \cos\beta_y & 0 & \sin\beta_y \\ 0 & 1 & 0 \\ -\sin\beta_y & 0 & \cos\beta_y \end{pmatrix} \leftrightarrow \mathcal{R}_y(\beta_y) = \exp(\beta_y\{J_y, \circ\}_{(3)}), \quad (5.18)$$

$$\mathbf{R}_z(\beta_z) = \begin{pmatrix} \cos\beta_z & -\sin\beta_z & 0 \\ \sin\beta_z & \cos\beta_z & 0 \\ 0 & 0 & 1 \end{pmatrix} \leftrightarrow \mathcal{R}_z(\beta_z) = \exp(\beta_z\{J_z, \circ\}_{(3)}). \quad (5.19)$$

Below we shall see that the Lie exponential operators of $\{J_k, \circ\}_{(3)}$ will generate concomitant rotations of the 3-space of positions, $\vec{q} \in \mathsf{R}^3$, which ensure the canonicity of the full phase space transformation.

Remark: Euler angles of rotation A commonly used parametrization for the group of rotations $\mathsf{SO}(3)$, which is represented by the 3×3 orthogonal matrices (5.17)–(5.19) is by *Euler angles* [21, Sect. 2.6]:

$$\mathbf{R}(\psi, \theta, \phi) = \mathbf{R}_z(\psi)\,\mathbf{R}_x(\theta)\,\mathbf{R}_z(\phi). \quad (5.20)$$

The Euler parametrization has counterparts for pseudo-orthogonal and pseudo-unitary groups. This representation is simple enough to be the *presentation* of the rotation group. •

Remark: Matrices $\leftrightarrow$ Lie exponential operators. When we examined 1-parameter Lie groups in Sect. 3.4, we noted that Lie exponential operators, such as $\mathcal{R}_k(\beta)$ above, have the virtue of acting on functions by 'jumping into' their arguments (as used for the root factors of refraction in Sect. 4.2). This implies that their association ($\leftrightarrow$ above) is such that the action of $\mathcal{R}_k(\mathbf{R}_k)$ on the column vector of coordinates $\vec{p}$ (or $\vec{q}$) will be through the *inverse* of their fundamental representation matrices $\mathbf{R}_k$ above, i.e.,

$$\mathcal{R}(\mathbf{R}) : \vec{p} = \mathbf{R}^{-1}\vec{p}. \quad (5.21)$$

Because only then the order of the group product is preserved under the association,

[4] *I.e.* $\mathbf{R}^\top = \mathbf{R}^{-1}$. See ahead, Sects. 9.3 and 11.7.

$$\left.\begin{array}{l}\mathbf{R}_1 \leftrightarrow \mathcal{R}(\mathbf{R}_1)\\ \mathbf{R}_2 \leftrightarrow \mathcal{R}(\mathbf{R}_2)\end{array}\right\} \Rightarrow \quad \mathbf{R}_1\,\mathbf{R}_2 \leftrightarrow \mathcal{R}(\mathbf{R}_1\,\mathbf{R}_2) = \mathcal{R}(\mathbf{R}_1)\,\mathcal{R}(\mathbf{R}_2). \tag{5.22}$$

We prove this in three lines:

$$\begin{aligned}\mathcal{R}(\mathbf{R}_1)\,\mathcal{R}(\mathbf{R}_2) : \vec{p} &= \mathcal{R}(\mathbf{R}_1)\left(\mathcal{R}(\mathbf{R}_2) : \vec{p}\right)\\ &= \mathcal{R}(\mathbf{R}_1) : \mathbf{R}_2^{-1}\,\vec{p} = \mathbf{R}_2^{-1}\left(\mathcal{R}(\mathbf{R}_1) : \vec{p}\right)\\ &= \mathbf{R}_2^{-1}\,\mathbf{R}_1^{-1}\,\vec{p} = (\mathbf{R}_1\,\mathbf{R}_2)^{-1}\,\vec{p} = \mathcal{R}(\mathbf{R}_1\,\mathbf{R}_2) : \vec{p}.\end{aligned} \tag{5.23}$$

•

Example: Rotations around the x-axis We can find the generating functions $J_k(\vec{p}, \vec{q})$ from (3.35). For example, in (5.17) we let $\beta_x \to 0$, so

$$\mathbf{R}_x(\beta)^{-1}\,\vec{p} = \begin{pmatrix} p_x \\ p_y \cos\beta + p_z \sin\beta \\ -p_y \sin\beta + p_z \cos\beta \end{pmatrix} \approx \vec{p} + \beta \begin{pmatrix} 0 \\ p_z \\ -p_y \end{pmatrix}. \tag{5.24}$$

Analytic functions $f(\vec{p})$ transforming through (5.24) can be expanded in a Taylor series (only terms linear in β are needed), and compared with the Lie exponential series of Poisson operators,

$$\begin{aligned}\mathcal{R}_x(\beta) : f(\vec{p}) &= f(p_x,\, p_y + \beta p_z + \cdots,\, p_z - \beta p_y + \cdots)\\ &\approx f(\vec{p}) + \beta\Big(p_z \frac{\partial}{\partial p_y} - p_y \frac{\partial}{\partial p_z}\Big) f(\vec{p}),\end{aligned} \tag{5.25}$$

$$\begin{aligned}\exp(\beta\{J_x, \circ\}_{(3)}) : f(\vec{p}) &\approx f(\vec{p}) + \beta\{J_x, \circ\}_{(3)} f(\vec{p})\\ &= f(\vec{p}) + \beta\Big(\frac{\partial J_x}{\partial q_x}\frac{\partial}{\partial p_x} + \frac{\partial J_x}{\partial q_y}\frac{\partial}{\partial p_y} + \frac{\partial J_x}{\partial q_z}\frac{\partial}{\partial p_z}\Big) f(\vec{p}).\end{aligned} \tag{5.26}$$

From (5.25)–(5.26) it follows that $\partial J_x/\partial q_x = 0$, $\partial J_x/\partial q_y = p_z$, and $\partial J_x/\partial q_z = -q_y$. We solve for $J_x = q_y p_z - q_z p_y = (\vec{q}\times\vec{p})_x$, plus any function of $\vec{p}$ that we set to zero[5] so that the position vectors $\vec{q}$ be properly *covariant* with the vectors $\vec{p}$ of momentum 3-space. •

Finding the generator functions J_y and J_z in the same way as J_x above, we recognize/define the 3-vector of *angular momentum* functions

$$\begin{pmatrix} J_x \\ J_y \\ J_z \end{pmatrix} = \vec{J} := \vec{q}\times\vec{p} = \begin{pmatrix} q_y p_z - q_z p_y \\ q_z p_x - q_x p_z \\ q_x p_y - q_y p_x \end{pmatrix}. \tag{5.27}$$

The three functions $J_k(\vec{p}, \vec{q})$ generate respective one-parameter Lie groups (5.17)–(5.19); their Poisson brackets [cf. (3.66)], are

$$\{J_x, J_y\}_{(3)} = J_z, \quad \{J_y, J_z\}_{(3)} = J_x, \quad \{J_z, J_x\}_{(3)} = J_y. \tag{5.28}$$

This is the presentation of the Lie algebra $\mathsf{so}(3)$, which generates the Lie group $\mathsf{SO}(3)$ of rotations in 6-dim phase space $(\vec{p}, \vec{q})$.

[5] Additive functions of the components of $\vec{p}$ generate (generic) spherical aberration. See Sect. 3.6.1.

5.4 Rotations of the screen

We now reduce the 6-dim realization of 3-dim rotations generated by the angular momentum functions $J_k(\vec{p},\vec{q})$ in (5.27), down to the 4-dim optical phase space $(\mathbf{p},\mathbf{q}) \in \wp$ referred to the screen at the plane $z=0$.[6] The corresponding generator functions $j_k(\mathbf{p},\mathbf{q}) := J_k(\vec{p},\vec{q})|_\wp$, under the restrictions (5.14)–(5.15), yield the *reduced* angular momentum functions and Poisson operators,

$$j_x = q_y\sqrt{1-|\mathbf{p}|^2}, \qquad \{j_x,\circ\} = p_z(\mathbf{p})\frac{\partial}{\partial p_y} + \frac{q_y}{p_z(\mathbf{p})}\mathbf{p}\cdot\frac{\partial}{\partial \mathbf{q}}, \tag{5.29}$$

$$j_y = -q_x\sqrt{1-|\mathbf{p}|^2}, \qquad \{j_y,\circ\} = -p_z(\mathbf{p})\frac{\partial}{\partial p_x} - \frac{q_x}{p_z(\mathbf{p})}\mathbf{p}\cdot\frac{\partial}{\partial \mathbf{q}}, \tag{5.30}$$

$$j_z = \mathbf{q}\times\mathbf{p}, \qquad \{j_z,\circ\} = -\mathbf{q}\times\frac{\partial}{\partial \mathbf{q}} - \mathbf{p}\times\frac{\partial}{\partial \mathbf{p}}. \tag{5.31}$$

The 2-dim Poisson brackets between the j_k's above, are the same as between their homonym J_k's in (5.28), because of their covariance through (5.16). They are distinct *realizations* of the same abstract algebra on different homogeneous spaces.

Remark: The Casimir invariant. In $(\vec{p},\vec{q}) \in \mathsf{R}^6$ under rotations, there are three independent functions which are invariant under $\mathsf{SO}(3)$: $|\vec{p}|^2$, $|\vec{q}|^2$, and $\vec{p}\cdot\vec{q}$ (because $\mathcal{R}: |\vec{p}|^2 = |\vec{p}|^2$, $\{\vec{J},|\vec{p}|^2\} = \vec{0}$, etc.). A quadratic function of these is the *Casimir* (or square angular momentum) invariant

$$J^2 := J_x^2 + J_y^2 + J_z^2 = |\vec{q}\times\vec{p}|^2 = |\vec{p}|^2\,|\vec{q}|^2 - (\vec{p}\cdot\vec{q})^2. \tag{5.32}$$

In the optical phase space realization $(\mathbf{p},\mathbf{q}) \in \wp$, this reduces to

$$j^2 := J^2|_\wp = j_x^2 + j_y^2 + j_z^2 = |\mathbf{q}|^2(1-|\mathbf{p}|^2) + |\mathbf{q}\times\mathbf{p}|^2 = |\mathbf{q}|^2 - (\mathbf{p}\cdot\mathbf{q})^2, \tag{5.33}$$

which is also invariant under the realization of $\mathsf{SO}(3)$, generated by (5.29)–(5.31). While in R^6 there are 3 invariants, in the 4-dim phase space $\wp$ only $j^2(\mathbf{p},\mathbf{q})$ is invariant; both are homogeneous spaces for $\mathsf{SO}(3)$ with the 3 remaining coordinates subject to effective rotation. •

Remark: Skewness of rays and the Petzval invariant. The generator function of rotations around the optical z axis, $j_z = \mathbf{q}\times\mathbf{p}$ in (5.31), has the form of the (skew-scalar) mechanical angular momentum on the 2-dim screen, and is called the *skewness* of the ray $(\mathbf{p},\mathbf{q})$. Meridional rays (i.e., those lying in a plane with the optical axis, $\mathbf{p} \parallel \mathbf{q}$) have zero skewness; the two saggital rays ($\pm\mathbf{p} \perp \mathbf{q}$) have maximal skewness (of either sign). Axis-symmetric optical media (i.e., those which are invariant under rotations

[6] For brevity we omit the sign $\sigma = \operatorname{sign} p_z$ from the coordinates of phase space $\wp$. This sign accompanies all square root terms as $\sigma\sqrt{1-|\mathbf{p}|^2}$ in the following expressions.

around the optical axis) will not change the skewness of the rays, as follows from Sect. 3.5, and conversely, when j_z is constant for every ray, the medium is axis-symmetric. For example, propagation in a homogeneous medium (2.10), conserves the skewness: $\mathbf{q}(z)\times\mathbf{p}(z) = \mathbf{q}(0)\times\mathbf{p}(0)$. Similarly, axis-symmetric refracting surfaces $z = \zeta(|\mathbf{q}|)$ have the surface normal $\boldsymbol{\Sigma}(|\mathbf{q}|) := \nabla_{\mathbf{q}}\zeta(|\mathbf{q}|) = 2\mathbf{q}\,\partial\zeta(|\mathbf{q}|)/\partial(|\mathbf{q}|^2)$; the root transformation (4.8)–(4.9) yields the following relation between the skewness of the incident, surface, and refracted rays:

$$\mathbf{q}\times\mathbf{p} = \frac{\bar{\mathbf{q}}\times\bar{\mathbf{p}}}{1-\partial(\zeta(|\bar{\mathbf{q}}|))^2/\partial(|\bar{\mathbf{q}}|^2)} = \mathbf{q}'\times\mathbf{p}'. \tag{5.34}$$

The conserved skewness $\mathbf{q}\times\mathbf{p}$ is often called the *Petzval* invariant of the rays.[7] •

The Poisson operators of the components of angular momentum (5.29)–(5.31), can be exponentiated to the group $\mathsf{SO}(3)$, realized in geometric optics by rotations of the standard screen (or equivalently, of the ambient 3-space that contains the rays). But summing the Lie exponential series of Poisson operators head-on can be difficult. There is of course the fail-safe traditional approach of 3-dim Euclidean geometry to calculate the way in which line intersections with a plane change when the plane rotates. A smarter procedure (from our point of view) to find the maps of rays on rotated screens, is to note that under $\mathsf{SO}(3)$, the components of $\{\vec{p},\circ\}_{(3)}$ are covariant with (transform in the same way as) those of $\vec{p}$, $\vec{q}$ and $\vec{J}$ [namely, as in (5.17)–(5.19)]; and [after the reduction (5.15)–(5.16), remain] covariant with the three components of $\vec{p}|_{\wp} = (p_x, p_y, p_z(\mathbf{p}))^\top$. And then there are the invariant functions: $j^2(\mathbf{p},\mathbf{q})$ in (5.33) and, for rotations around a particular axis k, its generator $j_k(\mathbf{p},\mathbf{q})$, p_k, and q_k. Together they will yield the map of rays on the screen, as in the following example.

Example: Rotations around the x-axis. Reducing momentum (5.24) to $\mathbf{p}$ in $\wp$, we note that p_x is invariant, while p_y transforms with its partner $p_z(\mathbf{p})$ as[8]

$$\exp(\beta\{j_x,\circ\})\begin{pmatrix} p_y \\ \sqrt{1-|\mathbf{p}|^2}\end{pmatrix} = \begin{pmatrix}\cos\beta & \sin\beta \\ -\sin\beta & \cos\beta\end{pmatrix}\begin{pmatrix} p_y \\ \sqrt{1-|\mathbf{p}|^2}\end{pmatrix} \tag{5.35}$$

From the invariant function $j_x(\mathbf{p},\mathbf{q}) = q_y\,p_z(\mathbf{p})$ we extract $q_y = j_x(\mathbf{p},\mathbf{q})/p_z(\mathbf{p})$, where we know the transformations of both the numerator and denominator,

$$\begin{aligned}\exp(\beta\{j_x,\circ\})\,q_y &= \exp(\beta\{j_x,\circ\})\frac{j_x(\mathbf{p},\mathbf{q})}{p_z(\mathbf{p})} = \frac{j_x(\mathbf{p},\mathbf{q})}{\exp(\beta\{j_x,\circ\})\,p_z(\mathbf{p})} \\ &= \frac{q_y\sqrt{1-|\mathbf{p}|^2}}{-p_y\sin\beta + \sqrt{1-|\mathbf{p}|^2}\cos\beta}.\end{aligned} \tag{5.36}$$

---

[7] Some authors define the Petzval as $|\mathbf{q}\times\mathbf{p}|^2$.

[8] As the Descartes sphere of ray directions rotates, the points of the two momentum disks (distinguished by $\sigma = \operatorname{sign} p_z$) are mapped among themselves. Although some regions of $\wp$ may flip σ; we choose to disregard them for comfort, and concentrate on the regions that map within the same chart.

To find the transformation of q_x, we use the covariance of $j_y(\mathbf{p},\mathbf{q}) = -q_x\, p_z(\mathbf{p})$ and $j_z(\mathbf{p},\mathbf{q}) = \mathbf{q}\times\mathbf{p}$ with p_y and $p_z(\mathbf{p})$ given in (5.35), namely $\exp(\beta\{j_x,\circ\})\, j_y = j_y\cos\beta + j_z\sin\beta$. We extract q_x from j_y and obtain

$$\exp(\beta\{j_x,\circ\})\, q_x = \frac{q_x\sqrt{1-|\mathbf{p}|^2}\cos\beta - \mathbf{q}\times\mathbf{p}\sin\beta}{-p_y\sin\beta + \sqrt{1-|\mathbf{p}|^2}\cos\beta}. \tag{5.37}$$

•

As the above example shows, rotations of the screen are 'very' nonlinear maps of the ray coordinates $(\mathbf{p},\mathbf{q}) \in \wp$. But since they are covariant with the 3×3 matrix representation of $\mathsf{SO}(3)$, they are no more difficult to treat than in their simplest presentation. Rotations around the y-axis can be obtained from (5.35)–(5.37) through rotating the screen around the optical axis by $\frac{1}{2}\pi$, so $x \mapsto y$ and $y \mapsto -x$. Finally, rotations around the optical z-axis follow *verbatim* from 3-space for the two coordinates of the standard screen.

The three generator functions of (commuting) translations $\vec{p}$, and the three generator functions of (noncommuting) rotations $\vec{\jmath}$ [see (5.9) and (5.29)–(5.31)], have the following Poisson brackets among them,

$$\{j_x, j_y\} = j_z, \qquad \text{and cyclically,} \tag{5.38}$$

$$\{j_x, p_y\} = p_z, \qquad \text{and cyclically,} \tag{5.39}$$

$$\{p_i, p_j\} = 0. \tag{5.40}$$

This set of Poisson brackets presents the *Euclidean* Lie algebra $\mathsf{iso}(3)$.

5.5 Euclidean and semidirect product groups

The Euclidean Lie algebra $\mathsf{iso}(3)$ generates the Euclidean Lie group of rigid motions in 3-space, denoted $\mathsf{ISO}(3)$. Its simplest presentation is through 4×4 block matrices containing the 3-vector of translation $\vec{\alpha}$ [see (5.7)–(5.10)] and the 3 parameters of rotation in the 3×3 orthogonal matrix $\mathbf{R}$ (in Euler or other angles), associated ($\leftrightarrow$) with the action on $\vec{q} \in \mathsf{R}^3$ in the following way,[9]

$$\mathbf{E}(\vec{\alpha},\mathbf{R}) := \begin{pmatrix} \mathbf{R} & \vec{\alpha} \\ 0 & 1 \end{pmatrix} \quad \leftrightarrow \quad \mathcal{E}(\vec{\alpha},\mathbf{R}) \tag{5.41}$$

$$= \begin{pmatrix} \mathbf{1} & \vec{\alpha} \\ 0 & 1 \end{pmatrix}\begin{pmatrix} \mathbf{R} & \vec{0} \\ 0 & 1 \end{pmatrix} \qquad := \mathcal{T}(\vec{\alpha})\,\mathcal{R}(\mathbf{R}) \tag{5.42}$$

$$= \begin{pmatrix} \mathbf{R} & \vec{0} \\ 0 & 1 \end{pmatrix}\begin{pmatrix} \mathbf{1} & \mathbf{R}^{-1}\vec{\alpha} \\ 0 & 1 \end{pmatrix}, \qquad = \mathcal{R}(\mathbf{R})\,\mathcal{T}(\mathbf{R}^{-1}\vec{\alpha}), \tag{5.43}$$

[9] The lower left element of the matrices, indicated 0, is properly a row vector of 3 zeros.

$$\mathbf{E}(\vec{\alpha},\mathbf{R})^{-1}\begin{pmatrix}\vec{q}\\1\end{pmatrix} \quad\leftrightarrow\quad \begin{aligned}\mathcal{E}(\vec{\alpha},\mathbf{R}):f(\vec{q}) &= f\Big(\mathcal{E}(\vec{\alpha},\mathbf{R}):\vec{q}\Big)\\ &= f\Big(\mathbf{R}^{-1}(\vec{q}-\vec{\alpha})\Big),\end{aligned} \tag{5.44}$$

$$= \begin{pmatrix}\mathbf{R}^{-1} & -\mathbf{R}^{-1}\vec{\alpha}\\ 0 & 1\end{pmatrix}\begin{pmatrix}\vec{q}\\1\end{pmatrix},\ \mathcal{E}(\vec{\alpha},\mathbf{R})^{-1} = \mathcal{E}(-\mathbf{R}^{-1}\vec{\alpha},\mathbf{R}^{-1}). \tag{5.45}$$

[Cf. (5.10) and (5.21)–(5.23).] The Euclidean group $\mathsf{ISO}(3)$ has thus its fundamental representation given by the block-triangular matrices $\mathbf{E}(\vec{\alpha},\mathbf{R})$ in (5.41), with the product law

$$\mathbf{E}(\vec{\alpha}_1,\mathbf{R}_1)\,\mathbf{E}(\vec{\alpha}_2,\mathbf{R}_2) = \mathbf{E}(\vec{\alpha}_1+\mathbf{R}_1\vec{\alpha}_2,\ \mathbf{R}_1\mathbf{R}_2). \tag{5.46}$$

The Euclidean algebra and the Euclidean group, presented by (5.38)–(5.40) and (5.46) respectively, and connected by (3.69), are the best examples of a *semidirect sum* Lie algebra and its semidirect *product* Lie group. The constituents of the example were the algebras and groups of translations and rotations; we must only generalize slightly the former by *not* assuming that these commute [as those in (5.40) do]. Two definitions follow:

Definition: *Let* t *and* r *be two Lie algebras*[10] *with a common Lie bracket* $\{\circ,\circ\}$ *in the direct sum of their vector spaces such that, as sets,*

$$\{\mathsf{t},\mathsf{t}\}\subset\mathsf{t},\quad \{\mathsf{r},\mathsf{t}\}\subset\mathsf{t},\quad \{\mathsf{r},\mathsf{r}\}\subset\mathsf{r}. \tag{5.47}$$

The ensemble constitutes a Lie algebra called the **semidirect sum** *of* t *and* r, *denoted*[11] $\mathsf{a}=\mathsf{t}\oplus\!\!\!\!\supset\mathsf{r}$, *and containing both as subalgebras (see page 46);* t *is called the* **invariant** *subalgebra and* r *the* **radical** *subalgebra. If* $\{\mathsf{r},\mathsf{t}\}=0$, *then the vector sum of* t *and* r *is a* **direct sum** *of algebras,* $\mathsf{t}\oplus\mathsf{r}$.

Definition: *Let* T *and* R *be two Lie groups, with products between its elements indicated* $t_1\sharp t_2\in\mathsf{T}$ *and* $r_1\flat r_2\in\mathsf{R}$ *respectively, and such that* T *is a homogenous space for the action of* $r\in\mathsf{R}$, *as* $r:t=D(r)\,t\in\mathsf{T}$. *Then the set of pairs* $g(t,r)$ *with the product rule*

$$g(t_1,r_1)\,g(t_2,r_2) = g\Big(t_1\sharp D(r_1)\,t_2,\ r_1\flat r_2\Big), \tag{5.48}$$

and satisfying the distribution implied by associativity

$$D(r_1)\,(t_2\sharp D(r_2)\,t_3) = D(r_1)\,t_2\sharp D(r_1\flat r_2)\,t_3, \tag{5.49}$$

forms a Lie group called the **semidirect product** *of* T *and* R, *indicated* $\mathsf{G}=\mathsf{T}\otimes\!\!\!\!\supset\mathsf{R}$. *The first,* T, *is called the* **invariant** *subgroup of* G; *the second,* R, *is called the* **factor** *(or* **radical**) *subgroup. When* $D(r)\,t=t$, *the composite group is a* **direct product**, $\mathsf{G}=\mathsf{T}\otimes\mathsf{R}$.

---

[10] Cf. (5.38)–(5.40). We may think of t and r as translations and rotations.

[11] There are many notations in the literature; we emphasize with '$\mathsf{t}\oplus\!\!\!\!\supset\mathsf{r}$' that the summand algebras of the semidirect sum cannot be exchanged.

Example: The aberration algebra. Denote by F_k the vector space of homogeneous polynomials of integer degree $k \geq 1$ in the phase space coordinates $(\mathbf{p}, \mathbf{q})$. Since the Poisson bracket involves two dreivatives, the elements of these spaces satisfy

$$\{F_k, F_{k'}\} = F_{k+k'-2}. \tag{5.50}$$

It follows that F_2 is a Lie algebra with this bracket (because $\{F_2, F_2\} = F_2$) and that every F_k serves as a homogeneous space for the action of the Poisson operators $\{F_2, \circ\}$. But the F_k's, do not close into a Lie algebra when $k \geq 3$; (5.50) shows that they 'spill over' into ever higher k's. Only the (direct) sum of all spaces, $\oplus_{k\geq 3} F_k$, forms a Lie algebra (with derivation) under the bracket (5.50). As we recall from Sect. 3.6, F_2 generates linear transformations of phase space while F_k, $k \geq 3$, generates nonlinear ones called *aberrations*; we reserve part IV for their presentation. The linear + aberration algebra of generators $\oplus_{k\geq 2} F_k$ is thus a semidirect sum of F_2 with the invariant subalgebra of aberrations $\oplus_{k\geq 3} F_k$. The latter takes the place of the subalgebra of translations above, and here the Poisson brackets are not zero. Yet, because the dimension of this algebra is (denumerable) infinity, it is an improper one; Lie algebras of infinite dimension cannot be trusted in many crucial properties, such as exponentiating to a proper Lie group. As with vector spaces whose dimension is made to 'pass to the continuum' into a Hilbert space of functions, the problems of series convergence can become deadly. They are often called *functional* Lie algebras. •

Example: Inhomogeneous groups. With groups of translations and rotations we can build many groups of inhomogeneous (linear + constant) transformations. Let $r \in \mathsf{R} = \mathsf{SO}(3)$ be a rotation as before, but now let its action on the homogeneous space (and group) of translations, $r : t \in \mathsf{T}$, be through its $(2\ell + 1) \times (2\ell + 1)$ Wigner representation matrices $\mathbf{D}^{\ell}(r)$ of spin ℓ (integer or half-integer) familiar from angular momentum in quantum mechanics [21]. The dimension of the subgroup of translations will be then $2\ell + 1$, with the 3×3 matrices in (5.41)–(5.45) being the special $\ell = 1$ case. The corresponding 'Euclidean' groups can be distinguished by $\mathsf{I}_{2\ell+1} \oplus \mathsf{SO}(3)$. •

5.6 Lorentz boost of light-like vectors

In the framework of special relativity, a four-vector $v = (v_x, v_y, v_z, v_0)^\top$ boosted from rest to rapidity γ in the direction of the z-axis, will suffer the *Lorentz transformation*

$$\mathcal{B}_z(\gamma) : \begin{pmatrix} \mathbf{v} \\ v_z \\ v_0 \end{pmatrix} = \begin{pmatrix} \mathbf{1} & \mathbf{0} & \mathbf{0} \\ \mathbf{0}^\top & \cosh\gamma & \sinh\gamma \\ \mathbf{0}^\top & \sinh\gamma & \cosh\gamma \end{pmatrix} \begin{pmatrix} \mathbf{v} \\ v_z \\ v_0 \end{pmatrix} =: \mathbf{B}_z(\gamma)\, v, \tag{5.51}$$

where $\mathbf{v} = (v_x, v_y)^\top$ as before, and $\mathbf{0} = (0,0)^\top$. The space part $\vec{v} = (\mathbf{v}, v_z)^\top$ is identified with the direction of momentum both in optics and in classical relativistic kinematics. An observer with a screen at rest and another with a screen in relative motion will generally differ in their measurements of ray directions. Here we investigate this *Gedankenexperiment* marriage between geometric optics and special relativity.

Remark: Rapidity is the best parameter for boosts. Rapidity $\gamma \in \mathsf{R}$ correponds to velocity $c \tanh \gamma$, and is the best parameter for boosts. Whereas relativistic velocities add nonlinearly not to exceed c, the composition of two boosts in the same direction with rapidities γ_1 and γ_2 will result in a boost with the sum of their rapidities, $\gamma_1 + \gamma_2$, as we can see multiplying two boost matrices (5.51). Boosts evidently form a 1-parameter Lie group: $\mathcal{B}_z(\gamma_1)\, \mathcal{B}_z(\gamma_2) = \mathcal{B}_z(\gamma_1 + \gamma_2)$. •

A ray of colatitude angle $\theta \in [0, \pi]$ with respect to the optical axis, i.e. $\tan \theta = |\mathbf{v}|/\mathbf{v_z}$, after the boost (5.51) will have a colatitude angle θ' that can be determined through

$$\begin{aligned} \tan \tfrac{1}{2}\theta' &= \mathcal{B}_z(\gamma) : \tan \tfrac{1}{2}\theta = \mathcal{B}_z(\gamma) : \frac{|\mathbf{v}|}{v_z + v_0} \\ &= \frac{|\mathbf{v}|}{(v_z + v_0)(\cosh \gamma + \sinh \gamma)} = e^{-\gamma} \tan \tfrac{1}{2}\theta. \end{aligned} \tag{5.52}$$

This is the *Bargmann* deformation [15] of the Descartes sphere.[12] See Fig. 5.1, where each meridian half-circle is a homogeneous space for the 1-parameter group of Lorentz boosts, whose action is effective (except for the two poles) and transitive within the colatitude interval $(0, \pi)$.

Light-like four-vectors lie on the cones $v_0 = \pm|\vec{v}\,|$; we identify their homogeneous space part with the optical 3-momentum,

$$\vec{p} = (p_x, p_y, p_z)^\top := n\big(v_x/v_0, v_y/v_0, v_z/v_0\big)^\top. \tag{5.53}$$

As on the Descartes sphere, $|\vec{p}| = 1$ and the third coordinate $p_z = p_z(\mathbf{p}) = \sigma\sqrt{1 - |\mathbf{p}|^2}$ is a function of the other two [recall (5.15)]. Following the same route as in (5.36)–(5.37), but with the Lorentz transformation (5.51) in the screen coordinates of momentum (5.53), we find the following realization of the z-boost on the homogeneous space provided by the Descartes sphere,

$$\mathcal{B}_z(\gamma) : \begin{pmatrix} \mathbf{p} \\ p_z(\mathbf{p}) \end{pmatrix} = \begin{pmatrix} \mathbf{p}(\gamma) \\ p_z(\mathbf{p}(\gamma)) \end{pmatrix} = \begin{pmatrix} \dfrac{\mathbf{p}}{\cosh \gamma + p_z(\mathbf{p}) \sinh \gamma} \\ \dfrac{p_z(\mathbf{p}) \cosh \gamma + \sinh \gamma}{\cosh \gamma + p_z(\mathbf{p}) \sinh \gamma} \end{pmatrix}. \tag{5.54}$$

Remark: Mixing of forward and backward rays. For $\gamma > 0$, forward rays ($p_z > 0$, $\sigma = +$) in the disk $\mathcal{D}_+$ of momenta $|\mathbf{p}| < 1$, map on a *sub-disk*

[12] In positional astronomy it is called the *stellar aberration* of the celestial sphere.

of forward rays $|\mathbf{p}(\gamma)| < e^{-\gamma}$. Backward rays ($p_z < 0$, $\sigma = -$) inside the disk $|\mathbf{p}| < e^{-\gamma}$ remain backward, while their complement in $\mathcal{D}_-$, i.e. in the annullus $e^{-\gamma} < |\mathbf{p}| < 1$, map on the forward ray annullus $1 > |\mathbf{p}| > e^{-\gamma}$. Again for comfort we make no further precision on the global matching of the two charts under Lorentz boosts. •

See in Fig. 5.1 that under the Bargmann deformation (5.52), the angle elements $d\theta$ change their size. Since we know that no rays can be lost nor created, we must look at the whole of phase space to understand what happens with the rays. That is, we must find the action of $\mathcal{B}_z(\gamma)$ on both momenta *and positions* on the screen $(\mathbf{p}, \mathbf{q}) \in \wp$, to uphold the canonicity of Lorentz transformations.

Remark: Descartes' distorsion and coma. The action of the Lorentz z-boost on the Descartes sphere (5.54), is a point-to-point *distorsion* of momentum. We conclude from the previous equations (3.23) and (3.59), that Lorentz z-boosts are members of the *comatic family* of canonical maps of $\wp$. For this reason we have called their optical realization *relativistic coma* in [5] and [160]. •

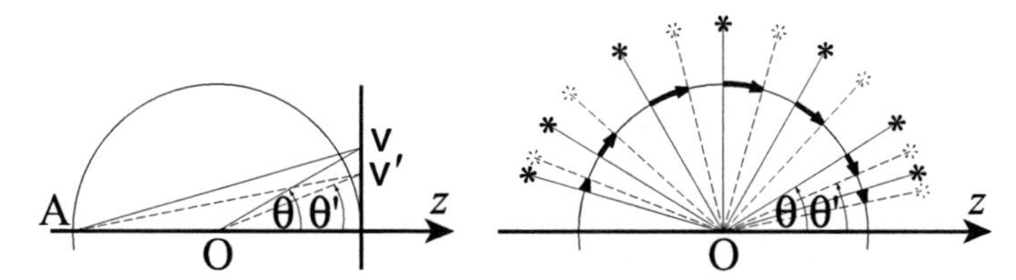

Fig. 5.1. Bargmann deformation of the circle and stellar aberration are Lorentz boosts acting on ray directions. *Left*: Contraction of the (vertical tangent) line $v \mapsto v' = e^{-\gamma}v$, maps the angle $\frac{1}{2}\theta = \widehat{z\mathrm{A}v}$ to $\frac{1}{2}\theta' = \widehat{z\mathrm{A}v'}$, and produces the Bargmann deformation of the circle $\theta \in \mathcal{S}^1$ (center at O). *Right*: Rays arriving from stars * to the observer at rest in O (continuous lines), when boosted, become rays appearing to come from the directions marked * (dotted). The action of the Lorentz boost on optical momentum $\mathbf{p}$ is the vertical component of the Bargmann deformation of the circle.

We now find the function $b_z(\mathbf{p}, \mathbf{q})$ that generates z-boosts on the screen through the Lie series $\mathcal{B}_z(\gamma) = \exp(\gamma\{b_z(\mathbf{p}, \mathbf{q}), \circ\})$, as we did for rotations in 3-space in the previous section. In this way we shall determine the Lorentz transformation of the position coordinates of phase space and guarantee that the transformation be canonical. We use (3.35) on (5.54) as in (5.25)–(5.26), carried to first order in the rapidity parameter $\gamma \ll 1$, to write

$$\mathcal{B}_z(\gamma) : \mathbf{p} = \mathbf{p}(\gamma) \approx \frac{\mathbf{p}}{1 + \gamma\, p_z(\mathbf{p})} \approx \mathbf{p} - \gamma\, p_z(\mathbf{p})\, \mathbf{p}, \tag{5.55}$$

$$\exp(\gamma\{b_z, \circ\})\, \mathbf{p} \approx \mathbf{p} + \gamma\Big(\frac{\partial b_z}{\partial q_x}\frac{\partial}{\partial p_x} + \frac{\partial b_z}{\partial q_y}\frac{\partial}{\partial p_y}\Big)\, \mathbf{p}. \tag{5.56}$$

Hence $\partial b_z/\partial \mathbf{q} = -p_z(\mathbf{p})\,\mathbf{p}$, up to an additive function of $\mathbf{p}$ (which we discard since it only contains spherical aberration). The solution is the generating function

$$b_z(\mathbf{p},\mathbf{q}) := -\sqrt{1-|\mathbf{p}|^2}\;\mathbf{p}\cdot\mathbf{q}, \tag{5.57}$$

The realization of the z-boost generator on $\wp$ is thus[13]

$$-\{b_z(\mathbf{p},\mathbf{q}),\circ\} = \sqrt{1-|\mathbf{p}|^2}\left(\mathbf{p}\cdot\frac{\partial}{\partial\mathbf{p}} + \mathbf{q}\cdot\frac{\partial}{\partial\mathbf{q}}\right) + \frac{\mathbf{p}\cdot\mathbf{q}}{\sqrt{1-|\mathbf{p}|^2}}\mathbf{p}\cdot\frac{\partial}{\partial\mathbf{q}}. \tag{5.58}$$

Below we shall see how to integrate this Poisson operator to a Lie group of transformations on optical phase space $\wp$.

5.7 Relativistic aberration of images

According to the previous arguments, a camera at rest focused on a nearby object (i.e. collecting cones of rays to form its image) will *not* produce the same image when the object and the camera are in relative motion. To find the aberrated image we should in principle exponentiate (5.58) acting on beam functions of $\wp$. The task is as straightforward as it is daunting. So we use the finer tools that we have developed.

Since boosts along an axis and rotations around the same axis commute, $\{j_z, b_z\} = 0$ [see (5.31)], both $j_z = \mathbf{q}(\gamma)\times\mathbf{p}(\gamma)$ and $-b_z = p_z(\mathbf{p}(\gamma))\,\mathbf{p}(\gamma)\cdot\mathbf{q}(\gamma)$ are invariants under z-boosts, as well as under rotations around the z-axis. We also know from (3.59) that $\mathbf{q}(\gamma)$ must be *linear* in the components of $\mathbf{q}$ because it must transform as a *vector* under rotations in the x–y plane, so it must have the general form

$$\mathbf{q}(\gamma) = R(|\mathbf{p}|;\gamma)\,\mathbf{q} + S(|\mathbf{p}|;\gamma)\mathbf{p}\cdot\mathbf{q}\,\mathbf{p}, \tag{5.59}$$

with two functions R and S of $|\mathbf{p}|$ and γ to be determined.[14] Then, from the conservation of j_z we determine $R(|\mathbf{p}|;\gamma)$ to be the denominator of $\mathbf{p}(\gamma)$ in (5.54). Finally, $S(|\mathbf{p}|;\gamma)$ can be found from the conservation of b_z,

$$-b_z = \left(1+|\mathbf{p}|^2\frac{S}{R}\right)p_z(\mathbf{p}(\gamma))\mathbf{p}\cdot\mathbf{q} \quad\Rightarrow\quad S = \frac{R}{|\mathbf{p}|^2}\left(\frac{p_z(\mathbf{p})}{p_z(\mathbf{p}(\gamma))} - 1\right). \tag{5.60}$$

Replacing this into (5.59), we find the map of ray positions on the screen under z-boosts, $\mathbf{q}(\gamma) := \mathcal{B}_z(\gamma) : \mathbf{q}$, to be

$$\mathbf{q}(\gamma) = (\cosh\gamma + p_z(\mathbf{p})\sinh\gamma)\left(\mathbf{q} - \frac{\sinh\gamma + \mathbf{p}\cdot\mathbf{q}\sinh\gamma}{p_z(\mathbf{p})\cosh\gamma}\mathbf{p}\right). \tag{5.61}$$

[13] We recall once more that the chart index $\sigma \in \{+,-\}$ can be introduced as a factor in every occurence of $\sqrt{1-|\mathbf{p}|^2}$.

[14] The second summand transforms as the *vector* $\mathbf{p}$ times the scalar $\mathbf{p}\cdot\mathbf{q}$; this term was absent in the 2-dim case (3.59).

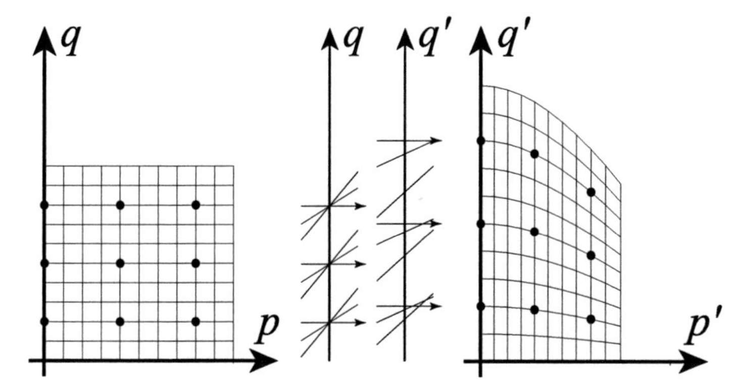

Fig. 5.2. Deformation of optical phase space under a Lorentz boost along the z-axis, in two dimensions. A patch of phase space $\wp$ at the standard screen (*left*), with several rays marked by bullets (•), is deformed comatically (vertical p-lines remain vertical) to another patch (*right*). This entails a map of the rays (*middle*), both in direction and position on the standard screen.

This transformation is the conjugate of that of ray momentum in (5.54), and ensures that Lorentz boosts on optical phase space are canonical. Figure 5.2 shows the 2-dim relativistic coma deformation of phase space (cf. Fig. 2.4), and Fig. 5.3 has the spot diagram of the 3-dim phenomenon (refer to page 41 for the interpretation of spot and spots diagrams).

Remark: Relativity near to the optical axis. When $|\mathbf{p}|$ and $|\mathbf{q}|$ are small, we can cut their Taylor series to some order in these quantities (see Sect. 4.4). Indicating by $^{[3]}$ the third-order approximation, (5.54) and (5.61) become

$$\mathbf{p}(\gamma)^{[3]} = e^{-\gamma}\mathbf{p} + \tfrac{1}{2}|\mathbf{p}|^2\mathbf{p}\, e^{-2\gamma}\sinh\gamma, \tag{5.62}$$

$$\mathbf{q}(\gamma)^{[3]} = e^{\gamma}\mathbf{q} - (\tfrac{1}{2}|\mathbf{p}|^2\mathbf{q} + \mathbf{p}\cdot\mathbf{q}\,\mathbf{p})\sinh\gamma. \tag{5.63}$$

The first-degree terms show there is *scaling* [cf. (3.60) with $\alpha = -\gamma$]: when beams are boosted, their directions 'squeeze' towards the optical axis; as a result, images must amplify.[15] The third-order terms of (5.62)–(5.63) approximate the straight part of the lines near to the apex of the spot diagram in Fig. 5.3. •

Remark: The apex and circles of coma. The spot in Fig. 5.3 shows why the aberration is called *coma* – we see a small *comet* directed outwards (for $\gamma > 0$) whose size increases towards the edges of the image field. To

[15] In quantum optics this transformation is called *squeezing* because the minimum uncertainty in the position of a photon can be reduced (for $\gamma < 0$) at the expense of its spread in momentum. Here, squeezing simply means scaling the image.

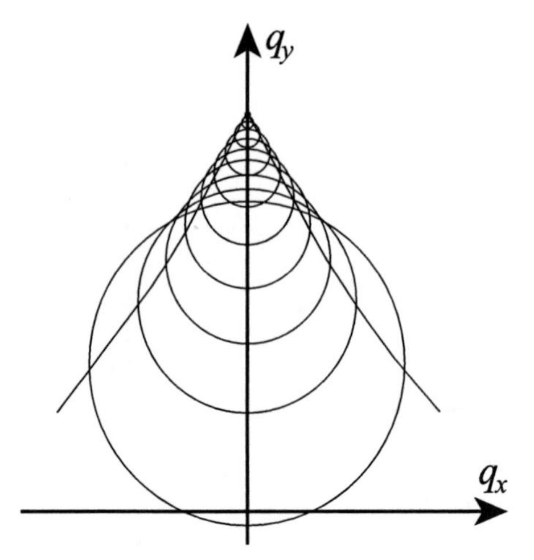

Fig. 5.3. Spot diagram of relativistic coma. A perfect imaging device which at rest would image a single point $(0, q_y)$, when boosted along the z-axis, suffers the relativistic coma aberration which yields a 'little comet' spot. The apex $(0, q_y)$ is the intersection of the ray parallel to the optical axis $\mathbf{p} = \mathbf{0}$; the circles are the images of cones of rays with the same (equally spaced) colatitude; the 4 lines issuing from the apex (two coincide with the q_y-axis) are the 2:1 images of 8 meridians on the Descartes sphere of ray directions.

analyze the spot further, we parametrize rays in polar coordinates by $\mathbf{q} = q(\cos\kappa, \sin\kappa)^\top$ and $\mathbf{p} = \sin\theta\,(\cos\phi, \sin\phi)^\top$, and trace out a family of rays on a cone of fixed colatitude θ. This will trace out an image that we find from (5.61). To third aberration order in $p = \sin\theta$ and q, (5.63) yields

$$\mathbf{q}(\mathbf{p}(p,\phi), \mathbf{q}(q,\kappa); \gamma) = \mathbf{q}_{\text{ctr}}(p^2, \mathbf{q}; \gamma) + \rho(p^2, q; \gamma)\,\mathbf{r}(\phi, \kappa), \quad \text{i.e.,} \tag{5.64}$$

$$\text{a circle} \quad \mathbf{r}(\phi,\kappa) = \begin{pmatrix} \cos(2\phi - \kappa) \\ \sin(2\phi - \kappa) \end{pmatrix}, \qquad -\pi < \phi \le \pi, \tag{5.65}$$

$$\text{with center at} \quad \mathbf{q}_{\text{ctr}} = (R + \tfrac{1}{2} S\,p^2)\mathbf{q} = (e^\gamma - p^2 \sinh\gamma - \cdots)\,\mathbf{q}, \tag{5.66}$$

$$\text{and radius} \quad \rho = \tfrac{1}{2} S\,p^2\,q = \tfrac{1}{2}\,p^2\,q \sinh\gamma + \cdots, \tag{5.67}$$

where the functions $R(p;\gamma)$ and $S(p;\gamma)$ are given in (5.59) and (5.60). The apex of the comets is at $\mathbf{q}_{\text{ctr}}(\mathbf{p} = \mathbf{0}, \mathbf{q}; \gamma)$, and as $p = \sin\theta$ grows, the centers of the circles move radially away from the apices by $\delta = |\mathbf{q}_{\text{ctr}} - e^\gamma \mathbf{q}| = p^2 q \sinh\gamma$, while their radii (5.67) increase as $\rho(q, p^2; \gamma) = \frac{1}{2} p^2 q \sinh\gamma$. The opening angle of the comet at the apex is twice $\delta/\rho = \frac{1}{2} = \sin 30°$, namely 60°. The circles on the screen, (5.65), contain the angle $2\phi - \kappa$; this means that a ray going around once in ϕ, draws out the image circle *twice*! The comets show thus a 2:1 map occurring between original and image rays at the 2-dim screen. Their boundaries are the *caustics* of the geometric image. Caustics can give trouble with the illumination on the screen because it becomes singular on a line, and markedly different of what it is in wave optics. One significant advantage of the phase space description of relativistic coma is that maps are canonical and 1:1. •

5.8 The Lorentz Lie algebra and group

The generators of boosts along the x- and y-directions, $b_x(\mathbf{p},\mathbf{q})$, $b_y(\mathbf{p},\mathbf{q})$, or along any direction $\vec{b}$ given by a vector of rapidities $\vec{\gamma} = (\gamma_x, \gamma_y, \gamma_z) \in \mathsf{R}^3$, can be found now by using the covariance under rotations of the 3-vector $\{\vec{b}, \circ\}$ with the 3-vector of generators of translations $\vec{p}$. The Poisson brackets of the functions $\vec{b}(\mathbf{p},\mathbf{q})$ with the generators of rotations $\vec{\jmath}$, will be as in (5.39), namely

$$\{j_x, b_y\} = b_z, \qquad \text{and cyclically,} \tag{5.68}$$

Thus from $b_x = \{j_y, b_z\}$ and $b_y = -\{j_x, b_z\}$, we complete the vector to find

$$b_x = q_x\, p_z^2 + p_y\, \mathbf{q}{\times}\mathbf{p} = q_x - p_x\, \mathbf{p}\cdot\mathbf{q}, \tag{5.69}$$

$$b_y = q_y\, p_z^2 - p_x\, \mathbf{q}{\times}\mathbf{p} = q_y - p_y\, \mathbf{p}\cdot\mathbf{q}, \tag{5.70}$$

$$b_z = -p_z\, \mathbf{p}\cdot\mathbf{q} = -\sigma\sqrt{1-|\mathbf{p}|^2}\, \mathbf{p}\cdot\mathbf{q}. \tag{5.71}$$

Remark: The global faces of coma. Because b_x and b_y are independent of the chart index $\sigma = \pm$ (see Sect. 2.1), boosts in these directions will respect the hemispheres of forward and backward rays; b_z however, contains $\sigma = \operatorname{sign} p_z$. From (5.61)–(5.64) we see that rays on the *forward pole*, $\mathbf{p} = \mathbf{0}$, $p_z = +1 = \sigma$, map $\mathbf{q} \in \mathsf{R}^2$ on the apices of the comets, $\mathbf{q}^+_{\text{apex}} = e^{\gamma}\mathbf{q}$ [see (5.66) for $\mathbf{p} = 0$], and that rays on the *backward pole*, $\mathbf{p} = \mathbf{0}$, $p_z = -1 = \sigma$, are mapped on paired points at $\mathbf{q}^-_{\text{apex}} = e^{-\gamma}\mathbf{q}$. The latter are apices of comets that open towards the origin, just as the former ones for negative rapidity ($\gamma \leftrightarrow -\gamma$ for $\sigma \leftrightarrow -\sigma$). On the other hand, boosts in x- and y-directions also produce their spot diagrams (of generally elliptic shape) with plenty of geometric properties. Indeed, *all* boosts globally map the Descartes sphere *twice* on the screen R^2. Aspects of these remarkable 4-dim faces of global relativistic coma were studied in [160]. •

The composition of two boosts is a boost only when they are along the same direction. If not, the result is found multiplying the 4×4 matrix presentation (fundamental representation) of boosts (5.51) and recalling the 3×3 submatrices of $\mathsf{SO}(3)$, (5.17)–(5.19), for rotation of the space components. The latter are orthogonal matrices which are the upper-left block in the larger Lorentz *pseudo-orthogonal* matrices.[16] The product of two such matrices also has this property, the unit and inverse matrices do also, so the full

[16] *I.e.*, $\mathbf{B}\mathbf{G}\mathbf{B}^\top = \mathbf{G}$ preserving the diagonal space-time metric matrix $\mathbf{G} = \operatorname{diag}(1,1,1,-1)$ for the scalar products of special relativity. See also Sects. 9.3 and 11.7.

set of these matrices presents a group. This is denoted $\mathsf{SO}(3,1)$ and called the *Lorentz* group. The composition of noncommuting boosts [cf. (3.63)] is found by computing the Poisson bracket of the generating functions, which can duly verified to be

$$\{b_x, b_y\} = -j_z, \quad \text{and cyclically.} \tag{5.72}$$

This set of Lie brackets (5.28), (5.68) and (5.72) present the Lorentz Lie algebra $\mathsf{so}(3,1)$ [cf. (5.38)–(5.40) for the Euclidean algebra $\mathsf{iso}(3)$]. The Lorentz group has 6 parameters, three for the axis and angle of rotation, and three for the direction and magnitude of the boost.

Remark: Pseudo-orthogonal algebras. Please note the minus sign in the $\mathsf{so}(3,1)$ Lie bracket (5.72); it makes the crucial difference between the Lorentz algebra and the 4-dim rotation algebra $\mathsf{so}(4)$ that follows the pattern of $\mathsf{so}(3)$ in (5.28) and (5.68), but replaces the sign in (5.72) with a plus. •

5.9 Other global optical transformations

What other Lie algebras and groups can be realized on optical phase space $\wp$? In this section we first learn why the *Poincaré* algebra and group (of translations, rotations and Lorentz boosts) can *not* be realized, although its Euclidean and Lorentz substructures can; and then we give generic and specific examples of other finite-dimensional algebras and groups which do admit this realization.

The Poincaré group is the semidirect product of spacetime translations with the Lorentz group. Its fundamental representation by 5×5 matrices has the right-triangular block form of (5.41)–(5.45) with 4×4 Lorentz pseudo-orthogonal matrices in place of the 3×3 rotation matrices, and a 5th column with 3-space translation parameters, one time translation, and a 1. While both the Euclidean and Lorentzian groups have 6 parameters each, the Poincaré group has 10. From the point of view of geometric optics, there appear two paradoxes (destined below to dissolve together). First, the time-translation parameter, which can be produced through boosting a space translation, is extraneous to our model on the screen; and second, the realizations of the Euclidean and Lorentzian Lie algebras by Poisson operators on $\wp$ do not fit. Let us examine the last impediment first, using the generator functions of translations [$\vec{p}$ with $p_z(\mathbf{p}) = \sqrt{1 - |\mathbf{p}|^2}$ as in (5.9)], of rotations [$\vec{\jmath}$ in (5.29)–(5.31)], and of boosts [$\vec{b}$ in (5.69)–(5.71)]. To merge the two algebras, the Euclidean with generators $\vec{p}$, $\vec{\jmath}$ and the Lorentzian with $\vec{\jmath}, \vec{b}$ (all $|_\wp$), we must be sure that all their Poisson brackets belong to the same set of functions. But when we calculate

$$\{b_j, p_k\} = \delta_{j,k} - p_j p_k, \qquad j, k \in \{x, y, z\}, \tag{5.73}$$

we see that the adjoint action of $\{\vec{b}, \circ\}$ on $\vec{p}$ is nonlinear, and the algebra will not close into the 10-dim vector space of the Poincaré algebra $\mathsf{iso}(3,1)$. So we inquire more closely about the dimensions of the algebras and the homogeneous spaces.

The dimension of the phase space manifold $(\mathbf{p},\mathbf{q}) \in \wp$ is 4; this is the same as the dimension of the Euclidean algebra spanned by the 6 generator functions $\vec{p}(\mathbf{p})$, $\vec{\jmath}(\mathbf{p},\mathbf{q})$ minus the 2 restrictions $|\vec{p}|^2 = 1$ and $\vec{p}\cdot\vec{\jmath} = 0$, which are consubstantial to their classical realization. This is also the same dimension of the space of Lorentzian generator functions $\vec{\jmath}(\mathbf{p},\mathbf{q}), \vec{b}(\mathbf{p},\mathbf{q})$ subject to the 2 other restrictions $|\vec{\jmath}|^2 - |\vec{b}\,|^2 = 0$ and $\vec{\jmath}\cdot\vec{b} = 0$. The adjoint action of the groups produce linear transformations of their generators that *respect* these restrictions, and which translate into the nonlinear transformations of phase space $(\mathbf{p},\mathbf{q})$ that we saw in the previous sections. However, one group of transformations does not respect the restrictions of the other, as (5.73) evinces for the algebras. If the two algebras and the groups realized on $\wp$ cannot be combined, space translations will not be boosted into the time translation of the first paradox. Our conclusion is that the full Poincaré group cannot be realized on the geometric optical phase space $\wp$; only its Euclidean and Lorentzian subgroups can.

The phase space of the model of 3-dim paraxial optics (to be seen further in chapters 9 and 10) and of 2-dim mechanics, is R^4; this is a convenient homogeneous space for the action of the 10-parameter *symplectic* group of linear canonical transformations, denoted $\mathsf{Sp}(4,\mathsf{R})$.[17] Here we show how the linear action of this group can be extended to nonlinear action on the coordinates $(\mathbf{p},\mathbf{q}) \in \wp$. We recall from page 31 the *opening coma* map of phase space, $\mathcal{C}_\gamma$ in (3.23) for $\gamma = \frac{1}{2}$, and shown in Fig. 3.3, that we indicate by $\mathcal{C}$ (see [92]). This maps the forward half of the Descartes momentum sphere onto the plane 'mechanical' momentum space; this map was extended canonically to the phase spaces $(\mathbf{p},\mathbf{q}) \in \wp_{(\sigma=+)}$ and $(\mathbf{P},\mathbf{Q}) = (\mathcal{C}:\mathbf{p},\ \mathcal{C}:\mathbf{q}) \in \mathsf{R}^4$. The inverse, $\mathcal{C}^{-1} = \mathcal{C}_{-1/2}$, maps R^4 back to $\wp$ in the open half-sphere. Hence, if $\mathcal{L} \in \mathsf{Sp}(4,\mathsf{R})$ is a linear canonical map of R^4 phase space, then

$$\mathcal{L}_{\text{opt}} := \mathcal{C}^{-1}\,\mathcal{L}\,\mathcal{C}. \tag{5.74}$$

will be a nonlinear canonical map of $\wp$ on $\wp$. A group of $\mathcal{L}$'s realized on R^4 will realize the same group of $\mathcal{L}_{\text{opt}}$'s in $\wp$, and the same holds for the algebras.

Example: The optical Fourier transform. We can define an *optical* fractional Fourier transformation from the linear map defined in (3.19)–(3.20) as

$$\mathcal{F}^\alpha_{\text{opt}} := \mathcal{C}^{-1}\,\mathcal{F}^\alpha\,\mathcal{C}. \tag{5.75}$$

We can write this map analytically using the composition property of Lie exponential operators (seen on page 36). For the case $\alpha = 1$ the optical Fourier map $\mathcal{F} : (\mathbf{P},\mathbf{Q}) \mapsto (\mathbf{Q},-\mathbf{P})$ on R^4 is [156]

[17] The structure of $\mathsf{Sp}(4,\mathsf{R})$ need not be detailed beyond pointing out that it contains all fractional Fourier transforms; part III is devoted to its study.

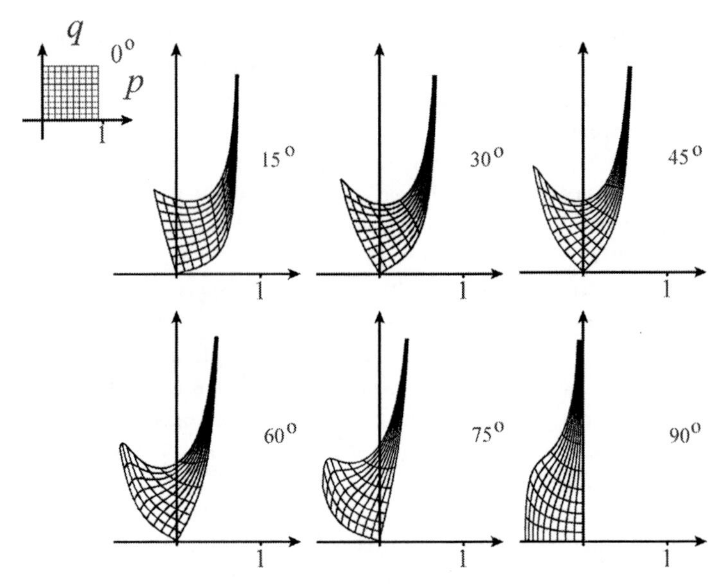

Fig. 5.4. Action of optical fractional Fourier transformations $\mathcal{F}^{\alpha}_{\rm opt}$ on a patch of optical phase space, for angles $\phi := \frac{1}{2}\pi\alpha$ at intervals of 15° from the identity ($\phi = 0$, *top center*) to the Fourier transform ($\phi = 90°$, *bottom right*). These are global canonical transformations from $(p, q) \in \wp(\sigma{=}+)$ on itself, which – as the ordinary fractional Fourier transforms $\mathcal{F}^{\alpha}$ – rotate a vanishing neighborhood of the origin rigidly by angles ϕ.

$$\mathcal{F}_{\rm opt} : \mathbf{p} = \mathcal{C}\,\mathcal{F}\frac{\mathbf{P}}{\sqrt{1-|\mathbf{P}|^2}} = \frac{\mathcal{C} : \mathbf{Q}}{\sqrt{1+(\mathcal{C} : \mathbf{Q})^2}}$$

$$= \frac{\sqrt{1-|\mathbf{p}|^2}}{\sqrt{1+(1+|\mathbf{p}|^2)([\mathbf{1}-\mathbf{pp}\cdot]\mathbf{q})^2}}(\mathbf{1}-\mathbf{pp}\cdot)\mathbf{q}, \tag{5.76}$$

$$\mathcal{F}_{\rm opt} : \mathbf{q} = -\frac{\sqrt{1+(1+|\mathbf{p}|^2)([\mathbf{1}-\mathbf{pp}\cdot]\mathbf{q})^2}}{\sqrt{1-|\mathbf{p}|^2}}$$

$$\times\,[\mathbf{p} + (1+|\mathbf{p}|^2)^2\mathbf{p}\cdot\mathbf{q}\,(\mathbf{1}-\mathbf{pp}\cdot)\mathbf{q}]. \tag{5.77}$$

For generic α the formulas are longer and we do not write them explicitly, but in Fig. 5.4 we show this as a map of 2-dim ($D = 1$) phase space. •

Remark: 2-Dim case. In Fig. 5.4 we plot the transformation (5.76)–(5.77) as a deformation of the 2-dim phase space of q and $p = \sin\theta$. This reduces to

$$p_{\rm F} = \frac{q\cos^3\theta}{\sqrt{1+q^2\cos^6\theta}}, \quad q_{\rm F} = -(1+q^2\cos^6\theta)^{3/2}\tan\theta. \tag{5.78}$$

The figure shows that a caustic will develop above a certain value of $q_{\rm F}$. •

6 Conformal optics – Maxwell fish-eyes

There is a very special optical medium, called the Maxwell *fish-eye*, where light enjoys the same symmetry and dynamics as free point masses on the surface of a sphere. Its guiding Hamiltonian is the quadratic invariant of the algebra that generates rotations of space. We use the stereographic projection – extended canonically to phase space – to map the mechanical free motion in conics, onto light in a remarkable *family* of optical media that we call *conformal*, because they are ruled by $\mathsf{SO}(4,2)$, the conformal Lie group.[1] This group underlies free motion on spheres and hyperboloids, and has long been known in connection with the bound and free orbits in the Kepler and Coulomb classical systems [57], as well as in the quantum Bohr atom [165]. The optical incarnation of the conformal group is the fish-eye family, which includes anisotropic media.

Many of the remarkable properties of the Maxwell fish-eye were beautifully analyzed by R.K. LUNEBURG in his posthumous book [90, Sect. 28] in 1964. Luneburg seems to have been the first to note that rays in this medium are circles, and the stereographic projection of maximal circles from a higher-dimensional sphere. The conserved quantities – essentially the generators of the Lie algebra – were found by H. BUCHDAHL in [27]. In our work [51] we applied what was known of the classical system [145] to complete the optical model.

6.1 On the eyes of fish and point rotors

Medical opticists of the middle of the nineteenth century were puzzled by the fact that the eyes evolved by fish are notoriously flat. In 1854, the Irish Academy offered a prize to determine the refractive index $n(\vec{q})$ of an inhomogeneous medium $\vec{q} \in \mathsf{R}^3$ which would bring parallel rays together to a point

[1] There is disagreement between mathematicians and physicists in the usage of the name *conformal* for the Lie group and algebra of this chapter. Mathematicians define conformal Lie operators L_i as those leaving invariant the Laplace equation $\Delta f(x) = 0$, so $[L_i, \Delta] = Q(x)\, \Delta$ (see *e.g.*, [74, Chap. 6]); for D-dim Laplacians this algebra is $\mathsf{iso}(D{+}1,1)$. The genesis of the physicists' interest lies in the spectrum of the hydrogen atom [19], where the Lie algebra of symmetry and transition operators is $\mathsf{so}(4,2)$. We follow the latter usage in this book.

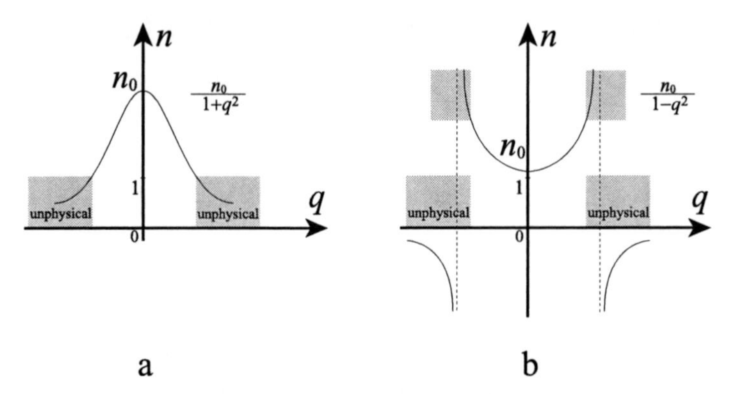

Fig. 6.1. Radial profiles of the refractive index of conformal fish-eyes. (**a**) Maxwell's (spherical) fish-eye. (**b**) Hyperbolic fish-eye. The shaded regions are unphysical because their refractive index would be less than 1 of vacuum, or impossibly high or negative (mathematically they are acceptable though.)

in the least distance. James Clerk Maxwell gave the solution [94]: he found that an optical medium of refractive index

$$n^+(\vec{q}) = \frac{n_o}{1+|\vec{q}|^2}, \tag{6.1}$$

cut in half by a plane through the center $\vec{q} = \vec{0}$, will bend the rays falling perpendicularly on its face so that they will follow circular paths, and that these all meet at a *fovea retinalis*. The refractive index of the *Maxwell fish-eye* (6.1) depends only on the radius $|\vec{q}|$ and is manifestly symmetric under $\mathsf{SO}(3)$ rotations around the center. In Fig. 6.1a we plot this index as a function of the radius $|\vec{q}|$.

It turns out that there is also a 'hyperbolic' fish-eye [33] with the radial refractive index profile,

$$n^-(\vec{q}) = \frac{n_o}{1-|\vec{q}|^2}, \tag{6.2}$$

which is shown in Fig. 6.1b, and also a member of equal standing in the conformal family. There, light follows arcs of circles which end perpendicularly at the singularity sphere $|\vec{q}| = 1$.

Remark: A glimpse of symmetry. We suspect that the refractive indices (6.1) and (6.2) are special when writing the ray equation in time, as in Sect. 1.6, (1.19), where $1/n(\vec{q})$ appears. For $n^\tau(\vec{q})$ as above, the ray equation is $d^2\vec{q}/d(ct)^2 = -\tau\vec{q}$, implying plane harmonic or exponential motion, according to the sign of τ. •

In this chapter we shall understand the Maxwell fish-eye through the *point rotor*. This is an ideal mechanical system consisting of a point of mass m constrained to move on a sphere, where its trajectories are great circles. To have a 3-dim optical medium we need the 3-dim surface of the sphere $\mathcal{S}^3$, in

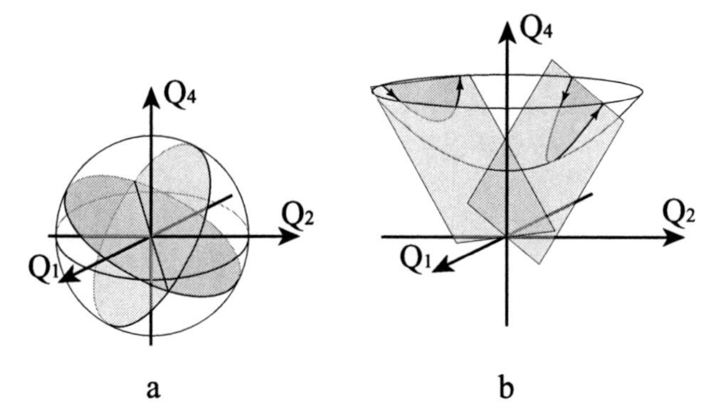

Fig. 6.2. Trajectories of free particles constrained to move on conics. (**a**) On the sphere $\mathcal{S}^2$ the trajectories are great circles (whose planes contain the origin). (**b**) On the hyperboloid $\mathcal{H}^2$ the trajectories are great hyperbolas (i.e., intersections of the upper sheet of a hyperboloid with planes that contain the origin).

a four-dimensional 'ambient' space. Since all essential features of conformal maps are D-dimensional, we shall mostly train our intuition in 2-dim optical media, as in Fig. 6.2, where the hyperbolic case appears too. In the latter, the trajectories of mass points are all 'great hyperbolas' contained in planes that pass through the origin. (Granted that the hyperbolic case is less intuitive, we will set aside its explicit treatment to occasional remarks and parentheses.) So, let the mass point moving on the sphere $\mathcal{S}^2$ of radius ρ, have vectors of position $\vec{Q}$, of momentum $\vec{P} = m\vec{v}$ (where $\vec{v}$ is velocity), and of angular momentum $\vec{J} := \vec{Q} \times \vec{P}$, which are mutually orthogonal, so $|\vec{J}| = \rho|\vec{P}|$. Since $|\vec{P}| = |\vec{J}|/\rho$ and the particle is free, its mechanical *energy* is purely kinetic:

$$H_{\text{rotor}} := \tfrac{1}{2}m|\vec{v}\,|^2 = |\vec{P}|^2/2m = |\vec{Q} \times \vec{P}|^2/2m\rho =: \iota|\vec{J}|^2, \qquad (6.3)$$

where $\iota := 1/2m\rho$, and $m\rho$ is the moment of inertia. Since energy is invariant under rotations and a constant of the motion, so is $|\vec{P}|^2$; hence $\vec{P}$ is constrained to its *own* sphere of radius $\tilde{r} = 2mH_{\text{rotor}}$. In this mechanical realization, the points on the $\vec{Q}$- and $\vec{P}$-spheres are further restricted by $\vec{P} \cdot \vec{Q} = 0$, hence the original manifold $(\vec{P}, \vec{Q}) \in \mathsf{R}^6$ is reduced to the 3-dim manifold of all points on all great circles of the sphere. In the next section we shall formalize these statements in terms of Poisson brackets and Lie algebras.

Remark: The manifold of great circles on the sphere. Points moving on great circles have three coordinates: two of these, $(\vartheta, \varphi) \in \mathcal{S}^2$, fix a unit axis and thus the plane of a trajectory; the third one, $\psi \in \mathcal{S}^1$, can fix an origin for each trajectory. Any such great circle can be obtained from the equator with the Greenwich origin by means of a unique $\mathsf{SO}(3)$ rotation, following Euler angles $\mathbf{R}(\psi, \vartheta, \varphi)$ in (5.20), so the manifold of trajectories with origins is $\mathsf{SO}(3) = \mathcal{S}^1 \times \mathcal{S}^2$, a 3-dim manifold. (On page 177 we generalize this parametrization for $\mathcal{S}^D$.) But if the elementary objects of our model are *whole* trajectories, ignoring thus $\psi \in \mathcal{S}^1 = \mathsf{SO}(2)$, their manifold is reduced to the 2-dim compact manifold of the sphere $\mathcal{S}^2$. [The group of rotations foliates into *spaces of cosets* $\mathcal{S}^2 = \mathsf{SO}(2)\backslash\mathsf{SO}(3)$.] Great circles are to conformal optics what straight lines are to the ordinary Euclidean optics of homogeneous spaces. •

6.2 Phase space and rotations

Let us indicate a point in *four*-dimensional space by a vector $Q = (\vec{Q}, Q_4)^\top$ with coordinates[2] $\{Q_\alpha\}_{\alpha=1}^4 \in \mathsf{R}$. Then, as we did in Sect. 5.2, but now in four dimensions, we consider functions $F(P, Q)$ of an R^8 (phase space) of position and a canonically conjugate momentum vector $P = (\vec{P}, P_4)^\top$, endowed with the natural Poisson bracket structure

$$\{F, \circ\}_{(4)} := \sum_{\mu=1}^{4} \left(\frac{\partial F(P,Q)}{\partial Q_\mu} \frac{\partial}{\partial P_\mu} - \frac{\partial F(P,Q)}{\partial P_\mu} \frac{\partial}{\partial Q_\mu} \right) \tag{6.4}$$

$$=: \frac{\partial F(P,Q)}{\partial Q} \bullet \frac{\partial}{\partial P} - \frac{\partial F(P,Q)}{\partial P} \bullet \frac{\partial}{\partial Q}, \tag{6.5}$$

where the last line defines the '•' inner product notation for 4-vectors (*versus* '·' for 3-vectors). In this way we have a Hamiltonian structure where Lie exponential operators $\exp(t\{F, \circ\}_{(4)})$ produce canonical Hamiltonian flows of 4+4-dim phase space.

The functions we met in Sects. 5.3 and 5.5 as generators of rotations ($\tau = +$) and Lorentz boosts ($\tau = -$), straightforwardly generalize to four dimensions:[3]

$$J_i := Q_j P_k - Q_k P_j, \quad i, j, k \text{ cyclic permutations of 1,2,3}, \tag{6.6}$$

$$B_i^\tau := Q_i P_4 - \tau Q_4 P_i, \qquad i = 1, 2, 3. \tag{6.7}$$

These functions close among themselves under Poisson brackets

$$\{J_i, J_j\} = J_k, \quad \{J_i, B_j^\tau\} = B_k^\tau, \quad \{B_i^\tau, B_j^\tau\} = \tau J_k. \tag{6.8}$$

For $\tau = +$, the J_i's and B_i^+'s are a basis for the four-dimensional rotation (orthogonal) Lie algebra $\mathsf{so}(4)$; in the $\tau = -$ case, it is the Lorentz (pseudo-orthogonal) Lie algebra $\mathsf{so}(3, 1)$. Since there are six linearly independent generators, the *vector* dimension of the two algebras is six.[4]

---

[2] In this section we use Q (capital, *without* arrow) to indicate 4-vectors, and $\vec{Q}$ (with arrow) for 3-vectors, to be consistent with the previous use of boldface for 2-vectors. After the stereographic projection we shall use lower-case coordinates, $\vec{p}$, $\vec{q}$, for optical phase space. (Generalizing, we may ascribe to $\vec{Q}$ dimension $D+1$ and the same formulas will hold.)

[3] Whenever i, j, k appear in the same equation, they will be understood to be cyclic permutations of 1,2,3.

We organize the six functions in (6.6)–(6.7) as components of the angular momentum *tensor*:

$$L^{\tau}_{\alpha,\beta} := \tau_\alpha Q_\alpha P_\beta - \tau_\beta Q_\beta P_\alpha, \tag{6.9}$$

where $\alpha, \beta = 1, 2, 3, 4$ and we indicate $\tau_1, \tau_2, \tau_3 = +$, and $\tau_4 = +$ (spherical case) or $\tau_4 = -$ (hyperbolic case). In the spherical case, the tensor is skew-symmetric, $L^{+}_{\alpha,\beta} = -L^{+}_{\beta,\alpha}$. In the hyperbolic case, while $L^{\tau}_{i,j} = J_k = -L^{\tau}_{j,i}$ is skew-symmetric, $L^{\tau}_{i,4} = B^{\tau}_i = -\tau_4 L^{\tau}_{4,1}$ is symmetric. One invariant Casimir function is built by 'exhausting indices':

$$L^{\tau\,2} := \sum_{1\leq\alpha<\beta\leq 4} \tau_\beta \, L^{\tau}_{\alpha,\beta}\, L^{\tau}_{\alpha,\beta} = |\,\vec{J}|^2 + \tau|\vec{B}^{\tau}|^2 = P^2\, Q^2 - (P \bullet Q)^2. \tag{6.10}$$

The last equality should be compared with the 3-dim case, (5.33), where also $|\vec{\jmath}|^2 = (\vec{p} \times \vec{q})^2 = |\vec{p}|^2\, |\vec{q}|^2 - (\vec{p}\cdot\vec{q})^2$ is twice the energy of the rotor.

Remark: Angular momentum in four dimensions. The short argument at the end of last section showed that the kinetic energy of the 3-dim point rotor is proportional to the square of its angular momentum. In more than three dimensions, angular momentum fully assumes its rôle as a second-order skew-symmetric tensor. The square of this tensor (contracting indices) is the invariant Casimir function, and this is the Hamiltonian of a point mass moving freely in the conic; its constant value is the energy. •

Remark: The second Casimir invariant. There is a *second* second-order Casimir invariant in $\mathsf{so}(4)$ and $\mathsf{so}(3,1)$, $C_2 := \vec{J}\cdot\vec{B}$. *But*, in the realization (6.9) of these algebras on phase space, it is identically zero. On other homogeneous spaces it need not be, though; we should be aware that the realization of these algebras and groups afforded by geometric optics is not the most general one allowed by Lie theory. •

Finally, we present the Lie brackets (6.8) of $\mathsf{so}(4)$ and $\mathsf{so}(3,1)$ in terms of a single expression with the $L^{\tau}_{\alpha,\beta}$'s, in a form suitable for all pseudo-orthogonal algebras $\mathsf{so}(D^+, D^-)$ that have D^+ coordinates with positive metric ($\tau_\alpha = +$, $\alpha = 1, 2, \ldots, D^+$) and D^- coordinates with negative metric ($\tau_\alpha = -$, $\alpha = D^+{+}1, \ldots, D^+{+}D^-$). It is

$$\{L^{\tau}_{\alpha,\beta},\, L^{\tau}_{\kappa,\lambda}\}_{(4)} = \tau_\alpha\, \delta_{\alpha,\kappa} L^{\tau}_{\beta,\lambda} + \tau_\alpha\, \delta_{\alpha,\lambda} L^{\tau}_{\kappa,\beta} + \tau_\beta\, \delta_{\beta,\kappa} L^{\tau}_{\lambda,\alpha} + \tau_\beta\, \delta_{\beta,\lambda} L^{\tau}_{\alpha,\kappa}. \tag{6.11}$$

Using the Leibniz derivation property, we can check straightforwardly that $\{L^{\tau\,2},\, L^{\tau}_{\alpha,\beta}\}_{(4)} = 0$, to confirm that the square angular momentum (i.e., Hamiltonian) is a constant of the motion.

---

[4] Compare the first of equations (6.8) with (5.28) and (5.38) replacing $1, 2 \mapsto z$, $2, 3 \mapsto x$, $3, 1 \mapsto y$; the second of these equations with (5.68); and the third again with (5.38) for $\tau = +$, or with (5.72) for $\tau = -$.

6.3 Restriction to conics

We shall now restrict the 4-dim angular momentum tensor functions (6.9) to a 3-dim subspace while preserving their Poisson brackets, to realize the Lie algebras $\mathsf{so}(4)$ and $\mathsf{so}(3,1)$ on the subspace $(\vec{P},\vec{Q}) \in \mathsf{R}^6$ where $|\vec{Q}|^2 + \tau Q_4^2 = \rho^2$, i.e., a sphere ($\tau = +$) or a hyperboloid ($\tau = -$).

A function L of $Q = (\vec{Q}, Q_4)^\top \in \mathsf{R}^4$, when restricted to the subspaces (surfaces in Fig. 6.2) $Q_4 = Q_4(\vec{Q})$, becomes a function $L_{\rm R}$ of $\vec{Q} \in \mathsf{R}^3$ given by[5]

$$L_{\rm R}(\vec{Q}) := L(Q)|_{\rm R} = L\Big(\vec{Q}, Q_4(\vec{Q})\Big). \tag{6.12}$$

Note carefully that the derivative of the restricted function with respect to its argument is

$$\frac{\partial L_{\rm R}(\vec{Q})}{\partial \vec{Q}} = \frac{\partial L(Q)}{\partial \vec{Q}}\bigg|_{\rm R} + \frac{\partial Q_4(\vec{Q})}{\partial \vec{Q}}\,\frac{\partial L(Q)}{\partial Q_4}\bigg|_{\rm R}. \tag{6.13}$$

To restrict the Poisson brackets, we shall need the space derivatives of functions $L(P,Q)$ that depend also on the four components of momentum P. These are

$$\frac{\partial L(P,Q)}{\partial \vec{Q}}\bigg|_{\rm R} = \frac{\partial L_{\rm R}(\vec{P},\vec{Q})}{\partial \vec{Q}} - \frac{\partial Q_4(\vec{Q})}{\partial \vec{Q}}\,\frac{\partial L(P,Q)}{\partial Q_4}\bigg|_{\rm R}, \tag{6.14}$$

$$\frac{\partial L(P,Q)}{\partial \vec{P}}\bigg|_{\rm R} = \frac{\partial L_{\rm R}(P,Q)}{\partial \vec{P}}, \quad \begin{aligned} L_{\rm R}(\vec{P},P_4,\vec{Q}) &= L(P,Q)|_{\rm R} \\ &= L\Big(\vec{P},P_4,\vec{Q},Q_4(\vec{Q})\Big). \end{aligned} \tag{6.15}$$

With these restrictions of the partial derivatives, the Poisson bracket of two functions F, G becomes

$$\begin{aligned} \{F,G\}_{(4)}\Big|_{\rm R} &= \left(\frac{\partial F}{\partial Q}\bullet\frac{\partial G}{\partial P} - \frac{\partial F}{\partial P}\bullet\frac{\partial G}{\partial Q}\right)\bigg|_{\rm R} = \frac{\partial F}{\partial Q}\bigg|_{\rm R}\bullet\frac{\partial G}{\partial P}\bigg|_{\rm R} - \{F\leftrightarrow G\} \\ &= \frac{\partial F}{\partial \vec{Q}}\bigg|_{\rm R}\cdot\frac{\partial G}{\partial \vec{P}}\bigg|_{\rm R} + \frac{\partial F}{\partial Q_4}\bigg|_{\rm R}\frac{\partial G}{\partial P_4}\bigg|_{\rm R} - \{F\leftrightarrow G\} \\ &= \left(\frac{\partial F_{\rm R}}{\partial \vec{Q}} - \frac{\partial Q_4}{\partial \vec{Q}}\,\frac{\partial F}{\partial Q_4}\bigg|_{\rm R}\right)\cdot\frac{\partial G_{\rm R}}{\partial \vec{P}} + \frac{\partial F}{\partial Q_4}\bigg|_{\rm R}\frac{\partial G}{\partial P_4}\bigg|_{\rm R} - \{F\leftrightarrow G\} \\ &= \frac{\partial F_{\rm R}}{\partial \vec{Q}}\cdot\frac{\partial G_{\rm R}}{\partial \vec{P}} - \frac{\partial F}{\partial Q_4}\bigg|_{\rm R}\left(\frac{\partial Q_4}{\partial \vec{Q}}\cdot\frac{\partial G_{\rm R}}{\partial \vec{P}} - \frac{\partial G_{\rm R}}{\partial P_4}\right) - \{F\leftrightarrow G\}. \end{aligned} \tag{6.16}$$

In the last line, the first term (and its $\{F\leftrightarrow G\}$) is the 3-dim Poisson bracket $\{F_{\rm R}, G_{\rm R}\}_{(3)}$ of the restricted functions; the restricted functions will form the same Lie algebra as the original ones when the second term vanishes.

Now we shall restrict the points Q to lie specifically on the sphere $\mathcal{S}^3$ of $Q^2 = \rho^2$ (or a two-sheeted hyperboloid $\mathcal{H}^3$ of $Q^2 = -\rho^2$). Actually, there will be *two* 1:1 maps; we distinguish them with the familiar sign $\sigma = \pm$, for

[5] Cf. the notation '$|_{\rm R}$' in Chap. 5, equation (5.15), which has a similar meaning.

the upper and the lower half-spheres $\tau = +$ (or for the two sheets of the hyperboloid $\tau = -$).[6] Importantly, the functions $L(P, Q)$ to be restricted are only those whose flows preserve the sphere (or hyperboloid). Then,

$$\left.\begin{array}{l} Q^2 := \vec{Q}\cdot\vec{Q} + \tau Q_4^2 = \rho^2 \\ Q_4(\vec{Q}) = \sigma\tau\sqrt{\rho^2 - \tau|\vec{Q}|^2} \end{array}\right\} \Rightarrow \frac{\partial Q_4(\vec{Q})}{\partial \vec{Q}} = \frac{-\tau\vec{Q}}{Q_4(\vec{Q})}, \tag{6.17}$$

$$\{L, Q^2\}_{(4)} = 0 \Rightarrow \frac{\vec{Q}}{Q_4}\cdot\frac{\partial L}{\partial \vec{P}} = -\tau\frac{\partial L}{\partial P_4}, \tag{6.18}$$

$$\text{hence for such } L(P,Q), \qquad \frac{\partial Q_4(\vec{Q})}{\partial \vec{Q}}\cdot\frac{\partial L}{\partial \vec{P}} = \frac{\partial L}{\partial P_4}, \tag{6.19}$$

and so the unwanted terms in (6.16) vanish. Therefore, for any two functions F, G satisfying (6.19) for a sphere or a hyperboloid,

$$\{F, G\}_{(4)}\Big|_{\mathrm{R}} = \{F_{\mathrm{R}}, G_{\mathrm{R}}\}_{(3)}. \tag{6.20}$$

The role of the coordinate P_4 is now that of a parameter: it will not participate in any 3-dim Poisson bracket nor change its value under transformations $\exp(\alpha\{F_{\mathrm{R}}, \circ\}_{(3)})$. Any constant P_4 may do, but we select $P_4 = 0$ and write this into the restriction,[7] so henceforth $Q^2|_{\mathrm{R}} = \tau\rho^2$ and

$$F(P,Q)|_{\mathrm{R}} := F\Big(\vec{P}, P_4 = 0; \vec{Q}, Q_4 = \sigma\sqrt{\rho^2 - \tau|\vec{Q}|^2}\Big). \tag{6.21}$$

The generators of 4-dim rotations $\mathsf{so}(4)$ of $\mathcal{S}^3$ [or Lorentz transformations $\mathsf{so}(3,1)$ of $\mathcal{H}^3$, (6.6) and (6.7)] restricted to the sphere, are thus

$$J_i|_{\mathrm{R}} = Q_j P_k - Q_k P_j, \quad B_i^\tau|_{\mathrm{R}} = -\sigma\tau\, P_i\sqrt{\rho^2 - \tau|\vec{Q}|^2}, \tag{6.22}$$

and the invariant Casimir Hamiltonian function (6.10) is restricted to

$$L^{\tau\,2}|_{\mathrm{R}} = \Big(P^2 Q^2 - (P\bullet Q)^2\Big)\Big|_{\mathrm{R}} = \tau\rho^2\,|\vec{P}|^2 - (\vec{P}\cdot\vec{Q})^2. \tag{6.23}$$

6.4 Stereographic map of phase space

The stereographic map [126, 121] is well known to be a 1:1 projection of the circle onto the line[8] as shown in Fig. 6.3a. In the $D = 1$ dimension of the figure, the ambient space is $Q = (Q_1, Q_4)^\top \in \mathsf{R}^2$ and the circle is $Q_1^2 + Q_4^2 = \rho^2$. Also, in Fig. 6.3b we show the *hyperbolic* stereographic projection from the upper sheet of a two-sheeted hyperbola to the interior of a segment bounded by two points that are the projections of the asymptotes.

[6] Formulas will agree for all four signs σ and τ, so we need not carry them as labels on the functions.

[7] We shall remark on the general case $P_4 \in \mathsf{R}$ in page 106.

[8] Under the map, the north pole of the circle is mapped to the point at infinity. The circle $\mathcal{S}^1$ is the *one-point compactification* of the real line $\mathsf{R}\cup\{\infty\}$ through the inverse stereographic map.

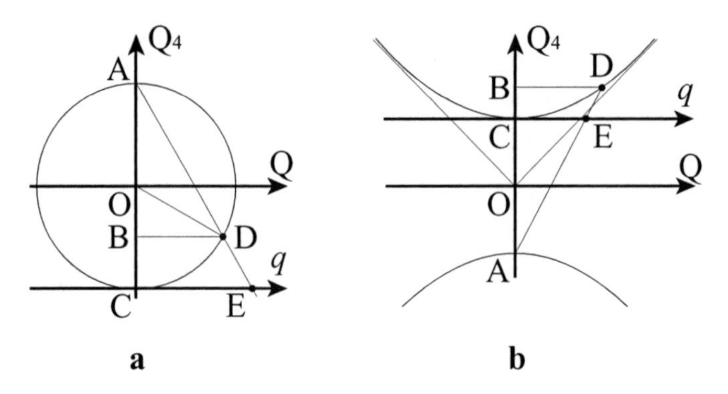

Fig. 6.3. Stereographic projection of conics to optical space (q-line in $D = 1$ dimension). (**a**) A point D on the circle, of coordinates (Q, Q_4), is projected from the north pole A on the q-axis at E; the angle $\widehat{\mathrm{CAD}}$ is half of $\widehat{\mathrm{COD}}$. (**b**) The point D on the upper branch of a hyperbola is projected from the south vertex A on the point E of the q-axis; similar hyperbolic angle relations hold.

For the cases both of spherical ($\sigma = \pm$, $\tau = +$) and hyperbolic ($\sigma \equiv +$, $\tau = -$) stereographic maps, Fig. 6.3 show the similarity of the triangles $\triangle$ABD and $\triangle$ACE. The ratio of their segments relates the two coordinates, $\vec{q}$ and $\vec{Q}$, through

$$\frac{|\vec{q}|}{2\rho} = \frac{\overline{\mathrm{CE}}}{\overline{\mathrm{AC}}} = \frac{\overline{\mathrm{BD}}}{\overline{\mathrm{AB}}} = \frac{|\vec{Q}|}{\rho - \tau Q_4}. \tag{6.24}$$

This is easily 'vectorized' to any dimension in the *stereographic map* of space, thus defined by

$$\vec{q} = \frac{2\rho\vec{Q}}{\rho - \sigma\tau\sqrt{\rho^2 - \tau|\vec{Q}|^2}}. \tag{6.25}$$

To find the *inverse* stereographic projection, note in Fig. 6.3a the angles $\chi = \widehat{\mathrm{CAD}}$ and $2\chi = \widehat{\mathrm{COD}}$, whose trigonometric functions are

$$\begin{aligned} \sin\chi &= |\vec{q}|/\sqrt{|\vec{q}|^2 + 4\rho^2}, & \sin 2\chi &= |\vec{Q}|/\rho, \\ \cos\chi &= 2\rho/\sqrt{|\vec{q}|^2 + 4\rho^2}, & \cos 2\chi &= -Q_4/\rho, \\ \tan\chi &= |\vec{q}|/2\rho, & \tan 2\chi &= -|\vec{Q}|/Q_4. \end{aligned} \tag{6.26}$$

From their 'vectorized' relations follows easily the *inverse* stereographic map from the $\vec{q}$-plane to the Q-sphere:

$$\vec{Q} = \frac{\vec{q}}{1 + \tau|\vec{q}|^2/4\rho^2}, \qquad Q_4(\vec{Q}) = \rho\frac{|\vec{q}|^2/4\rho^2 - \tau 1}{1 + \tau|\vec{q}|^2/4\rho^2}. \tag{6.27}$$

[There is a hyperbolic counterpart to the functions (6.26) which validates this result for $\tau = -$.] Since $Q_4(\vec{0}) = -\tau\rho$, the $\vec{q}$-plane osculates the conics from below. Following Fig. 6.2, it helps intuition to see the stereographic

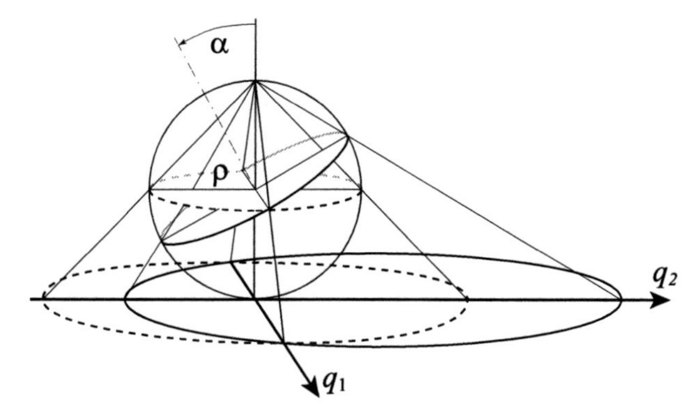

Fig. 6.4. Great circle on a sphere of radius ρ tilted by angle α and its stereographically projected circle (*continuous line*); the equator and its similarly projected *minimal* circle on the optical plane (q_1, q_2) (*dashed line*). Pairs of antipodal points on the sphere map on conjugate points in optical space. This figure shows the stereographic map to optical space $\vec{q}$ in one dimension more than the previous figure, but in one dimension less than that of the Maxwell fish-eye.

map of the spherical rotor trajectories in $D = 2$ dimensions, as shown in Fig. 6.4, on trajectories in the $\vec{q}$ plane, which will come to be recognized below as the Maxwell optical fish-eye medium. We now remark some well-known geometrical properties of the spherical stereographic projection (and some not-so well-known ones of its hyperbolic sibling).

Remark: Circles on the sphere map on circles on the plane. As shown in Fig. 6.4, great circles on the Q-sphere $\mathcal{S}^2$ map on circles $\mathcal{S}^1$ on the $\vec{q}$-plane. But moreover, *any* circle on the Q-sphere stereographically maps, in *any* dimension, on a circle $(\vec{q} - \vec{c})^2 = r^2$, with center at $\vec{c}$ and radius r. To prove this, we consider the intersection of the sphere $Q^2 = \rho^2$ with the cone $A \bullet Q =$ constant, whose axis A is a fixed vector on the sphere,

$$A = \begin{pmatrix} \vec{A} \\ A_4 \end{pmatrix} = \rho \begin{pmatrix} \vec{a} \sin\alpha \\ \cos\alpha \end{pmatrix}, \quad |\vec{a}\,| = 1, \quad \text{and} \tag{6.28}$$

$$A \bullet Q = \vec{A} \cdot \vec{Q} + A_4 Q_4 = \frac{\frac{1}{4}|\vec{q}|^2 + \rho\, \vec{q} \cdot \vec{a}\, \tan\alpha - \rho^2}{1 + |\vec{q}|^2/4\rho^2} \cos\alpha. \tag{6.29}$$

Let κ be the opening angle of the cone, $A \bullet Q = \rho^2 \cos\kappa$ ($|\kappa| \le \frac{1}{2}\pi$). We can now bring (6.29) to the form

$$|\vec{q}|^2(\cos\alpha - \cos\kappa) + 4\rho\, \vec{q} \cdot \vec{a}\, \sin\alpha = 4\rho^2(\cos\alpha + \cos\kappa), \tag{6.30}$$

where it describes a circle in the $\vec{q}$-plane, with

$$\text{center} \quad \vec{c} = \frac{2\rho \sin\alpha}{\cos\kappa - \cos\alpha}\vec{a}, \quad \text{and radius} \quad r = \frac{2\rho \sin\kappa}{\cos\kappa - \cos\alpha}. \tag{6.31}$$

•

Remark: Maximal and minimal circles. When we take $\kappa = \frac{1}{2}\pi$ in (6.28)–(6.31), the cone becomes the plane $A \bullet Q = 0$ and the Q-circle is then maximal on the sphere. On the $\vec{q}$-plane containing the origin, it is the circle

$$|\vec{q}|^2 + 4\rho\,\vec{q}\cdot\vec{a}\,\tan\alpha = 4\rho^2, \quad \text{of} \begin{cases} \text{center } \vec{c} = -2\rho\,\vec{a}\,\tan\alpha, \\ \text{radius } r = \quad 2\rho\sec\alpha. \end{cases} \tag{6.32}$$

We note that $|\vec{c}| < r$, so the origin lies in the interior of every circle, and $r \geq 2\rho$. The minimal circle ($r = 2\rho$, $\vec{c} = \vec{0}$, for $\alpha = 0$), is the projection of the equator of the sphere. In four dimensions ($D = 3$), the equatorial great sphere maps on a corresponding minimal sphere in the optical fish-eye space. Finally, note that the singularities in (6.31) at $\kappa \to \alpha$ occur when the projected circles degenerate into straight lines. •

Remark: Angles on the sphere equal angles on the plane. The stereographic map conserves angles. To prove this, consider two vectors on the sphere, A and B in the form of (6.28), that subtend an angle γ as shown in Fig. 6.5a. Then,

$$\begin{aligned} \frac{A \bullet B}{\rho^2} &= \begin{pmatrix} \vec{a}\,\sin\alpha \\ \cos\alpha \end{pmatrix} \bullet \begin{pmatrix} \vec{b}\,\sin\beta \\ \cos\beta \end{pmatrix} \\ &= \vec{a}\cdot\vec{b}\,\sin\alpha\sin\beta + \cos\alpha\cos\beta = \cos\gamma, \end{aligned} \tag{6.33}$$

Their projection on the plane are two circles whose centers and radii are given by (6.32) and shown in Fig. 6.5b (omitting the common factor of 2ρ). Thus we form the triangle between the two centers and either intersection point of the two circles, which has sides $\vec{a}\tan\alpha$, $\vec{b}\tan\beta$ and

$$\begin{aligned} \left|\vec{b}\tan\beta - \vec{a}\tan\alpha\right|^2 &= \tan^2\alpha + \tan^2\beta - 2\,\vec{a}\cdot\vec{b}\,\tan\alpha\tan\beta \\ &= \sec^2\alpha + \sec^2\beta - 2\sec\alpha\sec\beta\cos\gamma, \end{aligned} \tag{6.34}$$

where in the last line we have replaced $\vec{a}\cdot\vec{b}$ from (6.33) and used some trigonometric identities. This coincides with the 'rule of cosines' for the triangle in Fig. 6.5b which shows that γ is indeed the angle formed between the intersecting circles on the plane, *quod erat demonstrandum*. The two intersection points $\vec{q}_\pm$ lie in a line with the origin because they satisfy (6.32) with A and B, and the difference of these equations means that $\vec{q}_\pm \perp (\vec{b}\tan\beta - \vec{a}\tan\alpha)$. •

Remark: Bipolar and bispherical coordinates. It is evident from the previous remarks that the geographical orthogonal coordinate grid of longitude and latitude on the Q-sphere of Fig. 6.4 will project to the polar coordinate grid on the plane, which is orthogonal. When the sphere is tilted, the geographical grid will map on the *bipolar* coordinate system of the plane (see, for example [99, Sect. 10.1 and 10.3]). In one more dimension this map yields the spherical and bispherical coordinates of space, respectively. •

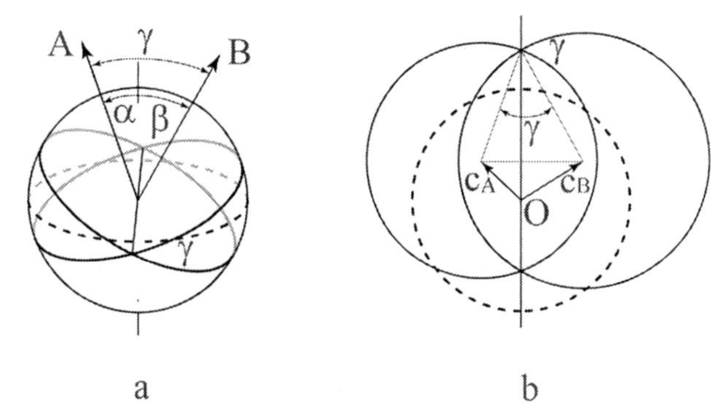

Fig. 6.5. The stereographic map conserves angles. (**a**) The angle γ between the two vectors A and B on the sphere is the angle between their normal great circles (*continuous lines*). (**b**) The two great circles project stereographically on circles in the optical plane with centers c_A and c_B, and radii given by (6.32); the equator maps on the minimal circle (*dashed lines*) with center O. The two circles intersect with the same angle γ. The proof uses the rule of cosines in the triangle formed between the two centers and either intersection point.

Remark: Conjugate points on sphere and plane. Two antipodal points on the sphere, $\pm(\vec{Q}, Q_4)$, will define two *conjugate points* in the fish-eye space through (6.25), $\vec{q}_\pm = \pm 2\rho\vec{Q}/(\rho \mp Q_4)$, whose directions are opposite and whose magnitudes are reciprocal, since $\vec{q}_+ \cdot \vec{q}_- = -4\rho^2$ and $|\vec{q}_+|\,|\vec{q}_-| = 4\rho^2$. Great circles on the sphere consist of antipodal points, so their projected circles in the fish-eye medium are composed of pairs of reciprocal points, and are thus self-reciprocal. The stereographic projection has a host of classical geometric properties, including harmonic quartets of points on all lines, coaxial circles, and inversions (see [126, Chaps. 4, 6 and 7].) •

Remark: Hyperbolic stereographic projection. In contradistinction to the spherical stereographic projection, the hyperbolic projection [equations (6.25) and (6.27) $\tau = -$ and $\sigma \equiv +$] maps the upper sheet of the hyperboloid to the bounded region $|\vec{q}| < 2\rho$ (see Fig. 6.2b). The boundary is the sphere projected from the asymptote cone, where the refractive index (6.2) becomes singular. The hyperbolic rotor motion takes place on great hyperbolas, which are the intersection of the hyperboloid $|\vec{Q}|^2 - Q_4^2 = -\rho^2$ $(Q_4 > 0,\, Q_4 > |\vec{Q}|)$ with planes passing through the origin, $A \bullet Q := \vec{A} \cdot \vec{Q} + A_4 Q_4 = 0$, and this implies that $|A_4| < |\vec{A}|$. The vectors A will then have a form distinct from that in (6.28); instead,

$$A = \begin{pmatrix} \vec{A} \\ A_4 \end{pmatrix} = \rho \begin{pmatrix} \vec{a}\cosh\alpha \\ \sinh\alpha \end{pmatrix}, \quad |\vec{a}\,| = 1, \;\; \alpha \in \mathsf{R}, \quad \text{and} \tag{6.35}$$

$$A \bullet Q = 0 \Rightarrow |\vec{q}|^2 + 4\rho\, \vec{q} \cdot \vec{a} \coth\alpha = -4\rho^2, \tag{6.36}$$

$$\text{i.e., \textbf{circle} arcs:} \quad \text{center } \vec{c} = -2\rho\,\vec{a}\coth\alpha, \quad \text{radius } r = 2\rho\,\mathrm{csch}\,\alpha. \tag{6.37}$$

Since $|\vec{c}| > r$, the origin lies in the *exterior* of every circle [cf. (6.32)]. Also, every circle intersects the boundary circle (which is projected from the asymptote cone) at right angles. The region of space *outside* the boundary circle can be given a mathematical existence by noting that it is the projection of the *lower* sheet of the hyperboloid (with $\sigma = -$). Many properties hold which are hyperbolic analogues of the spherical ones. •

Now that we have written the stereographic projection $\vec{Q} \mapsto \vec{q}$ between 'position' spaces, we recall the distorsion aberration map from Sect. 3.6.2, (3.58); thus we recognize the stereographic transformation as a *distorsion* of phase space, of which we know only its action on position. We must now determine the concomitant transformation of momentum before we can claim to have a *canonical* stereographic map. We first remark on the strategy to find it; the method is the Fourier transform of the one we used for coma in plane phase space, (5.59)–(5.61).

Remark: Canonical form of axis-symmetric distorsion. Technically, the task of finding the partner map of momentum which promotes a given distorsion of position to a canonical transformation, is probably the most difficult part of cruising into phase space. In (3.58) we gave the complete answer (with *caveata*), but only in one dimension. Now we consider maps between position subspaces that are axis-symmetric:

$$\vec{q} = U(|\vec{Q}|^2)\,\vec{Q}, \tag{6.38}$$

with a differentiable scalar function U. Then, to have a conjugate momentum $\vec{p}(\vec{P},\vec{Q})$ that preserves the Poisson bracket $\{Q_i, P_j\} = \{q_i, p_j\} = \delta_{i,j}$, it follows that $\vec{p}$ must be linear in $\vec{P}$. The momentum map canonically conjugate to (6.38) must then be of the form

$$\vec{p} = \frac{1}{U(|\vec{Q}|^2)}\vec{P} + V(|\vec{Q}|^2)\,\vec{Q}\vec{Q}\cdot\vec{P}, \tag{6.39}$$

where $V(x)$ will be determined below, and where we note the dyadic $\vec{Q}\vec{Q}$ which was absent in the $D = 1$ case. This form is sufficient to ensure the conservation of angular momentum:

$$\vec{q}\times\vec{p} = \vec{Q}\times\vec{P}. \tag{6.40}$$

•

Remark: Finding the companion function. To compute the Poisson brackets between position (6.38) and their presumed canonically conjugate momentum (6.39), we need $\partial|\vec{Q}|^2/\partial\vec{Q} = 2\vec{Q}$, so that $\partial U/\partial\vec{Q} = 2\vec{Q}\,U'$ with $U'(x) = dU(x)/dx$. Then,

$$\{q_i, p_j\} = \nabla_{\vec{Q}}\, q_i \cdot \nabla_{\vec{P}}\, p_j = \delta_{i,j} + 2Q_iQ_j\Big(U'/U + \tfrac{1}{2}U\,V + |\vec{Q}|^2U'V\Big). \tag{6.41}$$

The Poisson bracket will close properly when the second term is zero, i.e.,

$$V(x) = -2\frac{U'(x)}{U(x)\,(U(x) + 2xU'(x))}. \tag{6.42}$$

Lastly (after a long but tedious calculation) we can check that $\{p_i, p_j\} = 0$ holds. •

The stereographic map can be extended to a canonical transformation of phase space, (6.38)–(6.39), by finding (6.42). We set $U(x) = 2\rho/(\rho - Q_4(x))$, and use $U'/U^2 = -1/4\rho Q_4$ to conclude that $V(x) = -1/2\rho^2$.[9] For all σ, τ cases (spheres and hyperboloids), the result is

$$\vec{p} = \frac{\rho - \sigma\sqrt{\rho^2 - \tau|\vec{Q}|^2}}{2\rho}\,\vec{P} - \frac{\vec{P}\cdot\vec{Q}}{2\rho^2}\,\vec{Q}. \tag{6.43}$$

The *inverse* transformation can be prodded out of (6.40), and is

$$\vec{P} = \left(1 + \tau\frac{|\vec{q}|^2}{4\rho^2}\right)\left(\vec{p} + \tau\frac{\vec{p}\cdot\vec{q}/2\rho^2}{1 - \tau|\vec{q}|^2/4\rho^2}\,\vec{q}\right). \tag{6.44}$$

6.5 Hidden symmetry and the Hamiltonian

We now project stereographically the Lie algebras and groups which describe free motion on the sphere and hyperboloid, onto the same structures on a phase space that we shall recognize as optical. The generator functions of $\mathsf{SO}(4)$ rotations of the 3-sphere, and of Lorentz transformations $\mathsf{SO}(3,1)$ for the hyperbolic case (6.22), become, through (6.27) and (6.44),

$$j_i := J_i|_{\mathrm{R}} = q_j p_k - q_k p_j, \tag{6.45}$$

$$b_i^\tau := B_i^\tau|_{\mathrm{R}} = \tau\rho\Big(1 - \tau\frac{|\vec{q}|^2}{4\rho^2}\Big)p_i + \frac{\vec{p}\cdot\vec{q}}{2\rho}\,q_i. \tag{6.46}$$

The restricted Poisson brackets of these functions, $\{\cdot,\cdot\}|_{\mathrm{R}}$, are the same as those in (6.8), and they close into the same Lie algebras (and are independent of σ, as they should).

We saw in Sect. 6.1 that the free motion of a point rotor has the mechanical Hamiltonian (6.3), which is basically the square of the angular momentum tensor – the Casimir invariant of the Lie algebra. This motion on the sphere (or the hyperboloid) will map stereographically on the motion of a point on the $\vec{q}$-plane of Figs. 6.3 and 6.4.[10] We should now inquire into whether the

[9] Something as special as the stereographic map must give a neat result.

[10] The free motion of a mass-point on the sphere under parallel projection ($Q \mapsto \vec{Q}$, as a shadow) is harmonic; its trajectories are generally ellipses. Under stereographic projection ($Q \mapsto \vec{q}$) the trajectories are all circles but the motion is not seen (*prima facie*) as harmonic.

rotor Hamiltonian projected on $\vec{q}$-space corresponds to a plausible *optical* Hamiltonian for 'points of light' in some medium whose refractive index will be thus specified. The stereographic image of the Casimir invariant (6.23) is

$$\ell^{\tau\,2} := |\vec{\jmath}|^2 + \tau|\vec{b}^\tau|^2 = \tau\rho^2 \left(1 + \tau\frac{|\vec{q}|^2}{4\rho^2}\right)^2 |\vec{p}|^2. \tag{6.47}$$

Can this be a *bona fide* optical Hamiltonian? The answer is in the affirmative (*of course*). We look back to Sect. 1.6 to see that 'geometric optics in time' has Hamiltonians of the appropriate generic form (1.20): $|\vec{p}|^2$ times a scalar function of $\vec{q}$. We thus compare

$$H_{\rm opt}(\vec{p},\vec{q}) = \frac{|\vec{p}|^2}{2n(\vec{q})^2} \quad \leftrightarrow \quad \tfrac{1}{2}\tau\ell^{\tau\,2} = \tfrac{1}{2}\rho^2\left(1 + \tau\frac{|\vec{q}|^2}{4\rho^2}\right)^2 |\vec{p}|^2. \tag{6.48}$$

The refractive index of this privileged system is the one found by Maxwell:

$$n^\tau(\vec{q}) = \frac{n_o}{1 + \tau|\vec{q}|^2/4\rho^2}, \qquad n_o := n(\vec{0}) \tag{6.49}$$

– as announced in (6.1)–(6.2).

The movement of the 'points of light' in the Maxwell fish-eye are the stereographic projection of the motion of the free mass-points on the conic. We start with the spherical $D = 2$-dim case of Figs. 6.2a and 6.4, because it is the simplest to present in words. In the fish-eye space, the point [with velocity $v = c/n(\vec{q})$] will move slowest when it is near to the origin, and will zoom around the farthest part of its trajectory, where the refractive index is smallest. As we argued in Sect. 6.1, the manifold of great circles on the sphere $\mathcal{S}^2$ of Fig. 6.4, is $\mathcal{S}^2$, and hence all light trajectories in the plane Maxwell fish-eye can be parametrized by the two coordinates of $\mathcal{S}^2$.

We now step up by one dimension, to the fish-eye in R^3; trajectories are plane, and there is an $\mathcal{S}^2$ manifold of planes.[11] Hence, trajectories in the $D = 3$ fish-eye form a 4-dim manifold $\mathcal{S}^2 \otimes \mathcal{S}^2$. It is not mere coincidence that the optical phase space $\wp$ introduced in chapter 2 has the same dimensionality: straight lines in R^3 also form a 4-dim manifold; all of them can be obtained from the *standard ray* (through the optical center and along the optical axis) with rigid Euclidean transformations of space. Correspondingly, all light trajectories in the fish-eye can be obtained from a single *standard circle*, which is the stereographic projection of the equator of the sphere (in the 1–2 plane), i.e., a circle of minimal radius and center at the origin.

The **manifest** symmetry group of the Maxwell fish-eye is $\mathsf{SO}(3)$. (In the $D = 2$ case of Fig. 6.4 it is the $\mathsf{SO}(2)$ group of rotations around the vertical axis.) But as explained here, the fish-eye medium has a larger, **hidden**

[11] Antipodal plane normals correspond to the same trajectory but opposite senses of motion.

symmetry group SO(4), which can be *seen* only by inverse stereographic projection back to the sphere $\mathcal{S}^3$. (The $\mathcal{S}^2$ sphere in Fig. 6.4 has a manifest SO(3) symmetry, which becomes *hidden* on the optical plane.) Correspondingly for the Lie algebras, the angular momentum functions j_i's in (6.45) are generators of the manifest symmetry of rotations, while the b_i's generate rotations of the $\mathcal{S}^3$-sphere into its fourth direction. The latter are 'very' nonlinear transformations that map the minimal centered circles onto all other off-center circles. Moreover, all these are *canonical* transformations of phase space. Their lines of flow are $b_i^\tau(\vec{p}, \vec{q}) = \text{constant}$. Now follow several further remarks on this very special optical system.

Remark: Can light really go in circles? The minimal circle of light in a fish-eye medium is centered and has radius 2ρ; to contain it, the refractive index of a fish-eye must be such that $n^+(2\rho) > 1$, before the medium becomes vacuum. From (6.49) it follows that at the center it should satisfy $n_o > 2$, so diamond would do – provided its refractive index could tapered off as needed. Actually, because of light injection/detection problems, the full fish-eye is not quite realizable. Half fish-eyes, or fish-eye *regions* in an inhomogeneous medium can be of interest. Or further afield, Bose–Einstein condensates which 'stop light' by exhibiting huge refractive indices n_o, could support a wave version of fish-eye trajectories. For reasons explained in the appendix to this part, there *does* exist a corresponding *wave version* of the fish-eye – because it is built on a Lie group: imagine the $D = 2$ projection of a superposition of spherical harmonics [particularly $Y_{l,l}(\theta, \phi)$'s] from the sphere of Fig. 6.4 to the optical plane, with all its rotations and tilts. Only a discrete spectrum of colors can be supported by the fish-eye in wave optics [51]. •

Remark: Perfect imaging between conjugate points. Two antipodal points on the sphere $\mathcal{S}^3$ define an $\mathcal{S}^2$ manifold of great circles passing through both. Projected on the R^3 fish-eye space, light trajectories issuing in all directions from one of these points, will all converge on the other, its *conjugate* point; the latter will be the *perfect image* of the first. Moreover, since on the sphere it takes the same time for a mass point to reach its antipode independently of the trajectory, the image in time is also perfect. •

Remark: Descartes spheres in the fish-eye. Both the manifest and hidden symmetry transformations conserve the Casimir/Hamiltonian (6.48), i.e., the value of $H_{\rm opt}$ is constant on and for all trajectories in the fish-eye space. We choose this constant to be $\frac{1}{2}\rho^2$ in order that

$$|\vec{p}| = n^\tau(\vec{q}) = \frac{n_o}{1 + \tau|\vec{q}|^2/4\rho^2}, \qquad n_o = \frac{1}{\rho}. \tag{6.50}$$

The conformal optical model then provides a field of Descartes spheres to which momentum is there constrained. Geometric relations exist in this phase space which would merit a separate study. •

Remark: Hamilton equations of the fish-eye. The Hamilton equations of motion can be found with the Hamiltonian function $H_{\rm opt}(\vec{p}, \vec{q})$ for a 'light particle' in the fish-eye (6.48). [See Sect. 1.6, (1.20)]. Suggestive forms for velocity and acceleration are

$$\frac{d\vec{q}}{c\,dt} = \{H_{\rm opt}, \vec{q}\} \Rightarrow \frac{d\vec{q}}{dt} = \frac{c}{n^\tau(\vec{q})}\frac{\vec{p}}{|\vec{p}|}, \qquad \frac{d\vec{p}}{c\,dt} = \{H_{\rm opt}, \vec{p}\} = -\tau\frac{n^\tau(\vec{q})}{\rho}\vec{q}. \tag{6.51}$$

•

Remark: Motion in hyperbolic fish-eyes. The 'hyperbolic rotor' points on the maximal hyperbolas in Fig. 6.2b, will move vertically with a velocity $\sim \sinh \upsilon t$ (instead of the harmonic motion $\sim \sin \omega t$ in the spherical rotor), so they escape exponentially and approach the asymptote with the inverse exponential. As we saw above, the stereographic projection turns the great hyperbolas into circle arcs which end at right angles to the boundary sphere of the hyperbolic fish-eye space; there, the 'light particle' slows to a halt. There is the manifest $\mathsf{SO}(3)$ symmetry of the medium, and a hidden $\mathsf{SO}(3,1)$ group of canonical transformations which maps all of these circle arcs amongst themselves. •

Remark: Had we taken $P_4 \neq 0$. The restriction $P_4 = 0$ in the definition of (6.21) is in a sense arbitrary. Had we taken any other $P_4 = \varpi$, the generators $\vec{b}^\tau$ of the hidden symmetry Lie algebra (6.46), would in turn have a new form $\vec{b}^\tau_\varpi = \vec{b}^\tau + \varpi\vec{q}$. The new functions, although more complicated, still close into the same Lie algebra, $\mathsf{so}(4)$ or $\mathsf{so}(3,1)$ under Poisson brackets. However, the new invariant Casimir/Hamiltonian function (6.47) will be now

$$\ell^{\tau\,2}_\varpi := |\vec{\jmath}|^2 + \tau|\vec{b}^\tau + \varpi\vec{q}|^2 = \tau\left(\rho\Big(1 + \tau\frac{|\vec{q}|^2}{4\rho^2}\Big)\vec{p} + \varpi\vec{q}\right)^2. \tag{6.52}$$

This function unfortunately does not have any of the standard forms of a Hamiltonian function [cf. (6.48)]. The situation reminds us of the selection $H = 0$ that we made on the Hamiltonian for general optical media in Chap. 1, (1.7); any other choice would not allow the Descartes sphere (1.11) to be a useful concept. We may disregard the many alternatives offered by mathematics to concentrate on those which turn out to produce useful geometric optical models. •

6.6 Dynamical Lie algebra of the fish-eye

The transformations of optical phase space which do *not* leave a system invariant are called (somewhat redundantly) *non-invariance* transformations. The generating functions of that group will exhibit nonzero Poisson brackets with the Casimir/Hamiltonian of the conformal system (6.48), and will destroy the Descartes sphere (6.50). It is generally an open question whether a given optical system posesses a neat and useful Lie group (i.e., with a finite number of parameters) of non-invariance transformations. Tinkering with the Poisson brackets between the generator functions of $\mathsf{so}(4)$ and $\mathsf{so}(3,1)$ algebras in the stereographic projection, we note [145] the following Poisson brackets between the b_i^{τ}'s of (6.46):

$$\{b_i^+, b_j^-\} = \delta_{i,j} d, \quad \text{where} \quad d := -\vec{p}\cdot\vec{q} \quad \text{generates scalings.} \tag{6.53}$$

Further Poisson brackets between these functions shows that

$$\{d, b_i^+\} = b_i^-, \quad \{d, b_i^-\} = -b_i^+, \quad i = 1,2,3. \tag{6.54}$$

Thus we have 3 Lie algebras $\mathsf{so}(2,1)$ of generators $\{d, b_i^+, b_i^-\}$. Moreover, all generator functions of $\mathsf{so}(4)$ and of $\mathsf{so}(3,1)$, together with d, close into the larger Lie algebra $\mathsf{so}(4,1)$. This is called the *de Sitter* algebra in relativistic classical and field-theoretical models [39, 52].

To organize the functions of phase space that generate $\mathsf{so}(4,1)$, it is helpful to label them by $\ell_{\alpha,\beta}$, as in (6.11), with the metric $\tau_1, \dots \tau_4 = +$ and $\tau_5 = -$. We define

$$\begin{array}{lll} \ell_{i,j} := j_k, & \text{Eq. (6.45) } i,j,k \text{ cyclic } 1,2,3, & \\ \ell_{i,4} := b_i^+, & \text{Eq. (6.46)} \qquad\quad i = 1,2,3, & \\ \ell_{i,5} := b_i^- = b_i^+ - 2\rho\, p_i, & \qquad\qquad\qquad\quad i = 1,2,3, & \\ \ell_{4,5} := d := -\vec{p}\cdot\vec{q}. & & \end{array} \tag{6.55}$$

We have the new generator function d [see Sect. 3.6, (3.60)], whose action on phase space is to scale $\vec{p}$ and $\vec{q}$ by reciprocal ratios, and thus it maps fish-eyes in the following way:

$$e^{s\{d,\circ\}} f(\vec{p},\vec{q}) = f(e^{-s}\vec{p}, e^{s}\vec{q}) \;\Rightarrow\; e^{s\{d,\circ\}} : \begin{bmatrix}\text{fish-eye of}\\ \text{radius } \rho\end{bmatrix} \mapsto \begin{bmatrix}\text{fish-eye of}\\ \text{radius } e^{-s}\rho\end{bmatrix}, \tag{6.56}$$

as we can see from (6.48) and (6.50). Scaling is a non-invariance transformation, but one that relates Maxwell fish-eyes of all possible radii $0 < \rho < \infty$ (both spherical and hyperbolic).

To explore the action of other new functions, we note the linear combination

$$p_i = \tfrac{1}{2}(b_i^+ - b_i^-)/\rho, \tag{6.57}$$

namely *momentum*: the non-invariance algebra $\mathsf{so}(4,1)$ thus contains the generators of translation. On fish-eye systems, thus

$$e^{\vec{r}\cdot\{\vec{p},\circ\}} f(\vec{p},\vec{q}) = f(\vec{p},\vec{q}-\vec{r}) \;\Rightarrow\; e^{\vec{r}\cdot\{\vec{p},\circ\}} : \begin{bmatrix}\text{fish-eye of}\\ \text{center } \vec{0}\end{bmatrix} \mapsto \begin{bmatrix}\text{fish-eye of}\\ \text{center } \vec{r}\end{bmatrix}. \tag{6.58}$$

The de Sitter non-invariance group $\mathsf{SO}(4,1)$ thus transforms between all Maxwell fish-eye systems (both spherical and hyperbolic) of different centers and radii, displaced and scaled versions of the same *standard* system (say, $\rho = 1$, $\vec{r} = \vec{0}$). Within the group $\mathsf{SO}(4,1)$ thus, the hidden symmetries of the spherical and hyperbolic medium are dynamical transformations of the other two, so $\mathsf{SO}(4,1)$ merits the appellation of *dynamical group* of conformal optical systems.

Remark: Homogeneous optical media are also conformal. Not only the spherical and hyperbolic fish-eyes are members of the conformal family, but also their common contraction (by scaling as $\rho \to \infty$), the *homogeneous medium*, when the conics degenerate to a flat manifold. We have noted that in (6.57) appear the momentum functions p_i; these, together with the generator functions of rotations $\ell_{i,j}$, generate the Euclidean symmetry group $\mathsf{ISO}(3) \subset \mathsf{SO}(4,1)$ of the homogeneous medium seen in Chap. 5. •

6.7 Conformal Lie algebra

We may suspect a *still* larger algebra in conformal optical systems – beyond their dynamical algebra—, because the Casimir/Hamiltonian function of the symmetry algebra, ${\ell^{\top}}^2(\vec{p},\vec{q})$ in (6.47)–(6.48), is a perfect square. In the manner of (6.55), we can now add a *sixth* column of $\ell_{\alpha,\beta}$'s defining the functions

$$\begin{aligned} \ell_{i,6} &:= \quad l_i := \quad |\vec{p}|\, q_i, && i = 1,2,3, \\ \ell_{4,6} &:= -\ell^- := -\rho|\vec{p}|\,(1 - |\vec{q}|^2/4\rho^2), && \\ \ell_{5,6} &:= \quad \ell^+ := \quad \rho|\vec{p}|\,(1 + |\vec{q}|^2/4\rho^2). && \end{aligned} \tag{6.59}$$

Since there are now a lot of functions, it will be very useful to place $\ell_{\alpha,\beta}$, $\alpha < \beta$, in the α–β box of the following array [145]:

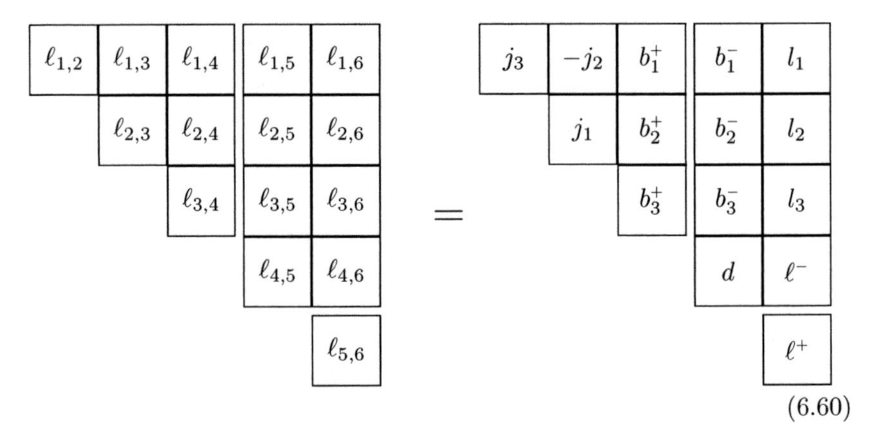

(6.60)

Since $\ell_{\beta,\alpha} = -\tau_\alpha \tau_\beta \ell_{\alpha\beta}$, we need indicate only $\alpha < \beta$. Also, there is a small gap in the pattern between the *compact* generators ℓ^+ and $\mathsf{so}(4)$, which generate rotations, and the *noncompact* ones, $\{\vec{b}^-, \vec{l}, d, \ell^-\}$, which can boost with *unbounded* rapidity.

Now we state that under Poisson brackets $\{\cdot,\cdot\}_{(3)}$, these functions close into the algebra $\mathsf{so}(4,2)$ defined in (6.11) with $\tau_1, \ldots \tau_4 = +$ and $\tau_5, \tau_6 = -$. This is called the *conformal* algebra; the corresponding conformal Lie group is $\mathsf{SO}(4,2)$.[12] The pattern (6.60) shows at a glance the structure (6.11) of the Lie algebra $\mathsf{so}(4,2)$: a generator $\ell_{\alpha,\beta}$ will have zero Lie bracket with all generators $\ell_{\kappa,\lambda}$ *outside* its row $\kappa \neq \alpha$ and column $\lambda \neq \beta$ (*and* their continuation, reflected at the diagonal, as column κ and row λ respectively). Thus we see that the 1-parameter Lie group of transformations generated by ℓ^+ will commute with the $\mathsf{SO}(4)$ symmetry subgroup of the spherical fish-eye and leave its Casimir/Hamiltonian function invariant (and similarly for ℓ^- and $\mathsf{SO}(3,1)$ in the hyperbolic case).

Remark: Conformal action on momentum. The new vector of functions $\vec{l} = |\vec{p}|\, \vec{q}$ in the conformal algebra $\mathsf{so}(4,2)$ [see (6.60)] acts rather simply on momentum space and under rotations transforms as a 3-vector. The repeated Poisson brackets between $\vec{r} \cdot \vec{l}$ (with constant $\vec{r}$ on a sphere $|\vec{r}| = 1$) and the coordinates of momentum p_k are as follows: $\{\vec{r} \cdot \vec{l}, p_k\} = r_k\, |\vec{p}|$, $\{\vec{r} \cdot \vec{l}, |\vec{p}|\} = \vec{r} \cdot \vec{p}$, $\{\vec{r} \cdot \vec{l}, \vec{r} \cdot \vec{p}\} = |\vec{p}|$, etc. Then, even and odd terms in the exponential series of the Lie-Poisson operators can be summed, to find the 1-parameter Lie group as

$$\exp(\chi\{\vec{r} \cdot \vec{l}, \circ\})\, \vec{p} = \vec{p} + [|\vec{p}| \sinh \chi + \vec{r} \cdot \vec{p}(\cosh \chi - 1)]\, \vec{r}. \tag{6.61}$$

Using now the terms starting from the second, we can also sum the Lie series of $\exp(\chi\{\vec{r}\cdot\vec{l}, \circ\})\, |\vec{p}|$, and fit the two transformations into a vector-and-matrix form:

$$\exp(\chi\{\vec{r} \cdot \vec{l}, \circ\}) \begin{pmatrix} \vec{r} \cdot \vec{p} \\ |\vec{p}| \end{pmatrix} = \begin{pmatrix} \cosh \chi & \sinh \chi \\ \sinh \chi & \cosh \chi \end{pmatrix} \begin{pmatrix} \vec{r} \cdot \vec{p} \\ |\vec{p}| \end{pmatrix}. \tag{6.62}$$

This shows that the field of Descartes spheres of momentum is mapped to ellipsoids of revolution around the direction of $\vec{r}$. Chap. 8 examines anisotropic media, in which such Descartes *ovoïds* occur. •

Remark: Conformal action on position. The transformation of one of the components of $\vec{q}$ under the conformal algebra generator $l_i = \ell_{i,6}$ can be found remembering that the function is conserved under its own flux, so $l_i = |\vec{p}|\, q_i$ is invariant. Hence,

$$\exp(\chi\{l_i, \circ\})\, q_i = \frac{q_i}{|\vec{p}| \cosh \chi + p_i \sinh \chi}. \tag{6.63}$$

This tactic can be generalized to include the other components q_j, $i \neq j$, noting from (6.11) that $\{l_i, l_j\} = -\ell_{i,j} = -j_k$, and so the exponential Lie-Poisson bracket series results in

$$\exp(\chi\{l_i, \circ\})\, l_j = l_j \cosh \chi - j_k \sinh \chi. \tag{6.64}$$

As before, knowing the transformation of $|\vec{p}|$ from (6.62) we find that of $q_j = l_j/|\vec{p}|$. In this way we can find the finite transformations of phase space under all generators of the conformal algebra. •

[12] Reader: please check. See that $|\vec{p}|$ is no longer on a Descartes spheres, but a function independent of $|\vec{q}|$; their Poisson bracket $\{|\vec{p}|, |\vec{q}|\}_{(3)}$ is nonzero.

Remark: The full family of conformal systems. In Sect. 6.6 we concluded that the 10-parameter dynamical group of the fish-eye, $\mathsf{SO}(4,1)$, maps any fish-eye into all other possible fish-eyes, scaled and translated with the 4 new parameters which the dynamical group provides in addition to the 6 parameters of the symmetry groups (manifest and hidden, $\mathsf{SO}(4)$ or $\mathsf{SO}(3,1)$). What is the rôle of the 15 parameters of the conformal group $\mathsf{SO}(4,2)$? To start with, note that the 3 new generators l_i do *not* commute with the fish-eye Hamiltonian $\ell^{\tau\,2}$. According to (6.11), $\{l_i, \ell^\tau\} = -\tau b_i^{-\tau}$, and the integrated group action on the square root of the Casimir/Hamiltonian will be

$$\begin{aligned}\exp(\chi\{l_i,\circ\})\,\ell^\tau &= \ell^\tau\cosh\chi - \tau b_i^{-\tau}\sinh\chi \qquad (6.65)\\ &= \rho\Big(1-\tau\frac{|\vec q|^2}{4\rho^2}\Big)(|\vec p|\cosh\chi + p_i\sinh\chi) - \tau\frac{\vec p\cdot\vec q}{2\rho}q_i\sinh\chi.\end{aligned}$$

This is no longer a 'true' fish-eye since anisotropy is present. Nevertheless, we welcome these new functions, their square Hamiltonian and pure angular momentum, into the *conformal* family of systems that live among the 15 conformal algebra generators. •

6.8 The Kepler system and its hidden rotor

We said at the beginning of this chapter that the Maxwell fish-eye is the Kepler (and Coulomb and Bohr) system of optics. The connection between both systems has been studied in the literature [137]; we shall review the hidden symmetry of the mechanical Kepler system.[13] The Kepler Hamiltonian is well known to be

$$H_{\text{Kepler}}(\vec P,\vec Q) := \frac{1}{2\mu}|\vec P|^2 - \frac{K}{|\vec Q|} = E = -\tau|E|, \qquad (6.66)$$

where μ is the reduced mass of the sun–planet system, K is a constant (including the gravity constant, or the charges of the electron and nucleus), and E is the energy. When $E<0$, $\tau=+$, and the orbits are bound; for $E>0$, $\tau=-$, and they are free.

When $E\neq 0$, we introduce the new coordinates $(\vec p,\vec q)\in\mathsf{R}^6$, rescaled by this energy,

$$\vec P = \vec p\,\sqrt{|E|}, \qquad \vec Q = \vec q\Big/\sqrt{|E|}. \qquad (6.67)$$

Dividing (6.66) by $|E|$ and multiplying it by $|\vec q|$, we bring the expression to the forms

$$\frac{1}{2\mu}|\vec p|^2\,|\vec q| - \frac{K}{\sqrt{|E|}} = \frac{E}{|E|}|\vec q| \quad\Rightarrow\quad |\vec q|\Big(1+\tau\frac{|\vec p|^2}{2\mu}\Big) = \frac{\tau K}{\sqrt{|E|}}. \qquad (6.68)$$

[13] The symbols $\vec P$, $\vec p$ and $\vec Q$, $\vec q$ for momentum and position used in this section are unrelated to their previous use for the point rotor phase space.

We compare this with $\ell^{\pm}$ in (6.59) [and, with an eye on the rotor Hamiltonian, with (6.47) and (6.48)]. To have a correspondence between the Kepler system and the fish-eye, we see it is still necessary to Fourier-transform the last expression in (6.68), i.e., $\vec{p} \mapsto \vec{q}$ and $\vec{q} \mapsto -\vec{p}$, and thence $\ell \mapsto \tilde{\ell}$. And *then* we can identify $\tilde{\ell}^{\tau\,2}/\rho^2$ there with $\tau K/\sqrt{|E|}$ here. We can further invert and square (6.68) to isolate the energy of the Kepler system and display it as

$$-\tau\,|E| = \left(\frac{\rho K}{\tilde{\ell}^{\tau}}\right)^2 = \frac{K^2}{|\vec{q}|^2\left(1+\tau\dfrac{|\vec{p}|^2}{2\mu}\right)^2}, \qquad \mu = 2\rho^2. \tag{6.69}$$

We conclude that the Kepler system also posesses hidden symmetry – but only *among orbits of the same energy.* The Kepler Hamiltonian function (6.66) is the inverse of the Casimir invariant of $\mathsf{SO}(4)$ for negative energy (bound orbits), and of $\mathsf{SO}(3,1)$ for positive energy (free orbits), while their common limit $E \to 0$ of parabolic orbits is under the ægis of $\mathsf{ISO}(3)$. The trajectory of the mechanical momentum in the Kepler system is thus the stereographic projection of a maximal circle trajectory on a sphere or hyperboloid of 4-dim momentum, whose radius depends now on the energy. The vector of functions $\vec{b}^{\tau}(\vec{p},\vec{q})$ in (6.46) is the (Fourier-transformed) *Runge-Lenz* vector $\vec{\jmath}\times\vec{p}+\vec{q}$, and the corresponding quantum mechanical operator is called the Pauli-Lubański vector.[14] This vector determines the (constant) orientation of the elliptical or hyperbolic orbits in the plane perpendicular to the angular momentum vector $\vec{\jmath}$, and their excentricity. The mechanical Kepler system is the prototype of symmetry in Nature; a recent book by Cordani [35] structures the group theoretical view on the music of the spheres.

Remark: The 'quantum number' operator. In quantum models, an operator $\hat{\ell}^{+}$ purportedly quantizing the classical $\ell^{+}(\vec{p},\vec{q})$ in (6.59), and with a lower-bound, discrete spectrum $n \in \{1, 2, \ldots\}$, serves to number the energy levels and is called the *number* operator. Rather unfortunately, the energy spectrum of the Bohr atom is $E_n \sim -1/n^2$, so the inverse square root of the Hamiltonian operator is needed. In the common Hilbert space of quantum mechanics, such operators can be defined only in terms of an appropriate eigenbasis, i.e., *weakly.* From the literature it follows that, in fact [145], the quantized $\mathsf{so}(4,2)$ operators do *not* close under commutators beyond the symmetry algebra. The energy-dependent rescaling of phase space presents a problem in the quantum system for the proper definition of a dynamical group. This has been surmounted in various ways to find algebraically the transition probabilities between levels of different energy and explicit sum rules [19], but the transformations themselves do not have a simple geometrical interpretation. That is the virtue of conformal optical systems alone. •

[14] In quantum mechanics, the matrix elements of the Pauli-Lubański operators obey three-term recursion relations that relate transition probabilities between states of different angular momenta.

7 Axial symmetry reduction

When the refractive index of an optical medium is invariant under rotations around a fixed (z-) axis, we call the system *axis-symmetric*. In the model of 3-dim geometric optics described with the 4-dim phase space $\wp$, axial symmetry entails the conservation of the single quantity $\mathbf{q}\times\mathbf{p}$. In this chapter we address the optical *guide* model defined in Sect. 2.3, that was solved for the elliptic-index-profile case. And so we further demand that the medium be invariant under *translations* along the the optical z-axis, and also (as is true for most optical materials) invariant under reflections across planes that contain the axis.[1]

Quantum models of Nature have often profited from axial and rotational symmetries, through separating the coordinate on which the dynamics of the system do *not* depend, i.e., the *ignorable* coordinate, and so the model reduces to its 'radial' part. Our study of 3-dim axis-symmetric z-invariant optical guides corresponds (see Sect. 1.5) to 2-dim mechanical models with a circularly-symmetric potential which is invariant in time. The results can be easily generalized to $(D-1)$-dim systems with spherical symmetry and one 'z'-axis. But we should not rely too much on the quantum theory of angular momentum with its stress on spherical harmonics; the geometric model admits a continuum of values for angular momentum and energy. Thus, axial symmetry reduction presents its own distinct face in geometric and other classical models.

7.1 Symmetry-adapted coordinates of phase space

A 3-dim optical guide whose refractive index $n(\mathbf{q})$ depends only on the distance $|\mathbf{q}|$ to the optical z-axis, will contain families of rays $\{\mathbf{q}(z), \mathbf{p}(z)\}_\phi$, $z \in \mathsf{R}$, $\phi \in \mathcal{S}^1$, obtained by rotation around the optical axis of any 'initial' $\phi = 0$ ray in the guide. This angle is the *ignorable* coordinate for evolution along z. The relevant coordinates for phase space $\wp$ are those *invariant* under rotations by ϕ, namely $|\mathbf{q}|^2$, $|\mathbf{p}|^2$ and $\mathbf{p}\cdot\mathbf{q}$ (or the angle θ between the two vectors,

[1] Reflection symmetry may be evident in optics; but when the theory is used for accelerator design, coaxial magnetic fields will contribute to the Hamiltonian with a term $\sim \mathbf{q}\times\mathbf{p}$ (and not its square) that changes sign with reflection.

$\cos\theta = \mathbf{p}\cdot\mathbf{q}/|\mathbf{p}||\mathbf{q}|$). Two of these coordinates appear in the Hamiltonian (2.7), *viz.*,

$$h = -\sqrt{n(|\mathbf{q}|)^2 - |\mathbf{p}|^2}, \qquad h^2 \le n(|\mathbf{q}|)^2. \tag{7.1}$$

The quadratic function

$$j_z := \mathbf{q}\times\mathbf{p}, \qquad j_z^2 = |\mathbf{q}|^2|\mathbf{p}|^2 - (\mathbf{p}\cdot\mathbf{q})^2 \ge 0, \tag{7.2}$$

is called the *skewness* of the ray, and is a constant of the motion because $\{h, j_z\} = 0$.

The *symmetry-adapted* coordinates that we choose are the quadratic functions

$$\begin{aligned} r_1 &:= \tfrac{1}{2}\mathbf{p}\cdot\mathbf{q}, & r_+ &:= r_3 + r_2 = \tfrac{1}{2}|\mathbf{p}|^2,\\ r_2 &:= \tfrac{1}{4}(|\mathbf{p}|^2 - |\mathbf{q}|^2), & r_0 &:= r_1 \qquad\;\; = \tfrac{1}{2}\mathbf{p}\cdot\mathbf{q},\\ r_3 &:= \tfrac{1}{4}(|\mathbf{p}|^2 + |\mathbf{q}|^2), & r_- &:= r_3 - r_2 = \tfrac{1}{2}|\mathbf{q}|^2. \end{aligned} \tag{7.3}$$

Because $|\mathbf{p}|^2, |\mathbf{q}|^2 \ge 0$ these coordinates cover only one quadrant of $\vec{r} \in \mathsf{R}^3$. Moreover, the coordinates of any given ray in the system are restricted to lie on a surface due to the constancy of the Hamiltonian (7.1),

$$0 \le h^2 = \nu(r_-)^2 - 2r_+ \le \nu(r_-)^2, \qquad \text{where} \quad \nu(r_-) := n(|\mathbf{q}|), \tag{7.4}$$

and to a second surface due to its skewness (7.2),

$$r_+ r_- - r_0^2 = -r_1^2 - r_2^2 + r_3^2 = \tfrac{1}{4}(|\mathbf{p}|^2|\mathbf{q}|^2 - (\mathbf{p}\cdot\mathbf{q})^2) = \tfrac{1}{4}j_z^2 \ge 0. \tag{7.5}$$

For each value of $0 \le |h| \le \nu(r_-)$, the surface (7.4) is a cylinder of section $r_+ = \frac{1}{2}\nu(r_-)^2$ and ruled along the r_0-axis, as shown in Fig. 7.1a. Similarly, for each value of $j_z^2 \ge 0$, the surface (7.5) is one sheet of an equilateral hyperboloid of revolution around the axis $r_3 = \frac{1}{2}(r_+ + r_-)$ shown in Fig. 7.1b. All hyperboloids lie 'inside' the *cone* $j_z = 0$ of zero-skewness rays; this solid cone we shall call the *hyperbolic onion* for short. The *reduced phase space* of axis-symmetric guides is the intersection of the two regions allowed by the previous two constraints, and we shall denote it by $\wp|_{\text{ax}}$. In Sect. 7.5 below shall justify that we use the name *phase* for this region of R^3. Ray trajectories in the guide thus map on trajectories in $\wp|_{\text{ax}}$ that lie in the intersection of an h- and a $\pm j_z$-constant surface.

Remark: Points in $\wp$ and in $\wp|_{\text{ax}}$. Each point $(\mathbf{p}, \mathbf{q}) \in \wp$ represents a ray in the guide that crosses the standard screen, while each point of $\vec{r} \in \wp|_{\text{ax}}$ represents *two* ϕ-families of rays in the guide which are mirror reflections of each other in a plane that contains the z-axis, namely

$$\vec{r} \leftrightarrow \left\{ \mathbf{R}(\phi)\begin{pmatrix} p_x \\ \pm p_y \end{pmatrix}, \mathbf{R}(\phi)\begin{pmatrix} q_x \\ \pm q_y \end{pmatrix} \,\middle|\, \mathbf{R}(\phi) := \begin{pmatrix} \cos\phi & -\sin\phi \\ \sin\phi & \cos\phi \end{pmatrix} \right\}. \tag{7.6}$$

The mirror reflection changes the sign of j_z, which is a *skew*-scalar. •

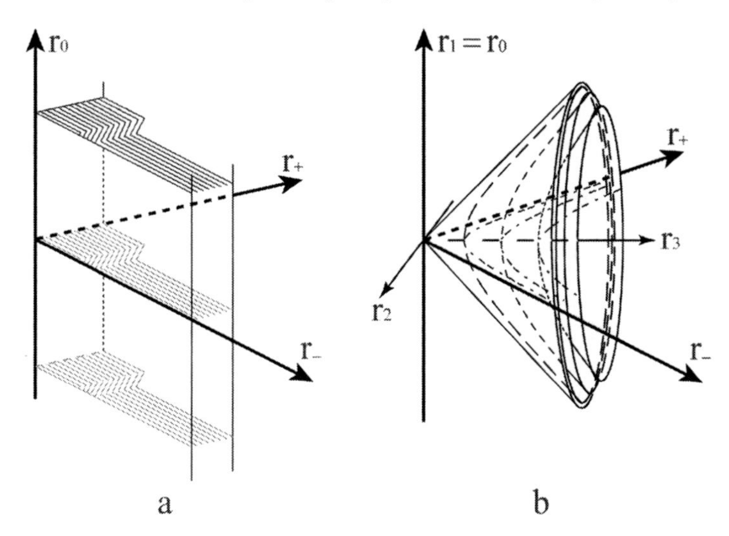

Fig. 7.1. Regions of $\vec{r}$ in first r_1–r_3 quadrant are further restricted by the Hamiltonian and angular momentum to $\vec{r}(z) \in \wp|_{\rm ax}$. (**a**) The value of the Hamiltonian h, $0 \le h \le \nu(r_-)$, foliates cylindrical surfaces along the $r_0 = r_1$ axis, bounded by $r_\pm \ge 0$ and $r_+ = \frac{1}{2}\nu(r_-)^2$. (**b**) The value of the skewness $\pm j_z$, $j_z^2 = 4(r_3^2 - r_1^2 - r_2^2) \ge 0$, foliates the interior of the equilateral cone into hyperbolic onion.

Remark: The 'energy' of trajectories. The Hamiltonian constant h in (7.1) can be called the *energy* of the ray trajectory in the guide of refractive index profile $n(|\mathbf{q}|)$, because it behaves as the mechanical energy of a particle in a potential $V(|\mathbf{q}|) \sim -n(|\mathbf{q}|)^2 + 1$.[2] The rays of *ground* energy in the guide, $h = -n(|\mathbf{q}_{\rm m}|)$, also called the *basal* or *design* rays, occur at the radius (or radii) $|\mathbf{q}_{\rm m}|$ where the refractive index is largest (usually the center of the guide), and among these, those rays which are parallel to the guide axis, $\mathbf{p} = \mathbf{0}$. The ϕ-families of ground-energy rays thus fall on the points $\vec{r} \in \wp|_{\rm ax}$ of coordinates $\{r_+{=}0,\ r_0{=}0,\ r_-{=}\frac{1}{2}|\mathbf{q}_{\rm m}|^2\}$, i.e., on the r_- axis of Figs. 7.1. On the other hand, rays of *maximal* energy occur at the value $h = 0$, where $|\mathbf{p}| = n(|\mathbf{q}|)$, i.e., at the points of the boundary surface $r_+ = \frac{1}{2}\nu(r_-)^2$ of Fig. 7.1a. As the reference screen progresses along the z-axis, the trajectory $\vec{r}(z) \in \wp|_{\rm ax}$ will remain on its $h = $ constant surface. •

Remark: Meridional, saggital and skew rays. The solid cone in Fig. 7.1b, whose points represent rays in the axially symmetric guide, is restricted by their angular momentum condition (7.5). The apex of this cone is the origin $\vec{r} = \vec{0}$, and corresponds to the ray along the center of the guide, $(\mathbf{p}, \mathbf{q}) = (\mathbf{0}, \mathbf{0})$, usually its design ray. The cone opens 45° and its axis is r_3; the r_+ axis corresponds to rays which pass the screen through its optical center $(\mathbf{p}, \mathbf{0})$, and the r_- axis to rays which at the screen are parallel to the

[2] See Sect. 1.5, (1.17), setting $E = 0$ so that vacuum $n = 1$ correspond to $V = 0$.

optical axis $(\mathbf{0}, \mathbf{q})$, in particular the ground-energy ray. The distance from the origin to the vertex of a hyperboloid (i.e., the *radius* of the hyperboloid) is $r_3 = \frac{1}{2}|j_z|$. When the screen moves along the optical z-axis, the trajectories of $\vec{r}(z) \in \wp|_{\text{ax}}$ are restricted to remain on a j_z = constant hyperboloid. *Meridional* rays, i.e., those which lie in a plane with the optical z-axis ($\mathbf{p} \parallel \mathbf{q}$, $j_z = 0$), correspond to the surface of the cone, $r_3^2 = r_1^2 + r_2^2$ (including the r_+ and r_- axes); thus they remain meridional under z-evolution. Rays of maximal angular momentum are *saggital* ($\mathbf{p} \perp \mathbf{q}$, $j_z = |\mathbf{p}|\,|\mathbf{q}|$), they correspond to the $r_1 = 0$ horizontal plane; the r_3 axis represents saggital rays which at the reference screen fulfill $|\mathbf{p}| = |\mathbf{q}|$. Under z-evolution, the trajectories $\vec{r}(z)$ of saggital rays generally move off the $r_1 = 0$ plane and then become *skew* rays; but they can never become meridional. •

7.2 Hamiltonian knife cuts hyperbolic onion

The Hamiltonian (7.1) is the generator of evolution along the z-axis of the guide, so we noted after (7.1) that $h(r_2, r_3)$ is a conserved quantity, and the Hamiltonian flow $\vec{r}(z)$ will respect the cylinder surfaces of Fig. 7.1a, determined by the radial profile of the refractive index $n(|\mathbf{q}|)$. Similarly, skewness j_z is an independent constant which restricts the flow to the hyperboloid surfaces of Fig. 7.1b. The flow $\vec{r}(z)$ is thus along the *lines* where the j_z = constant hyperbolic onion is cut by the h = constant Hamiltonian knife. We detail with two examples how to find and interpret the intersection of these two figures.

Example: Homogeneous medium. The homogeneous medium n = constant is also an axis-symmetric guide medium, and the evolution of the screen coordinates $(\mathbf{p}(z), \mathbf{q}(z)) \in \wp$ along the guide is given by (2.10). The corresponding trajectories for $\vec{r}(z) \in \wp|_{\text{ax}}$ are determined by $\vec{r}(0)$ and the constants n, $h_{\circ} = n - 2r_+$,

$$r_+ = \tfrac{1}{2}(n^2 - h_{\circ}^2), \quad r_0(z) = r_0 + z r_+/h_{\circ}, \quad r_-(z) = r_- + 2z r_0/h_{\circ} + z^2 r_+/h_{\circ}^2. \tag{7.7}$$

Then, as shown in Fig. 7.2, the Hamiltonian restricts the trajectory to the *plane* r_+ = constant, and angular momentum to the intersections with the hyperboloids $r_+ r_-(z) - r_0(z)^2 = \frac{1}{4}j_z^2$. Replaced in (7.7) this yields the nested *parabolas* in the vertical (r_0, r_-) plane of the figure. •

Example: Elliptic index-profile guide. When the radial profile of the refractive index of the guide is an ellipse $n_{\text{e}}^2(\mathbf{q}) := n_{\circ}^2 - \kappa^2|\mathbf{q}|^2$, of width $\sim 1/\kappa$ [see Sect. 2.3, equations (2.15), (2.18)–(2.19), and Fig. 2.5], then the Hamiltonian knife $h_{\text{e}}(\vec{r})$ is straight: (7.1) for $\nu(r_-)^2 = n_{\circ}^2 - 2\kappa^2 r_-$, leads to

$$h_{\text{e}}^2 = n_{\circ}^2 - 2\kappa^2 r_- - 2r_+, \quad \text{i.e.,} \quad (1-\kappa^2)r_2 + (1+\kappa^2)r_3 = \tfrac{1}{2}(n_{\circ}^2 - h_{\text{e}}^2) \geq 0. \tag{7.8}$$

These are the planes shown in Fig. 7.3 whose intersection with the hyperboloids of the j_z-onion $r_1^2 + r_2^2 = r_3^2 - \frac{1}{4}j_z^2$ are generically nested ellipses. Let

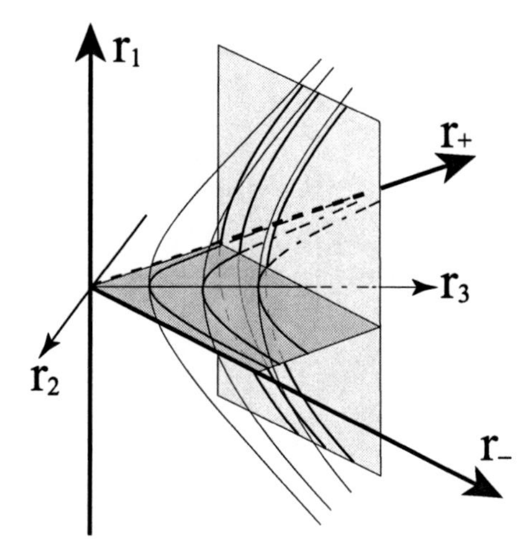

Fig. 7.2. Trajectories of rays in a homogeneous medium, represented in the reduced phase space coordinates $\vec{r}(z) \in \wp|_{\rm ax}$. The intersections of the planes r_+ = constant with the hyperbolic onion (*light lines*) are a set of nested parabolas (*heavy lines*).

$\kappa = 1$, so the planes are simply $0 < r_3 = \frac{1}{4}(n_{\rm o}^2 - h_{\rm e}^2) \leq \frac{1}{4}n_{\rm o}^2$, the trajectories are circles, and the r_3-planes of constant energy rotate rigidly,

$$\begin{pmatrix} r_1(z) \\ r_2(z) \end{pmatrix} = \begin{pmatrix} \cos 2\phi_h(z) & \sin 2\phi_h(z) \\ -\sin 2\phi_h(z) & \cos 2\phi_h(z) \end{pmatrix} \begin{pmatrix} r_1(0) \\ r_2(0) \end{pmatrix}, \quad \phi_h(z) := \frac{z}{h_{\rm e}}, \tag{7.9}$$

as follows from (2.18)–(2.19), with the double of their angle $\phi_h(z)$ of rotation in $\wp$ (cf. Fig. 2.5, which was drawn for $D = 2$-dim guides). The degenerate point-circle at the center of each nested set corresponds to saggital rays which move along the guide in *circular* spirals; these are the *fixed and stable rays* in the guide. When the guide widens, κ decreases and the straight Hamiltonian knife cuts the cone at smaller angles to the r_- axis, until at $\kappa = 0$ we are back in homogeneous medium where the closed ellipses degenerate into open parabolas. •

7.3 Stability of trajectories and critical rays

The previous two examples will be now combined into the more realistic model of an elliptic-profile guide surrounded by vacuum, i.e., an axially-symmetric medium with radial refractive index profile $n_{\rm ev}(|\mathbf{q}|)$ which is bounded from below by 1,

$$n_{\rm ev}(|\mathbf{q}|) := \begin{cases} \sqrt{n_{\rm o}^2 - \kappa^2|\mathbf{q}|^2}, & \kappa|\mathbf{q}| \leq \sqrt{n_{\rm o}^2 - 1}, \\ 1, & \kappa|\mathbf{q}| > \sqrt{n_{\rm o}^2 - 1}, \end{cases} \tag{7.10}$$

$$\Rightarrow \quad h_{\rm ev}(\vec{r})^2 := \begin{cases} n_{\rm o}^2 - 2\kappa^2 r_- - 2r_+, & \kappa^2 r_- \leq \frac{1}{2}n_{\rm o}^2, \\ 1 - 2r_+, & \kappa^2 r_- > \frac{1}{2}n_{\rm o}^2. \end{cases} \tag{7.11}$$

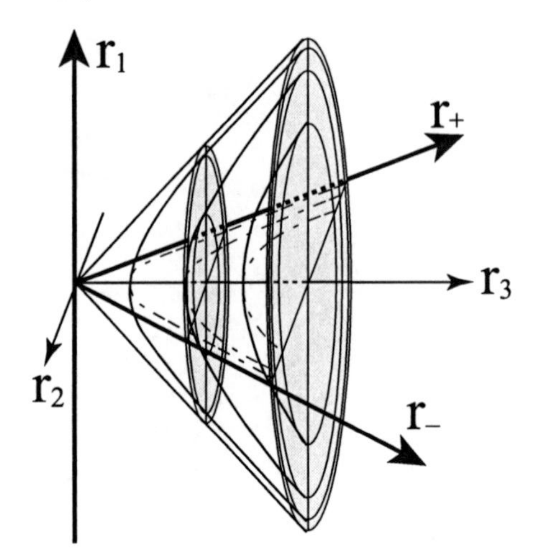

Fig. 7.3. Rays in an elliptic-index-profile guide in the reduced phase space. The trajectories $\vec{r}(z) \in \wp|_{\rm ax}$ are concentric ellipses in the intersection between the hyperbolic onion and parallel vertical planes.

Based on this composite example, we introduce a triplet of 2-dim diagrams to replace the 3-dim cut hyperbolic onions of Figs. 7.2 and 7.3, and which will serve for arbitrary refractive index profiles (and are easier to draw) [70].

The radial refractive index of the guide (7.10) is shown in Fig. 7.4a; Fig. 7.4b is the $r_0 = 0$ plane of Fig. 7.1, with the boundary section $h_{\rm ev} = 0$ of the Hamiltonian knife $[r_+ = \frac{1}{2}(\nu(r_-)^2 - h_{\rm ev}^2)$, where $\nu(r_-) = n_{\rm ev}(|\mathbf{q}|)$ is 1 beyond the point where the guide becomes vacuum], cutting some of the j_z = constant hyperbolas of the onion in this plane. Finally, Fig. 7.4**c** shows the trajectories $\vec{r}(z) \in \wp|_{\rm ax}$ which correspond to these j_z-hyperbolas, projected on the r_-–r_0 plane; the outermost trajectory consists of meridional rays $j_z = 0$, and the $h_{\rm ev} = 0$ section shown is that of highest energy. This guide-in-vacuum is composed of two regions, and all trajectories $\vec{r}(z)$ (for $0 \geq h_{\rm ev} \geq -n_\circ$ and $j_z \in \mathsf{R}$) are ellipse arcs inside the guide [(7.8) for $r_- \leq \frac{1}{2}n_\circ^2/\kappa^2$] and arcs of parabola (7.7), outside it; of course, they are *connected* lines. This divides all the trajectories of the system into bound (complete ellipses), free (either composed of ellipse and parabola when the rays impact the guide, or only a parabola when they miss it), and critical. And among the critical points, we have stable rays (those that advance in a spiral around a circular cylinder within the guide) and unstable ones (connected to the separatrix between bound and free trajectories).

Let us now generalize this example to guide media of arbitrary radial refractive index profiles. Points of tangency between the h = constant cylinders and the j_z = constant hyperboloids occur only on the $r_0 = 0$ plane (see Fig. 7.4b). In the r_+–r_- coordinates, these curves [see (7.4) and (7.5)] have the normal 2-vectors

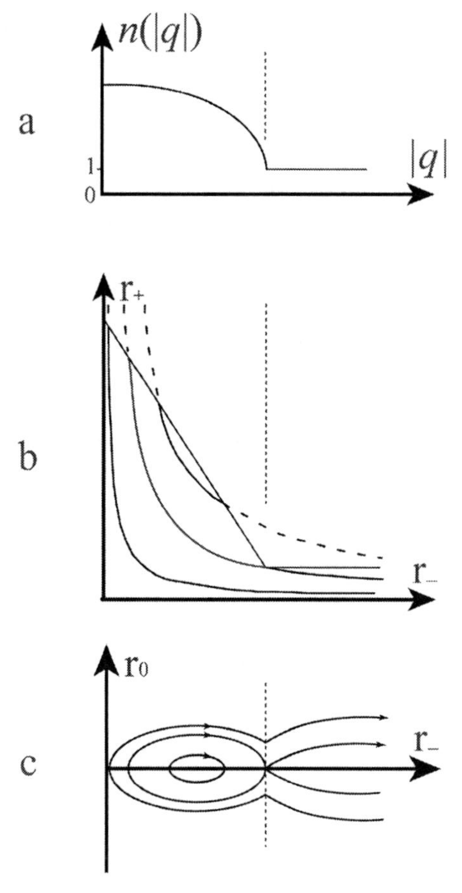

Fig. 7.4. Rays in an axis-symmetric guide of elliptic-index-profile surrounded by vacuum. (**a**) Radial profile of the refractive index of the guide, $n_{\rm ev}(|\mathbf{q}|) = \nu(r_-)$. (**b**) Hamiltonian knife $r_+ = \nu(r_1)^2$ cuts the hyperbolic onion in the plane $r_0 = 0$. The continuous line composed of two straight segments is the upper bound $h_{\rm ev} = 0$ of ray 'energy'; other cuts are r_+-displaced lines parallel to the one shown; dashed lines correspond to unphysical rays in $\wp|_{\rm ax}$. (**c**) Trajectories projected on the r_-–r_0 plane (i.e., $|\mathbf{q}|^2$ *vs.* $\mathbf{p}\cdot\mathbf{q}$). They are ellipses when the rays are trapped in the guide, parabolas for free rays that miss the guide (not shown here), connected ellipse and parabola arcs for free rays that impact and leave the guide, or critical (one unstable shown here).

$$\nabla_{r_\pm} h = \frac{1}{h}\begin{pmatrix} -1 \\ \frac{1}{2}\dfrac{d\nu(r_-)^2}{dr_-}\end{pmatrix}, \quad \nabla_{r_\pm} j_z = \frac{2}{j_z}\begin{pmatrix} r_- \\ r_+ \end{pmatrix}, \qquad \nabla_{r_\pm} := \begin{pmatrix} \partial/\partial r_+ \\ \partial/\partial r_- \end{pmatrix}. \tag{7.12}$$

Tangency occurs when they are collinear, i.e., $(\nabla_{r_\pm} h)\times(\nabla_{r_\pm} j_z) = 0$ on a *critical* point $\vec{r}^{\,\rm cr}$; this implies

$$2r_+^{\rm cr} + r_-^{\rm cr} \left.\frac{d\nu(r_-)^2}{dr_-}\right|_{r_-^{\rm cr}} = 0, \qquad r_0^{\rm cr} = 0. \tag{7.13}$$

While the hyperboloids are fixed geometric objects, the Hamiltonian cylinder depends on the refractive index profile of the guide. The two surfaces can osculate each other from 'outside' (see Figs. 7.4b and **c**) so that the critical point is a *stable* equilibrium point surrounded by closed trajectories; or from the 'inside', and then the critical point is *unstable*, i.e., a saddle point from which four arcs issue, and which will separate the bound from the free trajectories. The criterion for stability given in [70] is in terms of the curvatures: when the function $r_-\,\nu(r_-)$ is *concave downward* at a critical point, the equilibrium is *stable*.

The guide-in-vacuum of Figs. 7.4 has, for every allowed energy $h_{\rm ev} < -1$, one stable equilibrium point consisting of circular-spiral rays trapped in the guide. It also has one unstable critical point at the boundary between the guide and the surrounding vacuum for $-1 \le h_{\rm ev} \le 0$; there, rays spiraling the guide with the *critical* skewnesses $j_z^{\rm cr} = \pm\sqrt{r_+^{\rm cr} r_-^{\rm cr}}$ could escape through any small imperfection in the guide. Rays which in R^3 miss the guide cylinder altogether, correspond to the parabolic trajectories which do not surround the critical points. In the generic case when the radial refractive index profile $n(|\mathbf{q}|)$ has more than one maximum, and/or when discontinuities at cylindrical refracting surfaces are present, the diagrams of Figs. 7.4 are useful to follow the families of rays which can be bound, critical or free at various radii and angular momenta, and to represent their total internal reflection at discontinuities.

7.4 The reduced phase space of axis-symmetric systems

To justify that the 3-dim solid cone manifold of Fig. 7.1 is indeed a kind of *phase* space, as we defined it in Sect. 7.1, we must show that a Lie-Poisson bracket formalism for a Hamiltonian system can be set up on the basis of these *three* coordinates $\vec{r} \in \wp|_{\rm ax}$ (plus some conditions), as we did for the 4-dim optical phase space $(\mathbf{p}, \mathbf{q}) \in \wp$ in Sect. 2.1.

The Poisson bracket between two functions f, g of the coordinates in (7.3), indicated $\vec{r} = (r_1, r_2, r_3) = \{r_+, r_0, r_-\}$, can be obtained from (3.15) with the chain rule,

$$\begin{aligned}\{f, g\} &= \frac{\partial f}{\partial q_x}\frac{\partial g}{\partial p_x} + \frac{\partial f}{\partial q_y}\frac{\partial g}{\partial p_y} - \frac{\partial f}{\partial p_x}\frac{\partial g}{\partial q_x} - \frac{\partial f}{\partial p_y}\frac{\partial g}{\partial q_y} \\ &= 4r_1\left(\frac{\partial f}{\partial r_2}\frac{\partial g}{\partial r_3} - \frac{\partial f}{\partial r_3}\frac{\partial g}{\partial r_2}\right) + r_2\left(\frac{\partial f}{\partial r_3}\frac{\partial g}{\partial r_1} - \frac{\partial f}{\partial r_1}\frac{\partial g}{\partial r_3}\right) \\ &\quad - r_3\left(\frac{\partial f}{\partial r_1}\frac{\partial g}{\partial r_2} - \frac{\partial f}{\partial r_2}\frac{\partial g}{\partial r_1}\right)\end{aligned} \tag{7.14}$$

$$= r_+ \left(\frac{\partial f}{\partial r_0} \frac{\partial g}{\partial r_+} - \frac{\partial f}{\partial r_+} \frac{\partial g}{\partial r_0} \right) + 2r_0 \left(\frac{\partial f}{\partial r_-} \frac{\partial g}{\partial r_+} - \frac{\partial f}{\partial r_+} \frac{\partial g}{\partial r_-} \right)$$
$$+ r_- \left(\frac{\partial f}{\partial r_-} \frac{\partial g}{\partial r_0} - \frac{\partial f}{\partial r_0} \frac{\partial g}{\partial r_-} \right). \tag{7.15}$$

The components of $\vec{r}$ also yield the following Poisson operators:

$$\begin{aligned} \{r_1, \circ\} &= -r_2 \frac{\partial}{\partial r_3} - r_3 \frac{\partial}{\partial r_2}, & \{r_+, \circ\} &= -r_+ \frac{\partial}{\partial r_0} - 2r_0 \frac{\partial}{\partial r_-}, \\ \{r_2, \circ\} &= \ \ r_3 \frac{\partial}{\partial r_1} + r_1 \frac{\partial}{\partial r_3}, & \{r_0, \circ\} &= \ \ r_+ \frac{\partial}{\partial r_+} - r_- \frac{\partial}{\partial r_-}, \\ \{r_3, \circ\} &= -r_1 \frac{\partial}{\partial r_2} + r_2 \frac{\partial}{\partial r_1}; & \{r_-, \circ\} &= \ 2r_0 \frac{\partial}{\partial r_+} + r_- \frac{\partial}{\partial r_0}. \end{aligned} \tag{7.16}$$

And so, the axis-symmetric Hamiltonian function (7.1)–(7.4) begets a Poisson operator on the reduced 'radial' space $\vec{r} \in \wp|_{\rm ax}$, given by

$$\begin{aligned} \{h, \circ\} &= \frac{\partial h}{\partial r_+} \{r_+, \circ\} - \frac{\partial h}{\partial r_-} \{r_-, \circ\} \\ &= \frac{r_0}{h} \frac{\partial \nu^2}{\partial r_-} \frac{\partial}{\partial r_+} + \frac{1}{h} \left(r_+ + \frac{r_-}{2} \frac{\partial \nu^2}{\partial r_-} \right) \frac{\partial}{\partial r_0} + \frac{2r_0}{h} \frac{\partial}{\partial r_-}. \end{aligned} \tag{7.17}$$

The equations for the evolution of trajectories $\vec{r}(z)$ corresponding to the screen moving along the axis of the guide, are thus of the Hamiltonian form (3.26), namely

$$\frac{d}{dz} \vec{r} = -\{h, \vec{r}\}. \tag{7.18}$$

Hamiltonian systems were introduced in Sect. 2.2, defined on a basis of canonically conjugate (*Darboux*) coordinates $(p_i, q_j) \in \mathsf{R}^{2D}$ with their basic Poisson brackets $\{q_i, p_j\} = \delta_{i,j}$, etc. [see (3.18)]. For axis-symmetric systems, we let their place be taken by the three coordinates of $\vec{r}$, (7.3), with the *new* basic Poisson brackets:

$$\begin{aligned} \{r_1, r_2\} &= \ \ r_3, & \{r_0, r_+\} &= \ \ r_+, \\ \{r_2, r_3\} &= -r_1, & \{r_0, r_-\} &= -r_-, \\ \{r_3, r_1\} &= -r_2; & \{r_+, r_-\} &= -2r_0. \end{aligned} \tag{7.19}$$

This set of Poisson brackets *presents* the Lie algebra $\mathsf{so}(2,1)$.[3] They are sometimes called *Berezin* brackets (when serving for the quantization of axially symmetric classical systems). In the next section we will set up a definition of Hamiltonian systems on Lie algebras; below, we remark on some other properties of axial symmetry reduction.

Remark: Conjugate angular and radial subalgebras. The Casimir invariant function of the Lie algebra $\mathsf{so}(2,1)$ is (7.5), namely $-r_1^2 - r_2^2 + r_3^2 =$

[3] For the notation, see the footnote in page 74 and Chap. 6.

$\frac{1}{4}j_z^2$. This is the footprint left by the axial symmetry of rotations by ϕ, $\mathsf{so}(2)$, which was separated from the radial space $\vec{r}$; it consists of the restriction of $\vec{r}$ to the surface of one given hyperboloid, and thus determines the homogeneous space on which $\mathsf{so}(2,1)$ acts. The two algebras are said to be *conjugate* within the larger algebra of quadratic functions of phase space, which is valid for dimension D, $\mathsf{sp}(2D,\mathsf{R}) \supset \mathsf{so}(D)_\phi \oplus \mathsf{so}(2,1)_{\vec{r}}$. •

Remark: Axial symmetry is ubiquitous. Although we have examined here only guides, most image-forming optical arrangements – those that include lenses and free space – have axial symmetry. The analysis, as we shall see in part IV, serves well for axis-symmetric refracting surfaces. Spherical symmetry in higher dimensions D follows suit. •

Remark: Finite radial transformations are canonical. To a function $f(\vec{r})$ corresponds the 'radial' Poisson operator $\{f,\circ\}$ [(7.14)–(7.15) with $\{f,g\}$ replaced by $\{f,\circ\}$] which contains only the partial derivatives $\partial/\partial r_i$ (no ϕ); this will act on the phase space coordinates $(\mathbf{p},\mathbf{q}) \in \mathsf{R}^4$ as

$$\{f(\vec{r}),\circ\}\begin{pmatrix}\mathbf{p}\\ \mathbf{q}\end{pmatrix} = \begin{pmatrix}\partial f/\partial \mathbf{q}\\ -\partial f/\partial \mathbf{p}\end{pmatrix} = \begin{pmatrix}\partial f/\partial r_0\,\mathbf{1} & \partial f/\partial r_-\,\mathbf{1}\\ -\partial f/\partial r_+\,\mathbf{1} & -\partial f/\partial r_0\,\mathbf{1}\end{pmatrix}\begin{pmatrix}\mathbf{p}\\ \mathbf{q}\end{pmatrix}, \tag{7.20}$$

where we have used the chain rule. Successive powers of $\{f,\circ\}$ follow this factorization into a 2×2 unit-block matrix depending only on r_i, acting on $\begin{pmatrix}\mathbf{p}\\ \mathbf{q}\end{pmatrix}$. Thus, the exponential series of the 1-parameter Lie group will have the form of (7.20),

$$\exp(\alpha\{f,\circ\})\begin{pmatrix}\mathbf{p}\\ \mathbf{q}\end{pmatrix} = \mathbf{M}(\alpha,f,\vec{r})\begin{pmatrix}\mathbf{p}\\ \mathbf{q}\end{pmatrix}. \tag{7.21}$$

Of course we cannot in general integrate (7.20) to find a closed form for $\mathbf{M}(\alpha,f,\vec{r})$ [although if we know this matrix, its generating f is easily found from (3.69)], but (7.21) shows that for radial transformations we can completely prescind of $2D$-dim vectors and work with $\begin{pmatrix}p\\ q\end{pmatrix} \in \mathsf{R}^2$. And of course, any transformation generated by Poisson brackets is canonical, as we saw in Chap. 3. •

7.5 Hamiltonian structure on reduced phase space

We have enough elements and examples to enlarge the definition of Hamiltonian systems that we presented in Sect. 2.2 and further refined in Sect. 3.3, which is grounded on the basic set of Poisson brackets (3.16) between the Darboux coordinates $(\mathbf{p},\mathbf{q}) \in \wp \subset \mathsf{R}^{2D}$ of phase space. When the system is axis-symmetric, above we have written everything about the system in terms of the 3 coordinates of a reduced space $\vec{r} \in \wp|_{\text{ax}} \subset \mathsf{R}^3$. We hesitate to call this a *phase space* because these must have even dimensions – in recent work (on image processing and Wigner functions of discrete quantum systems [7, 2]) it has been called *meta*-phase space –, so here we conservatively name it *reduced* phase space, and define the basic Lie brackets to be the Berezin brackets (7.19) for $\mathsf{so}(2,1)$ – or any other useful Lie algebra a. We can now generalize the definition of Hamiltonian system of page 19 to the following:

Definition: *A* **Hamiltonian system** *on a Lie algebra* a *with derivation, is a model of Nature endowed with:*

(i) *An essential set of D coordinates r_i (reduced phase space), whose Lie bracket (3.67) is a presentation of the Lie algebra a of dimension D.*
(ii) *An* **evolution** *parameter $z \in \mathsf{R}$, and a* **generating** *function (Hamiltonian) $h(\vec{r})$.*
(iii) *A set of D* **Hamilton equations** *with the structure of (7.18), extended to beam functions $f(\vec{r})$ through the linearity and derivation properties of the Lie bracket.*

Transformations $\vec{r} \mapsto \vec{r}\,'(\vec{r})$ will be said to be **canonical** *when they preserve the structure of the Lie algebra (3.67).*

It would seem that we have replaced the definition of Hamiltonian system with that of a Lie group acting on its Lie algebra; this is so, except for the Hamiltonian function $h(\vec{r})$, which gives the system a 'dynamical' behavior which may (or may not) correspond to a useful model of Nature. Recalling the examples treated in this and the previous chapters, we see that the Hamiltonian function may relate to the algebra a in several ways. It can be:

(i) an **element** of the algebra (*e.g.* in the elliptic-index profile guide);
(ii) a **Casimir** invariant function of the algebra (*e.g.* the square angular momentum for point rotors and the Maxwell fish-eye); or
(iii) an element of the **covering** algebra (recall Sect. 3.7). In this case fall the Hamiltonians of axis-symmetric optical media with refractive indices of 'arbitrary' radial profile.

7.6 Reconstruction of the ignored coordinate

Once we have found the solutions to the Hamilton equations in reduced phase space (7.18) as $\vec{r}(z)$, we go back to optical space to follow the rays, so we need to reintroduce the hitherto ignored angle ϕ of the symmetry in the x–y plane, and reconstruct the position coordinates

$$\mathbf{q}(z) = \rho(z) \begin{pmatrix} \sin\phi(z) \\ \cos\phi(z) \end{pmatrix}, \qquad \rho := |\mathbf{q}|, \quad \phi := \arctan\frac{q_x}{q_y}. \tag{7.22}$$

We can compute directly that the Poisson bracket between the angle $\phi(\mathbf{q})$ and the skewness j_z is

$$\{\phi, j_z\} = \left\{\arctan\frac{q_x}{q_y}, \mathbf{q}\times\mathbf{p}\right\} = 1, \tag{7.23}$$

as if angle and angular momentum were canonically conjugate coordinates of a phase space, as q and p are.

To show that this is the case, we search for a canonical transformation between the Cartesian Darboux coordinates of phase space on the screen (p_x, p_y, q_x, q_y) [where (p_x, q_x) and (p_y, q_y) are two independent canonically conjugate pairs], and four coordinates $(p_\rho, j_z, \rho, \phi)$, where (p_ρ, ρ) and (j_z, ϕ) be also conjugate pairs under the same Poisson bracket; only the radius $\rho = |\mathbf{q}|$ still needs *its* canonical conjugate 'radial' momentum – some p_ρ. We take the hint of classical mechanics to decompose the square momentum into its radial and angular parts, and write

$$p^2 = p_\rho^2 + j_z^2/\rho^2, \quad p_\rho := \sqrt{p^2 - j_z^2/\rho^2} = \sqrt{|\mathbf{p}|^2 - \frac{(\mathbf{q}\times\mathbf{p})^2}{|\mathbf{q}|^2}}, \quad \{\rho, p_\rho\} = 1. \tag{7.24}$$

Poisson brackets between one (p_ρ, ρ) and one (j_z, ϕ) are zero, so $(p_\rho, j_z, \rho, \phi)$ is also a canonical set of (locally R^4) coordinates for the Hamiltonian system. The Hamiltonian function (7.1) in the new radius-and-angle coordinates is

$$h = -\sqrt{n(\rho)^2 - p_\rho^2 - j_z^2/\rho^2}. \tag{7.25}$$

Hence, we can write the missing Hamilton equation for $\phi(z)$: as

$$\frac{d\phi(z)}{dz} = -\{h, \phi\} = \frac{\partial h}{\partial j_z} = -\frac{j_z}{h}\frac{1}{\rho(z)^2} = -\frac{j_z}{h}\frac{1}{2r_-(z)}. \tag{7.26}$$

So once we have a trajectory $\vec{r}(z) \in \wp|_{\mathrm{ax}}$ in reduced phase space, the ray evolving along the z-axis of optical 3-dim space will be $\mathbf{q}(z)$ as given in (7.22) with

$$\rho(z) = \sqrt{2r_-(z)}, \qquad \phi(z) = -\frac{j_z}{h}\int^z \frac{dz'}{2r_-(z')}. \tag{7.27}$$

where j_z and h are constants.

Remark: Poisson operators in $(p_\rho, j_z, \rho, \phi)$ coordinates. The Poisson operator of a function $f(\mathbf{p}, \mathbf{q}) = F(p_\rho, j_z, \rho, \phi)$ was defined in (3.25). After using the chain rule, we can bring it to the form

$$\{f, \circ\} = \frac{\partial F}{\partial \rho}\frac{\partial}{\partial p_\rho} - \frac{\partial F}{\partial p_\rho}\frac{\partial}{\partial \rho} + \frac{\partial F}{\partial \phi}\frac{\partial}{\partial j_z} - \frac{\partial F}{\partial j_z}\frac{\partial}{\partial \phi} =: \{F, \circ\}. \tag{7.28}$$

So when F is independent of ϕ, then j_z is a constant of the motion that foliates space into surfaces which are respected by the Hamiltonian flow. This form of the Poisson bracket is useful when the Hamiltonian function does depend on the angular momentum of the ray, as in the case of particle accelerator design with coaxial magnetic fields. •

Remark: Angle coordinate and 'operator'. The change of variables in (7.22) suggests *prima facie* that out of the three generators $\{\vec{r}, \circ\}$ of the $\mathsf{so}(2,1)$ algebra (7.19) and one extra number j_z, we could produce a 4-dim algebra of phase space translations. Not so, *except* in this realization of the algebras by Poisson brackets, where multiplication of functions is commutative. When phase space translations do *not* commute (as in quantum mechanics or wave optics), the square root and arc tangent functions in (7.22) and (7.24) cannot be extended unambiguously. *Square* radial momentum can be handled, but not a 'phase' or 'angle' operator $\widehat{\phi}$. Much work has been expended on these since Carruthers and Nieto [29] called attention to their fundamental problems. Fortunately, various ways around the main difficulty are practiced in quantum optics and work in experiment [123]. Geometric optics is free of all this, *ab initio*. •

8 Anisotropic optical media

Certain natural crystals, stressed plastics and organic compounds are anisotropic optical materials that exhibit birefringence. This is a physical phenomenon where light with different directions of propagation is transmitted differently, in a medium that has fixed, distinguished axes. The medium splits unpolarized beams into an *ordinary* beam that propagates as if the medium were isotropic, and an *extraordinary* beam, which manifests the anisotropy of the medium. Since physical anisotropy is of electromagnetic origin, the proper treatment follows from the Maxwell wave equations, whose small-wavelength limit leads to the geometric formulation of anisotropic optics given in the book by Kravtsov and Orlov, [83].

Here we shall present a model of anisotropic geometric optics based on a Hamiltonian formulation where light rays are still abstracted to be lines in space (i.e., no polarization vectors), but where we allow the refractive index – hitherto a scalar function of the position coordinates only – to depend also on *direction* of the ray in the medium. The corresponding mechanical system (according to Sect. 1.5) would have a potential dependent on the velocity vector. By the end of the chapter we shall see the further restrictions which Nature will impose on the Hamiltonian model, but the proper place to start is again with the two postulates of geometric optics that we introduced in Sects. 1.1 and 1.2.

8.1 Direction and momentum of rays

The first Hamilton equation is geometric, and defines the momentum 3-vector of a ray, $\vec{p}(s)$ [see (1.1)], as tangent to the ray trajectory $\vec{q}(s)$. The scale of the momentum vector [determined from the Hamiltonian in (1.11)] gives the Descartes sphere of ray directions a radius $|\vec{p}| = n(\vec{q})$ for each point in the medium. The second Hamilton equation contains the dynamics of the system (1.5); it states that the changes in the momentum vector are due to the *space gradient* $\nabla_{\vec{q}}\, n(\vec{q})$ of the refractive index. The two equations must continue to hold for anisotropic media if the system is to be Hamiltonian.

We thus let the refractive index of the medium $n(\vec{q}, \dot{\vec{q}})$ depend not only on the point $\vec{q}$ in the medium, but also on the *direction* of the light rays,

$$\dot{\vec{q}} := \frac{d\vec{q}(s)}{ds}, \qquad |\dot{\vec{q}}| = 1. \tag{8.1}$$

A vector embodying this anisotropy must be inserted into the Hamilton equations, which are themselves vectorial. The extra vector now available is the *gradient of direction* of the refractive index, namely

$$\vec{A}(\vec{q}, \dot{\vec{q}}) := \nabla_{\dot{\vec{q}}}\, n(\vec{q}, \dot{\vec{q}})\Big|_{|\dot{\vec{q}}|=1} = \left(\frac{\partial n}{\partial \dot{q}_x}, \frac{\partial n}{\partial \dot{q}_y}, \frac{\partial n}{\partial \dot{q}_z}\right)^{\top}\Bigg|_{|\dot{\vec{q}}|=1}. \tag{8.2}$$

When the medium is isotropic, $\vec{A}$ is zero and we recover previous results. But when not, $\vec{A}(\dot{\vec{q}})$ is tangent to the unit sphere of directions $\dot{\vec{q}}$, i.e.,

$$\dot{\vec{q}} \perp \vec{A} \quad \Rightarrow \quad \dot{\vec{q}} \cdot \frac{\partial n}{\partial \dot{\vec{q}}} = |\dot{\vec{q}}|\, \frac{\partial n}{\partial |\dot{\vec{q}}|} = 0. \tag{8.3}$$

We now **postulate** that the *geometric* Hamilton equation (1.3) be modified for anisotropic media, through redefining *optical momentum* as

$$\vec{p} := n(\vec{q}, \dot{\vec{q}})\, \dot{\vec{q}} + \vec{A}, \qquad \text{(in anisotropic media)}. \tag{8.4}$$

This equation, rewritten as $n\dot{\vec{q}} = \vec{p} - \vec{A}$, is a vector with norm

$$|\vec{p} - \vec{A}| = n(\vec{q}, \dot{\vec{q}}). \tag{8.5}$$

Thus, while the direction vector $\dot{\vec{q}}$ ranges over the sphere, the momentum vector $\vec{p}$ will range over a closed surface determined by (8.5), that we call the *Descartes ovoïd* of the medium. See Fig. 8.1. This takes the place of the Descartes sphere in momentum space (1.11), whose radius also depends on the point of the medium, $\vec{q}$. We distinguish this surface from another important surface, $n(\vec{q}, \dot{\vec{q}})\, \dot{\vec{q}}$, which we call *ray surface*.

8.2 Hamilton equations for anisotropic media

With the redefined optical momentum in (8.4), the two postulates of Sects. 1.1 and 1.2 [(1.3) and (1.5)] lead to the following form of the Hamilton equations for rays in anisotropic media:

$$\frac{d\vec{q}(s)}{ds} = \dot{\vec{q}}(s) = \frac{\partial H_{\text{A}}}{\partial \vec{p}}, \qquad \frac{d\vec{p}(s)}{ds} = \nabla_{\vec{q}}\, n(\vec{q}, \dot{\vec{q}}) = -\frac{\partial H_{\text{A}}}{\partial \vec{q}}, \tag{8.6}$$

with the anisotropic Hamiltonian function [cf. (1.6)]:

$$H_{\text{A}}(\vec{p}, \vec{q}) := |\vec{p} - \vec{A}| - n(\vec{q}, \dot{\vec{q}}) = 0. \tag{8.7}$$

At surfaces of discontinuity of the refractive index, we proceed from the finite-difference form of the Hamilton equations (1.2) and (1.4), with no

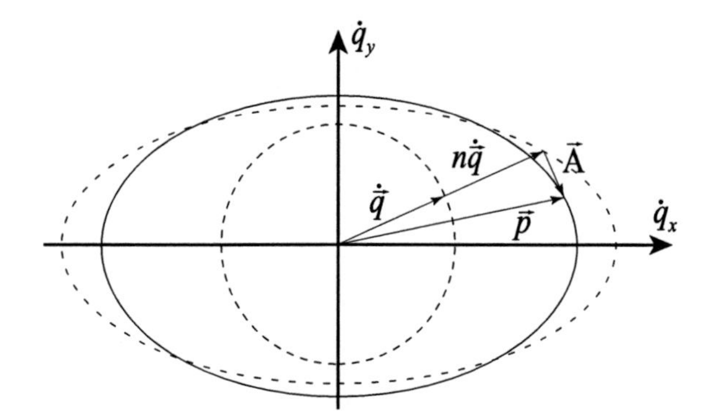

Fig. 8.1. The ray direction vector $\dot{\vec{q}}$ ranges on a unit sphere (*dotted circle*); in an anisotropic medium, where the refractive index $n(\vec{q}, \dot{\vec{q}})$ depends on position and direction, $n\dot{\vec{q}}$ draws out the *ray surface* (*dashed closed curve*), and the momentum vector $\vec{p} = n\dot{\vec{q}} + \vec{A}(\dot{\vec{q}})$ ranges over the Descartes ovoïd (*continuous line*).

change except for the redefinition of $\vec{p}$ in (8.4). We apply the same tactics of Sect. 1.4 to describe finite refraction at surfaces $S(\vec{q}) = 0$ between distinct anisotropic media n and n'. The conservation of the momentum component which is tangent to the surface, i.e., normal to $\vec{\Sigma} = \nabla_{\vec{q}} S|_{S=0}$ [see (1.9)], leads to the vector form of the Ibn Sahl–Descartes law of sines, $\Delta\vec{p}\times\vec{\Sigma} = 0$, with the redefined $\vec{p}$. In Fig. 8.2 we draw the new situation for plane optics (cf. Fig. 4.1 *right*), with Descartes ovoïds in both media. To find graphically the refracted ray, we proceed as follows:

1. Given the direction of the incoming ray $\dot{\vec{q}}$, we find its intersection with the ray surface $n\,\dot{\vec{q}}$; there we draw its perpendicular $\vec{A}$, so that $\vec{p} = n\,\dot{\vec{q}} + \vec{A}$ is a point on the Descartes ovoïd.
2. We project $\vec{p}$ on the plane tangent to the refracting surface. This defines $\vec{\bar{p}}$, the *conserved* momentum component.
3. We reproduce the same component $\vec{\bar{p}}$ in the second medium n', raising a perpendicular to a second momentum vector $\vec{p}'$ on the Descartes ovoïd of the second medium.
4. On this second ovoïd, we find a right triangle with hypotenuse $\vec{p}$, and right vertex on the ray surface $n'\dot{\vec{q}}'$ of the second medium; this yields the outgoing ray direction, $\dot{\vec{q}}'$.

This recipe is applicable for models of arbitrary dimension, but is most useful in 2-dim diagrams.

8.3 Angular dependences of the refractive index

Thus far we have not imposed any restriction on the dependence of the refractive index on ray direction; the ray surface $n(\dot{\vec{q}})$ and its Descartes ovoïd

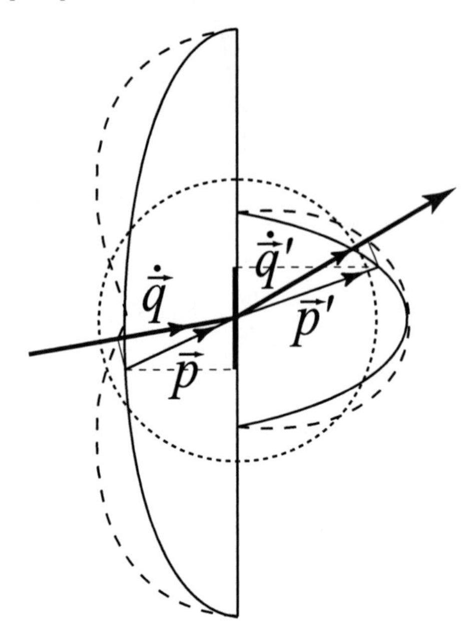

Fig. 8.2. Refraction of rays at the interface between two anisotropic plane media ($D = 1$). Initial and final rays (*thick lines*) have directions $\dot{\vec{q}}$ and $\dot{\vec{q}}'$ on the dotted ray direction circle, and the local tangent to the interface is the vertical line. The ray surfaces $n(\dot{\vec{q}})\,\dot{\vec{q}}$ are indicated by dashed curves, and the Descartes ovoïds – where lie the momentum vectors $\vec{p} = n(\dot{\vec{q}})\,\dot{\vec{q}} + \vec{A}(\dot{\vec{q}})$ – by solid lines. The projections of momentum on the plane tangent to the surface is the component that is conserved under refraction (*thick lines*).

could have any *star* shape. (To concentrate here on the effects of anisotropy, we shall simplify media to be homogeneous.)

Example: Dipolar anisotropic media. First we describe media where the anisotropy is a constant vector $\vec{d}$,[1] so the refractive index depends linearly on ray direction $\dot{\vec{q}}$, namely

$$n^{(1)}(\dot{\vec{q}}) = n_0 + n_1(\dot{\vec{q}}), \qquad n_1(\dot{\vec{q}}) = \vec{d}\cdot\dot{\vec{q}}. \tag{8.8}$$

Using the nomenclature of electrostatics, the scalar part n_0 would be that of a *monopole* medium, and we can call n_1 the *dipole* index of the medium. The anisotropy vector (8.2), being tangent to the direction sphere, can be found subtracting the radial part from the 3-dim direction gradient. Using dyadic notation, we write

$$\begin{aligned} \vec{A}_1(\dot{\vec{q}}) &= \left(\nabla_{\dot{\vec{q}}} - \dot{\vec{q}}\dot{\vec{q}}\cdot\nabla_{\dot{\vec{q}}}\right) n_{(1)}(\dot{\vec{q}}) = (1 - \dot{\vec{q}}\dot{\vec{q}}\cdot\,)\,\vec{d} \\ &= \vec{d} - n_1(\dot{\vec{q}})\,\dot{\vec{q}} = (\dot{\vec{q}}\times\vec{d})\times\dot{\vec{q}}. \end{aligned} \tag{8.9}$$

[1] Actually, we *invent* such media – because Nature does not provide them.

The momentum vector is then $\vec{p} = n_0\dot{\vec{q}} + \vec{d}$, i.e., ray direction plus the constant vector. The Descartes ovoïd is thus an off-center sphere. •

Let the refractive index of the medium be *quadrupolar*, i.e., its ray surface be an ellipsoid characterized by a symmetric and traceless matrix $\mathbf{Q} = ||Q_{i,j}||$,

$$n^{(2)}(\dot{\vec{q}}) = n_0 + n_2(\dot{\vec{q}}), \qquad n_2(\dot{\vec{q}}) = \sum_{i,j} Q_{i,j}\,\dot{q}_i\dot{q}_j = \dot{\vec{q}}\cdot\mathbf{Q}\dot{\vec{q}}. \tag{8.10}$$

The anisotropy vector of this medium is found from the general relation $\partial(\vec{v}\cdot\mathbf{Q}\vec{v})/\partial\vec{v} = 2\mathbf{Q}\vec{v}$ for any vector $\vec{v}$; it is

$$\vec{A}_2(\dot{\vec{q}}) = 2(1 - \dot{\vec{q}}\dot{\vec{q}}\cdot\,)\mathbf{Q}\dot{\vec{q}} = 2\Big(\mathbf{Q} - n_2(\dot{\vec{q}})\Big)\,\dot{\vec{q}}. \tag{8.11}$$

The momentum vector in quadrupole media is thus

$$\vec{p} = \left(n_0 + 2\mathbf{Q} - \dot{\vec{q}}\cdot\mathbf{Q}\dot{\vec{q}}\right)\dot{\vec{q}}, \tag{8.12}$$

and the Descartes ovoïd is $\vec{p}^{\,2} = n(\dot{\vec{q}})^2 - \vec{A}_2(\dot{\vec{q}})^2$.

Remark: Quadrupole media contain only two parameters. The number of independent parameters of the symmetric 3×3 quadrupole matrix $\mathbf{Q}$ in (8.10) appears to be five (the trace belongs to the monopole part). However, the number of *essential* parameters is *two*, because through rotation of the axes (by three angles), we can always bring the symmetric matrix to the diagonal form $\mathbf{Q} = \mathrm{diag}\,(Q_x, Q_y, Q_z)$, with $Q_x + Q_y + Q_z = 0$. •

Example: Uniaxial quadrupole media. A closed result can be found for the particular but important case of *uniaxial* quadrupole media. These media exhibit anisotropy only along one axis, their ray surfaces are revolution elipsoids, and their defining matrix (after diagonalization) has the generic form $\mathbf{Q} = \mathrm{diag}\,(\nu, \nu, -2\nu)$. When the ray direction is given in spherical coordinates with colatitude θ, the refractive index (8.10) is

$$n_{\mathrm{u}}^{(2)} = n_0 + \nu\,(\dot{q}_x^2 + \dot{q}_y^2) - 2\nu\dot{q}_z^2 = n_0 + \nu - 3\nu\cos^2\theta. \tag{8.13}$$

The momentum 3-vector can then be written explicitly in these coordinates as

$$\begin{pmatrix} \sqrt{p_x^2 + p_y^2} \\ p_z \end{pmatrix} = \begin{pmatrix} (n_0 + 4\nu - 3\nu\sin^2\theta)\sin\theta \\ (n_0 - 2\nu - 3\nu\sin^2\theta)\cos\theta \end{pmatrix}, \tag{8.14}$$

so the Descartes ovoïd differs from an ellipsoid by ν^2 terms; it is oblate in the z-direction when $\nu > 0$, and prolate when $\nu < 0$. •

8.4 Comparison with Maxwellian anisotropy

The two postulates of geometric optics that opened Chap. 1, with their present reformulation for anisotropic media, are insufficient to account for the two rays, the ordinary and extraoradinary, that must coexist in physical anisotropic media. (As we noted in Sect. 4.3, these postulates are also insufficient to predict the reflected ray at a refracting surface.) This is not necessarily a shortcoming of the geometric theory, but must be borne in mind. Nature is stricter on this matter: only quadrupolar anisotropy occurs. In this closing section we recall the physical reasons that restrict our Hamiltonian presentation of anisotropy.

The electric vector field in a medium without currents nor charges is commonly factored into a vector independent of time, times a complex phase

$$\vec{E}(\vec{q},t) = \vec{e}(\vec{q}) \exp i\Big(k\Phi(\vec{q}) - \omega t\Big), \tag{8.15}$$

where the wavenumber is $k = 2\pi/\lambda$ (λ is the wavelength), $\omega = ck$ is the angular frequency, and $\Phi(\vec{q})$ is the *eikonal* function [58, Sect. 9–8]; and similarly for the magnetic field vector $\vec{H}(\vec{q},t)$. The surfaces $\Phi(\vec{q})$ = constant are the *geometric wavefronts* and their normal *wave vector* is

$$\vec{p} := \nabla_{\vec{q}}\, \Phi(\vec{q}), \quad |\vec{p}| =: n_p = c/v_p, \qquad \text{(in the wave model).} \tag{8.16}$$

This we identify with the momentum vector $\vec{p}$ proposed in (8.4); its magnitude n_p is the *phase index* of the medium, which in turn is related to the *phase velocity* of the waves in the field, v_p. In birefringent media and for the extraordinary ray, this velocity is different from the *group* velocity (of energy and information transport) $v_r = c/n_r$, where n_r is the *ray index*. Energy propagates along the direction of the Poynting vector $\vec{S} = (c/4\pi)\vec{E}\times\vec{H}$, which coincides with the ray direction $\dot{\vec{q}}$ in (8.1).

In Fig. 8.3 we show the vectors of the Hamilton and Maxwell theories. The angle between the momentum and direction vectors, $\vec{p}$ and $\dot{\vec{q}}$, is the same as between the displacement and electric vectors. The latter are bound through the *material relation* $\vec{D} = \hat{\varepsilon}\vec{E}$ by the *permitivity tensor* (matrix) $\hat{\varepsilon} = |\varepsilon_{i,j}|$ of the physical medium. (The induction and magnetic vectors are bound by

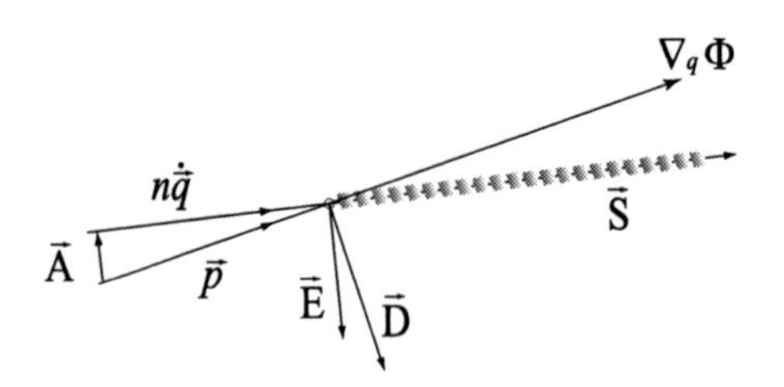

Fig. 8.3. Hamiltonian vectors (*left*) and electromagnetic vectors (*right*) in an anisotropic medium, where the electric and displacement vectors, $\vec{E}$ and $\vec{D}$, form an angle. (The magnetic and induction vectors point out of the page.) The wavefronts $\nabla_{\vec{q}}\, \Phi(\vec{q})$ = constant are perpendicular to the momentum vector $\vec{p}$, but wave packets and energy move along the Poynting vector $\vec{S}$, in the direction of $\dot{\vec{q}}$.

$\vec{B} = \mu\vec{H}$, but the magnetic permitivity μ is practically a constant.) The phase and ray refractive indices are reciprocal and fulfill $n_p\, n_r = \mu|\hat{\varepsilon}\vec{e}\,|/|\vec{e}\,|$. Finally, the permitivity tensor is symmetric (due to energy conservation) so the medium can be rotated to its principal axes, where

$$\hat{\varepsilon} = \mathrm{diag}\,(\varepsilon_0 + \eta_x,\ \varepsilon_0 + \eta_y,\ \varepsilon_0 + \eta_z) = \varepsilon_0 \hat{1} + \hat{\eta}, \tag{8.17}$$

$$\eta_x + \eta_y + \eta_z = 0, \quad |\eta_i| \ll \varepsilon_0. \tag{8.18}$$

In these components we can determine the ray refractive index n_r as a function of ray direction $|\dot{\vec{q}}| = 1$ by the *Fresnel ray equation*, which reads

$$\frac{\dot{q}_x^2}{n_r^2 - \mu(\varepsilon_0 + \eta_x)} + \frac{\dot{q}_y^2}{n_r^2 - \mu(\varepsilon_0 + \eta_y)} + \frac{\dot{q}_z^2}{n_r^2 - \mu(\varepsilon_0 + \eta_z)} = 0. \tag{8.19}$$

This equation is quadratic in n_r^2 and has two solution branches over the sphere.

Remark: Uniaxial media. Uniaxial materials aligned with the z-axis exhibit a permitivity tensor (8.18) with a single free parameter $\eta = \eta_x = \eta_y = -\frac{1}{2}\eta_z$. One can show directly [120] that the refractive indices for the ordinary and extraordinary rays are

$$n_{\mathrm{ord}} = \sqrt{\mu\,(\varepsilon_0 + \eta)}, \qquad n_{\mathrm{ext}}(\theta) = n_{\mathrm{ord}} - (3\mu\eta/2n_{\mathrm{ord}})\sin^2\theta, \tag{8.20}$$

respectively. The two rays propagate equally along the z-axis, and differ most at right angles to it. We compare (8.20) and (8.13) to relate the generic parameters in the latter with the physical ones in the former through

$$n_0 = \sqrt{\mu(\varepsilon_0 - 2\eta)} \approx (1 - \eta/\varepsilon_0)\sqrt{\mu\varepsilon_0}, \qquad \nu = -\mu\eta/2n_{\mathrm{ord}} \approx -\tfrac{1}{2}\eta\sqrt{\mu/\varepsilon_0}. \tag{8.21}$$

For example, at a wavelength of $\lambda = 404.7$ nm and $\mu = 1$, quarz exhibits ray indices $n_{\mathrm{ord}} = 1.557\,16$ and $n_{\mathrm{ext}} = 1.566\,71$. Comparing with (8.21), we see this corresponds to monopolar and quadrupolar refractive indices with values $n_0 = 1.563\,53$ and $\nu = 0.003\,18$. The approximation is thus of order $\nu/n_{\mathrm{ord}} \approx 2 \times 10^{-3}$. •

Remark: Why only quadrupoles? The spirit of special relativity suggests that we use the metric tensor $g^{i,j}$ to write the trajectory element as

$$c\,dt = \sqrt{g^{i,j}\,dq_i\,dq_j} =: n\,ds \tag{8.22}$$

(with summation implied for repeated indices). Dividing by ds and using the definition of ray direction in (8.1), we find the *square* of the refractive index to be

$$n(\dot{\vec{q}})^2 = g^{i,j}\,\dot{q}_i\,\dot{q}_j. \tag{8.23}$$

The metric tensor $g^{i,j}$ is symmetric and contains six independent quantities. When the off-diagonal elements of $g^{i,j}$ are *small* (and they are in quarz for example), its trace will approximate the monopole refractive index as $n_0 \approx \sqrt{g^{i,i}}$, and [in (8.10)] $Q_{i,j} \approx (1/2n_0)g^{i,j}$, so only quadrupolar anisotropy is allowed. This is a hybrid argument which nevertheless agrees with the observed restriction economically. •

B Euclidean optical models

The geometric and wave models of light in an optically homogeneous and isotropic medium are contained in a more general model based on the Euclidean group – together with many other models. In this appendix we shall connect these two models through a process which is similar in several aspects to the quantization of classical mechanics, but based on the Euclidean group of rotations and translations which properly *defines* optical homogeneity and isotropy. We call this the *Euclidean* model of optics; and to the strategy of connection we shall refer as the *wavization* of (the global, '4π' régimes of) geometric to wave optics. We shall use the results of Chap. 5 on the structure of the Euclidean Lie group $\mathsf{ISO}(3)$ and its representation by 4×4 matrices in Sect. 5.5.

B.1 Manifolds of rays, planes and frames

In a 3-dim homogeneous medium, the geometric optical model represents rays of light by oriented straight lines, all of which can be obtained by rotations and translations of the standard ray along the $+z$-axis. This we call the *standard object* of the geometrical model, denoted $\mathcal{O}_{\rm G}$; $\mathcal{O}_{\rm G}$ in turn is determined by the *symmetry* subgroup $\mathsf{T}_z \otimes \mathsf{SO}(2)_z \subset \mathsf{ISO}(3)$ of translations along and rotations around the z-axis. The beam functions of geometric optics (see Sect. 3.1) will thus depend on the parameters of the Euclidean transformations that carry $\mathcal{O}_{\rm G}$ to all rays, *modulo* those transformations which leave $\mathcal{O}_{\rm G}$ invariant. We recall that the number of parameters in $\mathsf{ISO}(3)$ is 6, while that of $\mathsf{T}_z \otimes \mathsf{SO}(2)_z$ is 2, so we reproduce the 4 coordinates of phase space $\wp$.

Remark: Signal and 'polarization' optics. There can be weaker versions of the geometric model that have smaller symmetry groups. The translation invariance is broken when the standard ray carries a signal $f(z)$, and leaves us with a model whose standard object has only $\mathsf{SO}(2)_z$-symmetry, that we can call *signal* geometric optics. On the other hand, when the rotation symmetry $\mathsf{SO}(2)_z$ of the standard object is broken, such as a standard ray with a 'flag' vector along the x axis, which we can imagine as an oriented ribbon, we define a model of geometric optics with *polarization*[1] based on T_z. When the translation distance and rotation angle in $\mathsf{T}_z \otimes \mathsf{SO}(2)_z$ depend linearly on the same parameter z, the standard object will be a ribbon screw. In all these cases, the enlarged manifolds of 'rays' are of dimension 5. •

The standard object for a model of scalar polychromatic wave optics is the oriented x–y plane that represents a Dirac-δ 'wavefront' pulse along the $+z$-axis; we denote it by $\mathcal{O}_{\mathrm{W}}$. Any wavefield in R^3-space can be expanded as a continuous complex superposition of such Dirac wavefronts, translated and rotated to all directions on the sphere $\mathcal{S}^2$, modulo the 3-parameter symmetry group $\mathsf{ISO}(2)_{xy}$ of translations and rotations within the x–y plane. This $6-3=3$-parameter manifold we denote by $\mathcal{W}$, noting its correspondence through Fourier analysis with the usual wavenumber space R^3. We can moreover *enlarge* the $\mathsf{ISO}(2)_{xy}$ invariance symmetry by the *covariance* symmetry T_z of the standard object with the phase e^{ikz}. This restricts the polychromatic wavefields into their monochromatic *Helmholtz* components $\mathcal{W}_k$ of wavenumbers $k \in \mathsf{R}$ (the *wavelength* is $\lambda := 2\pi/k \neq 0$), each of dimension 2.

Remark: Periodic and 'polarized' wave optics. Since $\mathsf{ISO}(2)_{xy} \otimes \mathsf{T}_z$ has many subgroups, there can be many 'stronger' models of polychromatic and Helmholtz optics. When T_z is restricted to discrete translations, so the standard object is a *stack* of equally spaced Dirac-wavefront planes, we represent wavefields composed of plane periodic signals. When $\mathsf{SO}(2)_z$ symmetry is violated we have (non-physical) versions of polarization optics. •

There can be as many models of 'optics' as subgroups $\mathsf{H} \subset \mathsf{ISO}(3)$ in the Euclidean group. For this reason we call $\mathsf{ISO}(3)$ the *mother* group of the sibling theories of geometric and wave optics, amongst others. The *strongest* model of this Euclidean family corresponds to the smallest symmetry $\mathsf{H} = \{h_o\}$ (the identity) of a standard Euclidean *frame* $\mathcal{O}_{\mathrm{E}}$. This standard frame is shown in Fig. B.1; its position is the origin $\vec{0} \in \mathsf{R}^3$ and its orientation is coincident with the (x, y, z) axes. From it, we obtain all frames through Euclidean transformations $\mathcal{E}(\vec{\alpha}, \mathbf{R})$ [see (5.41)–(5.45)]. This is the 6-dim manifold of $\mathsf{ISO}(3)$ itself, $\mathcal{E} = \mathsf{R}^3 \times \mathcal{S}^3/\mathcal{Z}_2$.[2] In the following sections, our strategy to subduce the geometric and wave models out of the Euclidean model, will consist in building their standard objects, $\mathcal{O}_{\mathrm{G}}$ and $\mathcal{O}_{\mathrm{W}}$, as *equivalence sets*

[1] This model is **different** from the physical, electromagnetic polarization theory, which involves both electric and magnetic fields. The polarization-flag model has some mathematical interest when *refraction* is introduced by assuming the conservation of the tangential momentum and flag vector components at warped interfaces between two media. Aberration expansions in this model have been examined in [159].

[2] Recall that the Euclidean group is a semidirect product; the R^3 coordinates correspond to translations, while the manifold of rotations $\mathsf{SO}(3)$ is the 3-sphere $\mathcal{S}^3 = \mathcal{S}^1 \times \mathcal{S}^2$, modulo the $\mathcal{Z}_2$-equivalence of the two rotations by $\pm\pi$ over every pair of antiparallel axes (see in Sect. 6.1 the remark on page 93).

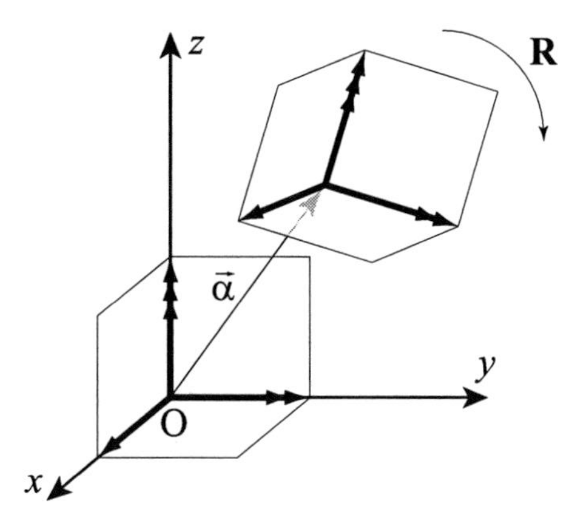

Fig. B.1. The standard frame $\mathcal{O}_{\rm E}$ (3 thick marked vectors at the origin), and the generic frame $(\vec{\alpha}, \mathbf{R})$ obtained through translation $\vec{\alpha}$ and rotation $\mathbf{R}$.

of frames $\mathcal{O}_{\rm H} := \{h : \mathcal{O}_{\rm E}\}$, $h \in \mathsf{H}$, determined by the symmetry subgroup $\mathsf{H} \subset \mathsf{ISO}(3)$.

The 4-dim manifold of straight lines $\wp$ in geometric optics, the 3-dim manifold of planes $\mathcal{W}$ in wave optics, and the 6-dim manifold of frames $\mathcal{E}$, are *vector bundles* [68]. We end this section remarking on their structure, to show the connection of their base spaces with the Descartes sphere $\mathcal{S}^2$ of momentum, and the way in which their fibers of position are complementary.

Remark: The manifold of rays $\wp$ in geometric optics. Consider first the manifold $\wp$ of oriented straight lines in space. With the tools of the mind we sort out all those lines that are oriented in a chosen direction and identify them by a point $\vec{p}$ on the Descartes sphere $\mathcal{S}^2$, which is the *base space* of the bundle (a projection $\pi : \wp \mapsto \mathcal{S}^2$). The inverse image of one such a point, $\pi^{-1}(\vec{p})$, is the set of paralell rays that can be brought one onto another by translations $\mathsf{T}_2^{(\perp\vec{p})}$ within the plane perpendicular to $\vec{p}$, such as shown in Fig. B.2a. $\mathsf{T}_2^{(\perp\vec{p})}$ is the typical *fiber* of $\wp$, and the *local screen*, because each is associated to *its* set of rays parallel to $\vec{p}$; two rays which are not paralell will have their screens oriented differently. The definition of a bundle [68] includes a group action (here rotations) which is effective on the base space $\vec{p} \in \mathcal{S}^2$, and in the fibers depends on their $\vec{p}$ [89, 28]. •

Remark: The manifold of planes $\mathcal{W}$ in wave optics. The manifold $\mathcal{W}$ of oriented Dirac-wavefront planes also has the structure of a vector bundle. Again, these are sorted by direction $\vec{p}$ on the Descartes sphere. But now the typical fiber [i.e., the inverse image $\pi^{-1}(\vec{p})$ of such a point], is a space-filling set of paralell planes: the 1-dim group and vector space $\mathsf{T}_1^{(\vec{p})}$, as shown in Fig. B.2b. Comparing the two models, we see that there is an obvious dimensional mismatch between the typical fibers of the geometric and polychromatic-wave models; thus, a 1:1 strict correspondence between beams and fields is excluded. But further, on the fibers of $\mathcal{W}$ one can perform Fourier

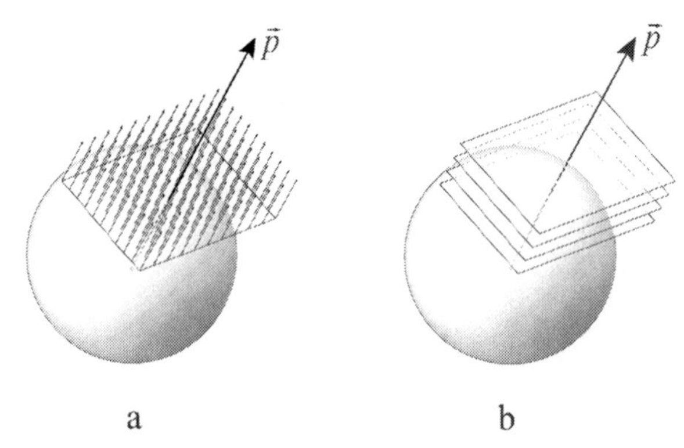

Fig. B.2. Geometric and wave optical models are distinct (Euclidean vector bundles), whose common base space is the Descartes sphere of momenta $\vec{p} \in \mathcal{S}^2$. (**a**) In the geometric model, for each $\vec{p}$ there is a local R^2-screen (fiber) of parallel rays. (**b**) In the polychromatic wave model, for each $\vec{p}$ there is a R-stack (fiber) of parallel planes; Fourier analysis further separates these into monochromatic components.

analysis to separate them into monochromatic Helmholtz components $\mathcal{W}_k$, each being invariant under – and a homogeneous space for – Euclidean transformations. •

Remark: The manifold of frames $\mathcal{E}$ in Euclidean optics. The manifold of elements $g(t,r)$ in a group which is a semidirect product [see (5.48)–(5.49)], such as the Euclidean group $\mathsf{ISO}(3) = \mathsf{T}_3 \circledast \mathsf{SO}(3)$, is also a bundle. The base space is here $\mathcal{S}^3/\mathcal{Z}_2$, the manifold of the factor subgroup $\mathsf{SO}(3)$, and the typical fiber is the normal subgroup, here the R^3 of T_3. When Euler angles (5.20) are used to parametrize rotations $R(\phi,\theta,\psi) \in \mathsf{SO}(3)$, equivalence under $\mathsf{SO}(2)_z$ turns ψ redundant and reduces the base space to the Descartes sphere $\vec{p}(\theta,\phi)$, while the remaining equivalence in the geometric and wave models, $\mathsf{T}_1^{(\vec{p})}$ and $\mathsf{T}_2^{(\perp\vec{p})}$ respectively, reduce the typical fiber from R^3 to an R^2 screen or a R signal. •

B.2 Coset spaces for geometric and wave models

Let G be a group and H one of its subgroups. We divide the manifold of G into disjoint equivalence subsets, one of which is H, through the following:

Definition: *For fixed $g \in \mathsf{G}$, the set of group elements*

$$g\,\mathsf{H} := \{g\,h\}, \quad h \in \mathsf{H} \subset \mathsf{G}, \tag{B.1}$$

is the **right coset** *of g by H. Two group elements g_1, g_2 belong to the same right coset when $g_1\,\mathsf{H} = g_2\,\mathsf{H}$* (i.e., *there exists an $h \in \mathsf{H}$ such that $g_1 = g_2\,h$*), *and this relation is denoted by $g_1 \equiv g_2$.*[3] *The manifold of right cosets is*

[3] This is an *equivalence* relation because it satisfies **a**: $g \equiv g$, **b**: $g_1 \equiv g_2 \Rightarrow g_2 \equiv g_1$, and **c**: $g_1 \equiv g_2$ and $g_2 \equiv g_3 \Rightarrow g_1 \equiv g_3$. Equivalence relations always divide a set into disjoint subsets.

indicated G/H. *(Left cosets and spaces of left cosets* $\mathsf{H}\backslash\mathsf{G}$ *are defined similarly by* $\mathsf{H}\,g := \{h\,g\}$, $h \in \mathsf{H}$, *etc.)*

Since $h\,\mathsf{H} = \mathsf{H} = \mathsf{H}\,h$, the coset of the identity g_o is H; all other right cosets $g\,\mathsf{H}$ are obtained by left translation from this distinguished coset. Thus, the space of right cosets G/H is a homogeneous space under the *left* action of the group G, *modulo* H.

Within each right coset we may choose a *representative* element $c \in \mathsf{G}$, to serve as point in the space of right cosets, so that every $g \in \mathsf{G}$ be written uniquely as a product

$$g(\kappa,\eta) = c(\kappa)\,h(\eta), \qquad h(\eta) \in \mathsf{H}, \quad c(\kappa) \in \mathsf{G}/\mathsf{H}. \tag{B.2}$$

We can thus parametrize the manifold of G with coordinates κ in the space of cosets, and η in the space of the coset H. These are referred to as *subgroup-adapted* coordinates for the group and its homogeneous spaces [97].

The manifold $\mathcal{E}$ of the Euclidean group $\mathsf{ISO}(3)$ was presented in (5.41)–(5.43) with a convenient set of coordinates and representation, namely

$$\mathcal{E}(\vec{\alpha},\,\mathbf{R}) := \mathcal{T}(\,\vec{\alpha})\,\mathcal{R}(\mathbf{R}) \leftrightarrow \mathbf{E}(\vec{\alpha},\,\mathbf{R}) := \begin{pmatrix} \mathbf{R} & \vec{\alpha} \\ 0 & 1 \end{pmatrix}, \tag{B.3}$$

$$\begin{array}{r} \mathcal{T}(\,\vec{\alpha}) \in \mathsf{T}_3, \\ \vec{\alpha} = (\alpha_x,\alpha_y,\alpha_z)^\top \in \mathsf{R}^3, \end{array} \qquad \begin{array}{l} \mathcal{R}(\mathbf{R}) \in \mathsf{SO}(3), \\ \mathbf{R} = \mathbf{R}(\phi,\theta,\psi) \\ \quad = \mathbf{R}_z(\phi)\,\mathbf{R}_x(\theta)\,\mathbf{R}_z(\psi). \end{array} \tag{B.4}$$

In these terms, our two models of light are defined through the symmetry of their standard object,

$$\begin{array}{cc} \text{GEOMETRIC MODEL:} & \text{WAVE MODEL:} \\ \mathsf{H} = \mathsf{T}_z \otimes \mathsf{SO}(2)_z & \mathsf{H} = \mathsf{ISO}(2)_{xy} \\ \mathcal{E}\Big((0,0,s)^\top,\,\mathbf{R}_z(\psi)\Big) : \mathcal{O}_{\mathrm{G}} = \mathcal{O}_{\mathrm{G}}, & \mathcal{E}\Big((t_x,t_y,0)^\top,\,\mathbf{R}_z(\psi)\Big) : \mathcal{O}_{\mathrm{W}} = \mathcal{O}_{\mathrm{W}}. \end{array} \tag{B.5}$$

The manifold of all objects in an optical model is the space of right cosets of the Euclidean group by its symmetry subgroup.

In the geometric model, we can extract the factor in the symmetry subgroup to the right, and choose the following coset representatives to the left:

$$\mathcal{E}\Big(\,\vec{\alpha},\,\mathbf{R}(\phi,\theta,\psi)\Big) = \mathcal{E}\Big((q_x,q_y,0)^\top,\,\mathbf{R}_z(\phi)\,\mathbf{R}_x(\theta)\Big)\,\mathcal{E}\Big((0,0,s)^\top,\,\mathbf{R}_z(\psi)\Big). \tag{B.6}$$

In this way the 2 coordinates in the space of the coset are $(s,\psi) \in \mathsf{R} \times \mathcal{S}^1$, and the 4 coordinates in the space of cosets include those 2 of the Descartes sphere of momentum, $\vec{p} = (p_x,p_y,p_z)^\top =: (\mathbf{p},p_z)^\top$, $|\vec{p}| = 1$,

$$\mathbf{P}(\phi,\theta) := \mathbf{R}_z(\phi)\,\mathbf{R}_x(\theta), \quad \vec{p}(\theta,\phi) := \mathbf{P}(\phi,\theta)\begin{pmatrix} \mathbf{0} \\ 1 \end{pmatrix} = \begin{pmatrix} \sin\theta\sin\phi \\ \sin\theta\cos\phi \\ \cos\phi \end{pmatrix} \tag{B.7}$$

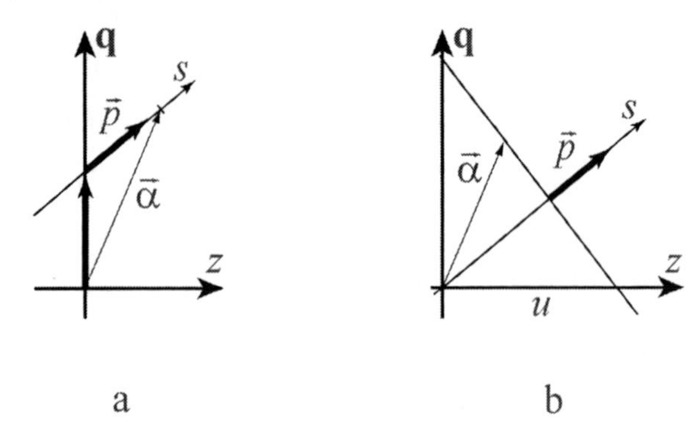

Fig. B.3. Coset coordinates for the geometric and wave models are $(\mathbf{p}, \mathbf{q})$ and $(\mathbf{p}, s)$ respectively. (**a**) The ray $\vec{\alpha}(s)$ of direction $\vec{p}$ (with optical length parameter s counted from the standard screen) has position $\mathbf{q} = \vec{\alpha}(0)$. (**b**) The plane $\vec{\alpha}(\mathbf{t})$ normal to $\vec{p}$ (parametrized by $\mathbf{t} \in \mathsf{R}^2$, not shown) intersects the z-axis at u and its distance to the origin is s.

[cf. (5.17)–(5.19)] and the 2 independent coordinates of position $\mathbf{q}$ obtained from the parameters of translation $\vec{\alpha} = (\boldsymbol{\alpha}, \alpha_z)^\top$, as shown in Fig. B.3a,

$$\begin{aligned} \alpha_x &= q_x + s\sin\theta\cos\phi, \\ \alpha_y &= q_y + s\sin\theta\sin\phi, \\ \alpha_z &= \qquad\quad s\cos\theta, \end{aligned} \qquad \text{i.e.,} \qquad \begin{aligned} \mathbf{q} &= \boldsymbol{\alpha} - s\mathbf{p}, \\ s &= \alpha_z/p_z = \alpha_z \sec\theta. \end{aligned} \tag{B.8}$$

The standard ray $\mathcal{O}_{\mathrm{G}}$ can be thus transformed to any other ray $(\mathbf{p}, \mathbf{q}) \in \wp$, as $\mathcal{O}_{\mathrm{G}}^{(\mathbf{p},\mathbf{q})} := \mathcal{E}(\vec{\alpha}, \mathbf{R}) : \mathcal{O}_{\mathrm{G}}$.

In the wave model on the other hand, the decomposition into a symmetry factor and a chosen right coset representative factor [cf. (B.6)] is

$$\mathcal{E}\Big(\vec{\alpha},\, \mathbf{R}(\phi,\theta,\psi)\Big) = \mathcal{E}\Big((0,0,u)^\top,\, \mathbf{R}_z(\phi)\,\mathbf{R}_x(\theta)\Big)\,\mathcal{E}\Big((t_x,t_y,0)^\top,\, \mathbf{R}_z(\psi)\Big). \tag{B.9}$$

Now the 3 coordinates in the space of the coset are $(\mathbf{t}, \psi) \in \mathsf{R}^2 \times \mathcal{S}^1$ (translation and rotation within the Dirac-wavefront planes), and the 3 remaining coordinates of the space of cosets (classifying the distinct planes) include again 2 of the Descartes sphere of momentum, with $\mathbf{P}(\phi, \theta)$ and $\mathbf{p}$ defined exactly as in (B.7), and the ubiquitous sign $\sigma := \operatorname{sign} p_z$ which must be kept in mind. The remaining coordinate u (and an s defined below) can be found from the composition of translations in (B.6), and are indicated in Fig. B.3b,

$$\begin{aligned} \begin{pmatrix} \alpha_x \\ \alpha_y \end{pmatrix} &= \begin{pmatrix} \cos\phi & \sin\phi \\ -\sin\phi & \cos\phi \end{pmatrix} \begin{pmatrix} t_x \\ t_y\cos\theta \end{pmatrix}, \\ \alpha_z &= \quad u - t_x \sin\theta, \end{aligned} \qquad \begin{aligned} t_x &= \boldsymbol{\alpha}\cdot\mathbf{p}/p_z|\mathbf{p}|, \\ t_y &= \boldsymbol{\alpha}\times\mathbf{p}/|\mathbf{p}|, \\ u &= \vec{\alpha}\cdot\vec{p}/p_z, \\ s &:= u\,p_z = u\cos\theta \\ &= \vec{\alpha}\cdot\vec{p}. \end{aligned} \tag{B.10}$$

The quantity s is the *optical distance* from the wavefront plane to the origin, and this is the natural coordinate on which to impress a signal and/or fix the wavenumber for a Helmholtz wavefield. When $\alpha_z = 0$ we are on the standard screen, where we interpret $\boldsymbol{\alpha} =: \mathbf{q}$; there, the wavefront plane cuts the line $\mathbf{p}\cdot\mathbf{q} = u\,p_z$. Finally, by Euclidean maps, the standard plane $\mathcal{O}_{\rm W}$ can be transformed to all other planes in $\mathcal{W}$, as $\mathcal{O}_{\rm W}^{(s,\mathbf{p})} := \mathcal{E}(\vec{\alpha}, \mathbf{R}) : \mathcal{O}_{\rm W}$.

B.3 Conservation of volume and structure

In Chap. 3 we introduced *beam* functions $\rho(\mathbf{p},\mathbf{q})$ as distributions over the phase space of geometric optics $\wp$, and saw that the 'total amount of light', $E := \int_\wp d\mathbf{p}\,d\mathbf{q}\,\rho(\mathbf{p},\mathbf{q})$ in (3.5) is conserved under canonical transformations; in Chap. 5, we detailed the Euclidean transformations of phase space, which in particular must also conserve E. In this section we show that this is a consequence of the conservation of the volume in the Euclidean *group* under Euclidean transformations; the corresponding consequence in the wave model will be thus obtained. Then, we shall broach the matter of 'Euclidean canonicity' in the conservation of the algebra realized on the frame, line and plane models of optics.

We first need a good measure for volume elements in the Euclidean group, $d\mathcal{E}$, to reproduce the 'total amount of objects' integral for E in geometric optics. The integral can be performed over the various coordinate sets that we have used above for $\mathcal{E}(\vec{\alpha}, \mathbf{R}) \in \mathsf{ISO}(3)$. These 6-dim volumes in the group should not change size when they are subject to $\mathsf{SO}(3)$ rotations and R^3 translations. For the latter, $d^3\,\vec{\alpha} := d\alpha_x\,d\alpha_y\,d\alpha_z$ is a good and natural invariant measure. The former is the manifold of Euler angles, where we know the measure $d^3R(\phi,\theta,\psi) := d^2S(\phi,\theta)\,d\psi$, with $d^2S(\phi,\theta) := d\phi\,\sin\theta\,d\theta$ being the usual rotation-invariant measure over the sphere. In the coordinates of this presentation, the Euclidean-invariant volume element is thus

$$d^6\mathcal{E} = d^3\,\vec{\alpha}\,d^3R = d\alpha_x\,d\alpha_y\,d\alpha_z\,d\phi\,\sin\theta\,d\theta\,d\psi. \tag{B.11}$$

This is the measure for the manifold of all frames [see Fig. B.1]; equal weight is given to all their positions and orientations.

Remark: The invariant Haar measure. The *Haar* measure of a Lie group manifold $g \in \mathsf{G}$ of dimension N is a volume element dg which is invariant under right and/or left action of the group, $g \mapsto g\,g'$ or $g \mapsto g'\,g$, so $d(g\,g') = dg$ and/or $d(g'\,g) = dg$. Of course the group elements are parametrized: $g(\gamma)$, $\gamma \in \mathsf{R}^N$, such that the identity is $g_o = g(0)$; the measure has thus the form

$$dg := \omega(\gamma)\,d\gamma, \tag{B.12}$$

where we are to search for the appropriate *weight function* $\omega(\gamma)$. Because G is a Lie group, there are composition functions $\gamma''(\gamma,\gamma')$ for $g'' = g\,g'$ which are analytic, so the N-differentials are bound through the Jacobian,

$$d\gamma'' = \left|\frac{\partial\gamma''(\gamma,\gamma')}{\partial\gamma}\right| d\gamma \tag{B.13}$$

(and similarly for left action). And moreover, it is sufficient to impose $dg = d(g\,g')$ only in a neighborhood of the identity, i.e., at $|_{g'=g_o} = |_{\gamma'=0}$. At the identity, dg is $d\gamma$ times a number $\omega(0)$ that fixes the overall normalization; this should equal $d(g\,g')$ everywhere, so

$$\omega(0)\,d\gamma = \omega(\gamma''(\gamma,\gamma'))\left|\frac{\partial\gamma''(\gamma,\gamma')}{\partial\gamma}\right|\Bigg|_{\gamma'=0} d\gamma \quad\Rightarrow\quad \omega(\gamma) = \frac{\omega(0)}{\left|\dfrac{\partial\gamma''(\gamma,\gamma')}{\partial\gamma}\right|_{\gamma'=0}}, \tag{B.14}$$

and similarly – but separately – for the left-invariant Haar measure. The two measures coincide for a large variety of group families, including translations, rotations, and their semidirect products (as well as all *compact*,[4] semisimple non-compact, etc.; but not, for example, in the *affine* group of translations and dilatiations.) •

Other parametrizations of $\mathcal{E}(\vec{\alpha},\mathbf{R})$ are those which follow from the geometric and the wave model [see (B.5), (B.6) and (B.9)]. We can find the corresponding measures through simply replacing[5] the coordinates $(\vec{\alpha},\mathbf{R})_{\rm E}$ in (B.11) by $(\mathbf{p},\mathbf{q};s,\psi)_{\rm G}$ from (B.8), and by $(\mathbf{p},s;\mathbf{t},\psi)_{\rm W}$ from (B.10):

$$\begin{aligned} d^6\mathcal{E} = d^4c_{\rm G}(\mathbf{p},\mathbf{q})\,d^2h_{\rm G}(s,\psi) \quad &= \quad d^3c_{\rm W}(\mathbf{p},s)\,d^3h_{\rm W}(\mathbf{t},\psi),\\ d^4c_{\rm G}(\mathbf{p},\mathbf{q}) := dp_x\,dp_y\,dq_x\,dq_y,\ & \ d^3c_{\rm W}(\mathbf{p},s) := \frac{dp_x\,dp_y}{p_z}\,ds,\\ d^2h_{\rm G}(s,\psi) := ds\,d\psi,\quad & \quad d^3h_{\rm W}(\mathbf{t},\psi) := dt_x\,dt_y\,d\psi. \end{aligned} \tag{B.15}$$

We remark below on the reason for these factorizations.

Remark: Measures split on coset spaces. The Jacobian in the weight function of the Haar measure [see (B.14)] is block-diagonal when the group is the direct product of two, because the cross-derivatives of their coordinate sets are zero. Thus the Haar measure of a direct product group *splits* into the Haar measures of the factors. It also splits in many other cases. For subgroup-adapted coordinates $\gamma := (\kappa,\eta)$, as in the coset decomposition $g(\gamma) = c(\kappa)\,h(\eta)$ in (B.2), the space of cosets with coordinates κ is a homogeneous space under the group, so $\kappa''(\kappa,g')$ is independent of the coset coordinates η; and $\eta''(\kappa,\eta,g')$ at $\kappa' = 0$ is the composition of elements at the coset of the identity, which are in the subgroup parametrized by η. Thus the Jacobian determinant also becomes block-diagonal with vanishing cross-derivatives between the κ and η coordinates, and splits into the product of two Jacobians. •

The 'total amount of light' contained in Euclidean, geometric and wave beams ρ, composed of frames, lines or planes respectively, is thus

[4] Compact groups G are those whose *volume*, computed with the Haar measure (B.12) as $V_G := \int_G dg$, is finite ($V_G < \infty$).

[5] Since we do not want to evaluate 6×6 Jacobian determinants, we recommended the reader to use the shorthand calculus of differential forms.

$$E_{\rm E}(\rho) := \int_{\mathsf{R}^3} d^3\,\vec{\alpha} \int_{\mathsf{SO}(3)} d^3R(\phi,\theta,\psi)\,\rho_{\rm E}\Big(\vec{\alpha}, R(\phi,\theta,\psi)\Big), \tag{B.16}$$

$$E_{\rm G}(\rho) := \sum_{\sigma\in\{+,-\}} \int_{|\mathbf{p}|<1} d^2\mathbf{p} \int_{\mathsf{R}^2} d^2\mathbf{q}\,\rho_{\rm G}(\mathbf{p},\sigma,\mathbf{q}), \tag{B.17}$$

$$E_{\rm W}(\rho) := \int_{\mathcal{S}^2} \sin\theta\,d\theta\,d\phi \int_{\mathsf{R}} ds\,\rho_{\rm W}\Big(\vec{p}(\theta,\phi), s\Big). \tag{B.18}$$

In particular we are reminded in (B.17) that the sign $\sigma := \operatorname{sign} p_z$ is included among the screen coordinates of $\wp$ [cf. (3.3)].[6] Finally, measures and integrals for the signal and 'polarization' models of optics that we mentioned above can be similarly treated from the measures in (B.15). These results generalize the 'Lavoisier–Liouville' conservation principle of light in Chap. 3 to a corresponding such principle in Euclidean optical models.

What is the *structure* preserved under Euclidean transformations? In Chap. 3 we had the Hamiltonian structure, and the transformations were required to preserve the basic Poisson brackets between the phase space coordinate functions $(\mathbf{p},\mathbf{q}) \in \wp$. Here, the structure preserved under Euclidean transformations is the Euclidean Lie *algebra* $\mathsf{iso}(3)$, which was presented in (5.38)–(5.40). The basis of translation and rotation generators is obtained differentiating the group action at the origin [see (3.69) and (5.25)]. On functions of the Euclidean group parameters $(\vec{\alpha}, \mathbf{R}(\phi,\theta,\psi))_{\rm E}$, the generators in these coordinates,[7] are

$$\widehat{T}^{\rm E}_x = -\partial_{\alpha_x}, \quad \widehat{T}^{\rm E}_y = -\partial_{\alpha_y}, \quad \widehat{T}^{\rm E}_z = -\partial_{\alpha_z}; \tag{B.19}$$

$$\widehat{J}^{\rm E}_x = -\alpha_y\,\partial_{\alpha_z} + \alpha_z\,\partial_{\alpha_y} + \cos\phi\cot\theta\,\partial_\phi + \sin\phi\,\partial_\theta - \frac{\cos\phi}{\sin\theta}\,\partial_\psi, \tag{B.20}$$

$$\widehat{J}^{\rm E}_y = -\alpha_z\,\partial_{\alpha_x} + \alpha_x\,\partial_{\alpha_z} + \sin\phi\cot\theta\,\partial_\phi - \cos\phi\,\partial_\theta - \frac{\sin\phi}{\sin\theta}\,\partial_\psi, \tag{B.21}$$

$$\widehat{J}^{\rm E}_z = -\alpha_x\,\partial_{\alpha_y} + \alpha_y\,\partial_{\alpha_x} - \partial_\phi. \tag{B.22}$$

The Lie bracket for this realization of $\mathsf{iso}(3)$ by first-order differential operators, is their commutator.

Example: Reduction to the arrow model. The standard object of the Euclidean model with $\mathsf{SO}(2)_z$ rotational symmetry is a unit arrow in the z direction. A beam or field of such oriented arrows is a measurable function of 5 coordiantes $\rho(\vec{\alpha};\phi,\theta)$ – with no ψ. On this linear space of functions, the generators of $\mathsf{iso}(3)$ are realized by the differential operators (B.19)–(B.22) replacing $\partial_\psi \mapsto 0$. The arrow field 'energy' integral is (B.16) suppressing $\int_{\mathcal{S}^1} d\psi$. •

---

[6] It is not uncommon to have to stitch together two or more regions of the space of cosets when the coordinates are not global, such as $(\mathbf{p},\mathbf{q})$, or those on two-sheeted hyperboloids.

[7] Recall the left *vs.* right action of Lie exponential operators in (5.10) and (5.21); also, we use the common notation $\partial_x := \partial/\partial x$.

The reduction from the Euclidean to the geometric model of optics is carried through changing coordinates from $(\vec{\alpha}, \mathbf{R})_{\mathrm{E}}$ to $(\mathbf{p}, \mathbf{q}; s, \psi)_{\mathrm{G}}$ by (B.8) and eliminating the coset coordinates (s, ψ) from all differentials and integrals. The result is the realization of the $\mathsf{iso}(3)$ generators $\vec{p}, \vec{\jmath}$ given in (5.9) and (5.29)–(5.31). The reduction to wave models of optics will be the subject of the rest of this appendix.

Remark: Recuperating the basic Poisson brackets. The differential operators $\widehat{T}_i^{\mathrm{E}}$, $\widehat{J}_k^{\mathrm{E}}$ in (B.19)–(B.22) reduce to *Poisson* operators $\{p_i, \circ\}$, $\{j_k, \circ\}$ only in the model of geometric optics. Only in this realization do we have a derivation with commutative product, so that we can define unambiguously two new functions of the generators: $q_x := -j_y/p_z$ and $q_y := j_x/p_z$ —see (5.29)–(5.30), that will close with p_x and p_y into the basic Poisson bracket set. A situation analogous to this was analyzed before in Chap. 7 for the Berezin brackets (7.19). So we can expect a canonical set of phase space Darboux coordinates satisfying a basic Poisson brackets only in the geometric model of Euclidean optics. Actually, the challenge is to extend the very intuitive concepts of geometric optical phase space $(\mathbf{p}, \mathbf{q}) \in \wp$ to the wave or other Euclidean models of optics.[8] •

B.4 Signal and Helmholtz models

We now realize the Lie algebra of the Euclidean group $\mathsf{ISO}(3)$ on the linear space of Dirac-δ plane wavefronts $\mathcal{W}$ (Sect. A.1); i.e., polychromatic superpositions of plane waves in the direction of $\vec{p}(\theta, \phi) \in \mathcal{S}^2$ and carrying signals $\rho_{\mathrm{W}}(\vec{p}(\theta, \phi), s)$ along $s \in \mathsf{R}$. As we did above in reducing to the geometric model, with the change of variables to the space of cosets by $\mathsf{ISO}(2)_{xy}$ in (B.10), and wiping out the coset coordinates $(\mathbf{t}, \psi)$, we find the following *wave* realization of the Euclidean algebra $\mathsf{iso}(3)$:

$$\begin{aligned} \widehat{T}_x^{\mathrm{W}} &= -\sin\theta\sin\phi\,\partial_s, & \widehat{J}_x^{\mathrm{W}} &= \cos\phi\cot\theta\,\partial_\phi + \sin\phi\,\partial_\theta, \\ \widehat{T}_y^{\mathrm{W}} &= -\sin\theta\cos\phi\,\partial_s, & \widehat{J}_y^{\mathrm{W}} &= \sin\phi\cot\theta\,\partial_\phi - \cos\phi\,\partial_\theta, \\ \widehat{T}_z^{\mathrm{W}} &= -\cos\theta\,\partial_s, & \widehat{J}_z^{\mathrm{W}} &= -\partial_\phi, \end{aligned} \tag{B.23}$$

with the same commutators as before.

There are two quadratic Casimir invariants in $\mathsf{iso}(3)$, which in this signal model are

$$(\widehat{T}_x^{\mathrm{W}})^2 + (\widehat{T}_y^{\mathrm{W}})^2 + (\widehat{T}_z^{\mathrm{W}})^2 = \frac{\partial^2}{\partial s^2}, \tag{B.24}$$

[8] The formulation of *paraxial* wave optics on phase space will be seen in part III, appendix C; it is the full analogue of the quantization of mechanical oscillator systems, and is done through the *Wigner* quasiprobability distribution function [144] for $(\mathbf{p}, \mathbf{q}) \in \mathsf{R}^{2D}$.

and $\sum_i \widehat{T}_i^{\rm W} \widehat{J}_i^{\rm W} = 0$. Thus we divide the polychromatic signals $\rho_{\rm W}(\vec{p}(\theta,\phi),s)$ of $\mathcal{W}$ into subspaces $\mathcal{W}_k$ of monochromatic wavetrains in all directions $\vec{p} \in \mathcal{S}^2$, with a corresponding distribution factor over the sphere

$$\rho_{{\rm W},k}(\vec{p}(\theta,\phi),s) := \rho_k^{\rm S}(\theta,\phi)\, e^{iks}, \qquad k \in \mathsf{R}, \tag{B.25}$$

where subspaces of different wavenumbers k will not mix under Euclidean transformations. Replacement of (B.25) into (B.23) yields the *complex* wave realization on a sphere which, because of the strict analogy between those coordinates in the spaces of cosets, we identify with the Descartes sphere of geometric optics,

$$\begin{array}{ll} \widehat{T}_x^{\rm S} = -ik\sin\theta\sin\phi, & \widehat{J}_x^{\rm S} = \cos\phi\cot\theta\,\partial_\phi + \sin\phi\,\partial_\theta, \\ \widehat{T}_y^{\rm S} = -ik\sin\theta\cos\phi, & \widehat{J}_y^{\rm S} = \sin\phi\cot\theta\,\partial_\phi - \cos\phi\,\partial_\theta, \\ \widehat{T}_z^{\rm S} = -ik\cos\theta, & \widehat{J}_z^{\rm S} = -\partial_\phi, \end{array} \tag{B.26}$$

and whose Casimir (B.24) is a number, $-k^2$.

A continuous linear superposition of plane waves $\exp(ik\,\vec{p}\cdot\vec{q})$ of wavenumbers k, by $f(\vec{p}) := \rho_k^S(\theta,\phi)$ over all directions in the sphere, $\vec{p} \in \mathcal{S}^2$, is a wavefield $F(\vec{q})$, $\vec{q} \in \mathsf{R}^3$, given by

$$F(\vec{q}) = \frac{k}{2\pi}\int_{\mathcal{S}^2} d^2S(\vec{p})\, f(\vec{p})\, \exp(ik\vec{p}\cdot\vec{q}), \quad d^2S(\vec{p}) = \frac{dp_x\, dp_y}{p_z}, \tag{B.27}$$

and under this Fourier-like transform, the Casimir invariant (B.24) becomes the *Helmholtz equation*:

$$\left(\frac{\partial^2}{\partial q_x^2} + \frac{\partial^2}{\partial q_y^2} + \frac{\partial^2}{\partial q_z^2}\right) F(\vec{q}) = -k^2 F(\vec{q}) \tag{B.28}$$

This equation can be written also in a Hamiltonian *evolution* form with first-order derivatives acting on two-component functions:

$$\begin{pmatrix} 0 & 1 \\ -\Delta_k & 0 \end{pmatrix} \begin{pmatrix} F(\vec{q}) \\ F_z(\vec{q}) \end{pmatrix} = \partial_{q_z} \begin{pmatrix} F(\vec{q}) \\ F_z(\vec{q}) \end{pmatrix}, \qquad \Delta_k := k^2 + \frac{\partial^2}{\partial q_x^2} + \frac{\partial^2}{\partial q_x^2}, \tag{B.29}$$

The first component of this equation defines $F_z(\vec{q})$ to be the q_z–derivative of $F(\vec{q})$, while the second reproduces the Helmholtz equation (B.28).

B.5 Hilbert space of Helmholtz wavefields

Quantum mechanics requires a Hilbert space for its wavefunctions; an inner product is needed to provide a norm for the probabilistic interpretation, in which only unitary evolution should take place. In wave models, the inner product is usually defined as an integral over R^3-position space as in quantum mechanics, and the absolute-square norm is associated to the total energy of the wavefield. Plane waves have infinite energy in this scheme, and free Helmholtz wavefields all exhibit infinite $\mathcal{L}^2(\mathsf{R}^3)$ norm. This need not thwart our Euclidean model of Helmholtz optics if we take instead the natural Hilbert space provided by square-integrable functions over the Descartes sphere, $\mathcal{L}^2(\mathcal{S}^2)$, given by (B.18) for the factor $f(\vec{p}) := \rho_k^{\mathrm{s}}(\theta,\phi)$ in (B.25) and (B.27), for each wavenumber k. Below we shall see that this leads to a Euclidean-invariant, *non-local* inner product over the standard screen – the $q_z = 0$ plane, which brings geometric and Helmholtz optics closer.

Working with Hilbert spaces we must have care with the global definition of functions on the sphere when projected on the *two sides* of the standard screen (recall Sect. 2.1). We shall indicate the functions of $\vec{p}$ on the Descartes sphere $\mathcal{S}^2$ equivalently as functions on the two σ-disks $\mathcal{D}$ where $|\mathbf{p}| \leq 1$ and $\sigma = \operatorname{sign} p_z$, as often before ignoring the limit circle between the two,

$$f(\vec{p}) := f_\sigma(\mathbf{p}) := f(\theta,\phi), \qquad \vec{p} = \begin{pmatrix} \mathbf{p} \\ \sigma\,|p_z| \end{pmatrix} = \begin{pmatrix} \sin\theta\sin\phi \\ \sin\theta\cos\phi \\ \cos\theta \end{pmatrix}. \tag{B.30}$$

The integral of the sesquilinear product of two such functions f, g over the sphere (i.e., over the *irreducible subspace* of cosets $\mathcal{W}_k$ with all plane wavetrains of the same wavenumber) is

$$\begin{aligned} (f,g)_{k,\mathcal{S}^2} &:= \int_{\mathcal{S}^2} d^2S(\vec{p})\, f(\vec{p})^*\, g(\vec{p}) \\ &= \int_{\mathcal{D}} \frac{d^2\mathbf{p}}{\sqrt{1-|\mathbf{p}|^2}} \big(f_+(\mathbf{p})^*\, g_+(\mathbf{p}) + f_-(\mathbf{p})^*\, g_-(\mathbf{p})\big). \end{aligned} \tag{B.31}$$

To write this $\mathcal{L}^2(\mathcal{S}^2)$-inner product in terms of the Helmholtz wavefields (B.27) *on the standard screen*, we must guarantee the unitarity of the integral transformation $F(\vec{q}) \leftrightarrow f(\vec{p})$, which is close to (but not coincident with) the Fourier integral transform. This we do by writing the functions of this – call it *wave*— transform at $q_z = 0$. Both the value of the field at the standard screen $F(\mathbf{q})$, and its *normal derivative* $F_z(\vec{q})$, are thus involved [cf. (B.29)!]:

$$F(\mathbf{q}) = F(\vec{q})\Big|_{q_z=0} = \frac{k}{2\pi}\int_{\mathcal{D}} \frac{d^2\mathbf{p}}{\sqrt{1-|\mathbf{p}|^2}} \big(f_+(\mathbf{p}) + f_-(\mathbf{p})\big)\exp(ik\mathbf{p}\cdot\mathbf{q}), \tag{B.32}$$

$$F_z(\mathbf{q}) = \frac{\partial F(\vec{q})}{\partial q_z}\bigg|_{q_z=0} = \frac{ik^2}{2\pi}\int_{\mathcal{D}} d^2\mathbf{p}\,\big(f_+(\mathbf{p}) - f_-(\mathbf{p})\big)\exp(ik\mathbf{p}\cdot\mathbf{q}). \tag{B.33}$$

The inverse map is obtained using a linear combination of the two 'initial' conditions of the Helmholtz wavefield, and the inverse Fourier transform,

$$f_\pm(\mathbf{p}) = \frac{k}{4\pi}\int_{\mathsf{R}^2} d^2\mathbf{q}\left(\sqrt{1-|\mathbf{p}|^2}\,F(\mathbf{q}) \pm \frac{1}{ik}F_z(\mathbf{q})\right)\exp(-ik\mathbf{p}\cdot\mathbf{q}). \tag{B.34}$$

Now the Parseval identity of the Fourier transform allows us to make the wave transform unitary, between the $\mathcal{L}^2(\mathcal{S}^2)$ Hilbert space with inner product (B.31), and a *Helmholtz* Hilbert space $\mathcal{H}^2_k(\mathsf{R}^2)$, whose inner product $(F^{\rm s}, G^{\rm s})_k := (f,g)_{\mathcal{S}^2}$ integrates the wavefields and their normal derivatives $F^{\rm s}(\mathbf{q}) := \{F(\mathbf{q}), F_z(\mathbf{q})\}$, over the standard screen.

We proceed to find the measure for $\mathcal{H}^2_k(\mathsf{R}^2)$. There is a triple integral, where the integral over the compact disk $\mathcal{D}$ can be moved to the right. We thus obtain the *Helmholtz* inner product

$$(F^{\rm s}, G^{\rm s})_k = \frac{k^2}{4\pi^2}\int_{\mathsf{R}^2} d^2\mathbf{q} \int_{\mathsf{R}^2} d^2\mathbf{q}' \begin{pmatrix} F(\mathbf{q}) \\ F_z(\mathbf{q}) \end{pmatrix}^{\dagger} \begin{pmatrix} \omega(|\mathbf{q}-\mathbf{q}'|) & 0 \\ 0 & \varpi(|\mathbf{q}-\mathbf{q}'|) \end{pmatrix} \begin{pmatrix} G(\mathbf{q}) \\ G_z(\mathbf{q}) \end{pmatrix}, \tag{B.35}$$

where $\dagger$ is transpose conjugation, and we have assimilated the $\mathbf{p}$–integration in two *non-local weight functions*, ω and ϖ,[9]

$$\begin{aligned} \omega(|\mathbf{q}-\mathbf{q}'|) &= \frac{1}{2}\int_{\mathcal{D}} d^2\mathbf{p}\, \sqrt{1-|\mathbf{p}|^2}\, \exp(-ik\mathbf{p}\cdot(\mathbf{q}-\mathbf{q}')) \\ &= \pi \frac{j_1(k|\mathbf{q}-\mathbf{q}'|)}{k|\mathbf{q}-\mathbf{q}'|}, \qquad \frac{j_1(\tau)}{\tau} = \frac{\sin\tau - \tau\cos\tau}{\tau^3}, \end{aligned} \tag{B.36}$$

$$\begin{aligned} \varpi(|\mathbf{q}-\mathbf{q}'|) &= \frac{1}{2k^2}\int_{\mathcal{D}} d^2\mathbf{p}\, \frac{1}{\sqrt{1-|\mathbf{p}|^2}} \exp(-ik\mathbf{p}\cdot(\mathbf{q}-\mathbf{q}')) \\ &= \frac{\pi}{k^2} j_0(k(\mathbf{q}-\mathbf{q}')), \qquad j_0(\tau) = \frac{\sin\tau}{\tau}, \end{aligned} \tag{B.37}$$

where j_0 and j_1 are the spherical Bessel functions [61].

Example: A plane wavetrain. A plane wave directed along $\vec{v} = (\mathbf{v}, v_z)^\top \in \mathcal{S}^2$ is the wave transform of a Dirac delta,[10] i.e.,

$$w_{\vec{v}}(\vec{p}) := \delta_{\mathcal{S}^2}(\vec{v}, \vec{p}) = \frac{1}{\sin\theta}\,\delta(\theta_v - \theta)\,\delta(\phi_v - \phi), \tag{B.38}$$

$$w_{\mathbf{v},\sigma}(\mathbf{p}) = \sqrt{1-|\mathbf{p}|^2}\,\delta(v_x - p_x)\,\delta(v_y - p_y)\,\delta_{\sigma,\mathrm{sign}\,v_z}; \tag{B.39}$$

$$W^{\rm s}_{\vec{v}}(\mathbf{q}) = \begin{pmatrix} W_{\vec{v}}(\mathbf{q}) \\ W_{\vec{v},z}(\mathbf{q}) \end{pmatrix} = \frac{k}{2\pi}\begin{pmatrix} 1 \\ ikv_z \end{pmatrix} \exp(ik\mathbf{v}\cdot\mathbf{q}). \tag{B.40}$$

The sign of the normal derivative distinguishes the plane wave along $\vec{v} = (\mathbf{v}, v_z)$ and its *reflection* in the standard screen, along $\vec{v}^{\,\rm R} = (\mathbf{v}, -v_z)$. A

---

[9] We used the integrals in [61, Eqs. 6.567.1, and 6.554.2]:

$$\int_0^n p\,dp\,(n^2 - |\mathbf{p}|^2)^\mu J_0(xp/n) = 2^\mu \Gamma(\mu+1) n^{2(\mu+1)} \frac{J_{\mu+1}(x)}{x^{\mu+1}}.$$

[10] Our first example concerns a wavefield which is actually *outside* the Hilbert space on the Descartes sphere $\mathcal{L}^2(\mathcal{S}^2)$, but belongs to a Gel'fand *triplet* whose larger space includes Dirac δ's. Through the wave transform this process can be followed through for the Helmholtz Hilbert space $\mathcal{H}^2_k(\mathsf{R}^2)$; in the place of the Dirac δ's will be plane waves.

Helmholtz field whose normal derivative vanishes on the screen is the sum of a a field and its reflection. •

Example: The narrowest Helmholtz wavefields. The weight functions $\omega(\mathbf{q})$ and $\varpi(\mathbf{q})$ in (B.36)–(B.37) are solutions of the Helmholtz equation with zero normal derivative; they relate through $\omega = \Delta_k \varpi$ [see (B.29)]. Note now that $\varpi(\mathbf{q})$ is the wave transform of a constant over the Descartes sphere, and $\omega(\mathbf{q})$ that of $\cos\theta$; and since these are the widest, smoothest functions on the sphere, ϖ and ω are expected to be the *narrowest* Helmholtz wavefunctions which can be collected on the screen. It is this minimal width which characterizes the non-locality of the Helmholtz Hilbert space, together with many other properties [155]. •

B.6 Euclidean algebra in Helmholtz optics

The generators of Euclidean translations and rotations on the sphere (B.26), are *skew-adjoint* in $\mathcal{L}^2(\mathcal{S}^2)$, so they generate *unitary* transformations[11]

$$(\widehat{S}^\dagger f, g) := (f, \widehat{S}\, g) = -(\widehat{S}\, f, g) \quad \Rightarrow \quad (f, [e^{\widehat{S}}]^\dagger g) = (f, e^{-\widehat{S}} g). \qquad \text{(B.41)}$$

Finite translations $\exp(\vec{\alpha}\cdot\widehat{\vec{T}})$ multiply the functions on the sphere by phases $\exp(-ik\,\vec{\alpha}\cdot\vec{p})$ while rotations move them around rigidly. After the wave transform, we have equivalent statements for the Helmholtz Hilbert space $\mathcal{H}^2_k(\mathsf{R}^2)$. The 'total amount of energy' of a Helmholtz wavefield, as well as the overlap between any two fields (B.35) measured on a plane screen, is *independent* of the location and orientation of that screen.

The realization of the generators of the Euclidean algebra $\mathsf{iso}(3)$ by skew-adjoint operators in the Helmholtz Hilbert space is

$$\begin{aligned}
\widehat{T}^{\mathrm{H}}_x &= \begin{pmatrix} \partial_{q_x} & 0 \\ 0 & \partial_{q_x} \end{pmatrix}, & \widehat{J}^{\mathrm{H}}_x &= \begin{pmatrix} 0 & q_y \\ -q_y\Delta_k - \partial_{q_y} & 0 \end{pmatrix}, \\
\widehat{T}^{\mathrm{H}}_y &= \begin{pmatrix} \partial_{q_y} & 0 \\ 0 & \partial_{q_y} \end{pmatrix}, & \widehat{J}^{\mathrm{H}}_y &= \begin{pmatrix} 0 & -q_x \\ q_x\Delta_k + \partial_{q_x} & 0 \end{pmatrix}, \\
\widehat{T}^{\mathrm{H}}_z &= \begin{pmatrix} 0 & 1 \\ -\Delta_k & 0 \end{pmatrix}, & \widehat{J}^{\mathrm{H}}_z &= \begin{pmatrix} q_x\partial_{q_y} - q_y\partial_{q_x} & 0 \\ 0 & q_x\,\partial_{q_y} - q_y\,\partial_{q_x} \end{pmatrix},
\end{aligned} \qquad \text{(B.42)}$$

where $\Delta_k := k^2 + \partial^2_{q_x} + \partial^2_{q_x}$ as given in (B.29). Evidently $\widehat{T}^{\mathrm{H}}_x$, $\widehat{T}^{\mathrm{H}}_y$ and $\widehat{J}^{\mathrm{H}}_z$ generate the $\mathsf{ISO}(2)_{xy}$ motions of the screen, while (B.29) shows that $\widehat{T}^{\mathrm{H}}_z$ advances the screen in the z direction, and reveals itself as the Hamilton evolution equation for this free system. The analysis of $\widehat{J}^{\mathrm{H}}_x$ and $\widehat{J}^{\mathrm{H}}_y$ is more

[11] Equation (B.41) defines only skew-*hermitian* operators, since the equality occurs between inner products, it is a *weak* definition for the operators. We use the terminology of *strong* equality without proof.

complicated; rotations of the screen out of its plane will generally mix the values of the field and its normal derivatives through an integral transform that will realize the Euclidean group.

Remark: Helmholtz and Klein-Gordon measures. Searching for Euclidean-invariant inner products with an in general nonlocal plane measure involving field values and normal derivatives, the authors of [134] set up boundary and differential conditions holding in the solution space of the Helmholtz equation, showing that it is *unique.* A similar 2-component treatment for the inner product in the (2+1)-dim *Klein-Gordon* free field (with Poincaré mother symmetry group $\mathsf{ISO}(2,1)$) shows *its* measure to be also unique, *local* and, in matrix form, antidiagonal; i.e.

$$(F,G)_{\rm KG} = \int_{\mathsf{R}^2} d^2\mathbf{q} \int_{\mathsf{R}^2} d^2\mathbf{q}' \begin{pmatrix} F(\mathbf{q}) \\ F_z(\mathbf{q}) \end{pmatrix}^{\dagger} \begin{pmatrix} 0 & \delta(\mathbf{q}-\mathbf{q}') \\ \delta(\mathbf{q}-\mathbf{q}') & 0 \end{pmatrix} \begin{pmatrix} G(\mathbf{q}) \\ G_z(\mathbf{q}) \end{pmatrix}. \tag{B.43}$$

This well-known Poincaré invariant [134] is the analogue of (B.35). •

Remark: Lorentz transformations of Helmholtz fields. The relativistic coma transformation of geometric optics seen in Sects. 5.5–5.7, has a wave version [6, 155]. Through a similar formulation one can build the vector of generators of boosts through the commutator $\vec{B} = (i/k)[J^2, \vec{T}]$, between the square angular momentum and the translation generators. (The i is put in to preserve skew-adjointness.) This done, we find

$$\widehat{\mathbf{B}}^{\rm H} = \frac{i}{k}\begin{pmatrix} \widehat{D}\,\partial_{\mathbf{q}}+k^2\mathbf{q} & 0 \\ 0 & (\widehat{D}+1)\,\partial_{\mathbf{q}}+k^2\mathbf{q} \end{pmatrix}, \quad \widehat{B}_z^{\rm H} = -\frac{i}{k}\begin{pmatrix} 0 & -\widehat{D} \\ (\widehat{D}+1)\Delta_k - k^2 & 0 \end{pmatrix}, \tag{B.44}$$

where $\widehat{D} := \frac{1}{2}(\mathbf{q}\cdot\partial_{\mathbf{q}} + \partial_{\mathbf{q}}\cdot\mathbf{q})$ is a symmetrized generator of dilatations. These matrix operators generate a Lorentz group of transformations in the Hilbert space of Helmholtz fields $\mathcal{H}_k^2(\mathsf{R}^2)$, and produces a relativistic coma phenomenon on beams, subject to aberration expansion [6, 155]. Nevertheless, since (as in the geometric optics model) there is no Doppler effect on the wavenumber k, this transformation, although unitary, cannot be physical. •

B.7 The recipe for wavization

Now that we have a Hilbert space for Helmholtz wavefields, and a set of generators which close into the Euclidean algebra $\mathsf{iso}(3)$, we can establish their correspondence ($\leftrightarrow$) with the generators of the same transformations in the geometric model of optics. The former (times i) are self-adjoint 2×2 matrix operators in $\mathcal{H}_k^2(\mathsf{R}^2)$, and the latter are functions of phase space $(\mathbf{p},\mathbf{q}) \in \wp$; their correspondence leads to the following *wavization* recipe:

$$\mathbf{p} \leftrightarrow \frac{1}{ik}\begin{pmatrix} \partial_{\mathbf{q}} & 0 \\ 0 & \partial_{\mathbf{q}} \end{pmatrix}, \qquad \mathbf{q}\,p_z \leftrightarrow \frac{1}{ik}\begin{pmatrix} 0 & \mathbf{q} \\ -\mathbf{q}\,\Delta_k - \partial_{\mathbf{q}} & 0 \end{pmatrix},$$
$$p_z \leftrightarrow \frac{1}{ik}\begin{pmatrix} 0 & 1 \\ -\Delta_k & 0 \end{pmatrix}, \qquad \mathbf{q}\times\mathbf{p} \leftrightarrow \frac{1}{ik}\begin{pmatrix} q_x\,\partial_{q_y} - q_y\,\partial_{q_x} & 0 \\ 0 & q_x\,\partial_{q_y} - q_y\,\partial_{q_x} \end{pmatrix}, \tag{B.45}$$

and we note the anticommutator $\frac{1}{2}(\mathbf{q}\Delta_k + \Delta_k\mathbf{q}) = \mathbf{q}\Delta_k + \partial_{\mathbf{q}}$.

The role of Planck's constant $h = 2\pi\,\hbar$ in quantization is taken by the wavelength $\lambda = 2\pi/k$ in Helmholtz optics; optical momentum $\mathbf{p} \leftrightarrow \mathbf{1}\partial_{\mathbf{q}}/ik$ has no units, and generates translations within the screen spaces of both the geometric and the wave models. The generation of translations out of the screen plane bestow upon p_z the rôle of a Hamiltonian, and its action on $\mathcal{H}_k^2(\mathsf{R}^2)$ gives the Hamilton evolution equations for the system. The wavization (B.45) does *not* give, however, any operator corresponding to the *position* $\mathbf{q}$ on the screen; there is no position operator with a Dirac-δ eigenfunction, the narrowest wavefunction is $\varpi(\mathbf{p},\mathbf{q})$, given in (B.37). Only $\mathbf{q}\,p_z = \mathbf{q}\cos\theta$ has a Helmholtz operator counterpart.[12]

Example: Wavization of Lorentz boost generators. The wavization recipe (B.45), applied to the vector of boost generators (5.69)–(5.71), yields

$$\vec{b} = \vec{p}\times\vec{R} = \begin{pmatrix} p_z\,(q_x p_z) + p_y\,(\mathbf{q}\times\mathbf{p}) \\ -p_x\,(\mathbf{q}\times\mathbf{p}) + p_z\,(q_y p_z) \\ -p_x\,(q_x p_z) - p_y\,(q_y p_z) \end{pmatrix} \Rightarrow \widehat{\vec{B}}^{\mathrm{H}} \text{ in Eqs. (B.44)}, \tag{B.46}$$

provided we do not break the factors $(\ldots)$ before substituting them with their wavization recipe (B.45). If we did write $p_z^2 = 1 - p_x^2 - p_y^2$ and 'wavize' $\mathbf{q} \leftrightarrow \mathbf{1}\,\mathbf{q}$, the result would be $(\widehat{D} + \frac{1}{2})\partial_{\mathbf{q}} + k^2\mathbf{q}$; this is in fact the *average* of the two diagonal matrix elements of $\widehat{\mathbf{B}}^{\mathrm{H}}$ in (B.44). •

As in the quantization process of classical mechanics, the Helmholtz wavization of geometric optics presents operator-ordering problems beyond second degree in the generators. In the (Gel'fand triplet of the) Helmholtz Hilbert space, there is the Dirac basis of plane waves [eigenfunctions of two of the three commuting momenta $(\widehat{p}_x, \widehat{p}_y)$], the cylindrical basis of invariant (i.e., 'nondiffracting') beams [eigenfunctions of $(\widehat{p}_z, \widehat{\jmath}_z)$], the basis of multipole fields [eigenfunctions of $(\widehat{\jmath}^2, \widehat{\jmath}_z)$], and generally the wave transforms of functions on the Descartes sphere separated in elliptic coordinates [75]; but no 'position' eigenfunctions because of the nonlocal measure. A Hilbert space is important because it gives the proper context to problems of interpolation and approximation by minimal-energy wavefields between a finite number of sensors (cf. [60]). Phase space in Helmholtz wavefields can be also explored with an appropriate Wigner quasiprobability distribution function [162], together with all radiometric observables of scalar monochromatic

[12] One *cannot* simply divide by p_z because the corresponding operator has 0 in its spectrum; if we did, we would find that $\mathbf{q}\,\mathbf{1}$ maps the momentum disk $\mathcal{D}$ and hence $\mathcal{H}_k^2(\mathsf{R}^2)$ out of itself.

fields [3]. Presently, there are also drawbacks: the evanescent-wave solutions of the Helmholtz equation have not been incorporated (and they do exist in diffracted fields); all inhomogeneity of the medium has been left out; and reflection *plus* refraction at curved interfaces requires a geometric theorem for a root transformation (see Chap. 4) that conserves field values and their tangential derivatives.

Part III

The paraxial régime

Introduction

The paraxial régime of optics is a model of light which is *tangent* to the global model of the previous two parts of this book. 'Paraxial' means *near to the axis*, where we disregard all but the first order – or *linear* part – of the optical transformations of phase space. As we remarked in several places (pages 23, 31, 39, 85), linear transformations cannot generally be optical maps, because while the range of position coordinates is unbounded, that of momentum is bounded by the refractive index.

Definition: *The* **paraxial régime** *is a model of D-dim geometric optics where rays are points of an R^{2D} phase space, and are subject to linear canonical transformations.*

The paraxial régime *replaces* the optical phase space $\wp$ with the simpler, flat R^4 space for our real 3-dim world with 2-dim screens, and generally R^{2D} for D-dim screens in any $(D+1)$-dim world. In particular, all our plane figures of phase space correspond to $D = 1$. In this régime, optical elements can be represented by matrices that satisfy the conditions for the system to be Hamiltonian; this constrains them to be *symplectic.*[13] In chapter 9 we set up optical systems composed of free spaces and refracting surfaces, and characterize their manifold with matrix, Bargmann and Iwasawa parameters. The structure of the group of symplectic matrices is very, very rich. In chapter 10 we decompose the generic paraxial system into distinct factors: fractional Fourier transformers, and imager systems; these are respectively the maximal compact unitary, and the solvable Iwasawa subgroups.

It should be stressed that our exposition of the symplectic group, algebra, and their optical realizations, are far from being the most general treatment

---

[13] SYMPLECTIC (sim**plek**tik), first appearance: 1839. [*Adaptation from Greek συμπλεκτικός formed on σύν* SYM– +πλέκειν, TO TWINE, PLAIT, WEAVE.] A. *Adjective* Epithet of a bone of the suspensorium in the skull of fishes, between the hyomandibular and the quadrate bones. B. *Substantive* The symplectic bone. – *The Oxford Universal Dictionary on Historical Principles*, 3[d] ed., 1955. The group-theoretical acceptance of the term seems to have originated in Hermann Weyl's book on the classical groups [142]. In 2003, a search of the word `symplectic` in the web produced *about 109,000 entries*, mostly in physics, mathematics and optics, and also a few in geology; vernacular dictionaries still list only the ichtyological meaning.

of symplectic geometry as practised by contemporary mathematicians. As in the rest of this volume, we strive to follow the Middle Path between *ad hoc* construction and *non plus ultra* abstraction. Classical models in mechanics and in optics do not seem to require the plethora of unitary irreducible representations of the unitary symplectic or real symplectic Lie groups, but only finite-dimensional, non-unitary matrices. The reader's mathematical interest in symplectic geometry may be whetted in works such as those of Siegel [127], O'Meara [111], Weinstein [139], and Guillemin and Sternberg [62].

Chapter 11 is of a mathematical nature; its aim is to present the classical Cartan Lie algebras. Starting with the general linear Lie groups and algebras, we restrict these to leave spheres invariant. For real, complex and quaternionic spheres, one obtains the orthogonal, unitary and unitary-symplectic structures. With Weyl's trick we bind the latter to the real symplectic algebras of the paraxial régime, writing explicitly the accidental equalities between algebras that are used in optics. Root and weight diagrams are given for all classical algebras, in particular the symplectic one, that will be used to classify aberrations in Part IV. In Chap. 12 we explore further the nature of the 10-dim symplectic groups and algebras, to resolve the equivalence classes of 3-dim paraxial Hamiltonians, and to integrate the ray trajectories in corresponding optical guides.

Again, the appendix serves to step out from the classical geometric models, into the wave/quantum models known as Fourier optics or Schrödinger oscillator mechanics. Symplectic transformations in the former turn into unitary integral transforms for the latter, with complex Gaussian kernels. The wave model actually realizes the *metaplectic* group, i.e., the *two-fold cover* of the geometrical symplectic group. The large dispersion of understanding evident in the research literature around fractional Fourier transforms and the 'metaplectic sign problem' openly needs a detailed clarification of the wavization process of paraxial optics.

9 Optical elements of the symplectic group

The basic elements in any optical workshop are free spaces and lenses; with these building blocks one can construct a very wide variety of systems. How many adjustable parameters can there be and what is the resulting manifold of transformations? These questions can be answered completely within the paraxial régime. We choose a *design ray* (also called basal, or standard ray) in the system and examine its immediate neighborhood in phase space $(\mathbf{p},\mathbf{q}) \in \wp$. This ray serves to fix the origin of a local phase space coordinate grid (optical center $\mathbf{q}=\mathbf{0}$ and optical axis $\mathbf{p}=\mathbf{0}$), on which we assume that any phase space map $(\mathbf{p}'(\mathbf{p},\mathbf{q}),\, \mathbf{q}'(\mathbf{p},\mathbf{q}))$ can be subject to a Taylor expansion in powers of the coordinates p_i, q_j, and where we can consistently keep the linear terms only.

9.1 Free spaces, thin lenses, and action on phase space

The program outlined above was followed in Chap. 1 for free displacements generated by (2.22) in a homogeneous medium [(2.11)–(2.12) and (3.51)], and for the root transformation [(4.25)–(4.26) and (3.55)]; the paraxial refracting surface transformation follows immediately [see (4.13)]. Using the Poisson operator (3.25), *viz.*,

$$\{f(\mathbf{p},\mathbf{q}),\circ\} := \frac{\partial f(\mathbf{p},\mathbf{q})}{\partial \mathbf{q}} \cdot \frac{\partial}{\partial \mathbf{p}} - \frac{\partial f(\mathbf{p},\mathbf{q})}{\partial \mathbf{p}} \cdot \frac{\partial}{\partial \mathbf{q}}, \tag{9.1}$$

and its Lie exponential operator $\exp\{f,\circ\}$ [see (3.31)–(3.32)], we fix the notation as follows:[1]

Free displacement by z in vacuum

$$\mathcal{D}_z := \exp(-z\{\tfrac{1}{2}|\mathbf{p}|^2,\circ\}), \qquad z \geq 0, \tag{9.2}$$

$$\mathcal{D}_z : \begin{pmatrix} \mathbf{p} \\ \mathbf{q} \end{pmatrix} = \begin{pmatrix} \mathbf{p} \\ \mathbf{q}+z\mathbf{p} \end{pmatrix} = \begin{pmatrix} \mathbf{1} & \mathbf{0} \\ z\mathbf{1} & \mathbf{1} \end{pmatrix} \begin{pmatrix} \mathbf{p} \\ \mathbf{q} \end{pmatrix}. \tag{9.3}$$

Free displacement is shown in Fig. 9.1, for rays in optical space (q,z), and its approximation as a paraxial map of phase space $(p,q) \in \mathsf{R}^2$ between two 1-dim

[1] In the paraxial régime, the phase space $(\mathbf{p},\mathbf{q})^\top \in \mathsf{R}^{2D}$ *is* a vector space; in the global régime, it was only a convenient notation.

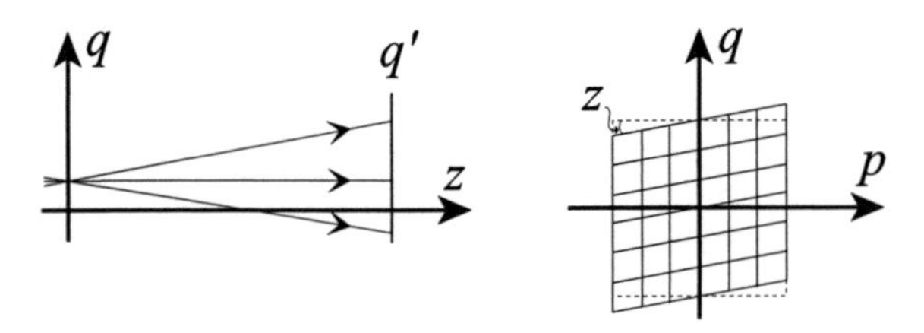

Fig. 9.1. Free displacement by z along the optical axis. *Left*: Rays propagating from screen q to screen q'. *Right*: Corresponding map of phase space (p, q); the flow is along the vertical lines $p =$ constant; horizontal lines acquire slope $\sim z$.

screens. This is the linear approximation to the 'true' free displacement of Fig. 2.4, when the angle θ between the ray and the optical axis is 'small', i.e., $|\mathbf{p}| = \sin\theta \approx \theta \approx \tan\theta$. When the medium has refractive index n, the relation between this angle and momentum $|\mathbf{p}|$ becomes $\theta := |\mathbf{p}|/n$, and the summand $z\mathbf{p}$ in (9.3) becomes the shortened optical distance $z\mathbf{p}/n$. The operator (9.3) of displacement in this medium is thus $\mathcal{D}_{z/n}$.

Thin lens of *Gaussian power* $\mathbf{G}$ (a symmetric $D \times D$ matrix),

$$\mathcal{L}_{\mathrm{G}} := \exp(-\{\tfrac{1}{2}\mathbf{q} \cdot \mathbf{G}\mathbf{q}, \circ\}), \qquad \mathbf{G}^\top = \mathbf{G}, \tag{9.4}$$

$$\mathcal{L}_{\mathrm{G}} : \begin{pmatrix} \mathbf{p} \\ \mathbf{q} \end{pmatrix} = \begin{pmatrix} \mathbf{p} - \mathbf{G}\mathbf{q} \\ \mathbf{q} \end{pmatrix} = \begin{pmatrix} \mathbf{1} & -\mathbf{G} \\ \mathbf{0} & \mathbf{1} \end{pmatrix} \begin{pmatrix} \mathbf{p} \\ \mathbf{q} \end{pmatrix}. \tag{9.5}$$

Figure 9.2 shows the action of a thin convex 1-dim lens of Gaussian power $G > 0$ on rays, and as a map of phase space between two screens, q and q', which are coincident with the lens. A convex lens has $G > 0$ and a concave lens $G < 0$.

Remark: Focal distance. Rays entering parallel to the optical axis of Fig. 9.2, $(p = 0, q)$, will refract to $(p', q') = (-Gq, q)$ and, after (paraxial) free displacement, to $(p'', q'') = (p', q' + zp') = (-Gq, q - zGq)$. When $q'' =$

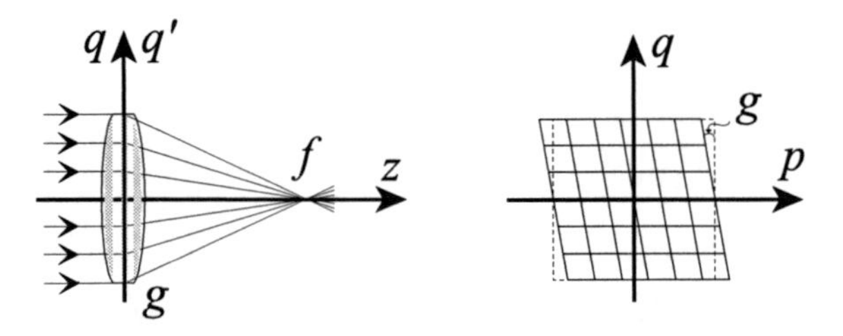

Fig. 9.2. Thin convex lens. *Left*: Action of a thin lens of Gaussian power g on rays parallel to the axis $(p = 0)$, refered to the two coincident screens q and q', and their intersection on the right focal point $f = 1/g$. *Right*: Action on phase space (p, q); the flow is along horizontal lines $q =$ constant and the tangent of the shear angle is $\sim g$.

$q(1 - zG) = 0$, they all converge to the *focal point* at the *focal distance* $z = F := 1/G$, which is simply the inverse of its Gaussian power. •

Remark: Gaussian power of the root, surface and lens maps. The paraxial approximation to the root and refracting surface transformations seen in Chap. 4, refers to quadratic surfaces

$$z = \zeta(\mathbf{q}) := \mathbf{q} \cdot \mathbf{Z}\mathbf{q} = \sum_{i,j} q_i Z_{i,j} q_j, \tag{9.6}$$

between media n and n' [see (4.11)–(4.13) and (4.25)–(4.26)]. Their Gaussian powers are, correspondingly,

$$\mathcal{R}_{n;\zeta} \leftrightarrow \mathbf{G} = -2n\mathbf{Z}, \tag{9.7}$$

$$\mathcal{S}_{n,n';\zeta} \leftrightarrow \mathbf{G} = 2(n' - n)\mathbf{Z}, \tag{9.8}$$

$$\mathcal{S}_{n,n';\zeta}\, \mathcal{S}_{n',n;\zeta'} \leftrightarrow \mathbf{G} = 2(n' - n)(\mathbf{Z} - \mathbf{Z}'). \tag{9.9}$$

The last operator realizes a thin lens of refractive index n' in a medium n, with left and right surfaces characterized by the matrices $\mathbf{Z}$ and $\mathbf{Z}'$ in (9.6) respectively, and of *zero* thickness. •

Remark: Spherical surfaces. A spherical surface of radius R and center on the optical axis at $z = R$, will be tangent to the standard screen $z = 0$ at the optical center $|\mathbf{q}|^2 = \mathbf{0}$. The surface is given by the equation

$$S(\vec{q}) := \sqrt{(R - z)^2 + |\mathbf{q}|^2} - R = 0. \tag{9.10}$$

In the paraxial régime we approximate this surface with its osculating parabola $z = |\mathbf{q}|^2/2R$, so the generic quadratic surface shape (9.6) is

$$Z_{i,j} = \frac{\delta_{i,j}}{2R}, \quad \text{i.e.} \quad \mathbf{Z} = \frac{1}{2R}\,\mathbf{1}, \quad \mathbf{G} = \frac{n' - n}{R}\,\mathbf{1}. \tag{9.11}$$

•

Out of free displacement and refracting surfaces we can produce composite systems, and the issue arises on how to multiply the operators and the matrices. To have a working class of cases, we first exemplify the composition in systems of a useful kind.

Example: Symmetric systems. Symmetric systems are paraxial optical systems represented by symmetric matrices. Since they will be composed of the free displacement and lens matrices [(9.3) and (9.5)] which have unit determinant, they too must have unit determinant:

$$\mathbf{H} = \begin{pmatrix} \lambda + \rho & \eta \\ \eta & \lambda - \rho \end{pmatrix}, \quad \det \mathbf{H} = \lambda^2 - \rho^2 - \eta^2 = 1. \tag{9.12}$$

In the $D = 1$ case, symmetric systems have only two independent parameters; two empty spaces determine the lens to be placed in between, yielding[2]

[2] Although the matrix expressions are formally correct because the systems are symmetric, we shall discuss below the left-to-right *vs.* right-to-left order of operators *vs.* matrices.

$$\begin{pmatrix} 1 & 0 \\ z_1 & 1 \end{pmatrix} \begin{pmatrix} 1 & \dfrac{z_1+z_2}{1-z_1z_2} \\ 0 & 1 \end{pmatrix} \begin{pmatrix} 1 & 0 \\ z_2 & 1 \end{pmatrix} = \frac{1}{1-z_1z_2} \begin{pmatrix} 1+z_2^2 & z_1+z_2 \\ z_1+z_2 & 1+z_1^2 \end{pmatrix}. \tag{9.13}$$

When $z_1 z_2 = 1$ there is a singularity that we must avoid. Note also that $z_1, z_2 > 0 \Rightarrow \eta \neq 0$, so pure magnifiers, represented by diagonal matrices, are excluded from this construction. •

Example: Fractional Fourier transformers. A set of useful systems is composed of one convex lens between two equal displacements,

$$\begin{pmatrix} 1 & 0 \\ z & 1 \end{pmatrix} \begin{pmatrix} 1 & \dfrac{-2z}{1+z^2} \\ 0 & 1 \end{pmatrix} \begin{pmatrix} 1 & 0 \\ z & 1 \end{pmatrix} = \frac{1}{1+z^2} \begin{pmatrix} 1-z^2 & -2z \\ 2z & 1-z^2 \end{pmatrix}. \tag{9.14}$$

Since free spaces are positive, $z > 0$, this setup realizes fractional Fourier transformers $\mathcal{F}^\alpha$ for $\sin \frac{1}{2}\pi\alpha = 2z/(1+z^2) > 0$, i.e. $0 \leq \alpha < 2$ [see (3.19)–(3.20), Fig. 3.2 and the remark on page 90]. With 3 such transformers we can cover the full range of α modulo 4. The representing matrix (9.14) is orthogonal ($\mathbf{O}^\top = \mathbf{O}^{-1}$) and its effect is to rotate phase space around the origin (see Fig. 3.2). For D-dim systems, the matrices are $2D \times 2D$ *ortho-symplectic*, as will be seen below in Sects. 9.3 and 10.3. •

Remark: Positive definite matrices. A complex N-dim[3] matrix $\mathbf{H}$ is called *positive definite* if it is self-adjoint (i.e., if it equals its transpose conjugate, $\mathbf{H} = \mathbf{H}^\dagger$), and $\mathbf{v}^\dagger \mathbf{H}\, \mathbf{v} > 0$ for all vectors $\mathbf{v} \in \mathsf{R}^N$. In particular, the 2×2 matrix $\mathbf{H}$ in (9.12)–(9.13) can be diagonalized (because it is symmetric and real, $\mathbf{H} = \mathbf{H}^\top = \mathbf{H}^*$); its determinant is unity, so $|\lambda| > \sqrt{\rho^2 + \eta^2}$, and its two eigenvalues are $\lambda_\pm := \lambda \pm \sqrt{\rho^2+\eta^2}$. When $\lambda > 0$, $\lambda_\pm > 0$ and $\mathbf{H}$ is positive definite, while when $\lambda < 0$, the matrix is *negative definite*, with $\lambda_\pm < 0$. In the latter case, $-\mathbf{H}$ will be positive definite. •

Remark: Bargmann factorization of optical systems. It is a well known theorem that any real matrix can be decomposed into the product of an orthogonal and a positive definite symmetric matrix [55, Sect. 15] (or, for complex matrices, a unitary and a self-adjoint one). This is called the *polar* decomposition of matrices. Consequently, all paraxial optical systems (this is evident at least for the 1-dim case above) can be composed with a Fourier transformer and a symmetric subsystems (the former form a group, the latter do not). This factorization was used by Bargmann [15] to unravel the manifold of the 3-dim Lorentz group, $\mathsf{SO}(2,1) \stackrel{1:2}{=} \mathsf{Sp}(2,\mathsf{R})$. •

Remark: Units. Position coordinates $\mathbf{q}$ have units of distance, as do the displacements z. In optical models, momentum $\mathbf{p}$ has no units; this implies that the Gaussian power of lenses, i.e., the matrix elements of $\mathbf{G}$, have units of inverse distance. When performing linear transformations therefore, the

[3] We speak of optical systems with D-dim screens and $2D$-dim phase spaces, the cases of optical interest being $D = 1$ and 2; but when we refer to group and matrix dimensions, we favor the letter N.

lower-left matrix elements [see (9.3)] will have units of distance, and the upper-right elements of inverse distance [see (9.5)]. •

When we arrange our optical elements $\mathcal{D}_z$ and $\mathcal{L}_G$ along a workbench – from left to right, as usual – the matrices (9.3) and (9.5) that carry their linear action on phase space will compose through multiplication in a way that merits close attention. In Sect. 3.4 and 4.2 we pointed out that operators act on beam functions $\rho(\mathbf{p},\mathbf{q})$ by 'jumping into' their arguments [see this chain of events in (3.40)–(3.43) and (4.14)] as

$$\mathcal{M} : \rho(\mathbf{p},\mathbf{q}) = \rho(\mathcal{M} : \mathbf{p},\, \mathcal{M} : \mathbf{q}) = \rho(\mathbf{p}'(\mathbf{p},\mathbf{q}), \mathbf{q}'(\mathbf{p},\mathbf{q})) = \rho'(\mathbf{p},\mathbf{q}). \qquad (9.15)$$

Any other operator coming next from the left will see the argument of the new function $\rho'(\mathbf{p},\mathbf{q})$ and act on it in turn. This happens automatically when these linear operators are realized by differential operators, such as the Lie exponentials $\mathcal{D}_z$ and $\mathcal{L}_G$ in (9.2) and (9.4). They act on the variables $(\mathbf{p},\mathbf{q})$ wherever they find them.

We thus require that the concatenation of two or more linear operators $\mathcal{M}(\mathbf{M}_1)$, $\mathcal{M}(\mathbf{M}_2)$, parametrized by the matrices $\mathbf{M}_1$ and $\mathbf{M}_2$ standing for optical sub-systems placed from left to right, should produce a linear maps of phase space which are consistent with the ordinary, ordered matrix multiplication. Below we show that

$$\begin{aligned} \mathcal{M}(\mathbf{M}) : \begin{pmatrix} \mathbf{p} \\ \mathbf{q} \end{pmatrix} = \begin{pmatrix} \mathcal{M}(\mathbf{M}) : \mathbf{p} \\ \mathcal{M}(\mathbf{M}) : \mathbf{q} \end{pmatrix} = \mathbf{M}^{-1} \begin{pmatrix} \mathbf{p} \\ \mathbf{q} \end{pmatrix} \\ \Rightarrow \quad \mathcal{M}(\mathbf{M}_1)\, \mathcal{M}(\mathbf{M}_2) = \mathcal{M}(\mathbf{M}_1\, \mathbf{M}_2). \end{aligned} \qquad (9.16)$$

We place no restrictions on $\mathbf{M}$ save $\det \mathbf{M} \neq 0$, because of the necessity of existence of $\mathbf{M}^{-1}$; the conditions of the Hamiltonian theory (paraxial canonicity), will be imposed in the next section. To prove (9.16) we use (9.15) twice,

$$\begin{aligned} \Big(\mathcal{M}(\mathbf{M}_1)\, \mathcal{M}(\mathbf{M}_2)\Big) : \begin{pmatrix} \mathbf{p} \\ \mathbf{q} \end{pmatrix} &= \mathcal{M}(\mathbf{M}_1) : \left(\mathcal{M}(\mathbf{M}_2) : \begin{pmatrix} \mathbf{p} \\ \mathbf{q} \end{pmatrix} \right) \\ &= \mathcal{M}(\mathbf{M}_1) : \left(\mathbf{M}_2^{-1} \begin{pmatrix} \mathbf{p} \\ \mathbf{q} \end{pmatrix} \right) \\ &= \mathbf{M}_2^{-1} \begin{pmatrix} \mathcal{M}(\mathbf{M}_1) : \mathbf{p} \\ \mathcal{M}(\mathbf{M}_1) : \mathbf{q} \end{pmatrix} \\ &= \mathbf{M}_2^{-1}\, \mathbf{M}_1^{-1} \begin{pmatrix} \mathbf{p} \\ \mathbf{q} \end{pmatrix} = (\mathbf{M}_1\, \mathbf{M}_2)^{-1} \begin{pmatrix} \mathbf{p} \\ \mathbf{q} \end{pmatrix} \\ &= \mathcal{M}(\mathbf{M}_1\, \mathbf{M}_2) : \begin{pmatrix} \mathbf{p} \\ \mathbf{q} \end{pmatrix}. \end{aligned} \qquad (9.17)$$

From this follows that $\mathcal{M}(\mathbf{1}) = \mathit{1}$ is the unit operator, that the inverse exists and is $[\mathcal{M}(\mathbf{M})]^{-1} = \mathcal{M}(\mathbf{M}^{-1})$, and that the concatenation of three or more factors will be associative.

The two basic optical elements, displacements (9.2)–(9.3) and thin lenses (9.4)–(9.5), will be henceforth labeled by their matrices in the following way:

$$\mathcal{D}_z =: \mathcal{M}\begin{pmatrix} \mathbf{1} & \mathbf{0} \\ -z\mathbf{1} & \mathbf{1} \end{pmatrix}, \qquad \mathcal{L}_{\mathrm{G}} =: \mathcal{M}\begin{pmatrix} \mathbf{1} & \mathbf{G} \\ \mathbf{0} & \mathbf{1} \end{pmatrix}. \tag{9.18}$$

Now we can pose again the question: paraxial optical systems built from free flights (one parameter each) and thin lenses [$\frac{1}{2}D(D+1)$ parameters each for symmetric $\mathbf{G}$], how many independent parameters can they have? And how do we realize each parameter set on the workbench? It is rather awkward to meet this problem head-on, so we shall follow group-theoretical tactics of encroachment. In the rest of this chapter we study the matrices $\mathbf{M}$ that represent all paraxial optical systems. Instead of writing their action fastidiously with $\mathbf{M}^{-1}$, we call the matrices simply by $\mathbf{M}$. In Chap. 10 we shall examine optical arrangements and write their operators as we did in (9.18).

9.2 Linear canonical maps and symplectic matrices

The most general linear transformation between the paraxial R^{2D} phase space coordinates has the form

$$\mathcal{M}(\mathbf{M}) : \begin{cases} p_i \mapsto p_i' = \sum_j A_{i,j} p_j + \sum_j B_{i,j} q_j, \\ q_i \mapsto q_i' = \sum_j C_{i,j} p_j + \sum_j D_{i,j} q_j. \end{cases} \tag{9.19}$$

We demand that the linear transformation be *canonical* (see the definition in Sect. 3.2), i.e., that it mantain the Hamiltonian structure of phase space. This implies that the Jacobian matrix of the transformation (3.11),

$$\mathbf{M} := \begin{pmatrix} \partial p_i'/\partial p_j & \partial p_i'/\partial q_j \\ \partial q_i'/\partial p_j & \partial q_i'/\partial q_j \end{pmatrix} = \begin{pmatrix} \|A_{i,j}\| & \|B_{i,j}\| \\ \|C_{i,j}\| & \|D_{i,j}\| \end{pmatrix} =: \begin{pmatrix} \mathbf{A} & \mathbf{B} \\ \mathbf{C} & \mathbf{D} \end{pmatrix}, \tag{9.20}$$

must preserve the Hamilton equations (2.5)–(2.6). These are written in vector form of (3.10) and (3.24) with a Hamiltonian function $h(\mathbf{p},\mathbf{q})$, as follows:

$$\frac{d}{dz}\begin{pmatrix} \mathbf{p} \\ \mathbf{q} \end{pmatrix} = \boldsymbol{\Omega}\begin{pmatrix} \partial/\partial\mathbf{p} \\ \partial/\partial\mathbf{q} \end{pmatrix} h(\mathbf{p},\mathbf{q}), \quad \boldsymbol{\Omega} := \begin{pmatrix} \mathbf{0} & -\mathbf{1} \\ \mathbf{1} & \mathbf{0} \end{pmatrix}. \tag{9.21}$$

For the transformed quantities thus,

$$\begin{aligned} \frac{d}{dz}\begin{pmatrix} \mathbf{p}' \\ \mathbf{q}' \end{pmatrix} &= \mathbf{M}\frac{d}{dz}\begin{pmatrix} \mathbf{p} \\ \mathbf{q} \end{pmatrix} = \mathbf{M}\boldsymbol{\Omega}\begin{pmatrix} \partial/\partial\mathbf{p} \\ \partial/\partial\mathbf{q} \end{pmatrix} h(\mathbf{p},\mathbf{q}) \\ &= \mathbf{M}\boldsymbol{\Omega}\mathbf{M}^{\top}\begin{pmatrix} \partial/\partial\mathbf{p}' \\ \partial/\partial\mathbf{q}' \end{pmatrix} h'(\mathbf{p}',\mathbf{q}'), \\ h(\mathbf{p},\mathbf{q}) &= h'(\mathbf{p}'(\mathbf{p},\mathbf{q}),\mathbf{q}'(\mathbf{p},\mathbf{q})). \end{aligned} \tag{9.22}$$

When the original system is Hamiltonian and $\mathbf{M}\,\boldsymbol{\Omega}\,\mathbf{M}^{\top}$ is again $\boldsymbol{\Omega}$, the $\mathcal{M}$-transformed system will be also Hamiltonian. As a consequence, the basic Poisson brackets (3.18) are recovered, namely $\{p'_i, p'_j\} = 0$, $\{q'_i, q'_j\} = 0$, and $\{q'_i, p'_j\} = \delta_{i,j}$. A canonical transformation matrix $\mathbf{M}$, representing the most general linear optical system, must therefore fulfill the following definition.

Definition: *A* $2D \times 2D$ *matrix* $\mathbf{M}$ *which satisfies*

$$\mathbf{M}\,\boldsymbol{\Omega}\,\mathbf{M}^{\top} = \boldsymbol{\Omega}, \tag{9.23}$$

is a **symplectic matrix***, and* $\boldsymbol{\Omega}$ *is called the symplectic* **metric** *matrix. The set of symplectic matrices under multiplication form a group, called the (real)* **symplectic group***, and is denoted* $\mathsf{Sp}(2D, \mathsf{R})$.

Remark: Symplectic matrices form a group. We should check that symplectic matrices, satisfying the condition (9.23), indeed form a group:

(i) Closure holds, because when $\mathbf{M}_1$ and $\mathbf{M}_2$ are symplectic, then so is their product,

$$(\mathbf{M}_1\mathbf{M}_2)\,\boldsymbol{\Omega}\,(\mathbf{M}_1\mathbf{M}_2)^{\top} = \mathbf{M}_1\,\mathbf{M}_2\,\boldsymbol{\Omega}\,\mathbf{M}_2^{\top}\,\mathbf{M}_1^{\top} = \mathbf{M}_1\,\boldsymbol{\Omega}\,\mathbf{M}_1^{\top} = \boldsymbol{\Omega}. \tag{9.24}$$

(ii) The unit matrix $\mathbf{1}$ is evidently symplectic.

(iii) Every symplectic matrix has an inverse,

$$\mathbf{M}^{-1} = \boldsymbol{\Omega}\,\mathbf{M}^{\top}\,\boldsymbol{\Omega}, \quad \text{i.e.,} \quad \begin{pmatrix} \mathbf{A} & \mathbf{B} \\ \mathbf{C} & \mathbf{D} \end{pmatrix}^{-1} = \begin{pmatrix} \mathbf{D}^{\top} & -\mathbf{B}^{\top} \\ -\mathbf{C}^{\top} & \mathbf{A}^{\top} \end{pmatrix}, \tag{9.25}$$

which is easily seen to be symplectic too.

(iv) Associativity holds, as it does for all matrices. •

Remark: Properties of $\boldsymbol{\Omega}$. The symplectic metric matrix $\boldsymbol{\Omega} := \begin{pmatrix} \mathbf{0} & -\mathbf{1} \\ \mathbf{1} & \mathbf{0} \end{pmatrix}$ is a symplectic matrix, because it satisfies (9.23) with several interesting properties:

$$\boldsymbol{\Omega}^{\top} = -\boldsymbol{\Omega} = \boldsymbol{\Omega}^{-1}, \quad \boldsymbol{\Omega}^2 = -\mathbf{1}, \quad \boldsymbol{\Omega}^4 = \mathbf{1}. \tag{9.26}$$

It defines the transformation of phase space $\mathcal{M}(\boldsymbol{\Omega}) : \begin{pmatrix} \mathbf{p} \\ \mathbf{q} \end{pmatrix} = \begin{pmatrix} \mathbf{q} \\ -\mathbf{p} \end{pmatrix}$, which we recognize as *the* Fourier transform [cf. (3.19)–(3.20)]. •

Remark: Determinant of symplectic matrices. The determinant of (9.23) is $(\det \mathbf{M})^2 = 1$. Of the two possibilities, only the $\det \mathbf{M} = +1$ component will be in the manifold connected to the unit matrix. The manifold of the symplectic group is itself connected (as we shall see, *multiply* connected), because in its polar decomposition (into an orthogonal matrix of even dimension and a positive-definite symmetric matrix) the two factors clearly have determinant $+1$ on their components connected to the identity. This excludes the orthogonal reflection matrices from the first factor because their determinant is -1, such as the matrix $\mathbf{K} = \operatorname{diag}(-\mathbf{1}, \mathbf{1})$, which reverses the sign of

p, and thus of the fundamental Poisson bracket $\{q_i, p_j\}$ of any Hamiltonian system. Although it is not a proper *canonical* transformations of phase space, we shall see below that this particular non-symplectic matrix serves well to study reflection. •

The free displacement and the thin lens matrices [(9.3) and (9.5)] are easily checked to be symplectic. To check generic matrices for symplecticity, we write them in 2×2 block form, so that (9.23) reads

$$\begin{pmatrix} -\mathbf{A}\mathbf{B}^\top + \mathbf{B}\mathbf{A}^\top & -\mathbf{A}\mathbf{D}^\top + \mathbf{B}\mathbf{C}^\top \\ -\mathbf{C}\mathbf{B}^\top + \mathbf{D}\mathbf{A}^\top & -\mathbf{C}\mathbf{D}^\top + \mathbf{D}\mathbf{C}^\top \end{pmatrix} = \begin{pmatrix} \mathbf{0} & -\mathbf{1} \\ \mathbf{1} & \mathbf{0} \end{pmatrix}. \tag{9.27}$$

From the diagonal elements [and again from (9.23) with $\mathbf{M} \mapsto \mathbf{M}^{-1}$], we are led to the following **symplectic conditions**:

$$\mathbf{A}\mathbf{B}^\top,\ \mathbf{A}^\top\mathbf{C},\ \mathbf{B}^\top\mathbf{D},\ \text{and}\ \mathbf{C}\mathbf{D}^\top \quad \text{are symmetric} \tag{9.28}$$

(of these four only two are independent conditions); from the off-diagonal elements, we find the 'determinant' condition

$$\mathbf{A}\mathbf{D}^\top - \mathbf{B}\mathbf{C}^\top = \mathbf{1}. \tag{9.29}$$

Remark: The accident of 2×2 matrices. In the particular case of flat optics ($D = 1$, as most of our figures) the submatrices are numbers, so (9.29) only requires that the determinant of **M** be unity, while the conditions (9.28) are vacuous. The determinant restricts by one the number of independent matrix parameters from 4 to 3. Thus $\mathsf{Sp}(2, \mathsf{R}) = \mathsf{SL}(2, \mathsf{R})$ is the 3-parameter Lie group of real matrices of unit determinant. This is an 'accidental' equality between low-dimensional Lie groups; in fact it happens that also $\mathsf{Sp}(2, \mathsf{R}) = \mathsf{SU}(1, 1)$, and $\mathsf{Sp}(2, \mathsf{R}) \overset{2:1}{=} \mathsf{SO}(2, 1)$, as will be seen below. •

Remark: Number of independent symplectic parameters. In general $2D \times 2D$ matrices there are $4D^2$ parameters; since (9.27) is a skew-symmetric matrix, it entails $\frac{1}{2}(2D)(2D-1)$ bilinear restrictions, and this determines that $\mathsf{Sp}(2D, \mathsf{R})$ is a $D(2D+1)$-dimensional manifold. The relevant group for linear transformations of the real paraxial world, $\mathsf{Sp}(4, \mathsf{R})$, has thus 10 parameters. •

One book dedicated to symplectic matrices and its application for first-order systems is given in [77]; applications to quantum mechanics and quantum optics are archived in [4]. The reader who is comfortable using only the matrix parameters of paraxial optical elements can proceed to the next chapter, and take on faith the group-theoretical results on subgroups and decompositions that are detailed and proven in the remainder of this one. However, to appreciate the metaplectic sign of canonical integral transforms in paraxial wave optics (appendix C), this is necessary background.

9.3 Orthogonal and unitary matrices

A real N-dim nondegenerate *metric* matrix $\mathbf{\Upsilon}$ defines a sesquilinear[4] *inner product* $(\mathbf{u},\mathbf{v})_\Upsilon := \mathbf{u}^\dagger\,\mathbf{\Upsilon}\,\mathbf{v}$ between complex N-vectors $\mathbf{u}$ and $\mathbf{v}$. The metric serves to distinguish otherwise equivalent N-dimensional vector spaces [56, Sect. 2.III]. Metrics of special interest are:

$$\begin{aligned} \text{symmetric, } (\mathbf{u},\mathbf{v})_\Upsilon &= (\mathbf{v},\mathbf{u})^*_\Upsilon & \mathbf{\Upsilon}^\dagger &= \mathbf{\Upsilon}, \text{ and} \\ \text{real skew-symmetric, } (\mathbf{u},\mathbf{v})_\Omega &= -(\mathbf{v},\mathbf{u})_\Omega & \mathbf{\Omega}^\top &= -\mathbf{\Omega}. \end{aligned}$$

The antidiagonal-block form of the symplectic metric matrix $\mathbf{\Omega}$ in (9.21) depends on our choice of (*Darboux*) coordinates, and its dimension must be even: $N = 2D$.

Symmetric metrics $\mathbf{\Upsilon}$ can be always diagonalized and rescaled to a form with $+1$'s and -1's, but only when $\mathbf{\Upsilon} = \mathbf{1}$ will the inner product lead to a *positive-definite norm* $[|\mathbf{u}|^2 = \mathbf{u}^\dagger\mathbf{u} \geq 0$ and $\mathbf{u}^\dagger\mathbf{u} = 0 \Rightarrow \mathbf{u} = \mathbf{0}]$. The symplectic metric cannot define a norm because $(\mathbf{u},\mathbf{u})_\Omega = 0$ for all $\mathbf{u} \in \mathsf{R}^{2D}$. But in every case, matrices that preserve an inner product $(\mathbf{u},\mathbf{v}) = (\mathbf{M}\,\mathbf{u},\,\mathbf{M}\,\mathbf{v})$ form a *group*, as can be shown easily by following the steps in (9.24).

Definition: *A* $(N{+}M)\times(N{+}M)$ *matrix* $\mathbf{O}$ *which satisfies*

$$\mathbf{O}\,\mathbf{\Upsilon}_{N,M}\,\mathbf{O}^\top = \mathbf{\Upsilon}_{N,M}, \quad \mathbf{\Upsilon}_{N,M} := \operatorname{diag}(\overbrace{1,\ldots,1}^{N},\overbrace{-1,\ldots,-1}^{M}), \;\; \det\mathbf{O} = +1. \tag{9.30}$$

is a (pseudo-) **orthogonal matrix**. *The matrix elements can be real, complex numbers, or quaternions; when the field is unspecified we assume it is* R, *and denote their group by* $\mathsf{O}(N,M)$. *Orthogonal matrices whose determinant is* $\det\mathbf{O} = 1$ *form a subgroup, called the* **special** *(pseudo-) orthogonal group,* $\mathsf{SO}(N,M)$. *When* $M = 0$, $\mathbf{\Upsilon} = \mathbf{1}$ *and* $\mathbf{O}\mathbf{O}^\top = \mathbf{1}$, *the group is then* $\mathsf{SO}(N)$ *and we drop the pseudo-.*

Remark: Disconnected components of the pseudo-orthogonal groups. Starting from the identity $\det\mathbf{1} = 1$, a continuous line through the orthogonal group manifold will reach all matrices in $\mathsf{O}(N,M)$ within the *component* of the identity. Now, a reflection of the kth axis is produced by a diagonal orthogonal matrix $\mathbf{O}^{\text{R}}_k$ with 1's and a single -1 in the kth position, whose determinant is -1. The groups $\mathsf{O}(N)$ have thus two components and only $\mathsf{SO}(N)$ is a subgroup because it contains the identity. For $M \geq 1$, when we multiply two reflections on axes k and k' with the same sign of the metric $\mathbf{\Upsilon}_{N,M}$, their product $\mathbf{O}^{\text{R}}_k\mathbf{O}^{\text{R}}_{k'}$ is a rotation by π in the k–k' plane, which is in

[4] This means literally *one-and-one-half* linear, i.e., linear in the second argument and antilinear (complex conjugate) in the first (some mathematicians prefer the opposite convention); $\mathbf{u}^\dagger$ is the *adjoint* (transpose complex conjugate) of $\mathbf{u}$, or simple transpose for real vectors.

the component of the identity. And if the axes k and k' have different metric sign, their product matrix will have determinant $+1$ but will be in a manifold disconnected from the identity. This feature is well known in the *Lorentz* group $\mathsf{O}(3,1)$ for special relativity: there are reflections of space $(^{\uparrow}_{-})$, of time $(^{\downarrow}_{-})$, and of space-time $(^{\downarrow}_{+})$. The orthogonal groups $\mathsf{O}(N,M)$ have similarly 4 components $\mathsf{O}(N,M)^{\updownarrow}_{\pm}$, only one of which, $\mathsf{O}(N,M)^{\uparrow}_{+}$ contains the identity; this is the component we mean when writing simply $\mathsf{SO}(N,M)$. •

Remark: The orthogonality conditions. An orthogonal matrix satisfies $\mathbf{O}\mathbf{O}^{\top} = \mathbf{1}$; when we write $\mathbf{O} \in \mathsf{SO}(2D)$ in block form, we find the orthogonality conditions

$$\begin{pmatrix} \mathbf{A}\mathbf{A}^{\top} + \mathbf{B}\mathbf{B}^{\top} & \mathbf{A}\mathbf{C}^{\top} + \mathbf{B}\mathbf{D}^{\top} \\ \mathbf{C}\mathbf{A}^{\top} + \mathbf{D}\mathbf{B}^{\top} & \mathbf{C}\mathbf{C}^{\top} + \mathbf{D}\mathbf{D}^{\top} \end{pmatrix} = \begin{pmatrix} \mathbf{1} & \mathbf{0} \\ \mathbf{0} & \mathbf{1} \end{pmatrix} \qquad (9.31)$$

[cf. (9.27); from $\mathbf{O}^{\top}\mathbf{O} = \mathbf{1}$ we find different but not independent relations].•

Remark: Number of independent orthogonal parameters. The matrix equality in (9.31) is symmetric and contains $\frac{1}{2}(2D)(2D+1)$ conditions on its $(2D)^2$ elements, so the number of independent parameters of $\mathsf{SO}(2D)$ is $D(2D-1)$. Moreover, since the rows and the columns of orthogonal matrices are vectors of unit norm due to (9.31), the range of these $\mathsf{SO}(2D)$ parameters is *compact* (i.e., finite), and so is the *volume* of the group with the *Haar* measure defined in (B.12) *et seq.* •

Of course, not all orthogonal matrices can be used to represent optical systems, because they have to be symplectic also. A matrix which satisfies both the symplectic conditions (9.23) and the orthogonality conditions (9.30), is characterized as an **ortho-symplectic** matrix. It has the form

$$\left.\begin{matrix} \mathbf{O}^{\top} = \mathbf{O}^{-1} \\ \mathbf{O}\,\boldsymbol{\Omega} = \boldsymbol{\Omega}\,\mathbf{O} \end{matrix}\right\} \Rightarrow \mathbf{O} = \begin{pmatrix} \mathbf{A} & -\mathbf{B} \\ \mathbf{B} & \mathbf{A} \end{pmatrix} \qquad \left\{\begin{matrix} \mathbf{A}\mathbf{B}^{\top} \text{ symmetric,} \\ \mathbf{A}\mathbf{A}^{\top} + \mathbf{B}\mathbf{B}^{\top} = \mathbf{1}. \end{matrix}\right. \qquad (9.32)$$

The set of ortho-symplectic matrices is the manifold $\mathsf{Sp}(2D,\mathsf{R}) \cap \mathsf{SO}(2D)$, and therefore a group that we proceed to identify.

Definition: *A* $(N+M) \times (N+M)$ *complex matrix* $\mathbf{U}$ *which satisfies*

$$\mathbf{U}\,\boldsymbol{\Upsilon}_{N,M}\,\mathbf{U}^{\dagger} = \boldsymbol{\Upsilon}_{N,M}. \qquad (9.33)$$

is a (pseudo-) **unitary matrix***; its group is denoted* $\mathsf{U}(N,M)$*. The determinant of this condition is* $|\det \mathbf{U}| = 1$*, and thus*

$$\mathbf{U} = e^{i\omega}\,\mathbf{V}, \quad \det \mathbf{V} := 1, \quad \omega := \frac{1}{D} \arg\det \mathbf{U}, \quad D = N + M. \qquad (9.34)$$

The matrices $\mathbf{V}$ *of unit determinant form the subgroup* $\mathsf{SU}(N,M) \subset \mathsf{U}(N,M)$ *of* **special** *unitary matrices, while* $e^{i\omega} \in \mathsf{U}(1)$*. Again, when* $M = 0$ *and* $\mathbf{U}\mathbf{U}^{\dagger} = \mathbf{1}$*, we drop the pseudo-.*

And now, out of the two real submatrices $\mathbf{A}$ and $\mathbf{B}$ in (9.32), we produce the complex linear combination

$$\mathbf{U}(\mathbf{O}) = \mathbf{A} + i\mathbf{B}, \quad \mathbf{U}\,\mathbf{U}^\dagger = \mathbf{A}\,\mathbf{A}^\top + \mathbf{B}\,\mathbf{B}^\top - i(\mathbf{A}\,\mathbf{B}^\top - \mathbf{B}\,\mathbf{A}^\top) = \mathbf{1}, \quad (9.35)$$

which is a $D \times D$ *unitary* matrix.

Remark: Counting parameters. The $2D^2$ free parameters in the ortho-symplectic blocks $\mathbf{A}$ and $\mathbf{B}$ of (9.32), are curtailed by the $\frac{1}{2}D(D-1)$ restrictions to form symmetric $\mathbf{A}\mathbf{B}^\top$, and by the $\frac{1}{2}D(D+1)$ equalities to $\mathbf{1}$; so there remain D^2 parameters with compact ranges. Compare this with the number of *real* independent parameters in $D \times D$ complex matrices, $2D^2$, satisfying (9.33) with $\frac{1}{2}D(D-1)$ real and $\frac{1}{2}D(D+1)$ pure imaginary restrictions; this confirms that the manifolds of $\mathsf{U}(D)$ and $\mathsf{Sp}(2D,\mathsf{R}) \cap \mathsf{SO}(2D)$ have the same number D^2 of independent and bounded parameters. •

Remark: The manifold of $\mathsf{U}(D)$ matrices. The factors $e^{i\omega}\mathbf{1}$ in (9.34) naturally commute with all matrices (they are the *center* of the group $\mathsf{U}(D)$). But note that among them, the D matrices $\{e^{2\pi i n/D}\mathbf{1}\}_{n=0}^{D-1} \in \mathcal{Z}_D$ belong to both $\mathsf{U}(1)$ and $\mathsf{SU}(D)$, so

$$\begin{array}{ll} \mathsf{U}(D) = \mathsf{U}(1) \times \mathsf{SU}(D)/\mathcal{Z}_D, & \text{as manifolds, and} \\ \mathsf{U}(D) = \mathsf{U}(1) \otimes \mathsf{SU}(D), & \text{as Lie groups.} \end{array} \quad (9.36)$$

The unitary groups are *compact*, and $\mathsf{SU}(D)$ is simply connected, except for $\mathsf{U}(1) = \mathsf{SO}(2)$, which is a circle, and thus infinitely connected. The decomposition (9.36) with $D = M + N$ also applies to the noncompact pseudo-unitary groups. •

Remark: On $\mathsf{Sp}(2D,\mathsf{R}) \cap \mathsf{SO}(2D) = \mathsf{U}(D)$. Ortho-symplectic matrices can be written as

$$\mathbf{O}(\mathbf{U}) = \begin{pmatrix} \mathrm{Re}\,\mathbf{U} & -\mathrm{Im}\,\mathbf{U} \\ \mathrm{Im}\,\mathbf{U} & \mathrm{Re}\,\mathbf{U} \end{pmatrix}, \qquad \mathbf{U} \in \mathsf{U}(D), \quad (9.37)$$

because the real and imaginary parts of the unitary conditions (9.33) coincide with the ortho-symplectic conditions in (9.32). We check that the product of $\mathbf{O}(\mathbf{U}_1)$ and $\mathbf{O}(\mathbf{U}_2)$ is $\mathbf{O}(\mathbf{U}_1\,\mathbf{U}_2)$, that $\mathbf{O}(\mathbf{1}) = \mathbf{1}$, that the inverse entails $\mathbf{O}(\mathbf{U})^{-1} = \mathbf{O}(\mathbf{U}^{-1}) = \mathbf{O}(\mathbf{U}^\dagger) = \mathbf{O}(\mathbf{U})^\top$, and that there is no problem with association. Conversely $\mathbf{U}(\mathbf{O})$, defined in (9.35), has the same group properties; for every matrix $\mathbf{U}$ there is one and only one ortho-symplectic matrix $\mathbf{O}$. We conclude that the group of $2D \times 2D$ real ortho-symplectic matrices and $\mathsf{U}(D)$ are *isomorphic*; the former is a *representation* of the latter by real $2D \times 2D$ matrices. Paraxial optical systems $\mathcal{M}\big(\mathbf{O}(\mathbf{U})\big)$ are $\mathsf{U}(D)$-*fractional* $\mathsf{U}(D)$-*Fourier transformers* [131] that will be presented in Sect. 10.3. •

Remark: $\mathsf{SO}(D) \subset \mathsf{U}(D)$. Real unitary matrices are orthogonal matrices, as can be seen restricting (9.33) to the real field, which then become the orthogonality conditions (9.30). When $\mathbf{R}$ is such a matrix, the corresponding ortho-symplectic matrix is $\mathbf{O}(\mathbf{R}) = \mathrm{diag}\,(\mathbf{R},\mathbf{R})$. •

9.4 Bargmann parameters and group covers

To study the structure of the symplectic group of paraxial optical systems, it would be uncomfortable to use the parameters afforded by the optical setup (composed of displacements and lenses), because these cut the group manifold into various pieces (one-, two-, and three-lens arrangements, to be seen in Sect. 10.5) which do not relate simply to the parameters in the matrix entries,[5] nor do they follow subgroups, cosets, nor conjugation classes – among the clever structures known and favored by group-theorists. Our journey therefore goes in the opposite direction: we start with a parametrization of indubitable mathematical value – *Bargmann*'s – and leave for the next chapter its optical interpretation. We look first at the 2×2 matrices of $\mathsf{Sp}(2, \mathsf{R})$ in flat optics to understand the $2D \times 2D$ case, particularly the connectivity and the cover of the symplectic group manifold.

The generalization of the polar parameters of the plane, $c = |c|\, e^{i\phi}$ – a positive number $|c|$ and a phase $e^{i\phi}$, is the *polar decomposition* of a complex matrix. This factors it uniquely into a product of a positive definite hermitian and unitary matrix, respectively [55, Sect. 15] and real matrices decompose into the product of a positive definite symmetric and an orthogonal matrix. This was used by Bargmann to study the structure of the symplectic groups [15], so we refer to it as the *Bargmann* parametrization of $\mathsf{Sp}(2D, \mathsf{R})$ optical systems. A generic 2×2 real matrix of unit determinant is characterized by an angle $\phi \in \mathcal{S}^1$ and a complex number $\mu \in \mathsf{C}$, as

$$\begin{pmatrix} a & b \\ c & d \end{pmatrix} = \begin{pmatrix} \cos\phi & -\sin\phi \\ \sin\phi & \cos\phi \end{pmatrix} \begin{pmatrix} \lambda + \operatorname{Re}\mu & \operatorname{Im}\mu \\ \operatorname{Im}\mu & \lambda - \operatorname{Re}\mu \end{pmatrix} \tag{9.38}$$

$$= \begin{pmatrix} \lambda\cos\phi + \operatorname{Re}(e^{i\phi}\mu) & -\lambda\sin\phi + \operatorname{Im}(e^{i\phi}\mu) \\ \lambda\sin\phi + \operatorname{Im}(e^{i\phi}\mu) & \lambda\cos\phi - \operatorname{Re}(e^{i\phi}\mu) \end{pmatrix}, \tag{9.39}$$

$$\phi \in \mathsf{R} \bmod 2\pi, \quad \lambda := +\sqrt{|\mu|^2 + 1} \geq 1, \quad \mu \in \mathsf{C}. \tag{9.40}$$

The left matrix factor is the fractional Fourier transform of angle ϕ given in (3.39) [see also (9.14)], while the right factor is the symmetric matrix (9.12) determined by μ. From (9.39) follows the relation between the Bargmann and matrix parameters,

$$e^{i\phi}\lambda = \tfrac{1}{2}[a + d - i(b - c)], \qquad e^{i\phi}\mu = \tfrac{1}{2}[a - d + i(b + c)], \tag{9.41}$$

and conversely, within the ranges specified in (9.40),

$$\phi = \arg\,[a + d + i(b - c)], \quad \mu = \tfrac{1}{2}e^{-i\phi}[a - d + i(b + c)]. \tag{9.42}$$

With the Bargmann parameters ϕ, μ we indicate the 2×2 matrices (9.39) by $\mathbf{M}(\phi; \rho, \eta)$, $\mu := \rho + i\eta$, $\rho, \eta \in \mathsf{R}$. It is convenient to keep λ even though in

[5] The simplest multiplication rule for the symplectic group is its matrix form, but it hides much – as we shall see.

the $D = 1$ case it depends on $|\mu|$ through (9.40). The $\mathsf{Sp}(2, \mathsf{R})$ product law for $\mathbf{M} = \mathbf{M}_1 \mathbf{M}_2$ can be found using (9.39) and (9.41); with elementary algebra we arrive at

$$e^{i\phi}\lambda = e^{i(\phi_1+\phi_2)}\lambda_1\lambda_2 + e^{i(\phi_1-\phi_2)}\mu_1\mu_2^* =: e^{i(\phi_1+\phi_2)}\lambda_1\nu\lambda_2, \tag{9.43}$$

$$e^{i\phi}\mu = e^{i(\phi_1+\phi_2)}\lambda_1\mu_2 + e^{i(\phi_1-\phi_2)}\mu_1\lambda_2, \tag{9.44}$$

with the complex factor

$$\nu := 1 + e^{-2i\phi_2}\,\lambda_1^{-1}\mu_1\mu_2^*\lambda_2^{-1}, \qquad |\nu - 1| < 1, \tag{9.45}$$

because $|\mu|/\lambda < 1$. This range is shown in Fig. 9.3: the complex number ν is strictly *inside* a unit circle with center at 1, which *excludes* the origin. Hence, ν has an unambiguous *phase* for all $\mu_1, \mu_2 \in \mathsf{C}$. Therefore the product (9.43)–(9.44) in the Bargmann parameters is unique, and

$$\phi = \phi_1 + \phi_2 + \arg\nu, \qquad \arg\nu \in (-\tfrac{1}{2}\pi, \tfrac{1}{2}\pi), \tag{9.46}$$

$$\mu = e^{-i\arg\nu}(\lambda_1\mu_2 + e^{-2i\phi_2}\mu_1\lambda_2), \tag{9.47}$$

$$\lambda = \lambda_1\,|\nu|\,\lambda_2 \geq 1. \tag{9.48}$$

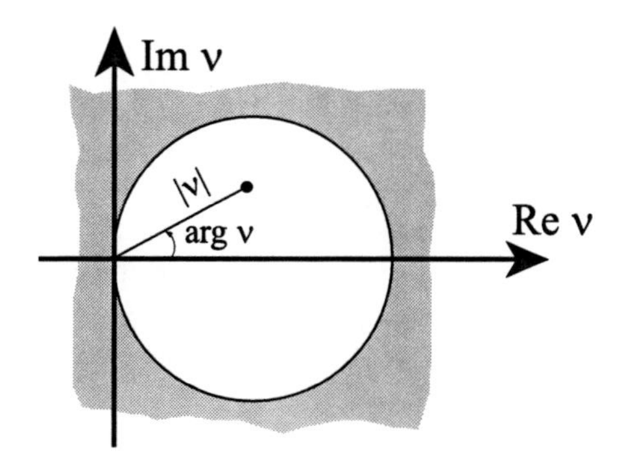

Fig. 9.3. Range of values of Bargmann's parameter ν in the complex plane. The phase $\arg\nu$ is thus uniquely determined in the range $(-\frac{1}{2}\pi, \frac{1}{2}\pi)$.

And now the connectivity of the symplectic group is revealed: when ϕ in (9.46) is counted modulo 2π, as in the range of (9.40), the $\mathsf{Sp}(2, \mathsf{R})$ matrices are uniquely specified by $\mathbf{M}(\phi, \rho, \eta)$, as originally intended. But the Bargmann presentation of the product allows us to consistently *define* the range of ϕ to be R modulo $2\pi n$, for $n = 1, 2, 3, \ldots$.

Definition: *The group whose elements are $M(\phi, \rho, \eta)$, with the parameters in the range $(\phi, \rho, \eta) \in \mathsf{R}^3$, and whose product is (9.46)–(9.47) (with $\mu = \rho + i\eta$), is the* **universal cover of** $\mathsf{Sp}(2, \mathsf{R})$, *denoted* $\overline{\mathsf{Sp}(2, \mathsf{R})}$.

Definition: *The group whose elements are $M(\phi, \rho, \eta)$, with ϕ modulo 4π, $(\rho, \eta) \in \mathsf{R}^2$, and with product law (9.46)–(9.47), is the two-fold cover of*

$\mathsf{Sp}(2,\mathsf{R})$; *it is called the* **metaplectic** *group,*[6] *denoted* $\mathsf{Mp}(2,\mathsf{R}) \stackrel{2:1}{=} \mathsf{Sp}(2,\mathsf{R})$. *The two groups are* $\mathcal{Z}_2$*-homomorphic (but not isomorphic). Similarly, counting* ϕ *modulo* 6π, 8π, *etc., we produce three-, four-, etc.-***covers** *of the symplectic group.*

The metaplectic group does *not* have a *faithful* finite-dimensional matrix representation (and neither does any other cover of $\mathsf{Sp}(2,\mathsf{R})$). This is probably the reason for its unpopularity with matrix-handlers. Yet with Bargmann parameters we see that distinct covers of the symplectic group involve only different ranges for ϕ. Further, the situation is exactly the same as the covering of a circle by the real line. And the same statement is true for arbitrary dimension D. It turns out that geometric optics does not *need* this discussion because $\mathsf{Sp}(2,\mathsf{R})$ acts faithfully with matrices on the phase space 2-vectors. The motivation for the following remarks is to prepare the background for the appendix of this part, where we wavize paraxial geometric to Fourier optics, and *this* requires the two-fold cover. We may think we have a perfect imaging device – a *unit* system, but it may turn out to be only the second metaplectic unit of the corresponding wave system.

Remark: The second metaplectic unit. In $\mathsf{Mp}(2,\mathsf{R})$ as in $\mathsf{Sp}(2,\mathsf{R})$, the (one and only) group unit is $M(0,0,0)$ in the Bargmann parameters (ϕ,ρ,η); for ϕ counted modulo 2π, the corresponding $\mathsf{Sp}(2,\mathsf{R})$ matrix is $\mathbf{1}$. It is descriptive to call $M(2\pi,0,0)$ the 'second metaplectic unit', because it also commutes with the rest of the group [check it in the presentation (9.45)–(9.48)]; and it *also* corresponds to the symplectic unit matrix $\mathbf{1}$. The only other matrix commuting with all of $\mathsf{Sp}(2,\mathsf{R})$ is $-\mathbf{1}$, which indicates that there should be a group covered twice by $\mathsf{Sp}(2,\mathsf{R})$ – to be seen below. A 2:1 homomorphism holds between the two metaplectic elements $M(\phi,\rho,\eta)$ and $M(\phi+2\pi,\rho,\eta)$, and the same symplectic matrix $\mathbf{M}(\phi,\rho,\eta)$. Within the metaplectic group, the square of the second metaplectic unit is the first (and only true) unit. Thus we write

$$\begin{aligned}\mathsf{Sp}(2,\mathsf{R}) &= \mathsf{Mp}(2,\mathsf{R})/\{M(0,0,0), M(2\pi,0,0)\}\\ &= \overline{\mathsf{Sp}(2,\mathsf{R})}/\{M(2n\pi,0,0),\ n\in\mathcal{Z}\}\end{aligned}$$

for the manifolds of the Lie groups $\mathsf{Sp}(2,\mathsf{R}) \stackrel{1:2}{=} \mathsf{Mp}(2,\mathsf{R}) \stackrel{1:\infty}{=} \overline{\mathsf{Sp}(2,\mathsf{R})}$. •

Remark: The isomorphism $\mathsf{Sp}(2,\mathsf{R}) = \mathsf{SU}(1,1)$. There is an isomorphism between real matrices of unit determinant $\mathbf{M} \in \mathsf{SL}(2,\mathsf{R}) = \mathsf{Sp}(2,\mathsf{R})$, and complex 2×2 pseudo-unitary matrices $\mathbf{V} \in \mathsf{SU}(1,1)$. It can be written as a similarity transformation by a unitary matrix $\mathbf{W} := \frac{1}{\sqrt{2}}\begin{pmatrix}1 & -i\\ 1 & i\end{pmatrix}$,

$$\mathbf{M} = \begin{pmatrix} a & b\\ c & d\end{pmatrix} = \mathbf{W}^\dagger\,\mathbf{V}\,\mathbf{W} = \begin{pmatrix}\mathrm{Re}\,(\alpha+\beta) & -\mathrm{Im}\,(\alpha-\beta)\\ \mathrm{Im}\,(\alpha+\beta) & \mathrm{Re}\,(\alpha-\beta)\end{pmatrix}, \tag{9.49}$$

$$\mathbf{V} = \begin{pmatrix}\alpha & \beta\\ \beta^* & \alpha^*\end{pmatrix} = \mathbf{W}\,\mathbf{M}\,\mathbf{W}^\dagger = \tfrac{1}{2}\begin{pmatrix}(a+d)-i(b-c) & (a-d)+i(b+c)\\ (a-d)-i(b+c) & (a+d)+i(b-c)\end{pmatrix}, \tag{9.50}$$

[6] METAPLECTIC (meh-ta**plek**tik), with μέτα BEYOND– the symplectic group.

where $ad - bc = 1 = |\alpha|^2 - |\beta|^2$. From (9.23) and (9.33) we see that $\mathbf{M}$ and $\mathbf{V}$ belong to the said groups, and it is immediate to verify that their correspondence is 1:1 and preserved under products and inversions. The two groups are (homomorphic and) *isomorphic*. •

Remark: How $\mathsf{Sp}(2,\mathsf{R})$ covers $\mathsf{SO}(2,1)$ twice. The original motivation of Bargmann [15] was to study the three-dimensional Lorentz group $\mathsf{SO}(2,1)$ and its unitary irreducible representations. In Chap. 7 we used coordinates of phase space that were adapted to systems with axial symmetry (7.3), which we now use. We note that when the 2-vector $(\mathbf{p},\mathbf{q})^\top$ undergoes the $\mathsf{Sp}(2,\mathsf{R})$ transformation $\mathbf{M}$ in (9.20), then the 3-vector of quadratic functions

$$\left(\tfrac{1}{2}pq,\ \tfrac{1}{4}(p^2 - q^2),\ \tfrac{1}{4}(p^2 + q^2)\right)^\top, \tag{9.51}$$

will undergo a linear transformation with the following 3×3 matrix:

$$\mathbf{M} = \begin{pmatrix} a & b \\ c & d \end{pmatrix} \in \mathsf{Sp}(2,\mathsf{R}) \ \stackrel{2:1}{\longleftrightarrow}\ \mathbf{D}(\pm\mathbf{M}) \in \mathsf{SO}(2,1), \tag{9.52}$$

$$\mathbf{D}(\pm\mathbf{M}) := \begin{pmatrix} ad+bc & ac-bd & ac+bd \\ ab-cd & \frac{1}{2}(a^2-b^2-c^2+d^2) & \frac{1}{2}(a^2+b^2-c^2-d^2) \\ ab+cd & \frac{1}{2}(a^2-b^2+c^2-d^2) & \frac{1}{2}(a^2+b^2+c^2+d^2) \end{pmatrix}. \tag{9.53}$$

It is straightforward to check that $\mathbf{D}$ is a pseudo-orthogonal matrix, satisfying (9.30) with $\boldsymbol{\Upsilon}_{2,1} = \operatorname{diag}(1,1,-1)$, and that $\mathbf{D}(\pm\mathbf{M}_1)\,\mathbf{D}(\pm\mathbf{M}_1) = \mathbf{D}(\pm\mathbf{M}_1\mathbf{M}_2)$. The $\mathbf{D}(\mathbf{M})$'s are a 3×3, 1:2 (*un*faithful) matrix representation of $\mathsf{Sp}(2,\mathsf{R})$, and the latter is thus the two-fold cover of the former:

$$\mathsf{SO}(2,1) = \mathsf{Sp}(2,\mathsf{R})/\{\mathbf{1},-\mathbf{1}\} = \overline{\mathsf{Sp}(2,\mathsf{R})}/\{M(n\pi,0,0),\ n \in \mathcal{Z}\}.$$

•

Remark: *Confer*: $\mathsf{SU}(1,1) \stackrel{2:1}{=} \mathsf{SO}(2,1)$ **and** $\mathsf{SU}(2) \stackrel{2:1}{=} \mathsf{SO}(3)$. The most famous homomorphism of all is that which occurs between the 'orbital' rotation group $\mathsf{SO}(3)$ and the spin group $\mathsf{SU}(2)$ [(9.49) and (9.53)]; the previous two remarks give the analogue homomorphism between their sibling noncompact groups, $\mathsf{SO}(2,1)$ and $\mathsf{SU}(1,1)$. We can compare directly with the case of spin [21, Eq. (2.22)], for $\alpha,\beta \in \mathsf{C}$ and $|\alpha|^2 - |\beta|^2 = 1$,

$$\mathbf{V} = \begin{pmatrix} \alpha & \beta \\ \beta^* & \alpha^* \end{pmatrix} \in \mathsf{SU}(1,1) \ \stackrel{2:1}{\longleftrightarrow}\ \mathbf{D}^{(2,1)}(\pm\mathbf{V}) \in \mathsf{SO}(2,1), \tag{9.54}$$

$$\mathbf{D}^{(2,1)}(\pm\mathbf{V}) := \begin{pmatrix} \operatorname{Re}(\alpha^2-\beta^2) & \operatorname{Im}(\alpha^2+\beta^2) & 2\operatorname{Im}(\alpha\beta) \\ \operatorname{Im}(\beta^2-\alpha^2) & \operatorname{Re}(\alpha^2+\beta^2) & 2\operatorname{Re}(\alpha\beta) \\ -2\operatorname{Im}(\alpha\beta^*) & 2\operatorname{Re}(\alpha\beta^*) & |\alpha|^2+|\beta|^2 \end{pmatrix}. \tag{9.55}$$

The corresponding compact group homomorphism holds also for $\alpha,\beta \in \mathsf{C}$, but is restricted to the 3-sphere $|\alpha|^2 + |\beta|^2 = 1$,

$$\mathbf{U} = \begin{pmatrix} \alpha & \beta \\ -\beta^* & \alpha^* \end{pmatrix} \in \mathsf{SU}(2) \quad \overset{2:1}{\longleftrightarrow} \quad \mathbf{D}^{(3)}(\pm\mathbf{U}) \in \mathsf{SO}(3), \tag{9.56}$$

$$\mathbf{D}^{(3)}(\pm\mathbf{U}) := \begin{pmatrix} \mathrm{Re}\,(\alpha^2 - \beta^2) & \mathrm{Im}\,(\alpha^2 + \beta^2) & -2\,\mathrm{Re}\,(\alpha\beta) \\ \mathrm{Im}\,(\beta^2 - \alpha^2) & \mathrm{Re}\,(\alpha^2 + \beta^2) & 2\,\mathrm{Im}\,(\alpha\beta) \\ 2\,\mathrm{Re}\,(\alpha\beta^*) & 2\,\mathrm{Im}\,(\alpha\beta^*) & |\alpha|^2 - |\beta|^2 \end{pmatrix}. \tag{9.57}$$

The R^3-submanifolds of the sibling groups $\mathsf{SO}(2,1)$ and $\mathsf{SO}(3)$ are shown in Fig. 9.4, with orthogonal axes $(\mathrm{Re}\,\alpha,\ \mathrm{Im}\,\alpha,\ \mathrm{Re}\,\beta)$, and $\pm\mathrm{Im}\,\beta$ foliating space by the conic surfaces

$$(\mathrm{Re}\,\alpha)^2 + (\mathrm{Im}\,\alpha)^2 - (\mathrm{Re}\,\beta)^2 = 1 + (\mathrm{Im}\,\beta)^2 \geq 1, \quad \text{for } \mathsf{SO}(2,1), \tag{9.58}$$

$$(\mathrm{Re}\,\alpha)^2 + (\mathrm{Im}\,\alpha)^2 + (\mathrm{Re}\,\beta)^2 = 1 - (\mathrm{Im}\,\beta)^2 \leq 1. \quad \text{for } \mathsf{SO}(3). \tag{9.59}$$

•

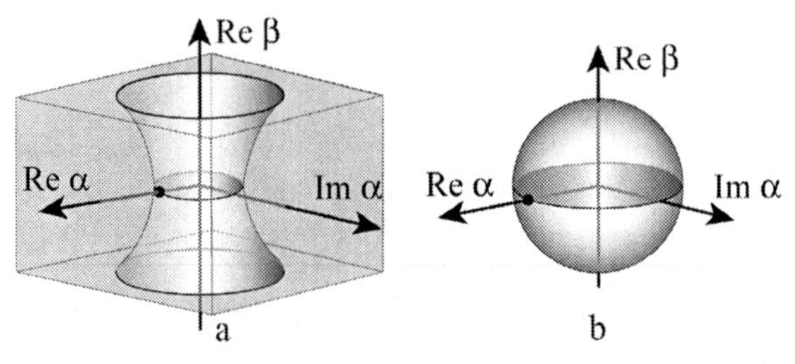

Fig. 9.4. Manifolds of two sibling Lie groups and their 2:1 covers. (**a**) $\mathsf{SO}(2,1)$ is the exterior (*shaded region*) of a one-sheeted revolution hyperboloid $|\alpha|^2 - |\beta^2| = 1$; this is a infinite (noncompact) volume which is multiply-connected, allowing for covering group manifolds; the dot represents the group unit at $(\alpha, \beta) = (1, 0)$, and the circle around the waist is the 1-parameter subgroup of fractional Fourier transforms. (**b**) $\mathsf{SO}(3)$ is the volume bounded by a 2-sphere of unit radius, $|\alpha|^2 + |\beta^2| = 1$; the dot represents the group unit $(1, 0)$. The elements of the 3×3 matrices [(9.55) and (9.57)] are quadratic in (α, β), while their 2:1 cover $\mathsf{SU}(1,1)$ and $\mathsf{SU}(2)$ are linear; thus, *two* of their elements (distinguished by $\pm\mathrm{Im}\,\beta$) correspond to each point in the above figures.

That spin is a *classical* rather than a quantum phenomenon was argued successfully by Lévy-Leblond [87] in the 1970s (although some colleagues still believe it arises only when Pauli $\boldsymbol{\sigma}$-matrices are present). Similarly, we argue here that symplectic (and metaplectic) group covers are basic features of geometric optics.

Remark: Bargmann parameters for the D-dim case. The symplectic groups of arbitrary dimension $2D$ were also subject to scrutiny by Bargmann [17]; they follow a $D \times D$ matrix version of (9.38)–(9.48), on which we remark briefly. There turns out to be again *one* distinguished phase parameter ϕ, and complex $D \times D$ matrices $\boldsymbol{\lambda}$ and $\boldsymbol{\mu}$ which are now independent but satisfy some restrictions. In terms of the block matrices of $\mathbf{M}$, (9.38) generalizes as follows:

$$\begin{pmatrix} \mathbf{A} & \mathbf{B} \\ \mathbf{C} & \mathbf{D} \end{pmatrix} = \begin{pmatrix} \cos\phi\,\mathbf{1} & -\sin\phi\,\mathbf{1} \\ \sin\phi\,\mathbf{1} & \cos\phi\,\mathbf{1} \end{pmatrix} \begin{pmatrix} \mathrm{Re}\,(\boldsymbol{\lambda}+\boldsymbol{\mu}) & \mathrm{Im}\,(-\boldsymbol{\lambda}+\boldsymbol{\mu}) \\ \mathrm{Im}\,(\boldsymbol{\lambda}+\boldsymbol{\mu}) & \mathrm{Re}\,(\boldsymbol{\lambda}-\boldsymbol{\mu}) \end{pmatrix}. \quad (9.60)$$

Corresponding to (9.41), now one has

$$e^{i\phi}\boldsymbol{\lambda} = \tfrac{1}{2}[\mathbf{A}+\mathbf{D}-i(\mathbf{B}-\mathbf{C})], \qquad e^{i\phi}\boldsymbol{\mu} = \tfrac{1}{2}[\mathbf{A}-\mathbf{D}+i(\mathbf{B}+\mathbf{C})]. \quad (9.61)$$

The crucial property $\lambda > 1$ in (9.40) is now $\det\boldsymbol{\lambda} > 1$, and ϕ can be found from the determinant of the first equation. The product rule of two matrices $M(\phi,\boldsymbol{\lambda},\boldsymbol{\mu}) \in \mathsf{Sp}(2D,\mathsf{R})$ in the Bargmann parameters uses a matrix and a phase corresponding to (9.45),

$$\boldsymbol{\nu} := \mathbf{1} + e^{-i\phi_2}\boldsymbol{\lambda}_1^{-1}\boldsymbol{\mu}_2\boldsymbol{\mu}_2^*\boldsymbol{\lambda}_2^{-1}, \quad \phi_\nu := \frac{1}{D}\arg\det\boldsymbol{\nu}. \quad (9.62)$$

Then it is shown that [7] $|\boldsymbol{\nu}-\mathbf{1}| < 1$, which again allows an unambiguous definition of the phase $\phi_\nu \in (-\frac{1}{2}\pi, \frac{1}{2}\pi)$. The product rule is

$$\phi = \phi_1+\phi_2+\phi_\nu, \quad \boldsymbol{\lambda} = e^{-i\phi_\nu}\boldsymbol{\lambda}_1\,\boldsymbol{\nu}\,\boldsymbol{\lambda}_2, \quad \boldsymbol{\mu} = e^{-i\phi_\nu}(\boldsymbol{\lambda}_1\boldsymbol{\mu}_2+e^{-2i\phi_2}\boldsymbol{\mu}_1\boldsymbol{\lambda}_2) \quad (9.63)$$

[cf. (9.46)–(9.48)]. Thus, for all symplectic groups, the *single* Bargmann phase ϕ reveals that the connectivity of $\mathsf{Sp}(2D,\mathsf{R})$ can be understood essentially as that of $\mathsf{Sp}(2,\mathsf{R})$, and similarly its cover $\overline{\mathsf{Sp}(2D,\mathsf{R})}$. •

Remark: Also $\mathsf{Sp}(4,\mathsf{R})$ covers $\mathsf{SO}(3,2)$ twice. As was the case for $D = 1$, where $\mathsf{Sp}(2,\mathsf{R}) \overset{2:1}{=} \mathsf{SO}(2,1)$, the physically important $D = 2$ case also leads to a homomorphism, $\mathsf{Sp}(4,\mathsf{R}) \overset{2:1}{=} \mathsf{SO}(3,2)$ (the latter is called the *anti-de Sitter* group [113] and used in '3 + 2' relativity models [115]). We shall write the homomorphism explictly below, in (11.151)–(11.152), and it will be the main tool for our analysis of Hamiltonian orbits in Chap. 12. Only in the $D = 1$ and 2 cases are there such *accidental* homomorphisms among the symplectic and other groups. •

9.5 The Iwasawa and other decompositions

Bargmann's factorization was used in the previous section to analyze the connectivity and cover of the symplectic groups. But there are reasons to work with other parametrizations of the symplectic groups,[8] in particular following the decomposition of the group into factors, each of which belongs to definite *sub*group of the full group; such is the well-studied Iwasawa decomposition

[7] The norm $|\mathbf{M}|$ of a matrix $\mathbf{M}$ is defined as the supremum $|\mathbf{M}|^2 \geq \mathbf{u}^\dagger\mathbf{M}^\dagger\mathbf{M}\mathbf{u}$ for all vectors of unit norm $\mathbf{u}$.

[8] The reader is invited to find the Bargmann parameters of the lower- and upper-triangular matrices (9.3) and (9.5) to see that the phases add and subtract when performing products in a complicated way. The practical ease of using the '*abc*' matrix parameters is evident.

[56, Chap. 7]. This applies not only to the symplectic groups, but for general Lie groups G, and is unique [73]. The closely related, *modified* Iwasawa factorization has been used to decompose an arbitrary paraxial system into convenient optical subsystems.

The Iwasawa and the modified-Iwasawa factorizations coincide in $\mathsf{Sp}(2,\mathsf{R})$; we shall see this case first. It is the following:[9]

$$\mathsf{G} = \mathsf{K}\,\mathsf{S} = \mathsf{K}\,\mathsf{A}\,\mathsf{N}, \qquad \text{where} \tag{9.64}$$

$$\mathsf{K}: \begin{array}{l}\text{maximal compact subgroup,}\\ \text{fractional Fourier transform,}\end{array} \qquad \begin{pmatrix} \cos\omega & -\sin\omega \\ \sin\omega & \cos\omega \end{pmatrix}, \tag{9.65}$$

$$\mathsf{S}: \begin{array}{l}\text{solvable subgroup}\\ \text{imager system,}\end{array} \qquad \begin{pmatrix} e^{\alpha} & e^{\alpha}G \\ 0 & e^{-\alpha} \end{pmatrix}, \tag{9.66}$$

$$\mathsf{A}: \begin{array}{l}\text{abelian subgroup}\\ \text{pure magnifier,}\end{array} \qquad \begin{pmatrix} e^{\alpha} & 0 \\ 0 & e^{-\alpha} \end{pmatrix}, \tag{9.67}$$

$$\mathsf{N}: \begin{array}{l}\text{nilpotent subgroup}\\ \text{thin lens,}\end{array} \qquad \begin{pmatrix} 1 & G \\ 0 & 1 \end{pmatrix}. \tag{9.68}$$

In the case of $\mathsf{Sp}(2,\mathsf{R})$, from the factorization

$$\begin{aligned}\begin{pmatrix} a & b \\ c & d \end{pmatrix} &= \begin{pmatrix} \cos\omega & -\sin\omega \\ \sin\omega & \cos\omega \end{pmatrix}\begin{pmatrix} e^{\alpha} & 0 \\ 0 & e^{-\alpha} \end{pmatrix}\begin{pmatrix} 1 & G \\ 0 & 1 \end{pmatrix} \\ &= \begin{pmatrix} \cos\omega\, e^{\alpha} & \cos\omega\, e^{\alpha}\, G - \sin\omega\, e^{-\alpha} \\ \sin\omega\, e^{\alpha} & \sin\omega\, e^{\alpha}\, G + \cos\omega\, e^{-\alpha} \end{pmatrix},\end{aligned} \tag{9.69}$$

we find the Iwasawa parameters (ω, α, G) to be

$$\omega = \arg(a + ic) \in \mathcal{S}^1, \quad e^{\alpha} = +\sqrt{a^2 + c^2} > 0, \quad G = \frac{ab + cd}{a^2 + c^2} \in \mathsf{R}. \tag{9.70}$$

Remark: Relation between the Bargmann and Iwasawa angles. The angle ω in the Iwasawa decomposition (9.70) equals the Bargmann parameter ϕ in (9.38)–(9.42) *only* when the symplectic matrix is a pure Fourier transform (with the solvable factor being the identity). To relate the two parameters for $\left(\begin{smallmatrix} a & b \\ c & d \end{smallmatrix}\right)$, draw the numbers $z_\omega = a - ic$, $z_\delta = d + ib$, and their vector sum $z_\phi = z_\omega + z_\delta$ on the complex plane. Then the Iwasawa angle ω is the argument of z_ω (i.e., the angle from the real axis to z_ω), while the Bargmann angle $\phi = \omega + \arg z_\delta$ is the argument of z_ϕ. •

[9] **Nilpotent** matrices $\mathbf{n}$ are such that $\mathbf{n}^P = \mathbf{0}$ for some power P; this is the case with upper-triangular matrices with zeros on the diagonal; they form Lie *algebras*. Matrices $\mathbf{N} = \exp \mathbf{n}$ with $\mathbf{n}$ nilpotent, form nilpotent Lie *groups*, such as upper-triangular matrices with 1's on the diagonal. **Solvable** matrices allow for arbitrary numbers on their diagonal; they also form *groups* and are generated by solvable *algebras*.

Remark: The Iwasawa decomposition of $\mathsf{Sp}(2D,\mathsf{R})$ matrices. Beyond the 2×2 case, ortho-symplectic matrices take the place at the left of the Iwasawa decomposition (9.65). On the right, (9.66) must be generalized nontrivially because triangular matrices cannot be symplectic; but when $\mathbf{M}$ is in the block form (9.20), $\mathbf{C}=\mathbf{0}$ and $\mathbf{A}$ upper-triangular, (9.29) requires that $\mathbf{D}^\top=\mathbf{A}^{-1}$; so $\mathbf{D}$ is lower-triangular. The Iwasawa decomposition is then

$$\begin{pmatrix}\mathbf{A} & \mathbf{B}\\ \mathbf{C} & \mathbf{D}\end{pmatrix}=\begin{pmatrix}\mathrm{Re}\,\mathbf{U} & -\mathrm{Im}\,\mathbf{U}\\ \mathrm{Im}\,\mathbf{U} & \mathrm{Re}\,\mathbf{U}\end{pmatrix}\begin{pmatrix}\mathbf{X} & \mathbf{Y}\\ \mathbf{0} & \mathbf{X}^{\top-1}\end{pmatrix},\qquad \mathbf{X}=\begin{pmatrix}e^{\chi_1} & x_{1,2} & \cdots & x_{1,D}\\ 0 & e^{\chi_2} & \cdots & x_{2,D}\\ \vdots & \vdots & \ddots & \vdots\\ 0 & 0 & \cdots & e^{\chi_D}\end{pmatrix}.$$
$$\mathbf{U}\in\mathsf{U}(D),\quad \mathbf{X}\mathbf{Y}^\top \text{ symmetric,} \tag{9.71}$$

There are D^2 real and bounded parameters in $\mathsf{K}=\mathsf{U}(D)$ and $D(D+1)$ independent unbounded parameters in the solvable matrix. One can factorize the latter into a diagonal matrix in the abelian subgroup $\mathsf{A}=\mathsf{T}_D$ with D independent positive diagonal parameters e^{χ_i}; the nilpotent matrix remaining to the right of (9.71) will be an element of the maximal nilpotent subgroup $\mathsf{N}\in\mathsf{Sp}(2D,\mathsf{R})$, and of the form (9.71) with 1's on the diagonal. •

Our interest in the D-dim case derives from the necessary 10-parameter $\mathsf{Sp}(4,\mathsf{R})$ for optical systems in real $D=2$ workbenches where the screens are planes. With physical realizability in mind, some authors have studied the symplectic groups with the **modified-Iwasawa** parameters [129, 130]. The generic $\mathsf{Sp}(2D,\mathsf{R})$ matrix is then factored in the following way:

$$\begin{pmatrix}\mathbf{A} & \mathbf{B}\\ \mathbf{C} & \mathbf{D}\end{pmatrix}=\begin{pmatrix}\mathrm{Re}\,\mathbf{U} & -\mathrm{Im}\,\mathbf{U}\\ \mathrm{Im}\,\mathbf{U} & \mathrm{Re}\,\mathbf{U}\end{pmatrix}\begin{pmatrix}\mathbf{S} & \mathbf{0}\\ \mathbf{0} & \mathbf{S}^{-1}\end{pmatrix}\begin{pmatrix}\mathbf{1} & \mathbf{G}\\ \mathbf{0} & \mathbf{1}\end{pmatrix}\quad \begin{cases}\mathbf{S}=\mathbf{S}^\top,\\ |\mathbf{S}|>0,\\ \mathbf{G}=\mathbf{G}^\top.\end{cases} \tag{9.72}$$

The first factor is again the ortho-symplectic matrix of a fractional $\mathsf{U}(D)$-Fourier transform (D^2 parameters); the second is a pure magnifier represented by a symmetric positive definite matrix; and the third is a thin lens [$\frac{1}{2}D(D+1)$ parameters each]. The factors can be found from

$$\mathbf{U}=(\mathbf{A}+i\mathbf{C})\mathbf{S}^{-1},\quad \mathbf{S}^\top\mathbf{S}=\mathbf{A}^\top\mathbf{A}+\mathbf{C}^\top\mathbf{C},\quad \mathbf{G}=(\mathbf{S}^\top\mathbf{S})^{-1}(\mathbf{A}^\top\mathbf{B}+\mathbf{C}^\top\mathbf{D}). \tag{9.73}$$

The modified-Iwasawa factorization is a *parametrization* of the group, but not a group *decomposition*, because symmetric matrices $\mathbf{S}$ do not form a group under multiplication. In $\mathsf{Sp}(4,\mathsf{R})$ the factors contain 4, 2, and 4 parameters, respectively.

Remark: Iwasawa and modified-Iwasawa parameters. The relation between Iwasawa and modified Iwasawa parameters is close because both have an ortho-symplectic matrix, $\mathbf{O}(\mathbf{U}_{\rm I})$ and $\mathbf{O}(\mathbf{U}_{\rm mI})$ respectively, factored to the left. A symplectic matrix written as (9.71) and as (9.72) relates the factor matrices through

$$\begin{pmatrix} \operatorname{Re} \mathbf{U}_{\mathrm{I}} & -\operatorname{Im} \mathbf{U}_{\mathrm{I}} \\ \operatorname{Im} \mathbf{U}_{\mathrm{I}} & \operatorname{Re} \mathbf{U}_{\mathrm{I}} \end{pmatrix} \begin{pmatrix} \mathbf{X} & \mathbf{Y} \\ \mathbf{0} & \mathbf{X}^{\top -1} \end{pmatrix} = \begin{pmatrix} \operatorname{Re} \mathbf{U}_{\mathrm{mI}} & -\operatorname{Im} \mathbf{U}_{\mathrm{mI}} \\ \operatorname{Im} \mathbf{U}_{\mathrm{mI}} & \operatorname{Re} \mathbf{U}_{\mathrm{mI}} \end{pmatrix} \begin{pmatrix} \mathbf{S} & \mathbf{S}\mathbf{G} \\ \mathbf{0} & \mathbf{S}^{-1} \end{pmatrix}, \tag{9.74}$$

The parameters $(\mathbf{U}_{\mathrm{mI}}, \mathbf{S}, \mathbf{G})$ on the left are given by (9.73), which also yield the parameters $(\mathbf{U}_{\mathrm{I}}, \mathbf{X}, \mathbf{Y})$ on the right replacing $\mathbf{S} \mapsto \mathbf{X}$, $\mathbf{S}\mathbf{G} \mapsto \mathbf{Y}$. Equating the two we obtain

$$\mathbf{X} = (\operatorname{Re} \mathbf{U})\, \mathbf{S}, \quad \mathbf{Y} = (\operatorname{Re} \mathbf{U})\, \mathbf{S}\, \mathbf{G} - (\operatorname{Im} \mathbf{U})\, \mathbf{S}^{-1}, \qquad \mathbf{U} := \mathbf{U}_{\mathrm{I}}^{\dagger}\, \mathbf{U}_{\mathrm{mI}}. \tag{9.75}$$

The first expression is the polar decomposition of $\mathbf{X}$ into an orthogonal and a positive definite symmetric matrix. •

Remark: Triangular decomposition of $\mathsf{Sp}(2D, \mathsf{R})$. The decomposition of symplectic matrices into two block-triangular and one block-diagonal matrices,

$$\begin{pmatrix} \mathbf{A} & \mathbf{B} \\ \mathbf{C} & \mathbf{D} \end{pmatrix} = \begin{pmatrix} \mathbf{1} & \mathbf{0} \\ \mathbf{C}\mathbf{A}^{-1} & \mathbf{1} \end{pmatrix} \begin{pmatrix} \mathbf{A} & \mathbf{0} \\ \mathbf{0} & \mathbf{A}^{\top -1} \end{pmatrix} \begin{pmatrix} \mathbf{1} & \mathbf{A}^{-1}\mathbf{B} \\ \mathbf{0} & \mathbf{1} \end{pmatrix}, \tag{9.76}$$

is valid for $\det \mathbf{A} \neq 0$ and $\mathbf{D}$ as determined from (9.29). This is the decomposition of a paraxial optical system into free anisotropic propagation $\mathbf{C}\mathbf{A}^{-1}$, a magnifier $\mathbf{A}$, and a lens $\mathbf{A}^{-1}\mathbf{B}$. •

Remark: Euler angle decomposition of $\mathsf{SO}(2,1) \stackrel{1:2}{=} \mathsf{Sp}(2, \mathsf{R})$. We remarked that (9.53) gave one 3×3 pseudo-orthogonal matrix for every two 2×2 symplectic matrices matrices. Parameters which generalize the $\mathsf{SO}(3)$ rotation Euler angles for all orthogonal *and* pseudo-orthogonal groups, are widely used [74, 97, 116, 146], and can be called *Euler parameters* for $\mathsf{Sp}(2, \mathsf{R})$. We denote the matrix of rotation through the angle θ in the j–k plane, by $\mathbf{R}_{j,k}(\theta)$,[10] the $\mathsf{SO}(3)$ Euler angle decomposition is $\mathbf{R}(\phi, \theta, \psi) = \mathbf{R}_{1,2}(\phi)\, \mathbf{R}_{2,3}(\theta)\, \mathbf{R}_{1,2}(\psi)$, and the matrices have trigonometric function entries. When the third direction has negative metric sign as in (9.53), then the range of θ is the real line,

$$\mathbf{R}_{1,2}(\phi) = \begin{pmatrix} \cos \frac{1}{2}\phi & -\sin \frac{1}{2}\phi \\ \sin \frac{1}{2}\phi & \cos \frac{1}{2}\phi \end{pmatrix} \quad \text{and} \quad \mathbf{R}_{2,0}(\theta) = \begin{pmatrix} e^{\theta/2} & 0 \\ 0 & e^{-\theta/2} \end{pmatrix}. \tag{9.77}$$

In this way we parametrize $\mathsf{Sp}(2, \mathsf{R})$ matrices, and flat paraxial systems, as two Fourier transforms with one magnifier in between,

$$\mathbf{M}\{\phi, \theta, \psi\}_{\text{Euler}} = \mathbf{R}_{1,2}(\phi)\, \mathbf{R}_{2,0}(\theta)\, \mathbf{R}_{1,2}(\psi), \quad \phi, \psi \in \mathsf{R} \bmod 4\pi,\ \theta \in \mathsf{R}. \tag{9.78}$$

It can be readily verified that the corresponding 3×3 matrices (9.53) contain trigonometric and hyperbolic functions of the double angles, ϕ and θ. •

Remark: Euler angles for $\mathsf{SO}(D)$. Euler angles can be generalized recursively for the D-dimensional rotation group $\mathsf{SO}(D)$, by factoring its elements into the $\mathsf{SO}(D-1)$ subgroup involving only the first $D-1$ coordinates,

---

[10] Cf. (5.17)–(5.19) where planes are labelled by normals: $x = 2, 3$, $y = 3, 1$, $z = 1, 2$.

times a '$(D-1)$-sphere' of rotations $\mathcal{S}^{D-1}$. We indicate the $D \times D$ matrices $\boldsymbol{\Theta}^{(\mathrm{E})}_{j,k} := \mathbf{R}_{j,k}(\theta^{(\mathrm{E})})$, $E = 1, 2, 3 \ldots, D-1$, to represent rotations through angles $\{\theta^{(\mathrm{E})}\}$ in the j–k planes of the E-dim subspace $1 \leq j < k \leq E+1$. We write

$$\mathbf{R}^{(2)}(\theta^{(1)}) = \boldsymbol{\Theta}^{(1)}_{1,2} \in \mathsf{SO}(2), \tag{9.79}$$
$$\mathbf{R}^{(3)}(\theta^{(1)}, \theta^{(2)}) = \mathbf{R}^{(2)}(\theta^{(1)})\, \boldsymbol{\Theta}^{(2)}_{2,3}\, \boldsymbol{\Theta}^{(2)}_{1,2} \in \mathsf{SO}(3), \tag{9.80}$$
$$\mathbf{R}^{(4)}(\theta^{(1)}, \theta^{(2)}, \theta^{(3)}) = \mathbf{R}^{(3)}(\theta^{(1)}, \theta^{(2)})\, \boldsymbol{\Theta}^{(3)}_{3,4} \boldsymbol{\Theta}^{(3)}_{2,3}\, \boldsymbol{\Theta}^{(3)}_{1,2} \in \mathsf{SO}(4), \tag{9.81}$$
$$\vdots \quad \vdots$$
$$\mathbf{R}^{(\mathrm{D})}(\theta^{(1)}, \ldots, \theta(D-1)) = \mathbf{R}^{(\mathrm{D}-1)}(\theta^{(1)}, \ldots, \theta^{(\mathrm{D}-2)})\, \mathbf{S}^{(\mathrm{D})}(\theta^{(\mathrm{D}-1)}) \in \mathsf{SO}(D), \tag{9.82}$$
$$\text{where } \mathbf{S}^{(D)}(\theta) = \boldsymbol{\Theta}_{D-1,D}\, \boldsymbol{\Theta}_{D-2,D-1} \cdots \boldsymbol{\Theta}_{1,2} \in \mathcal{S}^{\mathrm{D}-1}. \tag{9.83}$$

As local manifolds therefore, the rotation groups are nested products of spheres,

$$\mathsf{SO}(D) = \mathsf{SO}(D-1) \times \mathcal{S}^{D-1} = \mathcal{S}^1 \times \mathcal{S}^2 \times \cdots \times \mathcal{S}^{D-1}. \tag{9.84}$$

The total number of parameters is $\frac{1}{2}D(D-1)$, as expected. When some directions have negative metric sign, hyperbolic functions will replace the trigonometric ones in boosts, and thus parametrize $\mathsf{SO}(D^+, D^-)$ for $D^+ + D^- = D$. •

10 Construction of optical systems

In this chapter we shall realize each of the three factors in the Iwasawa decompositions of $\mathsf{Sp}(2,\mathsf{R})$ and $\mathsf{Sp}(4,\mathsf{R})$, the groups of linear canonical transformations of paraxial phase space on 1- and 2-dim screens, respectively. They will be realized with lenses, magnifiers and fractional Fourier transformers; their composition yields the set of all paraxial optical systems. Setups with a minimal number of lenses will be designed for the $D=1$-dim case, which also applies for the aligned, axially-symmetric $D=2$ case. Often we find that we can work in D dimensions at no extra cost.

10.1 Plane optical systems

Any plane paraxial optical system, represented by a 2×2 symplectic matrix, can be factored into subsystems following the Iwasawa decomposition (9.69) of the group $\mathsf{Sp}(2,\mathsf{R}) = \mathsf{KAN}$,

$$\begin{pmatrix} a & b \\ c & d \end{pmatrix} = \begin{pmatrix} \cos\omega & -\sin\omega \\ \sin\omega & \cos\omega \end{pmatrix} \begin{pmatrix} m & 0 \\ 0 & m^{-1} \end{pmatrix} \begin{pmatrix} 1 & G \\ 0 & 1 \end{pmatrix}. \tag{10.1}$$

The rightmost factor (nilpotent subgroup N) is a thin lens of Gaussian power G [*cf.* (9.5) and (9.18)] in contact with the screen to its right. This factor will not affect the image at all, since $\mathbf{q}'=\mathbf{q}$ [from (9.5)], but only change the direction of the rays falling on each point of the screen. The middle factor (abelian subgroup A) is a magnifier of coefficient $m>0$, and the leftmost factor (the maximal compact subgroup K) is a fractional Fourier transformer of angle $\omega\in\mathcal{S}^1$ (on a circle).

Example: Iwasawa decomposition of free displacement. Free displacements $\mathcal{D}_z = \mathcal{M}\begin{pmatrix} 1 & 0 \\ -z & 1 \end{pmatrix}$ in (9.3) and (9.18) can be decomposed into their Iwasawa factors KAN as follows:

$$\begin{pmatrix} 1 & 0 \\ -z & 1 \end{pmatrix} = \begin{pmatrix} \dfrac{1}{\sqrt{1+z^2}} & \dfrac{z}{\sqrt{1+z^2}} \\ \dfrac{-z}{\sqrt{1+z^2}} & \dfrac{1}{\sqrt{1+z^2}} \end{pmatrix} \begin{pmatrix} \sqrt{1+z^2} & 0 \\ 0 & \dfrac{1}{\sqrt{1+z^2}} \end{pmatrix} \begin{pmatrix} 1 & \dfrac{-z}{1+z^2} \\ 0 & 1 \end{pmatrix}. \tag{10.2}$$

The Fourier rotation angle is $\omega = \arg(1 - iz) \in (-\frac{1}{2}\pi, 0]$ for $z \geq 0$.[1] When z grows without bound, the free displacement is approximated by the concatenation of an inverse Fourier transformer ($\omega = -\frac{1}{2}\pi$), times a growing magnifier, and a Gaussian lens of power approaching -1. •

An optical setup consisting of a thin lens between two free displacements, will be called a *DLD configuration*; it is represented by the product

$$\begin{aligned}\mathcal{D}_{z_1}\,\mathcal{L}_G\,\mathcal{D}_{z_2} &= \mathcal{M}\begin{pmatrix} 1 & 0 \\ -z_1 & 1 \end{pmatrix}\mathcal{M}\begin{pmatrix} 1 & G \\ 0 & 1 \end{pmatrix}\mathcal{M}\begin{pmatrix} 1 & 0 \\ -z_2 & 1 \end{pmatrix} \\ &= \mathcal{M}\begin{pmatrix} 1 - Gz_2 & G \\ -z_1 - z_2 + z_1 G z_2 & 1 - z_1 G \end{pmatrix}.\end{aligned} \tag{10.3}$$

[Keep in mind the left-to-right order of correspondence between matrices and optical elements, insisted upon in page 161.] The *inverse* of this matrix acts on the phase space 2-vectors $\binom{p}{q}$, namely

$$\begin{pmatrix} p' \\ q' \end{pmatrix} = \mathcal{M}\begin{pmatrix} a & b \\ c & d \end{pmatrix} : \begin{pmatrix} p \\ q \end{pmatrix} = \begin{pmatrix} d & -b \\ -c & a \end{pmatrix}\begin{pmatrix} p \\ q \end{pmatrix} = \begin{pmatrix} dp - bq \\ -cp + aq \end{pmatrix}. \tag{10.4}$$

The system is an **imager** when its output is a sharp image of the object, i.e., when the position of a ray at the screen, $q'(q)$, is independent of its initial direction p. The lower-left (2–1) element c of the matrix (10.3) must then be zero. This occurs when the *focal condition* is satisfied:[2]

$$\begin{aligned} &z_1 + z_2 = z_1 G z_2, & \mu := 1 - Gz_2 &= (1 - z_1 G)^{-1} \\ &\text{i.e., } G = 1/z_1 + 1/z_2 = 1/F, & &= -z_2/z_1, \end{aligned} \tag{10.5}$$

where μ is the *magnification factor* of position, $q'(q) = \mu\, q$. Momentum is also magnified, but through the inverse factor, and with a *shear* of phase space, $p'(p, q) = \mu^{-1}p - Gq$ (see Fig. 9.2). An imager with no shear (or *pure* imager), $\mathcal{M}\begin{pmatrix} \mu & 0 \\ 0 & \mu^{-1} \end{pmatrix}$, we call a **magnifier**.[3]

We remarked before (page 160) that magnifiers, represented by diagonal matrices, cannot be realized with *DLD* configurations. But we can correct a *DLD* imager by adding a thin lens in contact with the screen, of Gaussian power G', to correct the ray directions at all image points. The resulting *DLDL* configuration is represented by

$$(\mathcal{DLD})_{\mu,G}\,\mathcal{L}_{G'} := \mathcal{M}\begin{pmatrix} \mu & G \\ 0 & \mu^{-1} \end{pmatrix}\mathcal{M}\begin{pmatrix} 1 & G' \\ 0 & 1 \end{pmatrix} = \mathcal{M}\begin{pmatrix} \mu & G + \mu G' \\ 0 & \mu^{-1} \end{pmatrix}, \tag{10.6}$$

---

[1] We shall usually exclude from consideration the trivial *identity* systems $\mathcal{M}(\mathbf{1})$, so $z > 0$ and the resulting inequalities will be strict.

[2] Recall from page 158 that a lens of Gaussian power G has focal distance $F = 1/G$. When $z_1 \to \infty$ then $z_2 \to F$; this is a telescopic system imaging stars.

[3] In turn, imagers are also called *impure* magnifiers.

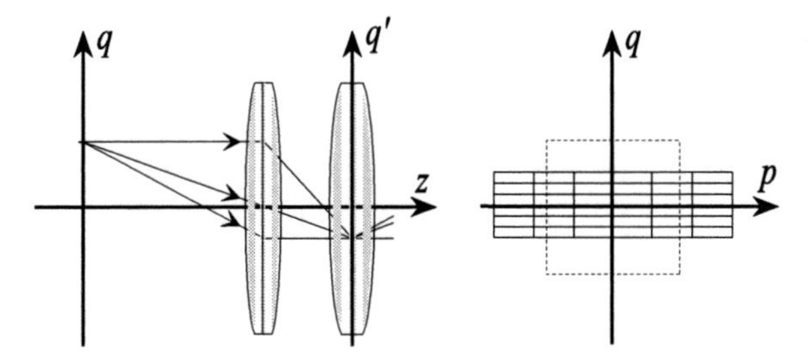

Fig. 10.1. Pure magnifier. Left: The magnifier is built with a $DLDL$ configuration. Right: The pure magnifier action on phase space scales positions (*horizontal lines*) and momenta (*vertical lines*) reciprocally, without shear nor rotation.

so when $G' = -G/\mu$, it is a magnifier; see Fig. 10.1. But *not all* magnifiers can be made from $DLDL$ systems: physically $z_1, z_2 > 0$, and the focal conditions (10.5) then imply that $\mu < 0$, so the resulting images will be inverted; and only convex lenses $G > 0$ are necessary. To obtain positive magnification (and fit the Iwasawa scheme), it is sufficient to concatenate two inverting magnifiers to obtain

$$\mathcal{A}_\zeta := \mathcal{M}\begin{pmatrix} e^\zeta & 0 \\ 0 & e^{-\zeta} \end{pmatrix}, \qquad \zeta \in \mathsf{R}, \quad \mu = e^\zeta > 0. \tag{10.7}$$

Actually, such a 4-lens setup is unnecessary; the minimal 3-lens configuration of magnifiers will be found in Sect. 10.5. With thin lenses and magnifiers we can cover the full two-parameter solvable group of paraxial imagers; they correspond to all upper-triangular matrices.

The leftmost factor in the Iwasawa decomposition (10.1) is a fractional Fourier transformer [see (9.39) and (9.65)], namely, a rigid *rotation* of the phase space plane in Fig. 3.2. A Fourier transformer is the *non-imaging* Iwasawa factor of any paraxial optical system – or, more positively, any $D = 1$ system can be decomposed into one imager preceded (or followed) by one fractional Fourier transformer.

Inverse fractional Fourier transformers can be built in a DLD configuration with one lens between two equal displacements, as shown in Fig. 10.2. Recalling the definition of $\mathcal{F}^\alpha$ on page 30, for $\omega = \frac{1}{2}\pi\alpha$, we write (10.3) as

$$\mathcal{F}^{-\alpha} := \mathcal{D}_z\,\mathcal{L}_G\,\mathcal{D}_z = \mathcal{M}\begin{pmatrix} \cos\frac{\pi}{2}\alpha & \sin\frac{\pi}{2}\alpha \\ -\sin\frac{\pi}{2}\alpha & \cos\frac{\pi}{2}\alpha \end{pmatrix}, \tag{10.8}$$

$$\begin{pmatrix} p' \\ q' \end{pmatrix} = \mathcal{F}^{-\alpha} : \begin{pmatrix} p \\ q \end{pmatrix} = \begin{pmatrix} \cos\frac{\pi}{2}\alpha & -\sin\frac{\pi}{2}\alpha \\ \sin\frac{\pi}{2}\alpha & \cos\frac{\pi}{2}\alpha \end{pmatrix} \begin{pmatrix} p \\ q \end{pmatrix}, \tag{10.9}$$

$$\text{where} \qquad z = \tan\tfrac{\pi}{4}\alpha > 0, \quad G = \sin\tfrac{\pi}{2}\alpha > 0. \tag{10.10}$$

Hence, with a single-lens DLD configuration we can cover the Fourier inverse power range $0 < \alpha < 2$. This includes *the* inverse Fourier transform $\alpha = 1$,

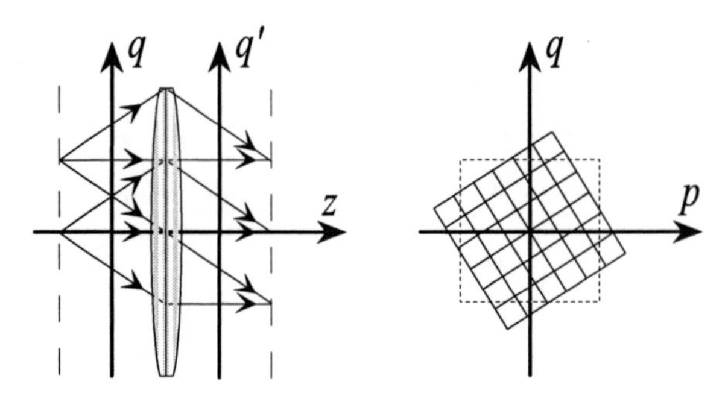

Fig. 10.2. Fractional Fourier transformer. *Left:* A fractional Fourier transformer $\mathcal{F}^\alpha$, for $0 > \alpha > -2$, is built in a left-right symmetric *DLD* configuration with a convex lens. Dashed vertical lines are screens at the focal distances from the lens; between them, the Fourier power is $\alpha = -1$. Continuous lines q, q' are screens between which the transformation is fractional. *Right:* The phase space rotation produced by the symmetric fractional Fourier transformer is rigid, i.e., has no shear nor magnification.

for which $z = 1$ and $G = 1$. The object and its inverse-Fourier image are at the two focal points of the convex thin lens, and in (10.9), $\binom{p'}{q'} = \mathcal{F}^{-1} : \binom{p}{q} = \binom{-q}{p}$. The configuration is said to be *symmetric* since it is invariant under reflection across the $z =$ constant plane of the lens.[4]

Remark: The Fourier cycle. Fractional Fourier transforms can be concatenated: $\mathcal{F}^{\alpha_1}\mathcal{F}^{\alpha_2} = \mathcal{F}^{\alpha_1+\alpha_2}$. Since with one symmetric *DLD* system we can only produce $0 > \alpha > -2$, with two *DLD* systems built into a *DLDLD* configuration, we can cover $0 > \alpha > -4$. The inversion of phase space is $\mathcal{I} := \mathcal{F}^{-2}$ (represented by $-\mathbf{1}$), which maps $(p, q) \mapsto (-p, -q)$, and is the only element that commutes with all paraxial optical systems. The Fourier transformation is $\mathcal{F} = \mathcal{F}^{-3}$ and can thus be obtained with a *DLDLD* configuration. To be able to go full circle back to the $\mathsf{Sp}(2, \mathsf{R})$ unit, $\mathcal{F}^0 := \mathit{1} = \mathcal{F}^4$, we need at least three lenses. But this will actually be the *second metaplectic unit* of $\mathsf{Mp}(2, \mathsf{R})$ in wave optics, rather than its identity element – recall the remark on page 170. •

Remark: Fractional Fourier transformation in guides. We saw in (2.18)–(2.19) and Sect. 2.4 that the paraxial approximation to a guide of refractive index $n(q, z) \approx n_o - \nu(q, z)$ along its length z corresponds to a mechanical system with potential $\nu(q, t)$ in time t. A harmonic guide has the maximum of its refractive index at the optical center $q = 0$; the second term of its Taylor expansion is a quadratic $\nu(q, z) \approx \frac{1}{2}k(z)\, q^2$, where $k(z) = -\partial n(q, z)/\partial z|_{q=0} \geq 0$ corresponds to the spring constant of a mechanical oscillator with generic time dependence. As we saw comparing (3.38) and

[4] Do not confuse configurations which are symmetric (invariant) under $z \leftrightarrow -z$, with systems whose representation *matrix* is symmetric, such as (9.12)–(9.13).

(3.39), displacement of the screen along the z-axis of a harmonic guide, generates an inverse fractional Fourier transform $\mathcal{F}^{-\alpha}$ for any power α modulo 4. We can fix units choosing $k = 1/n_o$, so that the trajectories are circles in phase space with a rotation angle $\omega = z/n_o = \frac{1}{2}\pi\alpha$. •

Example: Negative displacements. We assumed originally that free displacements could only be positive (or zero). But since the Fourier transform exchanges position and momentum, a similarity Fourier transformer can intertwine the transformations of free displacements and of Gaussian lenses,

$$\mathcal{F}\,\mathcal{L}_G\,\mathcal{F}^{-1} = \mathcal{M}\begin{pmatrix} 0 & -1 \\ 1 & 0 \end{pmatrix}\begin{pmatrix} 1 & G \\ 0 & 1 \end{pmatrix}\begin{pmatrix} 0 & 1 \\ -1 & 0 \end{pmatrix} = \mathcal{D}_G \tag{10.11}$$

Hence a concave lens $G < 0$ between Fourier transformers is a negative free displacement; systems which freely displace *backwards* can be thus built with three lenses. •

Example: Hyperbolic expanders. We noted below (10.6) that fractional Fourier transformers could be built only with convex lenses $G > 0$. A symmetric DLD system (10.3) with a *concave* lens is interesting because we obtain a *hyperbolic expander*:

$$\mathcal{H}_\zeta := \mathcal{D}_z\,\mathcal{L}_G\,\mathcal{D}_z = \mathcal{M}\begin{pmatrix} \cosh\zeta & -\sinh\zeta \\ -\sinh\zeta & \cosh\zeta \end{pmatrix}, \quad \begin{array}{c} G = -\sinh\zeta < 0, \\ z = \tanh\frac{1}{2}\zeta > 0. \end{array} \tag{10.12}$$

The optical setup and its action on phase space are shown in Fig. 10.3. With a one-lens arrangement, only the parameter range $\zeta > 0$ is available; but again, by similarity with Fourier transformers we can build $\mathcal{H}_{-\zeta} = \mathcal{F}\,\mathcal{H}_\zeta\,\mathcal{F}^{-1}$, and obviate this restriction. Hyperbolic expanders form the only subgroup $\mathsf{SO}(1,1) \subset \mathsf{Sp}(2,\mathsf{R})$ of symmetric matrices. Finally, note that these expanders belong to the same class of systems as the magnifiers $\mathcal{A}_\zeta$ in (10.7), because $\mathcal{A}_\zeta = \mathcal{F}^{-\frac{1}{2}}\,\mathcal{H}_\zeta\,\mathcal{F}^{\frac{1}{2}}$. The subgroup of hyperbolic expanders is thus conjugate to the subgroup of magnifiers by the square root of the Fourier transform. •

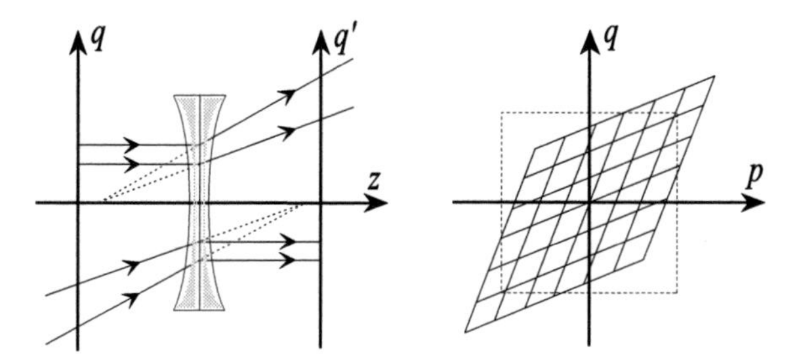

Fig. 10.3. Hyperbolic expander. *Left: DLD* arrangement between the two screens q, q' with a concave lens and two symmetric free spaces. *Right:* The action on phase space expands and contracts along the bisectrices of the p–q axes.

10.2 Astigmatic lenses and magnifiers

In the real 3-dim world with $D = 2$-dim screens, the set of paraxial optical systems form the 10-parameter group $\mathsf{Sp}(4,\mathsf{R})$ of linear canonical transformations of 4-dim phase space $(\mathbf{p},\mathbf{q})^\top \in \mathsf{R}^4$. Recalling (9.16), (10.4) and (10.9), their action is written

$$\mathcal{M}\begin{pmatrix}\mathbf{A} & \mathbf{B}\\ \mathbf{C} & \mathbf{D}\end{pmatrix} : \begin{pmatrix}\mathbf{p}\\ \mathbf{q}\end{pmatrix} = \begin{pmatrix}\mathbf{A} & \mathbf{B}\\ \mathbf{C} & \mathbf{D}\end{pmatrix}^{-1}\begin{pmatrix}\mathbf{p}\\ \mathbf{q}\end{pmatrix} = \begin{pmatrix}\mathbf{D}^\top\mathbf{p} - \mathbf{B}^\top\mathbf{q}\\ -\mathbf{C}^\top\mathbf{p} + \mathbf{A}^\top\mathbf{q}\end{pmatrix}. \tag{10.13}$$

We refer to the modified Iwasawa parametrization of $\mathsf{Sp}(4,\mathsf{R})$ given in (9.72), which we repeat here:

$$\begin{pmatrix}\mathbf{A} & \mathbf{B}\\ \mathbf{C} & \mathbf{D}\end{pmatrix} = \begin{pmatrix}\mathrm{Re}\,\mathbf{U} & -\mathrm{Im}\,\mathbf{U}\\ \mathrm{Im}\,\mathbf{U} & \mathrm{Re}\,\mathbf{U}\end{pmatrix}\begin{pmatrix}\mathbf{S} & \mathbf{0}\\ \mathbf{0} & \mathbf{S}^{-1}\end{pmatrix}\begin{pmatrix}\mathbf{1} & \mathbf{G}\\ \mathbf{0} & \mathbf{1}\end{pmatrix}, \tag{10.14}$$

where $\mathbf{S}$ and $\mathbf{G}$ are symmetric ($\det\mathbf{S} \neq 0$) and $\mathbf{U}$ is unitary [cf. (10.1)]. In the following, we discuss the three factors in (10.14) – lenses, magnifiers and Fourier transformers, as well as a few other remarkable arrangements. The results will apply straightforwardly to dimension D (cf. [91]).

10.2.1 Lenses

We start with a thin *cylindrical* lens $\mathcal{L}_{\mathrm{G}}^{(x)}$ which refracts rays in the x–z plane (changes p_x), leaving their y–z component (p_y) invariant. The generatrix of the cylinder is parallel to the y-direction, and its Gaussian power matrix is $\mathbf{G} = \mathrm{diag}\,(G_x, 0)$ [see (9.5) and (9.18)]. Such lenses exist only for $D \geq 2$, and for historical reasons are generally called *astigmatic.*[5] This x-cylinder lens can be rotated through the similarity transformation $\mathcal{R}(\phi)$ by angles $\phi \in \mathcal{S}^1$, to obtain

$$\mathcal{L}_{\mathrm{G}(\phi)} = \mathcal{R}(\phi)\,\mathcal{L}_{\mathrm{G}}^{(x)}\,\mathcal{R}(\phi)^{-1} = \mathcal{M}\begin{pmatrix}\mathbf{1} & \mathbf{G}(\phi)\\ \mathbf{0} & \mathbf{1}\end{pmatrix}, \tag{10.15}$$

$$\mathcal{R}(\phi) := \mathcal{M}\begin{pmatrix}\mathbf{R}(\phi) & \mathbf{0}\\ \mathbf{0} & \mathbf{R}(\phi)\end{pmatrix}, \quad \mathbf{R}(\phi) := \begin{pmatrix}\cos\phi & -\sin\phi\\ \sin\phi & \cos\phi\end{pmatrix}, \tag{10.16}$$

$$\mathbf{G}(\phi) = \mathbf{R}(\phi)\,\mathbf{G}\,\mathbf{R}(\phi)^{-1} = G\begin{pmatrix}\frac{1}{2}(1+\cos 2\phi) & \frac{1}{2}\sin 2\phi\\ \frac{1}{2}\sin 2\phi & \frac{1}{2}(1-\cos 2\phi)\end{pmatrix}. \tag{10.17}$$

We meet again the orthogonal matrix $\mathbf{R} \in \mathsf{SO}(D)$ of rotations in the plane of the screen (5.19); $\mathbf{R}(\phi)^\top = \mathbf{R}(\phi)^{-1} = \mathbf{R}(-\phi)$. Since $\mathbf{G}(\phi)$ depends on trigonometric functions of 2ϕ, the angle of rotation ϕ of astigmatic lenses counts modulo π.

---

[5] Astigmatic literally means *without stigma* or *blemish*. Paradoxically, the common usage of this word refers to systems whose action is precisely **not** invariant under rotations around the optical axis. Worse, the aberration called *astigmatism*, is a different thing altogether.

The set of lens transformations $\mathcal{L}_{\mathrm{G}}$ forms an abelian (i.e., commutative) group T_3, $\mathcal{L}_{\mathrm{G}_1}\mathcal{L}_{\mathrm{G}_2} = \mathcal{L}_{\mathrm{G}_1+\mathrm{G}_2}$, and a 3-dim vector space of 2×2 symmetric matrices. This allows us to concatenate (i.e., superpose) two mutually rotated cylindical lenses to obtain one astigmatic lens of power $\mathbf{G}_1+\mathbf{G}_2$. The manifold of astigmatic lenses $\mathcal{L}_{\mathrm{G}}$ has dimension 3 and is a homogeneous space under the adjoint action (10.17) of rotations around the optical z-axis. Every symmetric matrix can be brought to diagonal form by a similarity transformation through a rotation matrix, and conversely, from the diagonal matrices (i.e., superposed x- and y-cylindrical lenses) we can produce all symmetric matrices (i.e., all astigmatic lenses) through rotations by a single angle.

Remark: Scissor arrangements of cylindrical lenses. Two equal cylindrical lenses $\mathbf{G}^{(x)} = \operatorname{diag}(G,0)$ rotated in opposite directions by $\pm\phi$, will add up to $\mathbf{G} = 2G\operatorname{diag}(\cos^2\phi, \sin^2\phi)$ as we can see from (10.17). This *scissor* arrangement is equivalent to two crossed cylindrical lenses of powers $G_x = 2G\cos^2\phi$ and $G_y = 2G\sin^2\phi$ along the bisectrices of the scissor, i.e., along the *principal axes* of the system. In particular, when the two equal thin cylinders are crossed ($\phi = \frac{1}{4}\pi$) they result in an axially-symmetric thin lens. Another scissor arrangement where the two cylindrical lenses have opposite powers $\pm G$ (i.e., one convex and one concave), has Gaussian $\mathbf{G} = G\operatorname{antidiag}(\sin 2\phi, \sin 2\phi)$; this is an equilateral *saddle* lens, whose principal axes x', y' are rotated by 45° from the previous ones, and with Gaussian powers $G_{x'} = -G_{y'} = -G\sin 2\phi$. •

Remark: 'Spherical' and 'saddle' lenses. We can decompose a generic astigmatic lens into an axially-symmetric (i.e., 'spherical') thin lens and a rotated equilateral saddle lens, through

$$\begin{aligned}\mathbf{G} = \begin{pmatrix} G_{xx} & G_{xy} \\ G_{xy} & G_{yy}\end{pmatrix} &= G\,\mathbf{1} + \mathbf{R}(\psi)\begin{pmatrix} G^\perp & 0 \\ 0 & -G^\perp\end{pmatrix}\mathbf{R}(-\psi) \\ &= \begin{pmatrix} G + G^\perp\cos 2\psi & G^\perp \sin 2\psi \\ G^\perp\sin 2\psi & G - G^\perp\cos 2\psi\end{pmatrix},\end{aligned} \tag{10.18}$$

$$\text{where}\quad G = \tfrac{1}{2}(G_{xx}+G_{yy}),\quad G^\perp = \tfrac{1}{2}\sqrt{(G_{xx}-G_{yy})^2+4G_{xy}^2}, \tag{10.19}$$

$$\text{and}\quad \tan 2\psi = 2G_{xy}/(G_{xx}-G_{yy}). \tag{10.20}$$

•

10.2.2 Magnifiers

The middle factor of the modified-Iwasawa parametrization of paraxial systems (10.14) is the set of pure imagers, i.e., magnifiers of phase space characterized by $D\times D$ nonsingular symmetric matrices $\mathbf{S}$,

$$\mathcal{A}_{\mathrm{S}} := \mathcal{M}\begin{pmatrix}\mathbf{S} & \mathbf{0} \\ \mathbf{0} & \mathbf{S}^{-1}\end{pmatrix}, \quad \mathcal{A}_{\mathrm{S}}: \begin{pmatrix}\mathbf{p} \\ \mathbf{q}\end{pmatrix} = \begin{pmatrix}\mathbf{S}^{-1}\mathbf{p} \\ \mathbf{S}\,\mathbf{q}\end{pmatrix}. \tag{10.21}$$

The tactics we used above for astigmatic lenses also serve us here to build astigmatic magnifiers out of magnifiers along each axis (10.6) and (10.7), and rotating them. In particular, the effect of rotating $\mathcal{A}_{\mathbf{S}}$ around the optical axis of the apparatus through $\mathcal{R}(\phi)$ [cf. (10.16)–(10.17)] yields

$$\mathcal{A}_{\mathrm{S}(\phi)} := \mathcal{R}(\phi)\,\mathcal{A}_{\mathrm{S}}\,\mathcal{R}(\phi)^{-1}, \qquad \mathbf{S}(\phi) = \mathbf{R}(\phi)\,\mathbf{S}\,\mathbf{R}(\phi)^{-1}. \tag{10.22}$$

The set of astigmatic magnifiers does not form a group, since the product of two symmetric matrices is generally not symmetric.[6]

Remark: Building astigmatic magnifiers. Whereas superposed lenses sum their Gaussian power matrices, the composition of two magnifiers built with cylindrical lenses along different screen axes must take into account that the respective *DLDL* arrangements (10.6) must share the total length along the workbench: $z_{1x} + z_{2x} = z_{1y} + z_{2y}$. This optical setup is shown in Fig. 10.4, where crossed cylindrical lenses of powers G_x and G_y obey the focal conditions (10.5). Since the two superposed magnifiers have negative scaling factors μ_x and μ_y [cf. (10.7)], their matrices $\mathbf{S} = \mathrm{diag}\,(\mu_x, \mu_y)$ will be negative-definite. Positive magnification can be obtained by concatenating two inverting magnifiers. And finally, through rotations (10.22) of the apparatus of Fig. 10.4, we cover the full 3-dim manifold of symmetric, astigmatic magnifiers (10.21). •

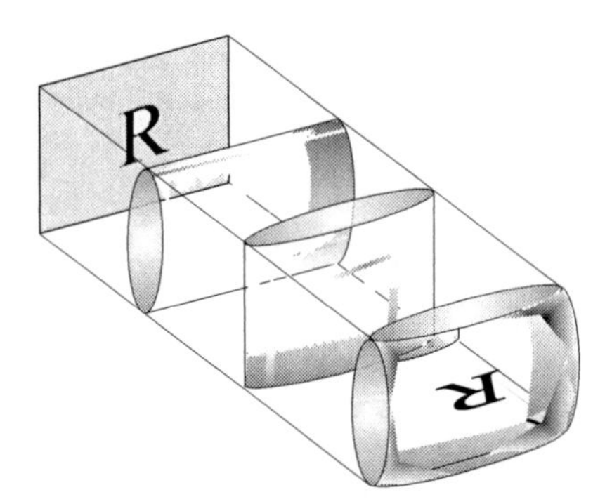

Fig. 10.4. Pure astigmatic magnifier built with *DLDL* arrangements of cylindrical lenses in the horizontal and the vertical directions, and sharing the same total length. The magnification factors are negative so the image is inverted across both axes, i.e., rotated by π; the range of these two factors and rotations of the setup around the optical axis yield the 3-dim manifold of pure astigmatic magnifiers.

[6] We recall that this is the *modified* Iwasawa parametrization, *not* the Iwasawa subgroup decomposition.

10.2.3 Reflectors and rotators

The manifold of astigmatic magnifiers is not a group, and the composition of two astigmatic magnifiers may take us out of that set. Thus we build *a posteriori* the putative *rotator* $\mathcal{R}(\phi)$ in (10.16). For this purpose we first need a special kind of astigmatic magnifier: a *reflector* of the x-axis,

$$\mathcal{I}_x := \mathcal{M}\Big(\mathrm{diag}\,(-1,1,-1,1)\Big), \quad \mathcal{I}_x : \begin{pmatrix} v_x \\ v_y \end{pmatrix} = \begin{pmatrix} -v_x \\ v_y \end{pmatrix}, \qquad v = p \text{ or } q. \tag{10.23}$$

Similarly for the y- or any other axis.

Remark: How to build a reflector. The x-reflector arrangement is shown in Fig. 10.5a; it has unit magnification on the x-axis and inverts the y-axis. The following $DLDLDL$ arrangement of spaces and x-cylindrical lenses, and $DLDL$ in y, achieves this purpose:

$$\begin{aligned} \mathcal{D}_{2F}\,\mathcal{L}^{(x)}_{1/F}\,\mathcal{D}_{4F}\,\mathcal{L}^{(x)}_{1/F}\,\mathcal{D}_{2F}\,\mathcal{L}^{(x)}_{2/F} &= \;\;\mathit{1}^{(x)}, \\ \mathcal{D}_{4F}\,\mathcal{L}^{(y)}_{1/2F}\,\mathcal{D}_{4F}\,\mathcal{L}^{(y)}_{1/2F} &= -\mathit{1}^{(y)}, \end{aligned} \tag{10.24}$$

where units are fixed by the focal distance F, and the total length of the apparatus is $8F$. Reflectors can be rotated [as in (10.18)],

$$\begin{aligned} \mathcal{I}(\phi) &:= \mathcal{R}(\phi)\,\mathcal{I}_y\,\mathcal{R}(\phi)^{-1} = \mathcal{M}\begin{pmatrix} \mathbf{I}(\phi) & 0 \\ 0 & \mathbf{I}(\phi) \end{pmatrix}, \\ \mathbf{I}(\phi) &= \mathbf{R}(\phi)\begin{pmatrix} -1 & 0 \\ 0 & 1 \end{pmatrix}\mathbf{R}(\phi)^{-1} = \begin{pmatrix} -\cos 2\phi & -\sin 2\phi \\ -\sin 2\phi & \cos 2\phi \end{pmatrix}, \end{aligned} \tag{10.25}$$

and act as a mirror across a line rotated by ϕ from the x-axis. •

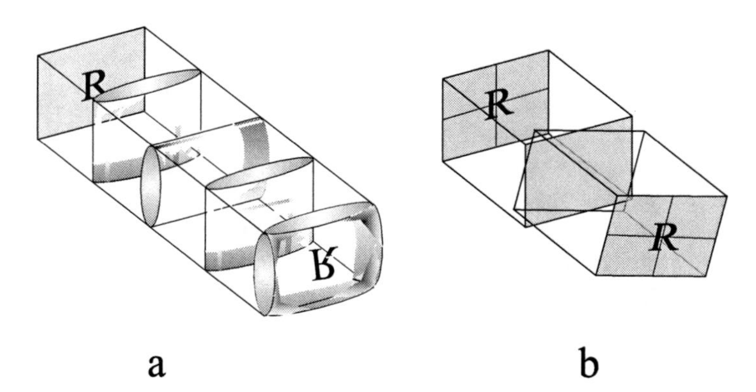

Fig. 10.5. Reflector and rotator. *Left*: Optical arrangement for $\mathcal{I}_x$, reflector of the vertical axis; this is a $DLDLDL$ unit magnifier in the horizontal direction, with a $DLDL$ inverting imager in the vertical direction. *Right*: Two such reflectors set at an angle yield a pure rotator.

Remark: How to build a rotator. With two reflectors at an angle we produce a pure rotation through the double angle, as shown in Fig. 10.5b. From the product of the matrices in (10.25) we see that

$$\mathcal{I}(\chi+\phi)\,\mathcal{I}(\chi)=\mathcal{I}(\chi)\,\mathcal{I}(\chi-\phi)=\mathcal{R}(2\phi). \tag{10.26}$$

Rotators actually belong to the leftmost Iwasawa factor of ortho-symplectic matrices (9.37), $\mathbf{O}(\mathbf{U})$, for real $\mathbf{U}=\mathbf{R}$. •

Remark: Seven dimensions of astigmatic imagers. Rotators and astigmatic magnifiers (including reflectors) are paraxial optical elements whose representing matrices have the generic block form $\mathcal{M}\Big(\operatorname{diag}(\mathbf{A},\mathbf{A}^{\top\,-1})\Big)$ with a nonsingular $\mathbf{A}\in\mathsf{GL}(2,\mathsf{R})\subset\mathsf{Sp}(4,\mathsf{R})$. They form the 4-parameter group of canonical phase space maps that keep the position and momentum subspaces separate: $\mathcal{M}(\mathbf{A}):\mathbf{q}=\mathbf{A}^{\top}\mathbf{q}$ and $\mathcal{M}(\mathbf{A}):\mathbf{p}=\mathbf{A}^{-1}\mathbf{p}$. In union with the 3-parameter manifold of thin astigmatic lenses, they yield the 7-parameter solvable group of astigmatic imagers, which will rotate and rescale the image, and/or change the directions of rays, but will keep the image *in focus*. Whereas in $\mathsf{Sp}(2,\mathsf{R})$ ($D=1$, page 181) 2 of the 3 parameters belonged to the manifold of imagers, in $\mathsf{Sp}(4,\mathsf{R})$ 7 out of 10 are. In dimension D, imagers have D^2+D+1 parameters: those of $\mathsf{Sp}(2D,\mathsf{R})$ minus those of $\mathsf{U}(D)$ plus those of $\mathsf{SO}(D)$. •

10.3 U(2) fractional Fourier transformers

The Iwasawa $\mathsf{U}(2)$ group of fractional Fourier transformers is the maximal compact subgroup of $\mathsf{Sp}(4,\mathsf{R})$, and is represented by all ortho-symplectic matrices,

$$\mathcal{F}(\mathbf{U}):=\mathcal{M}\begin{pmatrix}\operatorname{Re}\mathbf{U} & -\operatorname{Im}\mathbf{U}\\ \operatorname{Im}\mathbf{U} & \operatorname{Re}\mathbf{U}\end{pmatrix},\qquad \begin{matrix}\mathbf{U}\in\mathsf{U}(2),\\ \mathbf{U}\,\mathbf{U}^{\dagger}=\mathbf{1}.\end{matrix} \tag{10.27}$$

These are the canonical rotations of R^4 phase space that preserve the $\mathcal{S}^3$-spheres $|\mathbf{p}|^2+|\mathbf{q}|^2=$ constant. In turn, the group $\mathsf{U}(2)$ contains the subgroup of rotations around the optical axis, $\{\mathcal{R}(\phi)\}_{\phi\in\mathcal{S}^1}=\mathsf{SO}(2)$, which does not unfocus images; the remaining 3-parameter manifold of Fourier transformers mixes position and momentum coordinates of phase space. Also, $\mathsf{U}(2)$ has a *central* (commuting) subgroup $\mathsf{U}(1)$ [see (9.36)], which leaves a factor group $\mathsf{SU}(2)\overset{2:1}{=}\mathsf{SO}(3)$ of rotations of a *Fourier sphere*. The literature on fractional Fourier transforms has grown very considerably; the dedicated book by Ozaktas *et al.* [112] lists more than 550 references, mostly within the model of $D=1$-dim wave systems, to which we can hardly list with justice. We shall proceed to examine the structure of this group of $\mathsf{U}(2)$-Fourier transforms in paraxial geometric optics as proposed in [131].

10.3.1 Central Fourier transforms

One evident 1-parameter group of fractional Fourier transformers is the direct generalization of the plane DLD arrangement of Fig. 10.2, with axis-symmetric lenses whose parameters fulfill the relations (10.10)–(10.8).

As in (9.34), the factorization of $\mathbf{U} \in \mathsf{U}(2)$ into a phase and matrix of unit determinant [$\mathbf{U} = e^{i\phi}\mathbf{V}$, where $\phi = \frac{1}{2}\arg\det\mathbf{U}$, so $e^{i\phi}\mathbf{1} \in \mathsf{U}(1)$, and $\mathbf{V} \in \mathsf{SU}(2)$ with $\det\mathbf{V} = 1$] leads to the factorization of ortho-symplectic matrices as

$$\begin{pmatrix} \operatorname{Re}\mathbf{U} & -\operatorname{Im}\mathbf{U} \\ \operatorname{Im}\mathbf{U} & \operatorname{Re}\mathbf{U} \end{pmatrix} = \begin{pmatrix} \cos\frac{\pi}{2}\alpha\,\mathbf{1} & -\sin\frac{\pi}{2}\alpha\,\mathbf{1} \\ \sin\frac{\pi}{2}\alpha\,\mathbf{1} & \cos\frac{\pi}{2}\alpha\,\mathbf{1} \end{pmatrix} \begin{pmatrix} \operatorname{Re}\mathbf{V} & -\operatorname{Im}\mathbf{V} \\ \operatorname{Im}\mathbf{V} & \operatorname{Re}\mathbf{V} \end{pmatrix}. \tag{10.28}$$

We can see directly that the 'overall phase' subgroup $\mathsf{U}(1)$ commutes with all of $\mathsf{U}(2)$, i.e., it is the *center* of the group of Fourier transforms. For this reason we call *central* Fourier transforms to

$$\mathcal{F}_0^\alpha := \mathcal{M}\begin{pmatrix} \cos\frac{\pi}{2}\alpha\,\mathbf{1} & -\sin\frac{\pi}{2}\alpha\,\mathbf{1} \\ \sin\frac{\pi}{2}\alpha\,\mathbf{1} & \cos\frac{\pi}{2}\alpha\,\mathbf{1} \end{pmatrix}. \tag{10.29}$$

In particular, central Fourier transforms commute with rotations around the z-axis, so they are axially symmetric systems.

10.3.2 Separable Fourier transforms

To build all other $\mathsf{U}(2)$ Fourier transforms, we again resort to the tactic of composing astigmatic optical elements out of x- and y-cylindrical ones, as in Fig. 10.5. An (inverse) x-Fourier transformer $\mathcal{F}_x^{\alpha_x}$ is a DLD arrangement as before, but now with a lens $\mathcal{L}_{G_x}^{(x)}$ of Gaussian power $\operatorname{diag}(G_x, 0)$ whose parameters fulfill (10.10), and a unit system in y with a $DLDLDL$ configuration,

$$\begin{aligned} &\mathcal{D}_{4F}\,\mathcal{L}_{G_x}^{(x)}\,\mathcal{D}_{4F} = \mathcal{F}_x^{-\alpha}, \quad G_x = \sin\tfrac{\pi}{2}\alpha_x =: \sin(2\arctan 4F) \\ &\mathcal{D}_{2F}\,\mathcal{L}_{1/F}^{(y)}\,\mathcal{D}_{4F}\,\mathcal{L}_{1/F}^{(y)}\,\mathcal{D}_{2F}\,\mathcal{L}_{2/F}^{(y)} = \mathit{1}_y. \end{aligned} \tag{10.30}$$

This apparatus rotates the (p_x, q_x) phase plane by an angle $0 < \theta_x := \frac{\pi}{2}\alpha_x < \pi$; the required Fourier power $0 < \alpha_x < 2$ determines its length, $8F$.

Similarly, we can build a y-Fourier transformer $\mathcal{F}_y^{\alpha_y}$ which rotates only the (p_y, q_y) phase plane, by rotating the former 90° around its axis. Concatenating these two systems, we obtain a set of Fourier transformers that we call x–y-*separable*, because they rotate the two planes, (p_x, q_x) and (p_y, q_y), separately. In the notation (10.27), they are

$$\mathcal{F}\begin{pmatrix} e^{i\frac{1}{2}\pi\alpha_x} & 0 \\ 0 & e^{i\frac{1}{2}\pi\alpha_y} \end{pmatrix} = \mathcal{M}\begin{pmatrix} \cos\frac{\pi}{2}\alpha_x & 0 & -\sin\frac{\pi}{2}\alpha_x & 0 \\ 0 & \cos\frac{\pi}{2}\alpha_y & 0 & -\sin\frac{\pi}{2}\alpha_y \\ \sin\frac{\pi}{2}\alpha_x & 0 & \cos\frac{\pi}{2}\alpha_x & 0 \\ 0 & \sin\frac{\pi}{2}\alpha_y & 0 & \cos\frac{\pi}{2}\alpha_y \end{pmatrix}. \tag{10.31}$$

This is a toroidal manifold of transformations $(\alpha_x, \alpha_y) \in \mathcal{S}^1 \times \mathcal{S}^1$ of transforms which, for $\alpha_x = \alpha_y$, contains the subset of central Fourier transforms.

10.3.3 SU(2)-Fourier transforms

When we rotate the object and image screens of an x–y-separable Fourier transformer (10.31), by ϕ and ψ respectively, the result can be expressed as a central times a separable Fourier transformer:[7]

$$\begin{aligned}\mathcal{F}(\mathbf{U}) &= \mathcal{R}(\phi)\,\mathcal{F}\begin{pmatrix} e^{i\frac{1}{2}\pi\alpha_x} & 0 \\ 0 & e^{i\frac{1}{2}\pi\alpha_y} \end{pmatrix}\mathcal{R}(\psi) \\ &= \mathcal{F}(e^{i\frac{1}{4}\pi(\alpha_x+\alpha_y)}\mathbf{1})\,\mathcal{R}(\phi)\,\mathcal{F}\begin{pmatrix} e^{i\frac{1}{4}\pi(\alpha_x-\alpha_y)} & 0 \\ 0 & e^{-i\frac{1}{4}\pi(\alpha_x-\alpha_y)} \end{pmatrix}\mathcal{R}(\psi).\end{aligned} \tag{10.32}$$

In the last expression, the leftmost factor is a central Fourier transform (10.29) of power $\alpha = \frac{1}{2}(\alpha_x + \alpha_y)$, while the parameters of the three last factors are the Euler angles (ϕ, θ, ψ), $\theta = \frac{1}{2}(\alpha_x - \alpha_y)$, of a *generic* matrix in $\mathsf{SU}(2) \overset{2:1}{=} \mathsf{SO}(3)$. The middle Euler angle θ represents a *saddle* fractional Fourier transform, whose separated x- and y- powers are inverse of each other. Hence, $\mathsf{U}(2)$ Fourier transformers are separable Fourier transformers, up to rotations of the input and output screens.

Example: The Bartelt–Brenner–Lohmann Wigner transformer. One astigmatic lens in a *DLD* configuration cannot lead to *pure* Fourier transformers, but to a system which we can decompose into a pure Fourier transformer followed by a solvable imager factor. In [18], Bartelt, Brenner and Lohmann proposed a setup which inverts the object in one direction of the screen and performs a Fourier transform in the orthogonal direction. This is shown in Fig. 10.6; in the middle it has two crossed cylindrical lenses of focal distances $F_x = \frac{1}{2}z = 1/G_x$ (an inverting imager in x) and $F_y = z = 1/G_y$ (a Fourier transformer in y), so $G_y = \frac{1}{2}G_x = 1/z = G$, and the total length of the apparatus is $2z > 0$. Their purpose was to display a 1-dim signal, such as a bar code, in both position and momentum space, and thus see its wave-optical *Wigner* function on the screen turned phase space. The Bartelt–Brenner–Lohmann system is built and represented as

$$\begin{aligned}\mathcal{D}_{1/G}\,\mathcal{L}_{\mathrm{diag}(2G,G)}\,\mathcal{D}_{1/G} &= \mathcal{M}\left(\begin{array}{cc|cc} -1 & 0 & 2G & 0 \\ 0 & 0 & 0 & G \\ \hline 0 & 0 & -1 & 0 \\ 0 & -G^{-1} & 0 & 0 \end{array}\right) \\ &= \mathcal{M}\left(\begin{array}{cc|cc} -1 & 0 & 0 & 0 \\ 0 & 0 & 0 & 1 \\ \hline 0 & 0 & -1 & 0 \\ 0 & -1 & 0 & 0 \end{array}\right)\left(\begin{array}{cc|cc} 1 & 0 & 0 & 0 \\ 0 & G^{-1} & 0 & 0 \\ \hline 0 & 0 & 1 & 0 \\ 0 & 0 & 0 & G \end{array}\right)\left(\begin{array}{cc|cc} 1 & 0 & -2G & 0 \\ 0 & 0 & 0 & 0 \\ \hline 0 & 0 & 1 & 0 \\ 0 & 0 & 0 & 1 \end{array}\right).\end{aligned} \tag{10.33}$$

The leftmost factor is a $\mathsf{U}(2)$-Fourier transformer represented by the unitary matrix

[7] The reader familiar with the quantum theory of angular momentum will recognize the exponentiated Pauli matrices σ_2 and σ_3 in the factors of (10.32).

$$\mathbf{U} = \begin{pmatrix} -1 & 0 \\ 0 & -i \end{pmatrix} = e^{-i\frac{3}{4}\pi} \begin{pmatrix} e^{-i\frac{1}{4}\pi} & 0 \\ 0 & e^{i\frac{1}{4}\pi} \end{pmatrix}, \quad \text{i.e.,} \quad \begin{aligned} \alpha_x &= -2, \\ \alpha_y &= -1. \end{aligned} \tag{10.34}$$

This is equivalent to a central Fourier transformer of power $\alpha = \frac{1}{2}(\alpha_x + \alpha_y) = -\frac{3}{2}$ together with a saddle transformer of power $\frac{1}{2}(\alpha_x - \alpha_y) = -\frac{1}{2}$. The magnifier and lens factors of the accompanying imager are characterized by the 2×2 blocks which can be read from the 4×4 matrices in (10.33). •

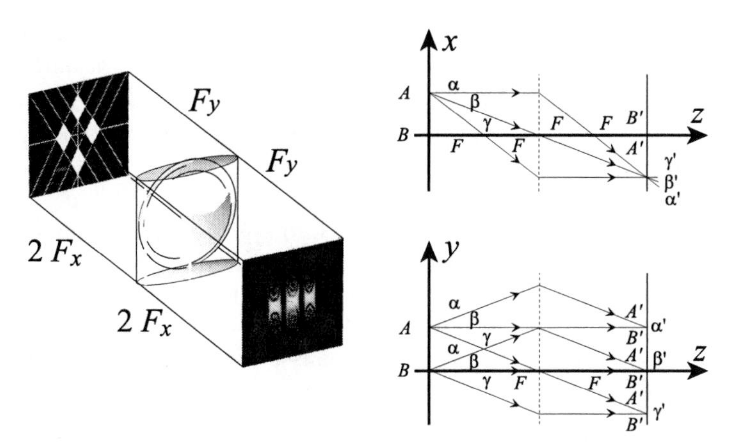

Fig. 10.6. *Left*: The Bartelt–Brenner–Lohmann setup with a spherical + cylindrical = astigmatic thin lens. *Right*: the arrangement inverts objects in the horizontal (x-) direction, and Fourier-transforms them in the y-direction. The depicted object is a pair of equal bar codes $f(q)$ superposed at an angle (opaque except for two transparent bands illuminated from behind); the image is essentially the *Wigner* transform of $f(q)$, which shows the bar code in phase space (p, q) [144].

10.4 Systems *cum* reflection

The reflection of a paraxial optical system in a mirror, $\mathcal{M} \mapsto \overline{\mathcal{M}}$, is again a paraxial system, even when the mirror is not flat but (paraxially) warped. This has interesting implications for building 'cat-eye' systems (which return a beam of light to its source) and fractional Fourier transformers thereby. We can work easily in D dimensions.

We introduced reflection as an antihomomorphism for the ordered product of optical transformations [Sect. 4.3, (4.34)] noting that free displacement in the mirror world is the same as in the real one [Fig. 4.3, (4.35)]; *thin* lenses coincident with the mirror are also invariant [as can be seen from (4.36) and (9.7)–(9.9), exchanging $n \leftrightarrow n'$ and $\mathbf{Z} \leftrightarrow -\mathbf{Z}$]. The reflected paraxial elements, indicated as in Sect. 4.6 with an overline, are thus represented by the matrices

$$\overline{\begin{pmatrix} 1 & 0 \\ -z1 & 1 \end{pmatrix}} = \begin{pmatrix} 1 & 0 \\ -z1 & 1 \end{pmatrix}, \qquad \overline{\begin{pmatrix} 1 & \mathbf{G} \\ 0 & 1 \end{pmatrix}} = \begin{pmatrix} 1 & \mathbf{G} \\ 0 & 1 \end{pmatrix}. \tag{10.35}$$

We state that the action of reflection on generic paraxial optical systems, represented on their matrices, is

$$\overline{\mathbf{M}} = \overline{\begin{pmatrix} \mathbf{A} & \mathbf{B} \\ \mathbf{C} & \mathbf{D} \end{pmatrix}} = \begin{pmatrix} \mathbf{D}^\top & \mathbf{B}^\top \\ \mathbf{C}^\top & \mathbf{A}^\top \end{pmatrix}. \tag{10.36}$$

To prove that this is the action of reflection, we check that it holds for free flights and lenses (10.35), and we verify that it is indeed an antihomomorphism: when $\mathbf{M}_1\,\mathbf{M}_2 = \mathbf{M}_3$, then $\overline{\mathbf{M}_3} = \overline{\mathbf{M}_2}\,\overline{\mathbf{M}_1}$.

Example: Looking through binoculars. The *DLD* configuration in (10.3) will reflect by exchanging the two displacements, $z_1 \leftrightarrow z_2$, and thereby exchanging its diagonal matrix elements: magnifications μ and μ^{-1}. An imaging system such as a camera will reflect into a projector, and binoculars will be seen through the 'wrong' end with inverse magnification. •

We also noted in Sect. 4.6 that reflection and inversion are two distinct antihomomorphisms, not homomorphically related. But in the paraxial régime, where optical elements $\mathcal{M}$ are approximated and represented by symplectic matrices $\mathbf{M} = \begin{pmatrix} \mathbf{A} & \mathbf{B} \\ \mathbf{C} & \mathbf{D} \end{pmatrix}$, reflection, inversion and transposition are linearly related by a similarity transformation:

$$\overline{\mathbf{M}} = \mathbf{K}\,\mathbf{M}^{-1}\,\mathbf{K} = \begin{pmatrix} \mathbf{D}^\top & \mathbf{B}^\top \\ \mathbf{C}^\top & \mathbf{A}^\top \end{pmatrix} = \begin{pmatrix} 0 & 1 \\ 1 & 0 \end{pmatrix} \mathbf{M}^\top \begin{pmatrix} 0 & 1 \\ 1 & 0 \end{pmatrix}, \tag{10.37}$$

$$\text{where} \quad \mathbf{K} := \begin{pmatrix} -1 & 0 \\ 0 & 1 \end{pmatrix} = \mathbf{K}^{-1}. \tag{10.38}$$

Remark: Reflection is an *outer* antihomomorphism. Although reflection $\mathbf{M} \mapsto \overline{\mathbf{M}}$ maps symplectic matrices on symplectic matrices, you will note that $\mathbf{K}$ is *not* a symplectic matrix; it does not satisfy (9.23). The similarity transformation by $\mathbf{K}$ in (10.37) is thus an *outer* homomorphism, and paraxial reflection an outer antihomomorphism of $\mathsf{Sp}(2D, \mathsf{R})$. •

Remark: Reflection of phase space. The matrix $\mathbf{K}$ introduced in (10.38) is not symplectic, but its action on rays is well defined: it inverts the momentum vector $\mathbf{K}\begin{pmatrix} \mathbf{p} \\ \mathbf{q} \end{pmatrix} = \begin{pmatrix} -\mathbf{p} \\ \mathbf{q} \end{pmatrix}$. This is evident for free propagation 'across' the mirror in Fig. 4.3: when the reflected ray is continued forward across the screen, again along $+z$, its momentum is $-\mathbf{p}$. We thus define a reflection *operator* whose action on functions of phase space is

$$\mathcal{K} : f(\mathbf{p}, \mathbf{q}) := f(-\mathbf{p}, \mathbf{q}). \tag{10.39}$$

By relating reflection with inversion through the similarity homomorphism by $\mathcal{K} = \mathcal{M}(\mathbf{K})$, we shall extrapolate its action to the aberration polynomials which lie the metaxial régime of Part IV. •

Imagine a paraxial optical system whose output screen has been substituted by a mirror. Rays go forth from the input z-plane through $\mathcal{M}(\mathbf{M})$, reflect (i.e., go into the mirror world), and then continue through what they see as the *reflected* system $\overline{\mathcal{M}}(\mathbf{M}) = \mathcal{M}(\overline{\mathbf{M}})$. The rays in fact arrive from the right at the original input plane, which now serves also to register the output rays of the *system cum reflection*, which is

$$\mathbf{M}^{\mathrm{II}} := \mathbf{M}\,\overline{\mathbf{M}} = \begin{pmatrix} \mathbf{A}\mathbf{D}^\top + \mathbf{B}\mathbf{C}^\top & 2\mathbf{A}\mathbf{B}^\top \\ 2\mathbf{C}\mathbf{D}^\top & (\mathbf{A}\mathbf{D}^\top + \mathbf{B}\mathbf{C}^\top)^\top \end{pmatrix}. \tag{10.40}$$

The two diagonal submatrices of the system *cum* reflection $\mathbf{M}^{\mathrm{II}}$ are transposes of each other, and the off-diagonal ones are symmetric because of the symplectic conditions (9.28).

Remark: Self-reflecting systems. Systems *cum* reflection are invariant under reflections because $\overline{\mathbf{M}^{\mathrm{II}}} = \overline{\mathbf{M}}^{\mathrm{II}} = \mathbf{M}^{\mathrm{II}}$. Conversely, self-reflecting systems,

$$\overline{\mathbf{M}} = \mathbf{M} \quad \Leftrightarrow \quad \mathbf{A} = \mathbf{D}^\top, \quad \mathbf{B} = \mathbf{B}^\top, \quad \mathbf{C} = \mathbf{C}^\top, \tag{10.41}$$

have submatrices with the symmetry properties noted under (10.40). Examples of self-reflecting systems are all astigmatic lenses and displacements in (10.35), central Fourier transforms $\mathcal{F}_0^\alpha$ in (10.29), and all separable Fourier transforms $\mathcal{F}_x^{\alpha_x}\mathcal{F}_y^{\alpha_y}$ in (10.31), rotated to any angle [see (10.32) with $\phi = -\psi$]. Systems which are **not** self-reflecting include the rotators $\mathcal{R}(\phi)$ in (10.16), and the magnifiers $\mathcal{A}_{\mathrm{S}}$ in (10.21); instead, reflection *inverts* these systems: $\overline{\mathcal{R}}(\phi) = \mathcal{R}(-\phi)$ and $\overline{\mathcal{A}_{\mathrm{S}}} = \mathcal{A}_{\mathrm{S}^{-1}}$. •

Remark: Specular roots of the identity. Only systems satisfying the self-reflection conditions (10.41) can have *specular roots*. These roots are not unique; many systems can yield the same action *cum* reflection. Consider first the specular roots of the unit system, i.e., $\mathbf{M}$'s such that $\mathbf{M}\,\overline{\mathbf{M}} = \mathbf{1}$; or, in the usual block form,

$$\overline{\mathbf{M}} = \mathbf{M}^{-1} \quad \Leftrightarrow \quad \mathbf{A} = \mathbf{D}^\top, \quad \mathbf{B} = \mathbf{0} = \mathbf{C}. \tag{10.42}$$

Specular roots of unity are thus all block-diagonal matrices $\operatorname{diag}(\mathbf{A}, \mathbf{A}^\top)$, $\det \mathbf{A} = 1$, i.e., pure astigmatic imagers and rotators. By multiplication, it follows that when $\mathbf{M}$ is a specular root of $\mathbf{M}^{\mathrm{II}}$, and $\mathbf{L}$ is any specular root of the unit, then any $\mathbf{M}\,\mathbf{L}$ will be also a specular root of $\mathbf{M}^{\mathrm{II}}$. Therefore, specular roots are *right coset* manifolds, modulo the subgroup $\mathsf{SL}(D, \mathsf{R})$ of roots of the identity. •

Remark: Reflection in a warped mirror. In Sect. 4.6 we saw that the reflecting transformation $\mathcal{S}^{\mathrm{R}}_{n;\zeta}$ [see (4.32)], due to a mirror $z = \zeta(\mathbf{q})$ in a medium n, is the product of the root transformation $\mathcal{R}_{n;\zeta}$ and its reflection, $\overline{\mathcal{R}_{n;\zeta}} = \mathcal{R}^{-1}_{n;-\zeta}$. As we saw in (9.7), the surface ζ is approximated in the paraxial régime by a quadratic polynomial (9.6) with a symmetric matrix $\mathbf{Z}$, and the root transformation is represented by an upper-triangular 'lens' matrix $\mathbf{R}_\zeta$ of Gaussian power $-2\mathbf{Z}$ (setting $n = 1$ for simplicity). The warped mirror is thus represented by $\mathbf{S}^{\rm R}_\zeta = \mathbf{R}_\zeta \overline{\mathbf{R}}_\zeta = (\mathbf{R}_\zeta)^2$, and a system $\mathbf{M}$ *cum* reflection in that warped mirror[8] will be represented by

$$\mathbf{M}^{\rm II}_\zeta = \mathbf{M}\,\mathbf{S}^{\rm R}_\zeta\,\overline{\mathbf{M}} = \mathbf{M}\,\mathbf{R}_\zeta\,\overline{\mathbf{M}\,\mathbf{R}_\zeta} = (\mathbf{M}\,\mathbf{R}_\zeta)^{\rm II} \tag{10.43}$$

This is still a system *cum* reflection, as before the warping. •

The previous three remarks were made in the direction of designing pure fractional Fourier transformers by systems cum reflection, such as a cat's eye with a curved, reflecting retina. Indeed, recall the modified Iwasawa parametrization of a symplectic matrix $\mathbf{M}$ in (9.72) and (10.14), which we write again for paraxial optical systems,

$$\mathcal{M}(\mathbf{M}) = \mathcal{F}(\mathbf{U})\,\mathcal{A}_{\rm S}\,\mathcal{L}_{\rm G}, \tag{10.44}$$

where $\mathcal{F}$ is a Fourier transformer such as (10.8)–(10.10), $\mathcal{A}_{\rm S}$ is an astigmatic magnifier by $\mathbf{S} \in \mathsf{SL}(D, \mathsf{R})$ as in (10.21), and $\mathcal{L}_{\rm G}$ is a lens of Gaussian power $\mathbf{G} = \mathbf{G}^\top$ as in (10.15). And we cap this system with a warped mirror which has the appropriate surface matrix $\mathbf{Z} = +\frac{1}{2}\mathbf{G}$ so that it *cancels* the last factor of (10.44). The next-to-last factor will also cancel in the system *cum* reflection, which according to (10.43) is now

$$\Big(\mathcal{M}(\mathbf{M})\,\mathcal{R}_\zeta\Big)^{\rm II} = \Big(\mathcal{F}(\mathbf{U})\,\mathcal{A}_{\mathbf{S}}\Big)^{\rm II} = \Big(\mathcal{F}(\mathbf{U})\Big)^{\rm II} = \mathcal{F}(\mathbf{U}\,\mathbf{U}^\top). \tag{10.45}$$

In particular, any impure imaging system, such as an astigmatic but focused slide projector ($\mathbf{U} = \mathbf{1}$) can be corrected with a warped mirror, so that the system *cum* reflection – back at the slide – is the identity paraxial optical system. In the general case when $\mathbf{U} \neq \mathbf{1}$ (such as an *un*focused projector), the warping can correct the system *cum* reflection only to $\mathbf{U}\,\mathbf{U}^\top$. Thus, *any* paraxial optical system, *cum* reflection in an appropriately warped mirror, is a fractional Fourier transformer between the coincident input and output screens. But for $D \geq 2$, since $\mathbf{U}\,\mathbf{U}^\top$ is a *symmetric* unitary matrix, these Fourier transformers exclude all rotations around the optical axis.

Example: Myopic cat's eye. We refer to the axis-symmetric cum-reflection cat-eye system of Fig. 10.7 in $D = 1$ dimensions. (We must assume this is *not* an imaging system, because then the 'Fourier transformation' produced cum reflection would be only the unit.) Let the 'air' and 'eye' refractive indices will be $n_1 = 1$ and $n_2 = \frac{3}{2}$ respectively, separated by a spherical surface of unit radius [this fixes the length scale so $\zeta(q) = \frac{1}{2}q^2$ and $G = \frac{1}{2} = Z$ in the paraxial approximation, see (9.6)–(9.8)]. We place the object at $z_1 = 6$ units to the left of the surface of the eye, and to its right we place the warpable mirror, at $z_2 = n_2\chi = \frac{3}{2}\chi$, where we use henceforth $\chi > 0$ for generic dis-

[8] We assume that the system is homogeneous in the region of space invaded or exposed by the warping.

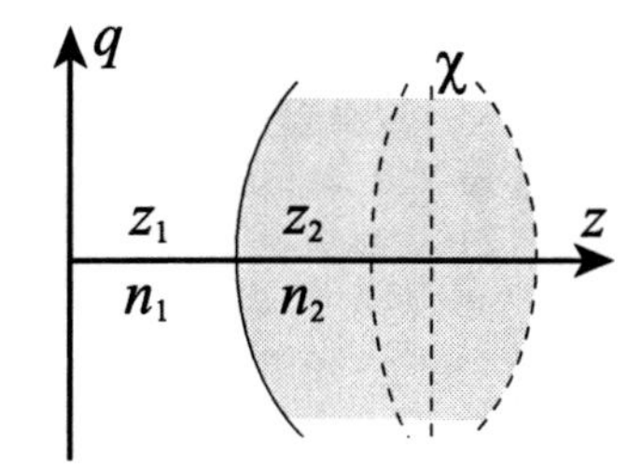

Fig. 10.7. Myopic cat-eye composed of a spherical interface between two media n_1, n_2, and a warpable mirror whose position z_2 and radius are parametrized by χ. The input and output screens q are coincident at a distance z_1 to the left the eye.

tance. Then, the matrix of the system $\mathbf{M}(\chi)$ is built and Iwasawa-decomposed as follows:

$$\begin{pmatrix} 1 & 0 \\ -6 & 1 \end{pmatrix}\begin{pmatrix} 1 & \frac{1}{2} \\ 0 & 1 \end{pmatrix}\begin{pmatrix} 1 & 0 \\ -\chi & 1 \end{pmatrix} = \begin{pmatrix} 1-\frac{1}{2}\chi & \frac{1}{2} \\ -6+2\chi & -2 \end{pmatrix}$$
$$= \begin{pmatrix} \cos\omega & -\sin\omega \\ \sin\omega & \cos\omega \end{pmatrix}\begin{pmatrix} L & 0 \\ 0 & L^{-1} \end{pmatrix}\begin{pmatrix} 1 & G \\ 0 & 1 \end{pmatrix}, \tag{10.46}$$

$$L\,e^{-i\omega} = (1+6i) - (\tfrac{1}{2}+2i)\chi, \qquad G = \frac{50-17\chi}{148-100\chi+17\chi^2}. \tag{10.47}$$

From here we see that at $\chi = 3$, the cat-eye system is an inverting imager of magnification $-\frac{1}{2}$ ($c = 0$, but impure since $b \neq 0$), while at $\chi = 2$ it is an impure Fourier transformer ($b \neq 1$, $d \neq 0$). The mirror that we place at $z_2 = \frac{3}{2}\chi$ should be warped to $z = \zeta_2 q^2$ (with $\zeta_2 = \frac{1}{2}G/n_2 = \frac{1}{3}G$ as given above), to cancel the nilpotent factor in the Iwasawa decomposition. The required warp coefficient $\zeta_2(\chi)$ is shown in Fig. 10.8a. The system *cum* warped

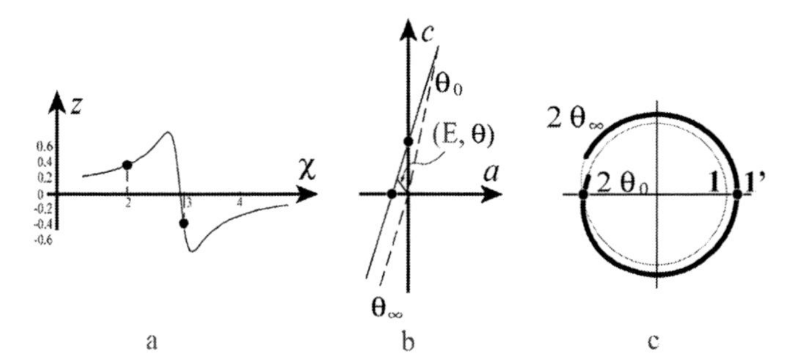

Fig. 10.8. (**a**) Warp coefficient ζ_2 of the mirror as function of $\chi = \frac{2}{3}z > 0$ for the cat-eye of the previous figure. We mark $\chi = 2$, where the unmirrored system is a Fourier transformer [$\zeta_2(2) = \frac{1}{3}$], and $\chi = 3$, where it is an imager [$\zeta_2(3) = -\frac{1}{3}$]. (**b**) Locus of χ-systems in the complex plane $L\,e^{-i\omega} = a+ic$. It is a line of slope 4; again we mark $\omega(2) = -\frac{1}{2}\pi$ and $\omega(3) = -\pi$. (**c**) Range of $\mathsf{U}(1)$-fractional Fourier transforms provided by the cat-eye system, with $-\omega^{\mathrm{II}}(\chi)$ (heavy line) counted modulo 4π.

reflection will be a *pure* inverse Fourier transformer of angle $-\omega^{\text{II}}(\chi) = 2\omega(\chi)$,[9] or power $4\omega/\pi$. In Fig. 10.8b we show the complex parameter range of $L\, e^{-i\omega}$ in (10.47) for $\chi > 0$. Finally, in Fig. 10.8c we draw the range of angles $\omega^{\text{II}}(\chi)$ modulo 4π, from $-\omega^{\text{II}}(0) = 2\arctan 6 \approx 162°$, through $-\omega^{\text{II}}(2) = 180°$ and $\omega^{\text{II}}(3) = 360°$, and up to $-\omega^{\text{II}}(\infty) = 2(\pi + \arctan 4) \approx 512° \equiv 152°$. (Be aware that at 360° lies the *second metaplectic unit* $\mathbf{1}'$ of paraxial wave optics – see remark on page 170.) •

10.5 Minimal lens arrangements

The construction of paraxial optical systems in this chapter has been supported mostly by the frame of the Iwasawa factorizations, which display the Lie group structure of the whole set. However, for the opticist at his proverbial workbench, the more immediate problem is to place adequately the least number of elements – lenses of fixed focal lengths – that will produce a required system. This yields a distinct parametrization of $\mathsf{Sp}(2,\mathsf{R})$ by upper- and lower-triangular matrices which is hardly ever found in the context of group theory. In this section we shall find the minimal lens arrangements that realize *all* paraxial optical systems; this will be done in the $D = 1$-dim case of plane optics, so it is applicable also for axis-symmetric optical systems. We have at our disposal the matrices of the elementary constituents, that we write again:

$$\text{displacements } z \geq 0, \qquad \mathcal{D}_z = \mathcal{M}\begin{pmatrix} 1 & 0 \\ -z & 1 \end{pmatrix} \quad \text{non-negative;} \tag{10.48}$$

$$\text{lenses } G \in \mathsf{R}, \qquad \mathcal{L}_G = \mathcal{M}\begin{pmatrix} 1 & G \\ 0 & 1 \end{pmatrix} \quad \begin{cases} \text{concave } G < 0, \\ \text{convex } G > 0. \end{cases} \tag{10.49}$$

[Cf., (9.3) and (9.5).]

10.5.1 The *abc*-parameters

We use the three independent parameters (a, b, c) of the generic symplectic matrix $\mathbf{M} = \begin{pmatrix} a & b \\ c & d \end{pmatrix}$, $ad - bc = 1$, in the role of Cartesian coordinates for an R^3-space of paraxial optical systems. Previously in Fig. 9.4a we drew the 3-dim $\mathsf{Sp}(2,\mathsf{R})$ manifold using the real matrix parameters of $\mathsf{SU}(1,1)$ to highlight the connectivity and cover of the group. Now we show the manifold of paraxial optical systems $\mathbf{M} \in \mathsf{Sp}(2,\mathsf{R})$ as in Fig. 10.9. The unit system $(a = 1\,, b = 0\,, c = 0)$ is at the Cartesian origin; $c \leq 0$ is the half-axis of displacements $\mathcal{D}_z$, b is the axis of lenses $\mathcal{L}_G$, and a is the magnification axis.

[9] The Iwasawa and Bargmann angle parameters are here the same – see the remark on page 174.

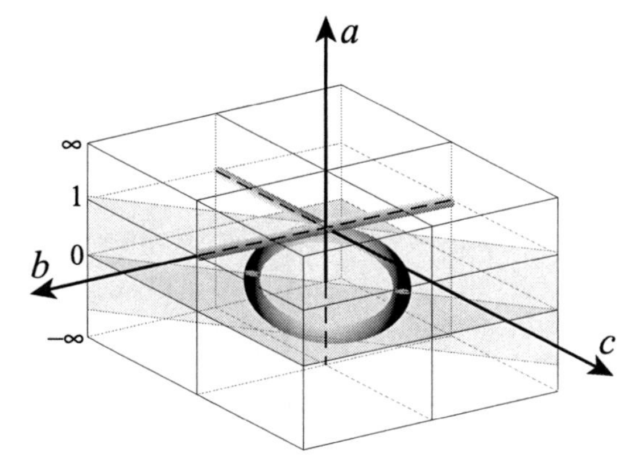

Fig. 10.9. The R^3 manifold of $\mathsf{Sp}(2,\mathsf{R})$ matrices in *abc*-coordinates. The unit optical element is at the origin of coordinates $a = 1$, $b = 0 = c$. Each point in the manifold (except the plane $a = 0$) corresponds to a paraxial optical system. On the plane $a = 1$ we mark positive free displacements and lenses. We also mark fractional Fourier transformers by a circle which lies in the $b = -c$ plane, between $a = \pm 1$.

Remark: Lines and planes in *abc*-space Positive displacements D and lenses L [(10.48) and (10.49)] are shown in Fig. 10.9; they are the basic, 1-element configurations, and are lines in *abc*-space. Magnifiers [(10.6) with $a = e^{\zeta} > 0$, $b = 0 = c$, but also with $a < 0$] are on the vertical axis. The plane $a = 0$ must be removed from the *abc* space because there the determinant dictates that $bc = -1$, promoting d to an independent parameter; this 2-dim submanifold must be treated separately. On the plane $b = -c$ we find the fractional Fourier transformers [(10.9), with $a = \cos\phi$, $c = -b = \sin\phi$], while on the plane $b = c$ of symmetric systems lie the hyperbolic expanders [(10.12) with $a = \cosh\zeta > 1$, $b = c = -\sinh\zeta < 0$, but also with $a < -1$]. Upper-triangular matrices representing generic imagers lie in the $c = 0$ plane; lower triangular matrices occupy the $b = 0$ plane. •

10.5.2 One-lens DLD configurations

We start with one lens and two displacements in a DLD configuration [see (10.3)]. Between the three parameters of such a system, z_1, G, z_2, and its *abc* coordinates, the following relations are found readily:

$$a = 1 - gz_2, \quad b = G, \quad c = -z_1 - z_2 + z_1 G z_2 = -z_1 a - z_2\,; \qquad (10.50)$$

$$z_1 = (1-d)/b = (a - bc - 1)/ab, \quad G = b, \quad z_2 = (1-a)/b. \qquad (10.51)$$

Of course, not all optical systems are realizable in a DLD configuration. For example, were we to use (10.51) for $\mathcal{M}\begin{pmatrix} 1 & 1 \\ 1 & 2 \end{pmatrix}$, the first of (10.50) would inform us that $z_1 = -1$, which is not physically realizable. There are two DLD cases to consider: concave and convex lenses.

A single **concave** lens $b = G < 0$ is the simplest case, because there $-Gz_2 \geq 0 \Rightarrow a \geq 1 \Rightarrow c \leq 0$; however, $c = 0 \Rightarrow z_1 = 0 = z_2 \Rightarrow a = 1$. Hence, the region of realizability by concave DLD configurations consists of an octant of R^3 space and a half-line:

$$\left.\begin{array}{l} a \geq 1,\, b < 0,\, c < 0, \\ a = 1,\, b < 0,\, c = 0. \end{array}\right\} \quad \begin{pmatrix} DLD \\ \text{concave} \end{pmatrix} \tag{10.52}$$

This is the upper solid block in Fig. 10.10a; the octant has a closed $a = 1$ face (the set includes the boundary of restricted DL systems with $z_2 = 0$), and open $b = 0$ and $c = 0$ faces (where the boundary is outside the set); the edge between the a- and c-faces is in the set and represents the lenses L.

A single **convex** lens in a DLD configuration will fill other regions of abc-space. In (10.50), $b = G > 0 \Rightarrow Gz_2 \geq 0 \Rightarrow a \leq 1$, and there are three cases to consider: $1 \geq a > 0 \Rightarrow 0 \leq Gz_2 < 1 \Rightarrow c \leq 0$, the singular plane $a = 0$, and $a < 0 \Rightarrow Gz_2 > 1$ which imposes no restriction on c. The coordinate plane $a = 0$ is singular, but not the parameters of the DLD configurations that realize it: $G = 1/z_2 = b > 0 \Rightarrow c = -1/b < 0$ and $d = 1 - z_1/z_2 < 1$; in Fig. 10.10a, the plane quadrant $(a = 0\,,\, b > 0\,,\, c < 0)$ thus bridges the $a > 0$ and $a < 0$ DLD regions. Hence, the parameter ranges of this configuration are:

$$\left.\begin{array}{rl} 0 < a \leq 1\,,\, b > 0,\, c \leq 0, & \\ a = 0,\, b > 0,\, c < 0, & d < 1, \\ a < 0,\, b > 0,\, c \in \mathsf{R}. & \end{array}\right\} \quad \begin{pmatrix} DLD \\ \text{convex} \end{pmatrix} \tag{10.53}$$

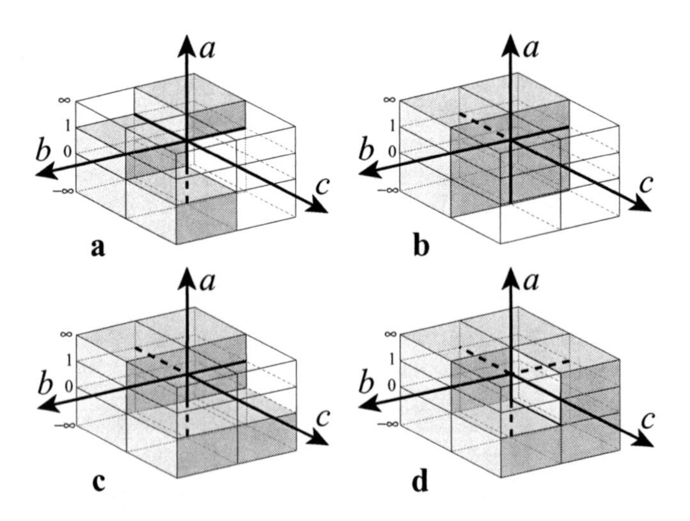

Fig. 10.10. Regions of the abc-manifold of paraxial optical systems that can be built with lens arrangements. (**a**) One-lens, DLD configuration. (**b**) Two-lens restricted LDL configuration. (**c**) The restricted configuration $DLDL$. (**d**) Two-lens generic $DLDLD$ configuration. The remaining unfilled region (first octant) corresponds to three-lens configurations.

This is the block in the lower part of Fig. 10.10a. Again, the $a = 1$ face contains the restricted DL systems and is closed on the edge of restricted L systems; the $c = 0$ face is also closed and contains *imaging* systems (*impure*, since $b \neq 0$), but magnifiers (*pure* imagers, $b = 0$) are excluded from this region. The other faces are open. The concave and convex DLD configurations are disjoint, and touch only at the common closing edge of systems with trivial lenses $G = 0$.

Example: First half of the Fourier cycle with DLD configurations. The region of convex-DLD configurations contains the first half of the inverse Fourier transform cycle $\mathcal{F}^{-\alpha}$, $0 \leq \alpha < 2$, with $\alpha = 0$ being the unit system (cf. Figs. 10.9 and 10.10a). The explicit expression for the parameters (a, b, c) in terms of α has been given in (10.10). •

Remark: Restricted DL and LD configurations. Boundary cases of the DLD configuration are the DL and LD configurations. From (10.50) with $z_2 = 0$, we find that the DL manifold is the horizontal half-plane ($a = 1\,,\, b \in \mathsf{R}\,,\, c \leq 0$). The case LD (with $z_1 = 0$), restricts $a = 1 + bc$ for $c \leq 0$; this is a ruled surface generated by lines parallel to the a-b plane and slope c. These two 2-dim manifolds are completely within the union of the concave- and convex-DLD regions shown in Fig. 10.10a. •

10.5.3 Two-lens configurations

With two lenses we build the generic $DLDLD$ configuration, which has five parameters z_1, G_1, z_2, G_2, z_3, to cover some regions of abc-space which cannot be realized with a single lens. To analyze this configuration, we start with an important restricted case: LDL optical systems, where two lenses are in contact with the input and the output screens. Then we shall add D's on both sides.

Restricted LDL configurations have parameters G_1, z, G_2, where the two lenses have Gaussian powers $G_1, G_2 \in \mathsf{R}$ and $z \geq 0$. They realize

$$\begin{pmatrix} 1 & G_1 \\ 0 & 1 \end{pmatrix} \begin{pmatrix} 1 & 0 \\ -z & 1 \end{pmatrix} \begin{pmatrix} 1 & G_2 \\ 0 & 1 \end{pmatrix} = \begin{pmatrix} a = 1 - G_1 z & b = G_1 a + G_2 \\ c = -z \leq 0 & d = 1 - zG_2 \end{pmatrix}, \tag{10.54}$$

in the half-space region shown in Fig. 10.10b. The $c = 0$ face is open because when $z = 0$, the system degenerates to L. The singular plane $a = 0$ is uneventful in the configuration parameters, and the LDL region is characterized by

$$\left.\begin{array}{l} a \in \mathsf{R},\, b \in \mathsf{R},\, c < 0, \\ a = 1,\, b \in \mathsf{R},\, c = 0. \end{array}\right\} \qquad (LDL) \tag{10.55}$$

With LDL configurations we thus fill the two new octants: ($a > 1\,,\, b \leq 0\,,\, c < 0$) and ($a < 1\,,\, b \geq 0\,,\, c < 0$). Still, systems in the one-lens DLD octant ($a < 0\,,\, b > 0\,,\, c > 0$) cannot be realized with restricted LDL two-lens systems. Systems in the intersection of the DLD and LDL regions can be realized in either configuration.

To reach the unfilled *abc*-region of paraxial optical systems we take an *LDL* configuration and separate by $z > 0$ one lens from its screen, forming a semi-restricted *DLDL* configuration. When (10.55) holds for the original system, then

$$\begin{pmatrix} 1 & 0 \\ -z < 0 & 1 \end{pmatrix} \begin{pmatrix} a \in \mathsf{R} & b \in \mathsf{R} \\ c < 0 & d \in \mathsf{R} \end{pmatrix} = \begin{pmatrix} a' = a \in \mathsf{R} & b' = b \in \mathsf{R} \\ c' = c - za & d' = d - zb \end{pmatrix}, \quad (10.56)$$

and $c' \geq 0$ can occur, but only when $a' = a < 0$; $c' < 0$ implies no restriction on a'. Erasing primes, we conclude that the *abc*-region that is realizable with *DLDL* systems, extends the previous *LDL* region (10.55) to

$$\left.\begin{array}{l} a \in \mathsf{R},\ b \in \mathsf{R},\ c < 0, \\ a = 1,\ b \in \mathsf{R},\ c = 0, \\ a < 0,\ b \in \mathsf{R},\ c \geq 0, \end{array}\right\} \qquad (DLDL) \qquad (10.57)$$

shown in Fig. 10.10c. As we shall remark below, the new quadrant of space $(a < 0\,,\ c \geq 0)$ includes now fractional Fourier transforms $\mathcal{F}^{-\alpha}$ for $2 \leq \alpha < 3$, and inverting magnifiers [$a < 0$, cf. (10.6)].

To reach the quadrant in Fig. 10.10c which is not realizable by restricted *DLDL* systems, we add a *D* to obtain the generic two-lens *DLDLD* configuration,

$$\begin{pmatrix} a & b \\ c & d \end{pmatrix} \begin{pmatrix} 1 & 0 \\ -z & 1 \end{pmatrix} = \begin{pmatrix} a' = a - zb & b' = b \\ c' = c - zd & d' = d \end{pmatrix}, \qquad (10.58)$$

where a, b, c are restricted by (10.57) and $z \geq 0$; note that $c' = c - zd = (ca' - z)/a$. Points in the 'missing' quadrant $(a' > 0\,,\ b' \in \mathsf{R}\,,\ c' \geq 0)$ can be obtained from (a, b, c)'s in the following ranges:

$$\begin{array}{r} a' > 0 \text{ when } a > 0,\ \ b \in \mathsf{R}, \text{ or } a < 0,\ \ b < 0; \\ a' > 0,\ \ c' \geq 0 \text{ when } a < 0,\ \ c \in \mathsf{R}, \text{ or } a > 0,\ \ c > 0. \end{array} \qquad (10.59)$$

Grouping the regions by the sign of a, we see that

$$\begin{array}{l} (\,a > 0,\ b' = b \in \mathsf{R},\ c > 0\,) \text{ but } (a, b, c) \text{ is } \mathbf{not} \text{ a } DLDL, \\ (\,a < 0,\ b' = b < 0,\ c \in \mathsf{R}\,) \Rightarrow (a' > 0\,,\ b' < 0\,,\ c' \geq 0) \\ \qquad\qquad\qquad\qquad\qquad \mathbf{is} \text{ a } DLDLD \text{ system.} \end{array} \qquad (10.60)$$

The region of optical systems realizable with generic two-lens configurations is shown in Fig. 10.10d; it adds to the previous *DLDL* region the new octant $(a \geq 0\,,\ b < 0\,,\ c \geq 0)$. Again, the plane $a = 0$ is uneventfully inside this region. Still, the remaining octant $(a > 0\,,\ b \geq 0\,,\ c \geq 0)$ **cannot** be realized with two lenses. The 2-lens region in Fig. 10.10d is open.

Remark: Completing the Fourier cycle – almost. We saw above that the first half of the (inverse) Fourier cycle $\mathcal{F}^{-\alpha}$, $0 \leq \alpha < 2$, can be realized with a single concave lens. [According to (10.55) and Fig. 10.10b, this range can be also realized with *LDL*'s.] Further along the cycle shown in Fig. 10.9, we see in Fig. 10.10c that inverse Fourier transformers of powers $2 \leq \alpha \leq 3$ are realized with a $DLDL$ configuration, and $3 \leq \alpha < 4$ with generic $DLDLD$'s (which are just two concatenated one-lens $\mathcal{F}^{-\alpha}$'s). But with two-lens systems we **cannot** reach $\alpha = 4$, the unit system;[10] neither can we produce magnifiers ($a \geq 1\,,\; b = 0\,,\; c = 0$), nor negative free displacements ($a = 1\,,\; b = 0\,,\; c \geq 0$). These will need three lenses. •

Remark: Extraction of lens parameters in LDL configuration. We analyze the correspondence between points in abc-space (optical systems) and the displacement and lens parameters by writing them explicitly. In the one-lens DLD case, their relation is 1:1 and given by (10.50) and (10.51). For the LDL configuration, the relations inverse to (10.54) are also 1:1,

$$G_1 = (a-1)/c, \quad z = -c, \quad G_2 = (d-1)/c = (1-a+bc)/ac, \tag{10.61}$$

and valid for $c < 0$. •

Remark: Freedom in restricted two-lens systems. In $DLDL$ configurations with parameters z_1, G_1, z_2, G_2 (from left to right), the inversion of (10.56) yields the relations

$$z_1 = (a - cG_1 - 1)/aG_1, \quad z_2 = (1-a)/G_1 = -az_1 - c, \quad aG_2 = b - G_1. \tag{10.62}$$

Since there are now four optical elements, we have a one-parameter freedom in choosing them; the two displacements must be non-negative and satisfy $az_1 + z_2 + c = 0$. As shown in Fig. 10.11, the allowed pairs (z_1, z_2) fall on lines with intercepts $-c/a$ and $-c$, in the regions ($a \neq 0\,,\; c < 0$) and ($a < 0\,,\; c \geq 0$) respectively. The parameter range in the $DLDL$ region is finite for $a > 0$, $c < 0$, and half-infinite in the region $a < 0$. •

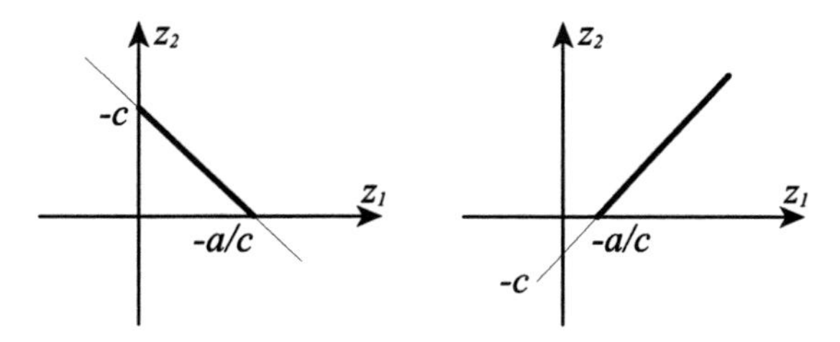

Fig. 10.11. The plane (z_1, z_2) of displacements for equivalent systems in $DLDL$ configuration; only the first quadrant is realizable. The lines are $az_1 + z_2 + c = 0$ for ($a \neq 0\,,\; c < 0$) and ($a < 0\,,\; c \geq 0$) respectively.

Remark: Freedom in generic two-lens systems. For the five-element generic $DLDLD$ configuration (10.58), invert to

$$z_1 = \frac{b - dG_1 - G_2}{bG_1}, \quad z_2 = \frac{G_1 + G_2 - b}{G_1 G_2}, \quad z_3 = \frac{b - G_1 - aG_2}{bG_2}. \tag{10.63}$$

[10] Actually, the second metaplectic unit system.

There are now two free parameters, so we may judiciously select lenses G_1 and G_2 from our collection (perhaps incomplete and surely finite) to build the required system. The parameter range of z_1, z_2, z_3 in (10.63) is the intersection of the plane $bG_1z_1 + G_1G_2z_2 + bG_2z_3 = 2b - dG_1 - aG_2$ with the octant $z_1, z_2, z_2 > 0$. It can be finite or 'half'-infinite. •

10.5.4 Three-lens configurations

Three-lens configurations may be expected to fill the remaining octant $(a \geq 0\,, b \geq 0\,, c \geq 0)$ of systems which cannot be made with two-lens configurations, namely the octant hole in Fig. 10.10**d**. We now place a third lens L at the front end of the semi-restricted two-lens setup $DLDL$, to build the restricted three-lens configuration $LDLDL$,

$$\begin{pmatrix} 1 & G \\ 0 & 1 \end{pmatrix} \begin{pmatrix} a & b \\ c & d \end{pmatrix} = \begin{pmatrix} a' = a + Gc & b' = b + Gd \\ c' = c & d' = d \end{pmatrix}. \tag{10.64}$$

Again (a, b, c) fulfil (10.57) and we look for (a', b', c') in the missing octant. It is evident that with $G \in \mathsf{R}$, a' can reach any value in the $c' = c > 0$ half-space. But note carefully that the value $c' = 0$ is *excluded* because then $a = a'$ is fixed. Therefore, with the restricted $LDLDL$ configuration we can realize all paraxial optical systems, *excepting* those in the quarter-*plane* $(a > 1\,, b \geq 0\,, c = 0)$ and the line segment $(0 < a < 1\,, b = 0\,, c = 0)$ [cf. relations (10.53) and (10.55)], to which we shall return soon.

Remark: Extraction of the three lens parameters. An $LDLDL$ configuration with parameters G_1, z_1, G_2, z_2, G_3 has again two redundant variables that we choose to be z_1 and z_2. In these terms, the inversion of the matrix product is

$$G_1 = \frac{c + az_1 + z_2}{cz_1}, \quad G_2 = \frac{c + z_1 + z_2}{z_1 z_2}, \quad G_3 = \frac{c + z_1 + dz_2}{cz_2}. \tag{10.65}$$

[Cf. (10.61), (10.62) and (10.63).] •

Remark: Negative displacements. An optical system was proposed in [136] that would emulate negative free flight. This is $\mathcal{L}_{4/z}\mathcal{D}_z\mathcal{L}_{2/z}\mathcal{D}_{2z}\mathcal{L}_{2/z} = \mathcal{D}_{-z}$.[11] This $3z$-long apparatus will act as a free space of negative length $-z$. Using (10.65) in $\begin{pmatrix} a & b \\ c & d \end{pmatrix} = \begin{pmatrix} 1 & 0 \\ z>0 & 1 \end{pmatrix}$ for $z_1 = z_2 = z$ we find $G_1 = G_2 = G_3 = 3/z$; that is [163],

$$\mathcal{L}_{3/z}\mathcal{D}_z\mathcal{L}_{3/z}\mathcal{D}_z\mathcal{L}_{3/z} = \mathcal{D}_{-z}. \tag{10.66}$$

This apparatus is more efficient because its total length is only $2z$. Moreover, being symmetric under reflections, some of the aberrations of this arrangement will be absent (see Part IV). •

---

[11] No hint was given there on how this system was found.

The systems which **cannot** be realized by one-, two-, nor restricted three-lens configurations $LDLDL$, lie in the $c = 0$ plane of Fig. 10.9. These are positive (upright) imagers ($a > 1$, $b \geq 0$, $c = 0$), which are pure for $b = 0$ and ($0 > a > 1$, $b = 0 = c$). That the $c = 0$ plane is singular (unattainable by $LDLDL$ configurations) is also evident in (10.65), because c appears in two of the three denominators while their numerators, $az_1 + z_2$ and $z_1 + dz_2$, are not zero. Again we 'place' one more displacement in front of the first lens, as we did in (10.56), and we see that from $a, c \neq 0$ we can pass to $c' = 0$ with an appropriate z. In this way we obtain all paraxial optical systems with semi-restricted $DLDLDL$ configurations.

Remark: Extraction of $DLDLDL$ parameters. The generic $DLDLDL$ configuration parameters $z_1, G_1, z_2, G_2, z_3, G_3$ can be solved for three of them in terms of the rest. We choose the lens powers to be independent because we may have only a small number of them available in the laboratory box. The three displacements are then

$$z_1 = \frac{1}{G_1} + \frac{G_2 + (d - cG_3)G_1}{(aG_3 - b)G_1}, \quad z_2 = \frac{1}{G_1} + \frac{1}{G_2} + \frac{aG_3 - b}{G_1 G_2}, \quad z_3 = \frac{1}{G_2} + \frac{G_1 + aG_2}{(aG_3 - b)G_2}, \tag{10.67}$$

and we only require that $aG_3 - b \neq 0$. •

Remark: Positive magnifiers with three equal lenses. We can construct the line of positive magnifiers ($a > 0$, $b = 0 = c$) on the positive vertical half-axis of Fig. 10.9 with three equal convex lenses. We set $G_1 = G_2 = G_3 = G > 0$ in (10.67) to find the distances at which we should place them:

$$z_1 = \frac{a^{-1} + 1 + a}{aG}, \quad z_2 = \frac{2 + a}{G}, \quad z_3 = \frac{1 + 2a}{aG}. \tag{10.68}$$

When $a = 1$, we have the unit system in a $(DL)^3$ configuration with the following arrangement:

$$\mathcal{D}_{3/G}\, \mathcal{L}_G\, \mathcal{D}_{3/G}\, \mathcal{L}_G\, \mathcal{D}_{3/G}\, \mathcal{L}_G = \mathit{1}, \tag{10.69}$$

of total length $F = 1/G$. This closes the fractional Fourier transform cycle (recall the footnote on page 201). •

With the preceding discussion we have explored the configurations with the minimum number of lenses that will produce all linear maps of phase space. However, if we only care about the images projected on a screen (independently of the direction of the ray's arrival), two systems differing in their b- and d-parameters become equivalent. A lens in contact with the output screen can be added or removed without affecting the performance of the system. This corresponds to projecting the abc-space on the a–c plane. With this widened criterion, all paraxial optical systems can be produced with configurations of no more than *two* lenses. And, except for the quadrant ($a > 0$, $c \geq 0$), with a single lens.

Remark: On the convenience of '*abc*...' parameters. We undertook the study of $DLD \cdots L$ configurations for completeness and use. The search method and the structure we found have a mathematical charm of their own, and are complementary to the strategy of Hamilton–Lie methods in the rest of this volume. However, the $DLD \cdots L$ factorization of optical systems becomes unwieldly in astigmatic systems; $\mathsf{Sp}(4, \mathsf{R})$ contains ten parameters and the corresponding 10-dim regions would be more difficult to visualize. Folklore says that five lenses, some of them astigmatic, are needed. •

11 Classical Lie algebras

In this chapter we present all classical Cartan families of Lie algebras from the perspective of geometric optics. They will be rooted in our previous developments, with emphasis on the real symplectic algebras of functions of phase space under Poisson brackets, and with the aim of placing the aberration multiplets of Part IV in a wider context. We start with the Lie algebras of the linear groups on the associative fields of real, complex and quaternion numbers, and the Lie algebras which preserve spheres in each field: the compact orthogonal, unitary and unitary-symplectic algebras. The Weyl trick allows us to know their noncompact broods; in particular, the real symplectic algebras of D-dim paraxial optics. The weight diagrams and the structure of the symmetric multiplets in all these algebras are developed concurrently.

11.1 Lie algebras of the linear groups

Let $\mathbf{A}$ be an $N \times N$ matrix; then, the exponential series

$$\mathbf{M}(\alpha) = \exp(\alpha\mathbf{A}) := \mathbf{1} + \alpha\,\mathbf{A} + \tfrac{1}{2!}\alpha^2\,\mathbf{A}^2 + \tfrac{1}{3!}\alpha^3\,\mathbf{A}^3 + \cdots \qquad (11.1)$$

exists and converges for all finite α.

Remark: Convergence of matrix series. The proof of convergence follows from the well-known convergence of the ordinary exponential series of a real number. Let the absolute maximum of the matrix elements of $\mathbf{A}$ be $A_{\max}$. Then, an upper bound for the elements of $\mathbf{A}^2$ is $NA^2_{\max}$, which is attained only when a row with maximum elements meets a similar column. By induction, an upper bound for the elements of $\mathbf{A}^n$ in the nth summand of the matrix series (11.1), is $N^{n-1}\,A^n_{\max}$, for $n = 2, 3, \ldots$. Excepting the $n = 0, 1$ terms, this is the series of $N^{-1}\exp(N\alpha\,A_{\max})$ that we know to be convergent, and an upper bound for the elements of the matrix $\exp(\alpha\,\mathbf{A})$ [98]. •

Remark: Non-exponential matrices. While every real matrix can be exponentiated, it is not true that every real matrix is the exponential of another. For example, the matrices

$$\begin{pmatrix} -e^{\zeta} & G \\ 0 & -e^{-\zeta} \end{pmatrix}, \qquad \begin{pmatrix} -\cosh\zeta & \sinh\zeta \\ \sinh\zeta & -\cosh\zeta \end{pmatrix}, \qquad (11.2)$$

do not have logarithms (but/because they are 'far' from $\mathbf{1}$). •

Example: Nilpotent matrices. Among the optical elements on the workbench, free displacements and lenses are represented by exponentials of nilpotent matrices, whose series terminates:

$$\begin{pmatrix} \mathbf{1} & \mathbf{0} \\ -z\mathbf{1} & \mathbf{1} \end{pmatrix} = \exp\left[-z\begin{pmatrix} \mathbf{0} & \mathbf{0} \\ \mathbf{1} & \mathbf{0} \end{pmatrix}\right], \qquad \begin{pmatrix} \mathbf{1} & \mathbf{G} \\ \mathbf{0} & \mathbf{1} \end{pmatrix} = \exp\begin{pmatrix} \mathbf{0} & \mathbf{G} \\ \mathbf{0} & \mathbf{0} \end{pmatrix}. \tag{11.3}$$

•

Example: Exponential and trigonometric matrices. Magnifiers, hyperbolic expanders and Fourier transformers involve exponential, trigonometric and hyperbolic series of matrices which can be summed as their numerical counterparts,

$$\begin{pmatrix} \exp\mathbf{A} & 0 \\ 0 & \exp(-\mathbf{A}^\top) \end{pmatrix} = \exp\begin{pmatrix} \mathbf{A} & 0 \\ 0 & -\mathbf{A}^\top \end{pmatrix}, \tag{11.4}$$

$$\begin{pmatrix} \cosh\mathbf{Z} & \sinh\mathbf{Z} \\ \sinh\mathbf{Z} & \cosh\mathbf{Z} \end{pmatrix} = \exp\begin{pmatrix} \mathbf{0} & \mathbf{Z} \\ \mathbf{Z} & \mathbf{0} \end{pmatrix}, \tag{11.5}$$

$$\begin{pmatrix} \cos\boldsymbol{\Theta} & -\sin\boldsymbol{\Theta} \\ \sin\boldsymbol{\Theta} & \cos\boldsymbol{\Theta} \end{pmatrix} = \exp\begin{pmatrix} \mathbf{0} & -\boldsymbol{\Theta} \\ \boldsymbol{\Theta} & \mathbf{0} \end{pmatrix}. \tag{11.6}$$

•

Consider now $N \times N$ nonsingular matrices which are 'close enough' to the unit matrix to be approximated by

$$\begin{aligned} \mathbf{M}(z) &\approx \mathbf{1} + \varepsilon\mathbf{A}(z) \qquad (\varepsilon^2 \approx 0) \\ &= \mathbf{1} + \varepsilon \sum_{\kappa,\lambda=1}^{N} z_{\kappa,\lambda}\, \mathbf{A}_{(\kappa,\lambda)}, \end{aligned} \qquad \mathbf{A}_{(\kappa,\lambda)} := \kappa \rightarrow \begin{pmatrix} 0 & \cdots & \cdots & 0 \\ \vdots & 0 & 0 & \vdots \\ \vdots & 1 & 0 & \vdots \\ 0 & \cdots & \cdots & 0 \end{pmatrix} \ (\text{column } \lambda\downarrow). \tag{11.7}$$

The N^2 matrices $\{\mathbf{A}_{(\kappa,\lambda)}\}_{\kappa,\lambda=1}^{N}$, with elements $(\mathbf{A}_{(\kappa,\lambda)})_{\mu,\nu} = \delta_{\mu,\kappa}\,\delta_{\nu,\lambda}$, form the *Kronecker* basis for the vector space R^{N^2} all $N \times N$ matrices, and the $z_{\kappa,\lambda}$'s are numerical coefficients. To order of ε, when $\mathbf{M} \approx \mathbf{1} + \varepsilon\mathbf{A}$, then $\mathbf{M}^{-1} \approx \mathbf{1} - \varepsilon\mathbf{A}$.

In Sect. 3.7, (3.64), we showed that, upon keeping terms to order of ε^2 in the exponential series (11.1), the infinitesimal group commutator determines the Lie bracket of the generators. The matrices $\mathbf{M}$ belong to the general linear group $\mathsf{GL}(N)$, and so their generator matrices (11.7) will be a vector basis for the corresponding Lie algebra $\mathsf{gl}(N)$. We can compute directly the Lie brackets that characterize this algebra through its matrix realization:

$$[\mathbf{A}_{(\mu,\nu)}, \mathbf{A}_{(\kappa,\lambda)}] := \mathbf{A}_{(\mu,\nu)}\mathbf{A}_{(\kappa,\lambda)} - \mathbf{A}_{(\kappa,\lambda)}\mathbf{A}_{(\mu,\nu)} = \delta_{\nu,\kappa}\,\mathbf{A}_{(\mu,\lambda)} - \delta_{\lambda,\mu}\,\mathbf{A}_{(\kappa,\nu)}. \tag{11.8}$$

This is the *canonical presentation* of the general linear Lie algebra $\mathsf{gl}(N)$.

Remark: Mnemonics for the $\mathsf{gl}(N)$ Lie brackets. The commutator (11.8) has the quartet of ordered indices $[\mu, \nu;\ \kappa, \lambda]$; in the right-hand side they are rearranged into two terms with $+\left[{}^{\mu \Longrightarrow \lambda}_{\nu \rightarrow \kappa}\right]$ and $-\left[{}^{\mu \longleftarrow \lambda}_{\nu \Leftarrow \kappa}\right]$, where the Kronecker δ's receive the indices in the order indicated by the single arrows, and the $\mathbf{A}$'s those of the double arrows. •

Vector spaces can be built with coefficients in any number field that distributes sum and multiplication, and is associative. Sum is commutative, but multiplication need not be. The distinct fields of our interest for the $z_{\kappa,\lambda}$'s in (11.7) are the reals R, the complex numbers C, and the quaternions Q.[1] We remark on the latter two.

Remark: Complex numbers. With $i^2 = -1$, complex numbers z can be represented by real 2×2 matrices $\mathbf{z}$ as follows:

$$z \in \mathsf{C}, \qquad \left\{ \begin{matrix} z = z^0 + i\, z^1 \\ z^0, z^1 \in \mathsf{R} \end{matrix} \right\} \longleftrightarrow \mathbf{z} = \begin{pmatrix} z^0 & z^1 \\ -z^1 & z^0 \end{pmatrix}. \tag{11.9}$$

Sums and products follow, complex conjugation corresponds to matrix transposition, and $\mathbf{z}\,\mathbf{z}^\top = |z|^2 \mathbf{1}$. •

Remark: Quaternions. It was also an idea of Hamilton to go beyond the complex field with one real and three noncommuting 'imaginary' units: e_α for $\alpha = 0$ and $1, 2, 3$. Using Latin letters for indices $j = 1, 2, 3$, their multiplication table is

$$e_0\, e_j = e_j = e_j\, e_0, \qquad e_k\, e_k = -e_0, \qquad e_j\, e_k = \epsilon_{j,k,\ell} e_\ell, \tag{11.10}$$

where $\epsilon_{j,k,\ell}$ is the Levi-Civita skew-symmetric symbol,[2] we do not distinguish between upper and lower indices, but we shall use the sum convention for repeated ones. Quaternion numbers can be represented by complex 2×2 matrices as follows:

$$z \in \mathsf{Q}, \qquad \left\{ \begin{matrix} z = z^\alpha e_\alpha \\ z^\alpha \in \mathsf{R}, \end{matrix} \right\} \longleftrightarrow \mathbf{z} = \begin{pmatrix} z^0 + i\, z^2 & z^1 + i\, z^3 \\ -z^1 + i\, z^3 & z^0 - i\, z^2 \end{pmatrix}. \tag{11.11}$$

Moreover, we can produce a representation of quaternions by 4×4 real matrices, using (11.9) to replace 1 by the 2×2 unit $\mathbf{1}$ and i by $\begin{pmatrix} 0 & 1 \\ -1 & 0 \end{pmatrix}$. In Sect. 11.3, we shall replace i's by 1's to build real symplectic matrices. •

---

[1] Octonions, whose multiplication is not associative, or other Cayley numbers which have nonzero divisors of zero, *cannot* be used, because a proper notion of linear independence of vectors must exist. Other subfields of the reals, such as that of rational numbers or their quadratic extensions (adding a term multiple of $\sqrt{2}$ for example), are not warranted by physical models.

[2] When j, k, ℓ are a cyclic permutation of $1, 2, 3$ then $\epsilon_{j,k,\ell} = 1 = -\epsilon_{k,j,\ell}$, otherwise $\epsilon_{j,j,k} = 0$.

Remark: Pauli matrices. To work comfortably with quaternions, we recall the well-known Pauli $\boldsymbol{\sigma}$-matrices,

$$\boldsymbol{\sigma}_1 := \begin{pmatrix} 0 & 1 \\ 1 & 0 \end{pmatrix}, \quad \boldsymbol{\sigma}_2 := \begin{pmatrix} 0 & -i \\ i & 0 \end{pmatrix}, \quad \boldsymbol{\sigma}_3 := \begin{pmatrix} 1 & 0 \\ 0 & -1 \end{pmatrix}, \quad \boldsymbol{\sigma}_0 := \begin{pmatrix} 1 & 0 \\ 0 & 1 \end{pmatrix}, \tag{11.12}$$

whose salient properties are $\boldsymbol{\sigma}_k^2 = \mathbf{1}$ and $\boldsymbol{\sigma}_j \boldsymbol{\sigma}_k = i\epsilon_{j,k,\ell}\boldsymbol{\sigma}_\ell$, while $\boldsymbol{\sigma}_0$ is a convenient symbol for the unit matrix. The multiplication table of the quaternion units e_α in (11.10) is satisfied by the following isomorphic ($\leftrightarrow$) set of matrices:

$$\begin{array}{llll} e_0 \leftrightarrow \mathbf{e}_0 := \boldsymbol{\sigma}_0 & e_1 \leftrightarrow \mathbf{e}_1 := i\,\boldsymbol{\sigma}_2 & e_2 \leftrightarrow \mathbf{e}_2 := i\,\boldsymbol{\sigma}_3 & e_3 \leftrightarrow \mathbf{e}_3 := i\,\boldsymbol{\sigma}_1 \\ = \begin{pmatrix} 1 & 0 \\ 0 & 1 \end{pmatrix}, & = \begin{pmatrix} 0 & 1 \\ -1 & 0 \end{pmatrix}, & = \begin{pmatrix} i & 0 \\ 0 & -i \end{pmatrix}, & = \begin{pmatrix} 0 & i \\ i & 0 \end{pmatrix}. \end{array} \tag{11.13}$$

•

Remark: Conjugation and norm. We shall write our numbers $z = z^\alpha e_\alpha$, understanding that $\alpha = 0, 1, 2, 3$ for quaternion numbers (and Latin letters for 1,2,3), $\alpha = 0, 1$ for complex numbers, and trivially $\alpha \equiv 0$ for reals. There is an operation of *conjugation* in the complex and quaternion fields, which maps z to $z^\dagger = z^0 e_0 - z^j e_j$. In the complex number case this operation transposes the matrix representation $\mathbf{z}$ in (11.9). In quaternion case, the matrix (11.11) is transposed and complex conjugated. Thus is defined a positive definite *norm* $|z|$ for z in these fields,

$$|z|^2 \mathbf{e}_0 := \mathbf{z}\,\mathbf{z}^\dagger = \mathbf{z}^\dagger\,\mathbf{z} = \mathbf{e}_0 \det \mathbf{z}, \qquad \begin{array}{l} \det \mathbf{z} = z^\alpha z^\alpha \geq 0, \\ |z| = 0 \Leftrightarrow \mathbf{z} = \mathbf{0}. \end{array} \tag{11.14}$$

This norm for the field in turn determines a positive definite norm for the N-dim vector spaces R^N, C^N, and Q^N. •

We return to the generic element $\mathbf{A}(z)$ in the Lie algebra $\mathsf{gl}(N)$ of matrices (11.7), with $z_{\kappa,\lambda}$ in the field F of real, complex or quaternion numbers, and of real components $z^\alpha_{\kappa,\lambda} \in \mathsf{R}$. This can be then written (with summation over repeated indices) as

$$\mathbf{A}(z) = z_{\kappa,\lambda}\,\mathbf{A}_{(\kappa,\lambda)} = z^\alpha_{\kappa,\lambda}\,\mathbf{A}^\alpha_{(\kappa,\lambda)}, \qquad \mathbf{A}^\alpha_{(\kappa,\lambda)} := e_\alpha\,\mathbf{A}_{(\kappa,\lambda)}. \tag{11.15}$$

The N^2 generators $\{\mathbf{A}_{(\kappa,\lambda)}\}^N_{\kappa,\lambda=1}$ are a vector basis for $\mathsf{gl}(N)$ with coefficients in F, while the $\mathbf{A}^\alpha_{(\kappa,\lambda)}$ are a basis for a Lie algebra with *real* coefficients, denoted $\mathsf{gl}(N,\mathsf{F})$. Counting $\dim \mathsf{F} = 1$, 2 and 4 for R, C and Q respectively, the real algebra of $\mathsf{gl}(N,\mathsf{F})$ has $N^2 \dim \mathsf{F}$ generators.

The commutation relations of the Kronecker basis of generators of $\mathsf{gl}(N,\mathsf{F})$ can be found from the Lie brackets of $\mathsf{gl}(N)$ in (11.8) and the quaternion multiplication table (11.10). They are

$$[\mathbf{A}^0_{(\mu,\nu)}, \mathbf{A}^\alpha_{(\kappa,\lambda)}] = \delta_{\nu,\kappa}\,\mathbf{A}^\alpha_{(\mu,\lambda)} - \delta_{\lambda,\mu}\,\mathbf{A}^\alpha_{(\kappa,\nu)}, \tag{11.16}$$

$$[\mathbf{A}^j_{(\mu,\nu)}, \mathbf{A}^j_{(\kappa,\lambda)}] = -\delta_{\nu,\kappa}\,\mathbf{A}^0_{(\mu,\lambda)} + \delta_{\lambda,\mu}\,\mathbf{A}^0_{(\kappa,\nu)}, \quad \text{(no sum)} \tag{11.17}$$

$$[\mathbf{A}^j_{(\mu,\nu)}, \mathbf{A}^k_{(\kappa,\lambda)}] = \epsilon_{j,k,\ell}(\delta_{\nu,\kappa}\,\mathbf{A}^\ell_{(\mu,\lambda)} + \delta_{\lambda,\mu}\,\mathbf{A}^\ell_{(\kappa,\nu)}). \tag{11.18}$$

All real Lie algebras $\mathsf{gl}(N, \mathsf{F})$ are accomodated by these Lie brackets, according to their field:

$$\begin{aligned} \text{real: } \mathsf{gl}(N,\mathsf{R}),&\quad \text{by (11.16) with } \alpha \equiv 0,\\ \text{complex: } \mathsf{gl}(N,\mathsf{C}),&\quad \text{by (11.16) with } \alpha = 0,1,\ \text{(11.17) with } j \equiv 1,\\ \text{quaternion: } \mathsf{gl}(N,\mathsf{Q}),&\quad \text{by (11.16)–(11.18) with } \alpha = 0,1,2,3. \end{aligned}$$

Remark: The $N = 1$ algebras. The algebra $\mathsf{gl}(1,\mathsf{R})$ has a single generator $\mathbf{A} \leftrightarrow 1$ stemming from the 1-parameter positive multiplicative group $\mathsf{GL}(1,\mathsf{R})$ of numbers e^{ζ}. The real Lie algebra $\mathsf{gl}(1,\mathsf{C})$ has two commuting generators, $\mathbf{A}^0 \leftrightarrow 1$ and $\mathbf{A}^1 \leftrightarrow i$, which generate the group of multiplication by complex numbers $e^{\zeta}\, e^{i\varphi}$; indicating the algebras by the group parameters, it decomposes as $\mathsf{gl}(1,\mathsf{C}) = \mathsf{gl}(1,\mathsf{R})_{\zeta} \oplus \mathsf{u}(1)_{\varphi}$. Finally, the real Lie algebra $\mathsf{gl}(1,\mathsf{Q})$ has four generators $\mathbf{A}^{\alpha} \leftrightarrow \mathbf{e}_{\alpha}$, whose Lie brackets are (11.16)–(11.18) with no subindices, so the first two commutators are null. The three traceless Pauli matrices $\boldsymbol{\sigma}_j$ in (11.12) generate the Lie algebra of *spin* $\mathsf{su}(2)$, so $\mathsf{gl}(1,\mathsf{Q}) = \mathsf{gl}(1,\mathsf{R}) \oplus \mathsf{su}(2)$. •

Remark: Determinant and invariant trace. Matrices near to the identity, $\mathbf{M}(z) \approx \mathbf{1} + \varepsilon \mathbf{A}(z)$, $\epsilon^2 \approx 0$ [see (11.7)], have determinant[3]

$$\det \mathbf{M}(z) \approx \prod_{\kappa=1}^{N} (1 + \varepsilon z_{\kappa,\kappa}) \approx 1 + \varepsilon \sum_{\kappa=1}^{N} z_{\kappa,\kappa} =: 1 + \varepsilon\, \mathrm{tr}\,(\mathbf{A}(z)). \qquad (11.19)$$

When $\det \mathbf{M}(z) = 1$, then $\mathrm{tr}\,(\mathbf{A}(z)) = 0$; so the matrix representing the generator will be traceless; and conversely, traceless matrices exponentiate to matrices whose determinant is unity. The commutator of two traceless matrices is traceless, as can be checked easily, so they form Lie subalgebras; correspondingly, the product of two matrices of unit determinant has also unit determinant, so they form subgroups. Matrices of unit determinant form a subgroup of $\mathsf{GL}(N)$, called the *special* linear group, and denoted $\mathsf{SL}(N)$. •

Remark: Special linear algebras $\mathsf{sl}(N,\mathsf{F})$. The Lie algebra $\mathsf{gl}(N,\mathsf{F})$ is a real vector space of dimension $N^2 \dim \mathsf{F}$. We distinguish the $\dim \mathsf{F}$ generators which are represented by (block-) diagonal matrices, which we call the *trace generators*:[4]

$$\mathbf{A}^{\alpha}_{\mathrm{tr}} := \sum_{\kappa=1}^{N} \mathbf{A}^{\alpha}_{(\kappa,\kappa)} \leftrightarrow \mathrm{diag}\,(\mathbf{e}_{\alpha}, \ldots, \mathbf{e}_{\alpha}) \qquad \alpha = 0,1,2,3. \qquad (11.20)$$

[3] A definition of the determinant of a matrix with noncommuting elements is needed for the quaternion field. In this case we replace the 2×2 complex matrix representation at each position and evaluate the determinant of the resulting $2N \times 2N$ matrix.

[4] When we use the representation of quaternions by 2×2 complex matrices, the $2N \times 2N$ matrices of the trace generators will be block-diagonal.

The trace generators close into the Lie algebra $\mathsf{gl}(1,\mathsf{F})$. For $\mathsf{F}=\mathsf{R}$ and C, the trace generators commute with the rest of $\mathsf{gl}(N,\mathsf{F})$. We can thus form the following generators represented by diagonal matrices

$$\widetilde{\mathbf{A}}^{\alpha}_{(\kappa)} := \mathbf{A}^{\alpha}_{(\kappa,\kappa)} - N^{-1}\mathbf{A}^{\alpha}_{\mathrm{tr}}, \qquad \begin{array}{l} \text{for } \alpha\equiv 0, \quad \text{when } \mathsf{F}=\mathsf{R}, \\ \text{or } \alpha = 0,1, \text{ when } \mathsf{F}=\mathsf{C}. \end{array} \tag{11.21}$$

The $(N-1)\dim\mathsf{F}$ independent 'diagonal' generators $\{\widetilde{\mathbf{A}}^{\alpha}_{(\kappa)}\}_{\kappa=1}^{N-1}$ and the 'off-diagonal' ones $\{\mathbf{A}^{\alpha}_{(\kappa,\lambda)}\}_{\kappa\neq\lambda=1}^{N}$ will close into the *special linear* Lie algebras $\mathsf{sl}(N,\mathsf{F})$. These will generate the linear groups of $N\times N$ matrices of unit determinant, which conserves the *volume* of the natural N-dim vector space. The algebra $\mathsf{sl}(N,\mathsf{R})$ has N^2-1 generators, and $\mathsf{sl}(N,\mathsf{C})$ has $2(N^2-1)$. In the quaternion case $\mathsf{gl}(N,\mathsf{Q})$, only $\mathbf{A}^{0}_{\mathrm{tr}}$ commutes with all generators, so the algebra $\mathsf{sl}(N,\mathsf{Q})$ has only 1 generator less, namely $4N^2-1$. •

Remark: Differential operator realization of $\mathsf{gl}(N,\mathsf{R})$. There is an extremely useful realization for the linear Lie algebras, and particularly for the classical Cartan subalgebras to be seen in the next section. Note that between the Kronecker basis matrices (11.7) of $\mathsf{gl}(N)$ and first-order homogeneous differential operators, there holds the isomorphism

$$\mathbf{A}_{\kappa,\lambda} \longleftrightarrow \widehat{A}_{\kappa,\lambda} := x_\kappa\partial_\lambda, \quad \text{where} \quad \partial_\lambda := \frac{\partial}{\partial x_\lambda}, \tag{11.22}$$

under linear combination (evidently) and under the commutation relations (11.8). The variables $x_\kappa\in\mathsf{F}$ can be seen as the Cartesian coordinates of an F^N vector space. •

Remark: Differential operator realization of $\mathsf{gl}(N,\mathsf{C})$. The real form of the generators of $\mathsf{gl}(N,\mathsf{C})$ can be obtained from (11.22) for the real coordinates x^α_κ, $\alpha=0,1$, by replacing $x_\kappa = x^0_\kappa + ix^1_\kappa$ (its complex conjugate $x^*_\kappa = x^0_\kappa - ix^1_\kappa$ is absent). Thus $\partial_\lambda = \frac{1}{2}(\partial^0_\lambda - i\partial^1_\lambda)$, and we find

$$\widehat{A}_{\kappa,\lambda} = \tfrac{1}{2}(x^0_\kappa + ix^1_\kappa)(\partial^0_\lambda - i\partial^1_\lambda) = \widehat{A}^0_{\kappa,\lambda} + i\widehat{A}^1_{\kappa,\lambda}, \tag{11.23}$$

$$\widehat{A}^0_{\kappa,\lambda} := \tfrac{1}{2}(x^0_\kappa\partial^0_\lambda + x^1_\kappa\partial^1_\lambda), \quad \widehat{A}^1_{\kappa,\lambda} := \tfrac{1}{2}(x^1_\kappa\partial^0_\lambda - x^0_\kappa\partial^1_\lambda). \tag{11.24}$$

Acting on functions of $x^\alpha_\kappa\in\mathsf{R}^{2N}$, these operators follow the commutation relations of the algebra, (11.16)–(11.17). •

Remark: Differential operator realization of $\mathsf{gl}(N,\mathsf{Q})$. The generators of the real Lie algebra $\mathsf{gl}(N,\mathsf{Q})$ follow the same pattern of (11.22) when acting on functions of quaternion variables $x_\kappa\in\mathsf{Q}^N$. But note that since quaternions do not commute, the differential ratio of $f(x+\Delta x)-f(x)$ and Δx can be made placing the increment either on the left or on the right of the function; so there are actually *two* realizations for the differential operators that will be distinguished below by the label $\pm$. The real generators of $\mathsf{gl}(N,\mathsf{Q})^\pm$ in the differential operator realizations are then

$$\widehat{A}^0_{\kappa,\lambda} := \tfrac{1}{2} x^\alpha_\kappa \partial^\alpha_\lambda, \quad \widehat{A}^{j,\pm}_{\kappa,\lambda} := \pm\tfrac{1}{2}(x^j_\kappa \partial^0_\lambda - x^0_\kappa \partial^j_\lambda) - \tfrac{1}{2}\epsilon_{j,k,\ell}\, x^k_\kappa \partial^\ell_\lambda, \tag{11.25}$$

[cf. (11.24)] and satisfy the commutation relations (11.16)–(11.18). •

11.2 The classical Cartan algebras

The existence of a positive definite norm $|x|$ for numbers $x \in \mathsf{F}$ in (11.14) leads to the definition of a *norm* for *vectors* $\mathbf{x} \in \mathsf{R}^N$, $\mathsf{C}^N = \mathsf{R}^{2N}$ or $\mathsf{Q}^N = \mathsf{C}^{2N} = \mathsf{R}^{4N}$,

$$|\mathbf{x}|^2 := \mathbf{x}^\dagger\, \mathbf{x} := \sum_{\kappa=1}^{N} |x_\kappa|^2 = \sum_{\kappa=1}^{N}\sum_{\alpha}(x^\alpha_\kappa)^2. \tag{11.26}$$

So now we can define *spheres* of dimension N in a field F as the submanifolds $|\mathbf{x}| = \rho$ of *radius* $\rho > 0$. The linear transformations in $\mathsf{GL}(N, \mathsf{F})$ which leave these spheres invariant – and only *rotate* them – form group *families* according to the field F. They are the three families of classical *Cartan* groups,

orthogonal	$\mathsf{SO}(N)$	rotates	R -sphere,
unitary	$\mathsf{U}(N)$		C -sphere,
unitary-symplectic	$\mathsf{USp}(N)$		Q -sphere.

The range of rotations of a sphere is finite, so all these groups are *compact.* The orthogonal and unitary groups were seen in chapter 9; the *unitary*-symplectic groups were called originally 'symplectic' in Weyl's nomenclature [142]. But since the 1970s, the *real* symplectic groups $\mathsf{Sp}(2N, \mathsf{R})$ that we use here became more popular, and are increasingly called *the* symplectic groups. The close relation between $\mathsf{USp}(N)$ and $\mathsf{Sp}(2N, \mathsf{R})$ will be the subject of the next section. Élie Cartan classified all *semisimple*[5] and *compact* Lie algebras. Beside the families named above, there are five *exceptional* Lie algebras, called G_2, F_4, E_6, E_7, and E_8 [138, 142].[6]

We consider the groups of matrices $\mathbf{R}$, $\det \mathbf{R} = +1$, that linearly transform the vectors $\mathbf{x} \in \mathsf{F}^N$ while preserving their norm,

$$\mathbf{x}'^\dagger\, \mathbf{x}' = (\mathbf{R}\,\mathbf{x})^\dagger\, \mathbf{R}\,\mathbf{x} = \mathbf{x}^\dagger\, \mathbf{x} \quad \Leftrightarrow \quad \mathbf{R}^\dagger\, \mathbf{R} = \mathbf{1}. \tag{11.27}$$

We find the generating Lie algebras by examining their behaviour near to the unit matrix, $\mathbf{R} \approx \mathbf{1} + \varepsilon \mathbf{\Lambda}$, $\epsilon^2 \approx 0$, as before. Replacing this in (11.27) to the order of ε, we characterize the Lie algebra of matrices $\mathbf{\Lambda}$ by

[5] Semisimple Lie algebras have no abelian invariant ideals, i.e., no semidirect translation subalgebra summands; they exhibit the 'most densely bonded' Lie brackets.

[6] The exceptional algebras do not seem to arise in optics –they do not have readily recognizable matrix realizations; indeed, they do quite seldom even in elementary particle physics [64].

$$(\mathbf{1} + \varepsilon\mathbf{\Lambda})^\dagger (\mathbf{1} + \varepsilon\mathbf{\Lambda}) = \mathbf{1}, \quad \Leftrightarrow \quad \mathbf{\Lambda}^\dagger = -\mathbf{\Lambda}. \tag{11.28}$$

The matrices $\mathbf{\Lambda}$ are skew-hermitian and, since the commutator of two skew-hermitian matrices is again skew-hermitian, we have Lie algebras represented by matrices which are:

orthogonal	$\mathsf{so}(N)$	real, skew-symmetric matrices,
unitary	$\mathsf{u}(N)$	skew-hermitian complex matrices,
unitary-symplectic	$\mathsf{usp}(N)$	skew-hermitian with quaternion conjugation.

A convenient real vector basis for these Lie algebras are the following linear combinations of the Kronecker basis in (11.7),

$$\mathbf{\Lambda}^\alpha_{(\kappa,\lambda)} := \mathbf{A}^\alpha_{(\kappa,\lambda)} - \mathbf{A}^{\dagger\,\alpha}_{(\lambda,\kappa)} = \begin{matrix} \\ \\ \kappa \to \\ \lambda \to \\ \\ \end{matrix} \begin{pmatrix} 0 & \cdots & \cdots & 0 \\ \vdots & 0 & \mathbf{e}_\alpha & \vdots \\ \vdots & -\mathbf{e}^\dagger_\alpha & 0 & \vdots \\ 0 & \cdots & \cdots & 0 \end{pmatrix}. \tag{11.29}$$

(with the columns κ and λ marked by arrows above the matrix)

We note the identities due to symmetry and the resulting dimension of the Lie algebras:

$$\begin{aligned} \mathbf{\Lambda}^0_{(\kappa,\lambda)} &= -\mathbf{\Lambda}^0_{(\lambda,\kappa)}, & 1 \le \kappa < \lambda \le N, &\quad \text{so } \dim \mathsf{so}(N) = \tfrac{1}{2}N(N-1), \\ \mathbf{\Lambda}^j_{(\kappa,\lambda)} &= +\mathbf{\Lambda}^j_{(\lambda,\kappa)}, & 1 \le \kappa \le \lambda \le N, &\quad \text{so there are } \tfrac{1}{2}N(N+1) \text{ of each}, \\ &\text{for } \mathsf{C}, & j \equiv 1, &\quad \text{hence } \dim \mathsf{u}(N) = N^2, \\ &\text{for } \mathsf{Q}, & j = 1,2,3, &\quad \text{and } \dim \mathsf{usp}(N) = N(2N+1). \end{aligned} \tag{11.30}$$

We organize the generators $\mathbf{\Lambda}^\alpha_{(\kappa,\lambda)}$ in a convenient pattern that generalizes the one used for the conformal algebra $\mathsf{so}(4,2)$ in Sect. 6.7, (6.60). Now we have a four-component scheme, for $\alpha = 0$ and $k = 1,2,3$:

$$\begin{array}{ccccc} 0 & \mathbf{\Lambda}^0_{1,2} & \mathbf{\Lambda}^0_{1,3} & \cdots & \mathbf{\Lambda}^0_{1,N} \\ & 0 & \mathbf{\Lambda}^0_{2,3} & \cdots & \mathbf{\Lambda}^0_{2,N} \\ & & 0 & \ddots & \vdots \\ & & & 0 & \mathbf{\Lambda}^0_{N-1,N} \\ & & & & 0 \end{array} \qquad \begin{array}{ccccc} \mathbf{\Lambda}^k_{1,1} & \mathbf{\Lambda}^k_{1,2} & \mathbf{\Lambda}^k_{1,3} & \cdots & \mathbf{\Lambda}^k_{1,N} \\ & \mathbf{\Lambda}^k_{2,2} & \mathbf{\Lambda}^k_{2,3} & \cdots & \mathbf{\Lambda}^k_{2,N} \\ & & \ddots & \ddots & \vdots \\ & & & \mathbf{\Lambda}^k_{N-1,N-1} & \mathbf{\Lambda}^k_{N-1,N} \\ & & & & \mathbf{\Lambda}^k_{N,N} \end{array}$$

generators of $\mathsf{so}(N)$, $\mathsf{u}(N)$, $\mathsf{usp}(N)$ of $\mathsf{u}(N)$, $k \equiv 1$; of $\mathsf{usp}(N)$, $k = 1,2,3$

(11.31)

Remark: On traces. The trace generators (11.20) of the subgroups of $\mathsf{GL}(N, \mathsf{F})$ which leave invariant the sphere in the field F, are

$$\mathbf{\Lambda}^{\alpha}_{\mathrm{tr}} := \sum_{\kappa=1}^{N} \mathbf{\Lambda}^{\alpha}_{(\kappa,\kappa)} = \begin{cases} \mathbf{0}, & \alpha = 0 \\ \operatorname{diag}(\mathbf{e}_j, \ldots, \mathbf{e}_j), & \alpha \equiv j = 1, 2, 3. \end{cases} \tag{11.32}$$

The real trace $\mathbf{\Lambda}^0_{\mathrm{tr}}$ is identically zero because all matrices (11.26) that leave the sphere invariant have determinant of absolute value 1. In $\mathsf{u}(N)$ however, the determinant *phase* generated by $\mathbf{\Lambda}^1_{\mathrm{tr}}$ is arbitrary; this phase can be subtracted directly from the algebra because its trace generator commutes with all others. The traceless generators form the unitary Lie algebras $\mathsf{su}(N)$, which have one generator less. Finally, in $\mathsf{usp}(N)$ there are three nonzero traces $\mathbf{\Lambda}^j_{\mathrm{tr}}$, $j = 1, 2, 3$, which form $\mathsf{su}(2)$ but do not commute with the rest of the algebra. •

The Lie algebras $\mathsf{so}(N)$, $\mathsf{u}(N)$, and $\mathsf{usp}(N)$ are characterized by commutation relations which are obtained from (11.16)–(11.18), (11.29), and (11.30). Notice the systematic changes of sign with α:

$$[\mathbf{\Lambda}^0_{(\mu,\nu)}, \mathbf{\Lambda}^0_{(\kappa,\lambda)}] = \delta_{\nu,\kappa}\mathbf{\Lambda}^0_{(\mu,\lambda)} + \delta_{\mu,\lambda}\mathbf{\Lambda}^0_{(\nu,\kappa)} + \delta_{\kappa,\mu}\mathbf{\Lambda}^0_{(\lambda,\nu)} + \delta_{\lambda,\nu}\mathbf{\Lambda}^0_{(\kappa,\mu)}, \tag{11.33}$$

$$[\mathbf{\Lambda}^0_{(\mu,\nu)}, \mathbf{\Lambda}^j_{(\kappa,\lambda)}] = \delta_{\nu,\kappa}\mathbf{\Lambda}^j_{(\mu,\lambda)} - \delta_{\mu,\lambda}\mathbf{\Lambda}^j_{(\nu,\kappa)} - \delta_{\kappa,\mu}\mathbf{\Lambda}^j_{(\lambda,\nu)} + \delta_{\lambda,\nu}\mathbf{\Lambda}^j_{(\kappa,\mu)}, \tag{11.34}$$

$$[\mathbf{\Lambda}^j_{(\mu,\nu)}, \mathbf{\Lambda}^j_{(\kappa,\lambda)}] = -\text{ (minus) the right-hand side of (11.33)}, \tag{11.35}$$

$$[\mathbf{\Lambda}^j_{(\mu,\nu)}, \mathbf{\Lambda}^k_{(\kappa,\lambda)}] = \text{r.h.s. of (11.33), with } \mathbf{\Lambda}^0_{(\kappa,\lambda)} \mapsto \epsilon_{j,k,\ell}\mathbf{\Lambda}^\ell_{(\kappa,\lambda)}. \tag{11.36}$$

These relations also hold for the special unitary algebra $\mathsf{su}(N)$, because $\mathbf{\Lambda}^1_{\kappa,\kappa}$ and $\mathbf{\Lambda}^1_{\kappa,\kappa} - N^{-1}\mathbf{\Lambda}^1_{\mathrm{tr}}$ commute equally in $\mathsf{u}(N)$. Note in particular that from (11.35), $[\mathbf{\Lambda}^j_{(\mu,\mu)}, \mathbf{\Lambda}^j_{(\kappa,\kappa)}] = 0$ (no sum), because of the skew-symmetry of the $\mathbf{\Lambda}^0$'s in (11.33).

Remark: Mnemonics for the commutators (11.33)–(11.36). To handle these commutators with ease, note that all left-hand sides have the same quartet of ordered indices $[\mu, \nu;\ \kappa, \lambda]$ and all signs in the right-hand side of (11.33) are $+$, but we must keep in mind the identities due to symmetry given in (11.30). It is sufficient to provide a mnemonic rule for the $\mathsf{so}(N)$ commutator (11.33), where the first two terms involve the 'parallel' left-to-right pairs and the last two terms lead to 'crossed' right-to-left pairs,

$$\begin{bmatrix} \mu \longrightarrow \lambda \\ \nu \Rightarrow \kappa \end{bmatrix} + \begin{bmatrix} \mu \Longrightarrow \lambda \\ \nu \rightarrow \kappa \end{bmatrix} + \begin{bmatrix} \mu \;\; \nwarrow\!\!\!\swarrow \;\; \lambda \\ \nu \qquad \kappa \end{bmatrix} + \begin{bmatrix} \mu \;\; \nwarrow\!\!\!\swarrow \;\; \lambda \\ \nu \qquad \kappa \end{bmatrix}. \tag{11.37}$$

As before, the δ's attach their indices in the order indicated by the single arrows, and the $\mathbf{\Lambda}$'s those of the double arrows. In the patterns (11.31), and as we explained following the $\mathsf{so}(4,2)$ pattern in (6.60), a generator $\mathbf{\Lambda}^\alpha_{\kappa,\rho}$ will *commute* with all those $\mathbf{\Lambda}^\beta_{\mu,\nu}$'s which lie *outside* its column κ and row ρ (and their continuation, beyond the diagonal, as row κ and column ρ respectively). The four patterns (for $\alpha = 0, \ldots, 3$) show the subalgebra chains $\mathsf{usp}(N) \supset \mathsf{u}(N) \supset \mathsf{so}(N)$, and the chain $\mathsf{usp}(N) \supset \mathsf{usp}(N-1) \supset \cdots \supset \mathsf{usp}(1)$ that we obtain by restricting the indices successively to $(\mu, \nu) \le N, N-1, \ldots, 1$. •

The maximum number of linearly independent generators that we can find in a Lie algebra is the dimension of the algebra as a *vector space*. There is also the *algebraic* dimension – or *rank* – of the algebra:

Definition: *The* **rank** *of a Lie algebra is the largest number of linearly independent and mutually commuting generators it contains (*i.e.*, the dimension of its maximal abelian subalgebra). These are called the* **weights**, *or weight generators, of the algebra.*

Cartan examined the structure of all classical Lie algebras classifying them first by rank $R \in \mathcal{Z}^+$, and setting up an eigenvalue problem in this R-dimensional vector space, as we shall see below.

Consider first the unitary algebra $\mathsf{u}(N)$, all of whose N 'diagonal' generators $\{\mathbf{\Lambda}^1_{\kappa,\kappa}\}_{\kappa=1}^N$ commute among themselves [see (11.35)]. By restricting their trace to zero, we are left with a maximal set of $R = N-1$ weight generators in the special unitary algebras $\mathsf{su}(N)$; these are called Cartan A_{N-1}-algebras. Next, in the quaternionic algebras $\mathsf{usp}(N)$ it is easy to see that $\{\mathbf{\Lambda}^j_{\kappa,\kappa}\}_{\kappa=1}^N$ with any fixed j, is a commuting set of generators; it is maximal since we cannot further subtract a trace. The unitary-symplectic algebras $\mathsf{usp}(N)$ thus have rank $R = N$, and are called Cartan C_N-algebras. Lastly, for the orthogonal algebras $\mathsf{so}(N)$ we select the commuting generators which are on the first diagonal of the generator pattern [(11.31) for $\alpha = 0$]: $\mathbf{\Lambda}_{1,2}$, $\mathbf{\Lambda}_{3,4}$, ..., $\mathbf{\Lambda}_{2n-1,2n}$, ... When N is even, the last generator in the set is $\mathbf{\Lambda}_{N-1,N}$, so the rank is $R = \frac{1}{2}N$, and the algebra is denoted D_R. When N is odd, the last generator is $\mathbf{\Lambda}_{N-2,N-1}$; the algebra has then rank $R = \frac{1}{2}(N-1)$, and is denoted B_R. We tabulate the classical Cartan families of Lie algebras of rank $R = 1, 2, 3, \ldots$:[7]

CARTAN FAMILY	CLASSICAL ALGEBRA	VECTOR DIMENSION	NONCOMPACT FORM
A_R	$\mathsf{su}(R+1)$	$(R+1)^2-1$	$\mathsf{sl}(R+1,\mathsf{R})$
B_R	$\mathsf{so}(2R+1)$	$R(2R+1)$	—
C_R	$\mathsf{usp}(R)$	$R(2R+1)$	$\mathsf{sp}(2R,\mathsf{R})$
D_R	$\mathsf{so}(2R)$	$R(2R-1)$	—

Remark: Accidental isomorphisms. There occur the following – called *accidental* – identities between some classical algebras of low rank:

$$\begin{aligned} A_1 &= B_1 = C_1, & \mathsf{su}(2) = \mathsf{so}(3) &= \mathsf{usp}(1), \\ B_2 &= C_2, & \mathsf{so}(5) &= \mathsf{usp}(2), \\ D_2 &= A_1 \oplus A_1, & \mathsf{so}(4) &= \mathsf{su}(2) \oplus \mathsf{su}(2), \\ A_3 &= D_3, & \mathsf{su}(4) &= \mathsf{so}(6). \end{aligned}$$

Several more accidental isomorphisms happen among noncompact forms of these algebras [56]. In particular, the case $\mathsf{so}(3,2) = \mathsf{sp}(4,\mathsf{R})$ is very relevant

[7] We annotate the noncompact forms that will be presented in the next section.

for paraxial astigmatic optics; it will be detailed in Sect. 11.7 and used in the next chapter. •

Remark: Noncompact sibling algebras. The natural line of generalization for the classical Cartan algebras (which are compact), is to preserve hyperboloids beside the spheres (11.26). Then, in place of (11.27), we have $\mathbf{R}^\dagger\,\mathbf{g}\,\mathbf{R} = \mathbf{g}$ with the $N\times N$ *metric* matrix $\mathbf{g} = \mathrm{diag}\,(+1_{\,(N^+\ \mathrm{times})},\ -1_{\,(N^-\ \mathrm{times})})$, $N^+ + N^- = N$. We thereby beget the classical noncompact algebras $\mathsf{so}(N^+, N^-)$, $\mathsf{su}(N^+, N^-)$ and $\mathsf{usp}(N^+, N^-)$, whose Lie brackets can be derived from (11.33)–(11.36) by replacing the $\delta_{\nu,\kappa}$'s by the appropriate metric $g_{\nu,\kappa} = \delta_{\nu,\kappa}\,\tau_\nu$, where τ_ν is the sign of the metric in the νth coordinate. •

Remark: Algebra realizations that generate deformations. There exist first-order differential operator realizations of the generators of the noncompact algebras that are distinct from (11.25), in that they can contain linear and quadratic terms in the x^α_κ's. These generate nonlinear *deformations* of the invariant sphere or conic manifold [24]. Examples of deformations that have crossed our way include the relativistic map of the Descartes sphere in Sects. 5.5–5.7, and the conformal symmetry obtained from the Euclidean algebra in Sects. 6.6 and 6.7. •

11.3 The Weyl trick for symplectic algebras

The *Weyl trick* is a mean way of calling a very useful transmutation between Lie algebras. Assume we have a Lie algebra L where we can divide the vector basis of generators into two sets, L^0 (forming a subalgebra of L) and L^1 (that we will rescale by a finite, nonzero constant c), such that their Lie brackets[8] *separate* in the following way:

$$\{\mathsf{L}^0, \mathsf{L}^0\} \subset \mathsf{L}^0, \quad \text{so } \mathsf{L}^0 \text{ is a subalgebra of } \mathsf{L}, \tag{11.38}$$

$$\{\mathsf{L}^0, \mathsf{L}^1\} \subset \mathsf{L}^1, \quad \text{is invariant under } \mathsf{L}^1 \mapsto c\,\mathsf{L}^1, \quad c \in \mathsf{C}, \tag{11.39}$$

$$\{\mathsf{L}^1, \mathsf{L}^1\} \subset \mathsf{L}^0, \quad \text{under } \mathsf{L}^1 \mapsto c\,\mathsf{L}^1, \text{ r.h.s. multiplies by } c^2. \tag{11.40}$$

Nothing essential is changed when c is real and nonzero: the structure of the Lie algebra L is the same, with only a re-scaled basis – which can be scaled back. However, when $c \to 0^\pm$ (or to $\pm\infty$), the structure of the algebra *changes*, and then the process is called a *contraction* of the algebra; this will be exemplified below. In this chapter we shall use mainly the change of structure brought about when $c = \pm i$, which is called the Weyl trick.

When $c^2 = -1$, the commutators (11.40) change sign and we give birth to a different real Lie algebra. The Weyl trick is the map

[8] The Lie brackets can be the Poisson brackets of functions of phase space; they can also be the commutators of Poisson (or other) operators, or of their matrix representations.

$$\mathsf{L} \stackrel{\text{Weyl}}{\longrightarrow} \mathsf{L}^{\text{W}} \quad \text{with} \quad \begin{cases} \mathsf{L}^0 \mapsto \mathsf{L}^0, \\ \mathsf{L}^1 \mapsto i\mathsf{L}^1, \end{cases} \tag{11.41}$$

and provides *noncompact forms* of the classical compact algebras, which will now contain the invariant L^0 as their maximal compact subalgebra. Several important examples follow.

Example: Rotation to Lorentz: $\mathsf{so}(3) \stackrel{\text{Weyl}}{\longrightarrow} \mathsf{so}(2,1)$. When exponentiated, the Lie algebra $\mathsf{so}(3)$ generates the group of rotations $\mathsf{SO}(3)$ [see Sect. 5.3, (5.27)–(5.28) and (5.29)–(5.31)]. The Weyl trick, with L^0 being the 1-dim subalgebra of generator J_z and L^1 spanned by J_x and J_y, is

$$\begin{array}{ccc} \mathsf{so}(3) & \stackrel{\text{Weyl}}{\longrightarrow} & \mathsf{so}(2,1) \\ \begin{cases} \{J_y, J_z\} = J_x \\ \{J_z, J_x\} = J_y \\ \{J_x, J_y\} = J_z \end{cases} & \begin{cases} \quad J_z \\ J_x = iB_x \\ J_y = iB_y \end{cases} & \begin{cases} \{B_y, J_z\} = \quad B_x \\ \{J_z, B_x\} = \quad B_y \\ \{B_x, B_y\} = -J_z. \end{cases} \end{array} \tag{11.42}$$

The Lie brackets on the right present the three-dimensional Lorentz algebra $\mathsf{so}(2,1)$ [cf. (5.68) and (5.72)]. This example is also a warning that it is dangerous to abuse the Weyl trick. It works readily for the commutation relations but says little about their representations, and less on the the new global group structure that corresponds to it. To appreciate the gap, compare the matrices of $\mathsf{SO}(3)$ [(5.17)–(5.19)], with those of its siblings $\mathsf{SO}(2,1) \stackrel{1:2}{=} \mathsf{Sp}(2,\mathsf{R})$ [(5.51), (9.52)–(9.53) and (9.54)–(9.57)]. There is no (simple) 'analytic continuation' between the properties of the two. •

Example: Compact to noncompact: $\mathsf{cl}(N^+ + N^-) \stackrel{\text{Weyl}}{\longrightarrow} \mathsf{cl}(N^+, N^-)$. There are several ways to perform Weyl tricks on algebras. Let us indicate by $\mathsf{L} = \mathsf{cl}(N)$ the three classical families $\mathsf{so}(N)$, $\mathsf{u}(N)$ or $\mathsf{usp}(N)$. Several separations of L into a L^0 and L^1 occur which satisfy (11.38)–(11.40). Invariant subalgebras are visible in the generator patterns (11.31) of size $N = N^+ + N^-$; they must form non-overlapping sub-patterns, of sizes N^+ and N^-, containing the generators of the subalgebras $\mathsf{cl}(N^+) \oplus \mathsf{cl}(N^-) \subset \mathsf{cl}(N^+ + N^-)$. In L^1 there are the remaining $N^+ \times N^-$ generators that will be multiplied by i under the trick. This separation respects (11.33)–(11.36) and decomposes the basis of $N \times N$ generator matrices as follows:

$$\begin{array}{lll} \text{in } \mathsf{L}^0 : \boldsymbol{\Lambda}_{(\kappa,\lambda)} & \left(\begin{array}{l} \kappa, \lambda \textbf{ both} \text{ in } \{1, \ldots, N^+\} \\ \quad \text{or both in } \{N^+{+}1, \ldots, N^+{+}N^-\} \end{array} \right) & \stackrel{\text{Weyl}}{\longrightarrow} \boldsymbol{\Lambda}_{(\kappa,\lambda)}, \\ \text{in } \mathsf{L}^1 : \boldsymbol{\Lambda}_{(\kappa,\lambda)} & \left(\begin{array}{l} \kappa, \lambda \textbf{ not} \text{ both in } \{1, \ldots, N^+\} \\ \quad \text{nor both in } \{N^+{+}1, \ldots, N^+{+}N^-\} \end{array} \right) & \stackrel{\text{Weyl}}{\longrightarrow} i\boldsymbol{\Lambda}_{(\kappa,\lambda)}. \end{array} \tag{11.43}$$

The Lie brackets for the new algebra L^{W} present the noncompact forms of the classical algebra $\mathsf{cl}(N)$ denoted $\mathsf{cl}(N^+, N^-)$. •

Remark: Turning coordinates imaginary. In the realization of the classical algebras $\mathsf{cl}(N, \mathsf{F})$ by differential operators [(11.22)–(11.25) and (11.30) substituted in (11.29)], the Weyl trick (11.43) ocurrs when we replace the coordinates as:

$$x^\alpha_\kappa \overset{\text{Weyl}}{\longrightarrow} \begin{cases} x^\alpha_\kappa, & \kappa \in \{1,\ldots,N^+\}, \\ i x^\alpha_{\bar\kappa}, & \bar\kappa \in \{N^+{+}1,\ldots,N^+{+}N^-\}. \end{cases} \tag{11.44}$$

It is convenient to mark those coordinates which 'go imaginary' by bars (such as $\bar\kappa$ above). Putting i's on some of the coordinates turns the invariant sphere (11.26) into an invariant hyperboloid of signature (N^+, N^-),

$$\sum_{\kappa=1}^{N} (x^\alpha_\kappa)^2 \overset{\text{Weyl}}{\longrightarrow} \sum_{\kappa=1}^{N^+} (x^\alpha_\kappa)^2 - \sum_{\bar\kappa=N^++1}^{N^++N^-} (x^\alpha_{\bar\kappa})^2. \tag{11.45}$$

•

Example: Complex to real: $\mathsf{su}(N) \overset{\text{Weyl}}{\longrightarrow} \mathsf{sl}(N,\mathsf{R})$. We recall the realization of $\mathsf{su}(N)$ in (11.29) by the basis of complex skew-hermitian matrices $\mathbf{\Lambda}^0_{(\kappa,\lambda)}$, which contain subsets of nonzero elements $\begin{bmatrix} 0 & 1 \\ -1 & 0 \end{bmatrix}$ in the κ–λ intersections and form $\mathsf{so}(N) \subset \mathsf{su}(N)$, and by the matrices $\mathbf{\Lambda}^1_{(\kappa,\lambda)}$, which contain nonzero submatrices $\begin{bmatrix} 0 & i \\ i & 0 \end{bmatrix}$ for $\kappa \neq \lambda$, or are diagonal and traceless for $\kappa = \lambda$. There are $N^2 - 1$ generators; this is the same number as in $\mathsf{sl}(N,\mathsf{R})$, which are real matrices. We perform the following Weyl trick [distinct from (11.43)], which makes all matrix elements real:

$$\mathbf{\Lambda}^0_{(\kappa,\lambda)} \overset{\text{Weyl}}{\longrightarrow} \mathbf{\Lambda}^{\text{w}\,0}_{(\kappa,\lambda)} = \mathbf{\Lambda}^0_{(\kappa,\lambda)}, \qquad \mathbf{\Lambda}^1_{(\kappa,\lambda)} \overset{\text{Weyl}}{\longrightarrow} \mathbf{\Lambda}^{\text{w}\,1}_{(\kappa,\lambda)} = -i\mathbf{\Lambda}^1_{(\kappa,\lambda)}. \tag{11.46}$$

The new matrices $\mathbf{\Lambda}^{\text{w}\,1}_{(\kappa,\lambda)}$, $\kappa \neq \lambda$, now contain the nonzero subsets $\begin{bmatrix} 0 & 1 \\ 1 & 0 \end{bmatrix}$, or are diagonal and traceless. Thus, $\mathbf{A}_{\kappa,\lambda} = \frac{1}{2}(\mathbf{\Lambda}^{\text{w}\,0}_{\kappa,\lambda} \pm \mathbf{\Lambda}^{\text{w}\,1}_{\kappa,\lambda})$, $\kappa \leq \lambda$, are real matrices that have a single off-diagonal 1 or a traceless diagonal form, i.e., we obtain the matrices (11.7) that realize the real linear Lie algebra $\mathsf{sl}(N,\mathsf{R})$.

•

Example: Contraction of rotations and boosts to Euclidean translations. Contraction is a distinct process which consists in scaling the generators of an algebra by various powers of a constant c, and then letting $c \to 0$ (or to ∞). As in the previous example (11.42), the division of $\mathsf{so}(3)$ into L^0 with generator J_z and L^1 with J_x and J_y, allows the following contraction of the rotation to the Euclidean algebra $\mathsf{iso}(2)$ [cf. (5.39)–(5.40)],

$$\begin{matrix} \mathsf{so}(3) & \overset{\text{Weyl}(c)}{\longrightarrow},\ \lim_{c\to\infty} & \mathsf{iso}(2) \\ \begin{cases} \{J_y, J_z\} = J_x \\ \{J_z, J_x\} = J_y \\ \{J_x, J_y\} = J_z \end{cases} & \begin{cases} J_z \\ J_x = c\,P_x \\ J_y = c\,P_y \end{cases} & \begin{cases} \{P_y, J_z\} = P_x \\ \{J_z, P_x\} = P_y \\ \{P_x, P_y\} = 0. \end{cases} \end{matrix} \tag{11.47}$$

For the groups, this contraction $\mathsf{SO}(3) \overset{c\to\infty}{\longrightarrow} \mathsf{ISO}(2)$ means that ever smaller rotations around the x- and y- axes of an ever larger sphere, become translations in its tangent x–y plane. An analogous contraction can be performed on the 3-dim Lorentz algebra [with J_x, J_y replaced by B_x, B_y from (11.42)]; this is the nonrelativistic limit of Lorentz transformations when the speed of light c is 'made' to increase without bound, $\mathsf{SO}(2,1) \overset{c\to\infty}{\longrightarrow} \mathsf{ISO}(2)$. In 3-dim, this contraction was first investigated by Inönü and Wigner [72] with the tools of group representation theory. •

Example: Contractions $\mathsf{cl}(N^+ + N^-) \overset{c\to\infty}{\longrightarrow} \mathsf{i}_{N^+\times N^-}[\mathsf{cl}(N^+) \oplus \mathsf{cl}(N^-)]$. The straightforward generalization of the previous example applies to all classical Lie algebras $\mathsf{L} = \mathsf{cl}(N)$ under separation into a subalgebra L^0 and an ideal L^1 satisfying (11.38)–(11.40). As in (11.47), when we rescale $\mathsf{L}^1 \mapsto c\mathsf{L}^1$ and let $c \to 0$, the last commutators vanish. The contracted algebra inherits the subalgebra L^0 while L^1 becomes its abelian ideal consisting of mutually commuting generators of translations. The resulting Lie algebras are the *inhomogeneous* siblings of $\mathsf{cl}(N)$. When the separation of L accords with (11.43) and $N = N^+ + N_-$, the patterns (11.31) indicate that the generators of the maximal compact subalgebra L^0 will be the sub-patterns $\mathsf{cl}(N^+) \oplus \mathsf{cl}(N^-)$, while the remaining $N^+ \times N^-$ generators belong to L^1. The latter are accomodated in a rectangular matrix, of which $\mathsf{cl}(N^+)$ transforms the rows and $\mathsf{cl}(N^-)$ the columns. Contracted algebras are semidirect sums [see (5.47)] between the invariant translation and the radical classical subalgebras, $\mathsf{T}_{N^+\times N^-} \,\oplus\!\!\!\!\supset [\mathsf{cl}(N^+) \oplus \mathsf{cl}(N^-)]$. The dimension of the translation subalgebra is rarely written, and most commonly we shall use and denote $\mathsf{icl}(N) := \mathsf{i}_N\mathsf{cl}(N)$. Combining the Weyl trick and contraction on different index sets, we can produce a variety of algebras, such as the original Inönü-Wigner contraction [72] $\mathsf{so}(3,1) \longrightarrow \mathsf{iso}(3)$, but also $\mathsf{so}(3,1) \longrightarrow \mathsf{iso}(2,1)$, etc. •

We now apply the Weyl trick to transmute unitary-symplectic algebras into *real* symplectic algebras, which are the relevant ones in the paraxial régime of geometric optics, $\mathsf{usp}(N) \overset{\text{Weyl}}{\longrightarrow} \mathsf{sp}(2N,\mathsf{R})$; both algebras have $N(2N+1)$ generators. We use the matrix representation of the unitary symplectic algebra (11.29) whose elements are quaternions $z_{\kappa,\lambda}$ that are in turn represented by the 2×2 complex submatrices $\mathbf{z}_{\kappa,\lambda}$ as in (11.11). In this representation, $\mathbf{\Lambda}^0_{(\kappa,\lambda)}$ and $\mathbf{\Lambda}^1_{(\kappa,\lambda)}$ are real, and form the subalgebra which will remain invariant, $\mathsf{L}^0 = \mathsf{u}(N) \subset \mathsf{usp}(N)$, while $\mathbf{\Lambda}^2_{(\kappa,\lambda)}$ and $\mathbf{\Lambda}^3_{(\kappa,\lambda)}$ are pure imaginary; the Weyl trick multiplies the latter two by i. Writing out the κ–λ submatrices of the $2N \times 2N$ matrices (11.29), $\mathsf{usp}(N) \overset{\text{Weyl}}{\longrightarrow} \mathsf{sp}(N,\mathsf{R})$ is

$$\mathbf{\Lambda}^0_{(\kappa,\lambda)} = \begin{pmatrix} \overset{\kappa}{\downarrow} & \overset{\lambda}{\downarrow} \\ 0 & \begin{smallmatrix} 1 & 0 \\ 0 & 1 \end{smallmatrix} \\ \begin{smallmatrix} -1 & 0 \\ 0 & -1 \end{smallmatrix} & 0 \end{pmatrix}, \qquad \mathbf{\Lambda}^1_{(\kappa,\lambda)} = \begin{pmatrix} \overset{\kappa}{\downarrow} & \overset{\lambda}{\downarrow} \\ 0 & \begin{smallmatrix} 0 & 1 \\ -1 & 0 \end{smallmatrix} \\ \begin{smallmatrix} 0 & 1 \\ -1 & 0 \end{smallmatrix} & 0 \end{pmatrix}, \tag{11.48}$$

$$\mathbf{\Lambda}^2_{(\kappa,\lambda)} = \begin{pmatrix} 0 & \begin{smallmatrix} i & 0 \\ 0 & -i \end{smallmatrix} \\ \begin{smallmatrix} i & 0 \\ 0 & -i \end{smallmatrix} & 0 \end{pmatrix} \overset{\text{Weyl}}{\longrightarrow} \mathbf{\Lambda}^{\text{w}\,2}_{(\kappa,\lambda)} = \begin{pmatrix} 0 & \begin{smallmatrix} -1 & 0 \\ 0 & 1 \end{smallmatrix} \\ \begin{smallmatrix} -1 & 0 \\ 0 & 1 \end{smallmatrix} & 0 \end{pmatrix}, \tag{11.49}$$

$$\mathbf{\Lambda}^3_{(\kappa,\lambda)} = \begin{pmatrix} 0 & \begin{smallmatrix} 0 & i \\ i & 0 \end{smallmatrix} \\ \begin{smallmatrix} 0 & i \\ i & 0 \end{smallmatrix} & 0 \end{pmatrix} \overset{\text{Weyl}}{\longrightarrow} \mathbf{\Lambda}^{\text{w}\,3}_{(\kappa,\lambda)} = \begin{pmatrix} 0 & \begin{smallmatrix} 0 & -1 \\ -1 & 0 \end{smallmatrix} \\ \begin{smallmatrix} 0 & -1 \\ -1 & 0 \end{smallmatrix} & 0 \end{pmatrix}. \tag{11.50}$$

Acting on R^{2N} column vectors, these matrices no longer preserve the quaternion sphere (11.14); instead, they satisfy the following relation:

$$\begin{array}{c}\mathbf{\Lambda}^{\mathrm{w}\,\alpha}_{(\kappa,\lambda)}\,\widetilde{\mathbf{\Omega}} = -\widetilde{\mathbf{\Omega}}\,(\mathbf{\Lambda}^{\mathrm{w}\,\alpha}_{(\kappa,\lambda)})^{\top}, \\ \alpha = 0,1,2,3,\end{array} \qquad \widetilde{\mathbf{\Omega}} := \begin{pmatrix} \begin{smallmatrix} 0 & -1 \\ 1 & 0 \end{smallmatrix} & & 0 \\ & \ddots & \\ 0 & & \begin{smallmatrix} 0 & -1 \\ 1 & 0 \end{smallmatrix} \end{pmatrix}. \tag{11.51}$$

We shall now identify the Lie algebra spanned by the $2N \times 2N$ real matrices $\mathbf{\Lambda}^{\mathrm{w}}$ that satisfy the condition (11.51). Having walked the path from sphere invariance to the classical algebras along (11.28), we can reverse our steps and compare (11.51) with the definition of the group of real symplectic matrices $\mathsf{Sp}(2N,\mathsf{R})$ in (9.23) for real $2N \times 2N$ matrices close to the unit,

$$\mathbf{M} = \mathbf{1} + \varepsilon\mathbf{m},\ \varepsilon^2 \approx 0, \mathbf{M}\,\mathbf{\Omega}\,\mathbf{M}^{\top} = \mathbf{\Omega} := \begin{pmatrix} \mathbf{0} & -\mathbf{1} \\ \mathbf{1} & \mathbf{0} \end{pmatrix} \tag{11.52}$$

$$\Rightarrow\ (\mathbf{1}+\varepsilon\mathbf{m})\,\mathbf{\Omega}\,(\mathbf{1}+\varepsilon\mathbf{m})^{\top} = \mathbf{\Omega} \quad \Rightarrow \quad \mathbf{m}\,\mathbf{\Omega} = -\mathbf{\Omega}\,\mathbf{m}^{\top}; \tag{11.53}$$

$$\text{writing}\quad \mathbf{m} = \begin{pmatrix} \mathbf{a} & \mathbf{b} \\ \mathbf{c} & \mathbf{d} \end{pmatrix} \text{ implies} \begin{cases} \mathbf{a} = -\mathbf{d}^{\top}, \\ \mathbf{b} = \mathbf{b}^{\top},\ \mathbf{c} = \mathbf{c}^{\top}. \end{cases} \tag{11.54}$$

Comparing $\mathbf{\Omega}$ with $\widetilde{\mathbf{\Omega}}$ in (11.51), we see that their connection is a permutation of the rows and of the columns (see below).

Remark: Renumbering the rows for phase space. We used the Weyl trick to substitute quaternion matrix elements by real 2×2 submatrices; this is supplemented by replacing the quaternion *vector* components $\{x_\kappa\}_{\kappa=1}^{N}$ with real two-vectors $\binom{p_\kappa}{q_\kappa}$. Equation (11.51) matches our previous designation of the momentum and position components of N-dimensional paraxial optics, $\mathbf{p} = \{p_\kappa\}_{\kappa=1}^{N}$ and $\mathbf{q} = \{q_\kappa\}_{\kappa=1}^{N}$, when the first are made to occupy the odd rows in the $2N$-vector, and the second the even rows. This renumbering is achieved through a similarity transformation with a $2N \times 2N$ permutation matrix $\mathbf{\Pi} = \|\mathit{\Pi}_{j,k}\|$, $\mathbf{\Pi}^{-1} = \mathbf{\Pi}^{\top}$, that we write as

$$\begin{pmatrix} \binom{p_1}{q_1} \\ \vdots \\ \binom{p_N}{q_N} \end{pmatrix} = \mathbf{\Pi} \begin{pmatrix} \begin{pmatrix} p_1 \\ \vdots \\ p_N \end{pmatrix} \\ \begin{pmatrix} q_1 \\ \vdots \\ q_N \end{pmatrix} \end{pmatrix}, \qquad \mathit{\Pi}_{j,k} := \begin{cases} 1, \text{ when } j = 2k-1 \\ \qquad \text{or } j = 2(k-N), \\ 0, \text{ otherwise.} \end{cases} \tag{11.55}$$

The two matrix sets are thus related by $\mathbf{A} = \mathbf{\Pi}\,\mathbf{m}\,\mathbf{\Pi}^{\top}$ and $\widetilde{\mathbf{\Omega}} = \mathbf{\Pi}\,\mathbf{\Omega}\,\mathbf{\Pi}^{\top}$. This renumbering is not really important when we abstract the elements of the algebra from their matrix realization. •

Definition: *A $2N \times 2N$ matrix* $\mathbf{m}$ *that satisfies (11.54),* i.e.,

$$\mathbf{m} = \begin{pmatrix} \mathbf{a} & \mathbf{b} = \mathbf{b}^{\top} \\ \mathbf{c} = \mathbf{c}^{\top} & -\mathbf{a}^{\top} \end{pmatrix}, \tag{11.56}$$

is called a **Hamiltonian** *(or infinitesimal symplectic) matrix. These matrices represent the Lie algebra* $\mathsf{sp}(2N, \mathsf{R})$.

Remark: Hamiltonian and symmetric matrices. A Hamiltonian matrix times the symplectic metric matrix $\mathbf{\Omega}$ is a symmetric matrix:

$$\mathbf{s} := \mathbf{m}\,\mathbf{\Omega}, \qquad \mathbf{m}\,\mathbf{\Omega} = -\mathbf{\Omega}\,\mathbf{m}^{\top} \quad \Leftrightarrow \quad \mathbf{s} = \mathbf{s}^{\top}. \tag{11.57}$$

Note that $\mathbf{\Omega}$ is also a Hamiltonian matrix, corresponding to the isotropic harmonic guide Hamiltonian function. In general, a symmetric matrix can be Hamiltonian only when it commutes with $\mathbf{\Omega}$. Products of Hamiltonian matrices are generally not Hamiltonian; only their commutators are. •

A particularly convenient vector basis for the Lie algebra $\mathsf{sp}(2N, \mathsf{R})$ of $2N \times 2N$ real symplectic matrices can be chosen to realize 'infinitesimal' (and astigmatic) optical operations. We may call it the *optical* presentation of the symplectic algebra, although it also merits other names: it is the infinitesimal version of the *triangular* decomposition in (9.76); and it will divide the generators of the algebra into *raising, weight and lowering*, in Sect. 11.5 below. We use the Kronecker basis of matrices $\mathbf{A}_{(\kappa,\lambda)}$ in (11.7), to define

$$\mathbf{a}_{(j,k)} := \mathbf{A}_{(j,k)} - \mathbf{A}_{(N+k,N+j)}, \quad 1 \le j,k \le N, \quad \text{MAGNIFIERS}, \tag{11.58}$$
$$\mathbf{b}_{(j,k)} := \mathbf{A}_{(j,N+k)} + \mathbf{A}_{(k,N+j)}, \quad 1 \le j \le k \le N, \quad \text{LENSES}, \tag{11.59}$$
$$\mathbf{c}_{(j,k)} := \mathbf{A}_{(N+j,k)} + \mathbf{A}_{(N+k,j)}, \quad 1 \le j \le k \le N, \quad \text{DISPLACEMENTS}. \tag{11.60}$$

The Lie brackets (commutators) of this optical basis separate as follows:

$$[\mathbf{a}_{(j,k)}, \mathbf{a}_{(m,n)}] = \delta_{k,m}\,\mathbf{a}_{(j,n)} - \delta_{j,n}\,\mathbf{a}_{(m,k)}, \quad \text{subalgebra } \mathsf{gl}(N, \mathsf{R}), \tag{11.61}$$
$$[\mathbf{a}_{(j,k)}, \mathbf{b}_{(m,n)}] = \delta_{k,m}\,\mathbf{b}_{(j,n)} + \delta_{k,n}\,\mathbf{b}_{(j,m)}, \quad [\mathbf{b}_{(j,k)}, \mathbf{b}_{(m,n)}] = 0, \tag{11.62}$$
$$[\mathbf{a}_{(j,k)}, \mathbf{c}_{(m,n)}] = -\delta_{j,m}\,\mathbf{c}_{(n,k)} - \delta_{j,n}\,\mathbf{c}_{(m,k)}, \quad [\mathbf{c}_{(j,k)}, \mathbf{c}_{(m,n)}] = 0, \tag{11.63}$$
$$[\mathbf{b}_{(j,k)}, \mathbf{c}_{(m,n)}] = \delta_{j,m}\,\mathbf{a}_{(k,n)} + \delta_{j,n}\,\mathbf{a}_{(k,m)} + \delta_{k,m}\,\mathbf{a}_{(j,n)} + \delta_{k,n}\,\mathbf{a}_{(j,m)}. \tag{11.64}$$

Remark: The $N = 1$ case. The optical representation of the Lie algebra $\mathsf{sp}(2, \mathsf{R})$ by 2×2 matrices[9] [cf. (9.18) and (10.7)] is

$$\mathbf{a} = \begin{pmatrix} 1 & 0 \\ 0 & -1 \end{pmatrix}, \quad \mathbf{b} = \begin{pmatrix} 0 & 1 \\ 0 & 0 \end{pmatrix}, \quad \mathbf{c} = \begin{pmatrix} 0 & 0 \\ 1 & 0 \end{pmatrix}. \tag{11.65}$$

Their commutators (11.62)–(11.64) are $[\mathbf{a}, \mathbf{b}] = 2\mathbf{b}$, $[\mathbf{a}, \mathbf{c}] = -2\mathbf{c}$, and $[\mathbf{b}, \mathbf{c}] = \mathbf{a}$. •

We summarize the road map around $\mathsf{Sp}(2N, \mathsf{R})$ as follows:

[9] We omit subindices (j, k) because they have the trivial range $\{1\}$.

$$\begin{array}{ccccccl} \mathsf{gl}(N,\mathsf{R}) & \supset & \mathsf{so}(N) & = & \mathsf{so}(N) & & \text{rotation generators} \\ \cap & & \cap & & \cap & & \\ \mathsf{gl}(N,\mathsf{C}) & \supset & \mathsf{u}(N) & = & \mathsf{u}(N) & & \text{Fourier generators} \\ \cap & & \cap & & \cap & & \\ \mathsf{gl}(N,\mathsf{Q}) & \supset & \mathsf{usp}(N) & \xrightarrow{\text{Weyl}} & \boxed{\mathsf{sp}(2N,\mathsf{R})} & \subset \mathsf{sl}(2N,\mathsf{R}) \subset \mathsf{gl}(2N,\mathsf{R}) & \\ \uparrow & & \uparrow & & \cup & & \\ \text{linear} & & \text{classical} & & \mathsf{gl}(N,\mathsf{R}) & & \text{magnifier generators.} \end{array} \tag{11.66}$$

11.4 Phase space functions, operators and matrices

With matrices we can represent all linear and classical Lie algebras, and they are useful in computation. But matrices are bulky for notation because they generally have more elements than there are linearly independent generators. For example, in real paraxial optics, the 4×4 Hamiltonian matrices (11.58)–(11.60) offer 16 element places, but $\mathsf{sp}(4, \mathsf{R})$ has only 10 generators. In fact, the most economic realization of the symplectic algebras is by Poisson operators of quadratic functions of phase space. [We have met Lie algebras of Poisson operators since Part I; in particular see (3.25), (9.2)–(9.3) and (9.4)–(9.5).] All classical algebras inherit this realization [67].

We extend language (and perhaps abuse it) by using the names of optical elements to designate their infinitesimal generators, as follows:

DISPLACEMENTS

$$\begin{pmatrix} \mathbf{0} & \mathbf{0} \\ -\mathbf{c} & \mathbf{0} \end{pmatrix} \qquad \begin{array}{c} \text{monomials } p_j p_k, \\ 1 \le j \le k \le N, \end{array} \qquad \{p_j p_k, \circ\} = -p_j \frac{\partial}{\partial q_k} - p_k \frac{\partial}{\partial q_j},$$

$$\{p_j p_k, \circ\} \begin{pmatrix} \mathbf{p} \\ \mathbf{q} \end{pmatrix} = \begin{pmatrix} 0 \\ -p_k \text{ in row } j, \; -p_j \text{ in row } k \end{pmatrix} = -\mathbf{c}_{(j,k)} \begin{pmatrix} \mathbf{p} \\ \mathbf{q} \end{pmatrix}; \tag{11.67}$$

MAGNIFIERS:

$$\begin{pmatrix} \mathbf{a} & \mathbf{0} \\ \mathbf{0} & -\mathbf{a}^\top \end{pmatrix} \qquad \begin{array}{c} \text{monomials } q_j p_k, \\ 1 \le j, k \le N, \end{array} \qquad \{q_j p_k, \circ\} = p_k \frac{\partial}{\partial p_j} - q_j \frac{\partial}{\partial q_k},$$

$$\{q_j p_k, \circ\} \begin{pmatrix} \mathbf{p} \\ \mathbf{q} \end{pmatrix} = \begin{pmatrix} p_k \text{ in row } j \\ -q_j \text{ in row } k \end{pmatrix} = \mathbf{a}_{(j,k)} \begin{pmatrix} \mathbf{p} \\ \mathbf{q} \end{pmatrix}; \tag{11.68}$$

LENSES:

$$\begin{pmatrix} \mathbf{0} & \mathbf{b} \\ \mathbf{0} & \mathbf{0} \end{pmatrix} \qquad \begin{array}{c} \text{monomials } q_j q_k, \\ 1 \le j \le k \le N, \end{array} \qquad \{q_j q_k, \circ\} = q_j \frac{\partial}{\partial p_k} + q_k \frac{\partial}{\partial p_j},$$

$$\{q_j q_k, \circ\} \begin{pmatrix} \mathbf{p} \\ \mathbf{q} \end{pmatrix} = \begin{pmatrix} q_k \text{ in row } j, \; q_j \text{ in row } k \\ 0 \end{pmatrix} = \mathbf{b}_{(j,k)} \begin{pmatrix} \mathbf{p} \\ \mathbf{q} \end{pmatrix}; \tag{11.69}$$

where in the vectors with text all unnamed entries are null and, when $j = k$ in (11.67) or (11.69), the two terms sum. This realization of $\mathsf{sp}(2N, \mathsf{R})$ is most economic because here are *all* $N(2N+1)$ quadratic monomials of phase space.

Hamiltonian matrices $\mathbf{m} \in \mathsf{sp}(2N,\mathsf{R})$ [see (11.56)] represent quadratic Hamiltonian functions $H(\mathbf{p},\mathbf{q})$; they generate evolution of phase space when exponentiated to $\mathbf{M} \in \mathsf{Sp}(2N,\mathsf{R})$. We recall once more from (9.16) that an optical system $\mathcal{M}(\mathbf{M})$ acts on the phase space vectors through the inverse matrix, i.e., $\mathcal{M}(\mathbf{M}) : \begin{pmatrix}\mathbf{p}\\ \mathbf{q}\end{pmatrix} = \mathbf{M}^{-1}\begin{pmatrix}\mathbf{p}\\ \mathbf{q}\end{pmatrix}$. When this system lies on a one-parameter subgroup close to the unit, $\mathbf{M}(z) = \exp(z\,\mathbf{m}) \approx \mathbf{1} + z\mathbf{m}$, with $z^2 \approx 0$ – so z we call ε,

$$\mathcal{M}(\mathbf{1}+\varepsilon\mathbf{m}) : \begin{pmatrix}\mathbf{p}\\ \mathbf{q}\end{pmatrix} = \begin{pmatrix}\mathbf{p}\\ \mathbf{q}\end{pmatrix} - \varepsilon\,\mathbf{m}\begin{pmatrix}\mathbf{p}\\ \mathbf{q}\end{pmatrix} \quad \Rightarrow \quad \mathcal{M}(\mathbf{m}) \leftrightarrow -\mathbf{m}. \qquad (11.70)$$

Quadratic functions of phase space $f(\mathbf{p},\mathbf{q})$ are linear combinations of the monomials $p_j p_k$, $q_j p_k$ and $p_j p_k$; they are associated to the realizations of $\mathsf{sp}(2N,\mathsf{R})$ by Hamiltonian matrices through (11.67)–(11.69). Paying close attention to the minus sign in (11.70), the generic D-dim case is

$$\begin{aligned} f(\mathbf{m};\mathbf{p},\mathbf{q}) &:= \tfrac{1}{2}\textstyle\sum_{j,k} c_{j,k}\, p_j p_k - \sum_{j,k} a_{j,k}\, q_j p_k - \tfrac{1}{2}\sum_{j,k} b_{j,k}\, q_j q_k \\ &\leftrightarrow \quad \mathbf{m} = \begin{pmatrix}\mathbf{a} & \mathbf{b}\\ \mathbf{c} & -\mathbf{a}^\top\end{pmatrix}. \end{aligned} \qquad (11.71)$$

The Poisson brackets of two such functions will be the function of the commutator of the matrices:

$$\{f(\mathbf{m}_1;\mathbf{p},\mathbf{q}), f(\mathbf{m}_2;\mathbf{p},\mathbf{q})\} = f([\mathbf{m}_1,\mathbf{m}_2];\mathbf{p},\mathbf{q}). \qquad (11.72)$$

In Sect. 9.5 we saw that fractional Fourier transformations form the maximal compact subgroup $\mathsf{U}(N) = \mathsf{Sp}(2N,\mathsf{R}) \cap \mathsf{SO}(2N) \subset \mathsf{Sp}(2N,\mathsf{R})$, and are represented by $2N \times 2N$ ortho-symplectic matrices (9.32). The (*infinitesimal Fourier*) Lie algebra $\mathsf{u}(N) = \mathsf{sp}(2N,\mathsf{R}) \cap \mathsf{so}(2N)$ can be found writing, near to the group identity, $\mathbf{U} \approx \mathbf{1} + \varepsilon\mathbf{u}$, $\varepsilon^2 \approx 0$, and proceeding in the same way as in (11.53). The ortho-symplectic conditions on the matrices in the group and algebra are:

$$\begin{array}{ccc} \text{SYMPLECTIC} & & \text{HAMILTONIAN} \\ \mathbf{U}\boldsymbol{\Omega}\mathbf{U}^\top = \boldsymbol{\Omega}, & \Rightarrow & \mathbf{u}\boldsymbol{\Omega} = -\boldsymbol{\Omega}\,\mathbf{u}^\top, \end{array} \quad \begin{pmatrix}\mathbf{a} & \mathbf{b}{=}\mathbf{b}^\top\\ \mathbf{c}{=}\mathbf{c}^\top & -\mathbf{a}^\top\end{pmatrix},$$

$$\begin{array}{ccc} \text{ORTHOGONAL} & & \text{SKEW-ADJOINT} \\ \mathbf{U}\mathbf{U}^\top = \mathbf{1} & \Rightarrow & \mathbf{u} = -\mathbf{u}^\top, \end{array} \quad \begin{pmatrix}\mathbf{a}{=}-\mathbf{a}^\top & \mathbf{b}\\ -\mathbf{b}^\top & \mathbf{d}{=}-\mathbf{d}^\top\end{pmatrix}, \qquad (11.73)$$

$$\text{ORTHO-SYMPLECTIC} \Rightarrow \mathsf{u}(N)\text{-FOURIER} \quad \begin{pmatrix}\mathbf{a}{=}-\mathbf{a}^\top & \mathbf{b}{=}\mathbf{b}^\top\\ -\mathbf{b} & \mathbf{a}\end{pmatrix}.$$

A good decomposition of the basis of $\mathsf{u}(N)$-Fourier generator matrices, and their corresponding quadratic polynomials of phase space, is thus

$$\mathbf{u} = \begin{pmatrix} \mathbf{l} & -\mathbf{f} \\ \mathbf{f} & \mathbf{l} \end{pmatrix}, \quad \begin{array}{l} \mathbf{l}_{(j,k)} := \mathbf{a}_{(j,k)} - \mathbf{a}_{(k,j)} \quad \leftrightarrow l_{j,k} := q_j p_k - q_k p_j, \\ \mathbf{f}_{(j,k)} := \mathbf{b}_{(j,k)} + \mathbf{b}_{(k,j)} \\ \qquad -\mathbf{c}_{(j,k)} - \mathbf{c}_{(k,j)} \leftrightarrow f_{j,k} := p_j p_k + q_j q_k. \end{array} \tag{11.74}$$

The subalgebra of the $l_{j,k}$'s is the subalgebra $\mathsf{so}(N) \subset \mathsf{u}(N)$ of angular momentum; it generates joint rotations of position and of momentum spaces. The $f_{j,k}$'s form an ideal of oscillator-like Hamiltonians which generate the 'true' Fourier transformations, i.e., those which *mix* position with momentum coordinates. They are realized as

PURE ROTATORS:

$$\begin{pmatrix} \mathbf{l} & \mathbf{0} \\ \mathbf{0} & \mathbf{l} \end{pmatrix}, \quad \mathbf{l} = -\mathbf{l}^\top \quad \{l_{j,k}, \circ\} = -q_j \frac{\partial}{\partial q_k} + q_k \frac{\partial}{\partial q_j} - p_j \frac{\partial}{\partial p_k} + p_k \frac{\partial}{\partial p_j},$$

$$\{l_{j,k}, \circ\} \begin{pmatrix} \mathbf{p} \\ \mathbf{q} \end{pmatrix} = \begin{pmatrix} p_k \text{ in row } j, & -p_j \text{ in row } k \\ q_k \text{ in row } j, & -q_j \text{ in row } k \end{pmatrix}; \tag{11.75}$$

'TRUE' FOURIER TRANSFORMS:

$$\begin{pmatrix} \mathbf{0} & -\mathbf{f} \\ \mathbf{f} & \mathbf{0} \end{pmatrix}, \quad \mathbf{f} = \mathbf{f}^\top \quad \{f_{j,k}, \circ\} = q_j \frac{\partial}{\partial p_k} - p_k \frac{\partial}{\partial q_j} + q_k \frac{\partial}{\partial p_j} + p_j \frac{\partial}{\partial q_k},$$

$$\{f_{j,k}, \circ\} \begin{pmatrix} \mathbf{p} \\ \mathbf{q} \end{pmatrix} = \begin{pmatrix} q_k \text{ in row } j, & q_j \text{ in row } k \\ -p_k \text{ in row } j, & -p_j \text{ in row } k \end{pmatrix}. \tag{11.76}$$

The generator functions of rotations $l_{j,k}$, are the components of the second-rank, traceless skew-symmetric tensor of *angular momentum* $\mathbf{l} = \mathbf{q} \wedge \mathbf{p}$; the $f_{j,k}$'s are a symmetric second-rank tensor composed of 1-dim harmonic guide (or oscillator) Hamiltonians $f_{k,k} = p_k^2 + q_k^2$ on the diagonal, and of 'cross harmonic' generators $f_{j,k} = p_j p_k + q_j q_k$ in the off-diagonal elements.

Remark: Central Fourier transforms and the trace. The trace generator [see (11.20) – divide by 2] and its associated matrix (11.71) are

$$f_{\mathrm{tr}} := \tfrac{1}{2} \sum_{k=1}^{N} f_{k,k} = \tfrac{1}{2}(|\mathbf{p}|^2 + |\mathbf{q}|^2) \quad \leftrightarrow \quad \begin{pmatrix} \mathbf{0} & -\mathbf{1} \\ \mathbf{1} & \mathbf{0} \end{pmatrix} = \mathbf{\Omega}. \tag{11.77}$$

This function is the Hamiltonian of an $\mathsf{SO}(N)$-isotropic harmonic guide; it Poisson-commutes with all $\mathsf{u}(N)$ generators (11.74), and so it generates the group $\mathsf{U}(1)$ of central Fourier transforms [cf. Sect. 10.3.1 and (10.29)]. The independent generators of $\mathsf{su}(N) \subset \mathsf{u}(N)$ include the $l_{j,k}$'s and the $f_{j,k}$'s with $j < k$ and, subtracting the trace, the $N-1$ independent traceless generators $\tilde{f}_{k,k} = f_{k,k} - N^{-1} f_{\mathrm{tr}}$ [cf. (11.21)]. The three partial trace functions $|\mathbf{p}|^2$, $\mathbf{p} \cdot \mathbf{q}$ and $|\mathbf{q}|^2$, Poisson-commute with all rotation generators of $\mathsf{so}(N)$. •

Remark: Separable Fourier transforms. The diagonal and mutually Poisson-commuting functions $f_{k,k} = p_k^2 + q_k^2$, generate the separable fractional Fourier transforms in the phase space planes (p_k, q_k), $k = 1, \ldots, N$. [This was seen for the $N = 2$ case in Sect. 10.3.2.] Their manifold is an N-dim torus $\mathcal{T}^N$, and every set of coordinates rotated by $\mathbf{R} \in \mathsf{SO}(N)$ will have *its* corresponding torus. Hence, the manifold of separable $\mathsf{U}(N)$-Fourier transforms is $\mathcal{T}^N \times \mathsf{SO}(N)$. In turn, the manifold of $\mathsf{SO}(N)$ is the product of spheres $\mathcal{S}^n$, $n = 1, \ldots, N$. [See (9.84).] •

Remark: The Iwasawa decomposition. In Sect. 9.5 we saw the Iwasawa decomposition of the symplectic groups $\mathsf{Sp}(2N, \mathsf{R})$ into the product of the $\mathsf{U}(N)$-Fourier subgroup, and the solvable group of matrices in (9.71), which in turn decomposes into an abelian and a nilpotent factor. There is a corresponding decomposition of the symplectic Lie algebras $\mathsf{sp}(2N, \mathsf{R})$ into a vector sum of three subalgebras. The $\mathsf{u}(N)$-Fourier subalgebra was seen above (11.74)–(11.76). The abelian algebra of N commuting diagonal matrices (11.58)–(11.68) corresponds, in the Poisson operator realization, to the *scaling* functions

$$d_\ell := \tfrac{1}{2} q_\ell p_\ell, \quad \ell = 1, 2, \ldots, N, \qquad \{d_\ell, \circ\} = \frac{1}{2}\left(p_\ell \frac{\partial}{\partial p_\ell} - q_\ell \frac{\partial}{\partial q_\ell}\right), \tag{11.78}$$

which generate reciprocal magnifications in each of the (p_ℓ, q_ℓ)-planes. Lastly, the nilpotent subalgebra in the Iwasawa decomposition is generated by the off-diagonal upper-triangular matrices $\mathbf{a}_{(j,k)}$, $j < k$, which correspond to the 'double-cross' functions $q_j p_k$, $j \neq k$; and by all 'infinitesimal lenses' given by the matrices $\mathbf{b}_{(j,k)}$, which correspond to $q_j q_k$. •

11.5 Roots and multiplets of the symplectic algebras

In this section we introduce the **Cartan roots** and their diagrams, which will serve to visualize the classification of their generators into *raising*, *lowering*, and *weight* subsets, and clarify further the structure of the classical Lie algebras. Their **multiplets** are finite sets of functions of phase space, arranged into patterns, which are raised, lowered and weighted by the algebra. In particular, this treatment will prepare the terrain for the classification of aberrations in Part IV. The symplectic algebras will be the first case to be studied, and the scaffold to the other Cartan families in the following two sections.

The set of Lie brackets which present a Lie algebra of rank N can be turned into a simultaneous set of N eigenvalue equations for the commuting *weight* generators. We recall the definition of rank from page 214, and recall that the weight generators of $\mathsf{Sp}(2N, \mathsf{R})$ are the scalings d_ℓ along the axes $\ell = 1, \ldots, N$, given in (11.78). The weights belong to the maximal abelian subalgebra, and determine the 'principal axes' of the algebra itself in an N-dim space. Following Cartan, we must search for the N-vector solutions ρ, called *roots*, of the simultaneous eigenvalue equations

$$\{d_\ell, \circ\}\,\rho := \{d_\ell, \rho\} = \lambda_\ell\,\rho, \quad \ell = 1, 2, \ldots, N, \tag{11.79}$$

$$\text{so we associate} \quad \rho \leftrightarrow \vec{\lambda} := (\lambda_1, \lambda_2, \ldots, \lambda_N). \tag{11.80}$$

The N-vector $\vec{\lambda}$ is the **weight vector** of the root ρ. For each root[10] ρ one draws its corresponding N-dim weight vector $\vec{\lambda}$ in the R^N space of a *Cartan diagram.* Here and in the following two sections we shall examine in detail the cases $N = 1, 2, 3$ of all classical Cartan families. Moreover, since any constant times a root is also a root (even if complex), Cartan diagrams will serve both for the compact and noncompact forms of an algebra, such as $\mathsf{u}(N)$ and $\mathsf{sl}(N, \mathsf{R})$, or $\mathsf{usp}(N)$ and $\mathsf{sp}(2N, \mathsf{R})$.

We start with the real symplectic algebras $\mathsf{sp}(2N, \mathsf{R})$ recalling their optical basis of quadratic monomials of phase space, $\{p_j p_k,\ q_j p_k,\ q_j q_k\}$ in (11.67)–(11.69), and realizing the Lie brackets by Poisson brackets. The weights $\{d_\ell, \circ\}$ in (11.78) count the number of p_ℓ's and subtract the number of q_ℓ's (and divide by 2):

$$\{d_\ell, \circ\}\ p_j p_k = \tfrac{1}{2}(\,\delta_{j,\ell} + \delta_{k,\ell}\,)\ p_j p_k, \tag{11.81}$$

$$\{d_\ell, \circ\}\ q_j p_k = \tfrac{1}{2}(-\delta_{j,\ell} + \delta_{k,\ell})\, q_j p_k, \tag{11.82}$$

$$\{d_\ell, \circ\}\ q_j q_k = \tfrac{1}{2}(-\delta_{j,\ell} - \delta_{k,\ell})\, q_j q_k, \tag{11.83}$$

and $\{d_\ell, \circ\}\, d_m = 0$. The correspondence between the roots of $\mathsf{sp}(2N, \mathsf{R})$ and their N-dim weight vectors of eigenvalues $\vec{\lambda}$ is thus

$$\begin{aligned} p_j p_k &\leftrightarrow \vec{\lambda}_+^{(j,k)} = (0, \ldots, 0, \overset{j}{\downarrow}{+\tfrac{1}{2}}, 0, \ldots, 0, \overset{k}{\downarrow}{+\tfrac{1}{2}}, 0, \ldots, 0), \\ q_j p_k &\leftrightarrow \vec{\lambda}_0^{(j,k)} = (0, \ldots, 0, -\tfrac{1}{2}, 0, \ldots, 0, +\tfrac{1}{2}, 0, \ldots, 0), \\ q_j q_k &\leftrightarrow \vec{\lambda}_-^{(j,k)} = (0, \ldots, 0, -\tfrac{1}{2}, 0, \ldots, 0, -\tfrac{1}{2}, 0, \ldots, 0). \end{aligned} \tag{11.84}$$

When $j = k$, the two components in the vector add, so $(\vec{\lambda}_\pm^{(j,j)})_k = \pm\delta_{j,k}$ and $\vec{\lambda}_0^{(j,j)} = \vec{0}$. Finally, since the d_ℓ's commute, the N weight generators can be thought as N points at the origin of R^N. In Fig. 11.1 we show the Cartan root diagrams for the algebras $C_N := \mathsf{sp}(2N, \mathsf{R})$ for $N = 1, 2, 3$, which we now proceed to explain.

Remark: Cartan root diagram for $\mathsf{sp}(2, \mathsf{R})$. For rank $N = 1$, the Cartan root diagram of $\mathsf{sp}(2, \mathsf{R})$ is 1-dimensional; the generators have no (j,k) subindices, and the eigenvalues are

$$p^2 \leftrightarrow \lambda = +1, \qquad qp \leftrightarrow \lambda = 0, \qquad q^2 \leftrightarrow \lambda = -1. \tag{11.85}$$

The Cartan diagram is shown in Fig. 11.1a. Since root eigenvectors of the algebra can be multiplied by any nonzero constants (including i's), this Cartan diagram describes *all* algebras of 3 generators and rank 1, i.e., $A_1 = \mathsf{su}(2)$ and $\mathsf{su}(1,1)$, $B_1 = \mathsf{so}(3)$ and $\mathsf{so}(2,1)$, and $C_1 = \mathsf{usp}(1)$. •

[10] To distinguish between the various solutions ρ and λ, these symbols will be later loaded with indices.

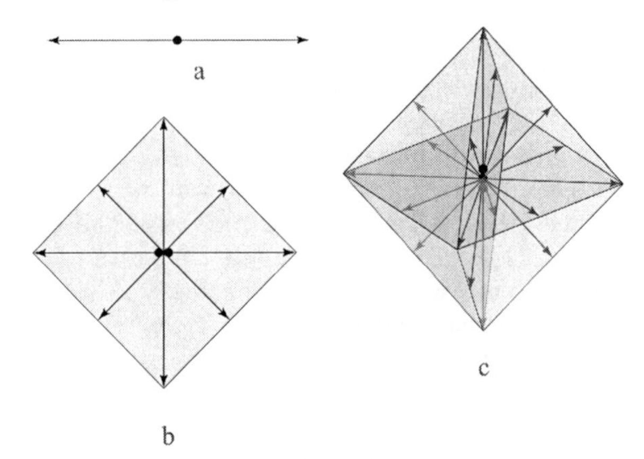

Fig. 11.1. Cartan diagrams for the symplectic algebras C_N, $N = 1, 2, 3$, where arrows represent the generators with eigenvalue N-vectors $\vec{\lambda}$ and dots stand for the N weight generators at origin. (**a**) $\mathsf{sp}(2, \mathsf{R})$ with 3 generators, (**b**) sp4 with 10 generators, and (**c**) $\mathsf{sp}(6, \mathsf{R})$ with 21 generators.

In the real paraxial optical world we are interested in $\mathsf{sp}(4, \mathsf{R})$, whose 8 roots and 2 weight generators are shown in Fig. 11.1b. The correspondence between root monomials and their weight vectors $\vec{\lambda} = (\lambda_1, \lambda_2)$, is

$$\begin{array}{ccc}
\boxed{\begin{array}{cc} p_1^2 & p_1p_2 \\ & p_2^2 \end{array}} & \boxed{\begin{array}{cc} q_1p_1 & q_1p_2 \\ q_2p_1 & q_2p_2 \end{array}} & \boxed{\begin{array}{cc} q_1^2 & q_1q_2 \\ & q_2^2 \end{array}} \\
\updownarrow & \updownarrow & \updownarrow \\
\boxed{\begin{array}{cc} (+1,0) & (+\frac{1}{2},+\frac{1}{2}) \\ & (0,+1) \end{array}} & \boxed{\begin{array}{cc} (0,0) & (-\frac{1}{2},+\frac{1}{2}) \\ (+\frac{1}{2},-\frac{1}{2}) & (0,0) \end{array}} & \boxed{\begin{array}{cc} (-1,0) & (-\frac{1}{2},-\frac{1}{2}) \\ & (0,-1) \end{array}}
\end{array} \tag{11.86}$$

The length of the roots, $|\vec{\lambda}|^2 = \lambda_1^2 + \lambda_2^2$ visible in Fig. 11.1b, distinguishes between *two* kinds of roots,

$$\begin{array}{lll} 4 \text{ `long' roots:} & |\vec{\lambda}| = 1, & \text{the } p_k^2\text{'s and } q_k^2\text{'s,} \\ 4 \text{ `short' roots:} & |\vec{\lambda}| = \frac{1}{\sqrt{2}}, & \text{all } j \neq k \text{ generators.} \end{array} \tag{11.87}$$

Remark: Lie brackets between roots. In the Cartan root diagram we can *see* the structure of the classical Lie algebras, because the weight vector of the commutator of two roots is the sum of the weight vectors of the roots, *viz.*,

$$\left.\begin{array}{l} \{\vec{d}, \rho\} = \vec{\lambda}\,\rho \\ \{\vec{d}, \tau\} = \vec{\mu}\,\tau \end{array}\right\} \quad \Rightarrow \quad \{\vec{d}, \{\rho, \tau\}\} = (\vec{\lambda} + \vec{\mu})\,\{\rho, \tau\} \tag{11.88}$$

where $\vec{d} = (d_1, d_2)$. So, when two roots in the Cartan diagram add to a root which is in the diagram, the latter is their Lie bracket (Poisson bracket or commutator). When the two roots add to zero, their bracket will be a linear combination of the weights, and if their sum does not fall on a root or weight, the two commute. •

Remark: Subalgebras of sp(4, R). Two opposite roots $\pm\rho$ of the Cartan diagram for sp(4, R), together with a linear combination of the weights d_ℓ, fall into a line which indicates a subalgebra $\mathsf{sp}(2,\mathsf{R}) \subset \mathsf{sp}(4,\mathsf{R})$, because of (11.88). In Fig. 11.1b, the two orthogonal lines through opposite vertices of the rhombus correspond to two mutually commuting sp(2, R) subalgebras of 'long' roots. There are also two sp(2, R)'s in the two pairs of 'short' roots, but these do not mutually commute. Subsets of sp(4, R) generators that form useful subalgebras are:[11]

SUBALGEBRA OF sp(4, R)	GENERATOR FUNCTIONS	
$\mathsf{sp}(2,\mathsf{R})_1 \oplus \mathsf{sp}(2,\mathsf{R})_2$	$p_1^2,\ p_1q_1,\ q_1^2;\quad p_2^2,\ p_2q_2,\ q_2^2;$	(11.89)
$\mathsf{so}(2)_\theta \oplus \mathsf{sp}(2,\mathsf{R})_r$	$l_{1,2} = q_1p_2 - q_2p_1;\quad \lvert\mathbf{p}\rvert^2,\ \mathbf{p}\cdot\mathbf{q},\ \lvert\mathbf{q}\rvert^2;$	(11.90)
FOURIER u(2)	$l_{1,2};\ f_{j,k} = p_jp_k + q_jq_k,\ 1\le j\le k\le 2.$	(11.91)

The subalgebra (11.89) is plainly visible in Fig. 11.1b; it generates *separable* astigmatic systems whose elements have axes aligned with the x–y Cartesian coordinates. But not every subalgebra of sp(4, R) can be readily seen in the Cartan root diagram: the other two subalgebras are built with linear combinations of roots in the diagram and do not coincide with the vertices of the rhombus or its midpoints. The subalgebra (11.90) generates paraxial axially-symmetric systems; when the system is symmetric under reflection, $(p_1, q_1) \mapsto (-p_1, -q_1)$, as optical systems are (but not magnetic ones), the generator $l_{1,2}$ of screen rotations must be removed from the list of generators, because it changes sign under reflection. Finally, the group of fractional U(2)-Fourier transformers is generated by (11.91). •

Remark: Cartan root diagram for sp(6, R). For sp(6, R) and its 21 generators, the 3-dim Cartan root diagram shown in Fig. 11.1c. We see a octahedron, with 6 vertices (length 1), 12 midpoints of edges (length $\frac{1}{\sqrt{2}}$), and the three weight operators at the center. The three orthogonal lines through the center and vertices indicate the reduction $\mathsf{sp}(6,\mathsf{R}) \supset \mathsf{sp}(2,\mathsf{R})_1 \oplus \mathsf{sp}(2,\mathsf{R})_2 \oplus \mathsf{sp}(2,\mathsf{R})_3$, while the three orthogonal planes that contain the vertices indicate the sp(4, R) subalgebras that are obtained when we supress the coordinate number 1, 2 or 3. There are four planes harboring u(3) subalgebras; one of them contains the roots $\{p_1p_2,\ p_1p_3,\ q_2p_3,\ q_3p_2,\ q_1q_2,\ q_1q_3\}$ arranged in a hexagon, and the three weights at the center. This shape must be known to the reader as the Eightfold Way of elementary particles [110], which asigns the eight baryions (and the controversial singlet) to the vertices and center of its Cartan root diagram. Since we disregard any i's in

[11] On the symbols of the subalgebras we add the subindices 1 and 2 refering to the Cartesian axes, while θ and r refer to angular and radial variables in polar coordinates.

these diagrams, we can be sure that in sp(6, R) these will be (see details below) noncompact u(2, 1)'s inside. Rotation-symmetric systems are treated with the reduction $\mathsf{sp}(6,\mathsf{R}) \supset \mathsf{so}(3)_{\theta,\phi} \oplus \mathsf{sp}(2,\mathsf{R})_r$, analogue of (11.90) in three dimensions. Finally, the reduction to infinitesimal orthosymplectic matrices $\mathsf{sp}(6,\mathsf{R}) \supset \mathsf{u}(3) \supset \mathsf{so}(3)$ in (11.74) has been used extensively in nuclear physics [100, 103]. •

The **multiplets** of a Lie algebra are closely related to its Cartan diagram of roots; they have the same axes to plot their weights. But the vectors to be represented, instead of being the roots of the algebra, are elements of a generally larger vector space, a homogeneous space where the algebra acts linearly. The name is borrowed directly from angular momentum and elementary particle theory, where multiplets are the (self-adjoint, irreducible) representations of the algebra [56]. Applications in geometric optics instead seem to require only *finite*-dimensional multiplets and, as other classical models, only the *totally symmetric* representations of the algebra – also called *most degenerate*, or *bosonic* representations.

The prototypical example of a multiplet occurs in the Lie algebra sp(2, R), whose generator functions are $\{\frac{1}{2}p^2,\ \frac{1}{2}pq,\ \frac{1}{2}q^2\}$ acting as Poisson operators on the space of homogeneous polynomials of fixed degree $\eta \in \{0,1,2,\ldots\} = \mathcal{Z}_0^+$ [i.e., $f(ap,aq) = a^\eta f(p,q)$]. There are $\eta+1$ monomials,

$$\{p^n q^m\}_{n,m\geq 0}^{n+m=\eta} = \{p^{\eta-m}q^m\}_{m=0}^{\eta} = \{p^\eta,\ p^{\eta-1}q,\ \ldots,\ pq^{\eta-1},\ q^\eta\}, \tag{11.92}$$

which form a natural basis for this vector space, and are a *multiplet* under sp(2, R) action. This action divides the generators (of *all* classical Cartan algebras) into three kinds:

$$\begin{aligned} \text{RAISING}: &\quad \tfrac{1}{2}\{p^2,\circ\}p^n q^m = -m\,p^{n+1}q^{m-1}, &(11.93)\\ \text{WEIGHT}: &\quad \tfrac{1}{2}\{pq,\circ\}p^n q^m = \tfrac{1}{2}(n-m)\,p^n q^m, &(11.94)\\ \text{LOWERING}: &\quad \tfrac{1}{2}\{q^2,\circ\}p^n q^m = n\,p^{n-1}q^{m+1}. &(11.95)\end{aligned}$$

Figure 11.2 shows the 1-dim weight diagram of the basis (11.92) obtained with the eigenvalues of (11.94).[12] We call $\{p^2,\circ\}$ a *raising* operator because it raises the weight $\lambda := \frac{1}{2}(n-m)$ of a monomial by one unit to $\lambda+1$; the multiplet, when finite, must have a member with the 'highest' weight $\lambda = \frac{1}{2}\eta$ for which $\{p^2,\circ\}p^\eta = 0$ prevents further raising. Correspondingly, $\{q^2,\circ\}$ is a *lowering* operator, since it lowers weights λ to $\lambda-1$, and down to a lowest weight $\lambda = -\frac{1}{2}\eta$. Generators which raise or lower the weights are also called *shift* generators; the Cartan basis separates weights from shifts, and shifts into raising and lowering generators. Finally, when there are no sub-multiplets invariant under the algebra, the multiplet is *irreducible*.

Remark: Symplectic spin. The finite irreducible multiplets of sp(2, R) and of su(2) and so(3) are the same. The $(\eta+1)$-dim multiplet of monomials in (11.93)–(11.95) will be said to have *symplectic spin* $s = \frac{1}{2}\eta$, –integer

[12] We usually omit – but do not forget – the rather trivial *singlet* $\eta = 0$.

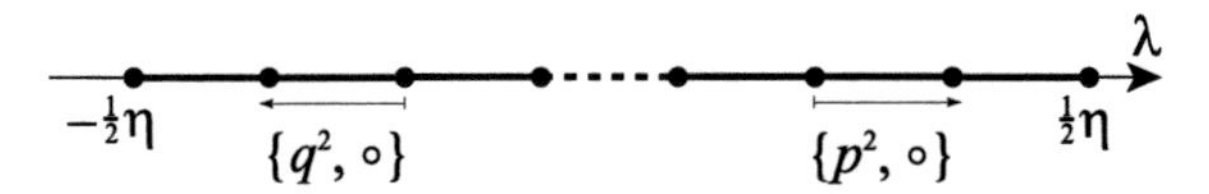

Fig. 11.2. Weight diagram for $\mathsf{sp}(2, \mathsf{R})$ multiplets. The axis marks the eigenvalues λ of the monomials (*heavy dots*) under the weight operator. The lowering and raising operators, $\{q^2, \circ\}$ and $\{p^2, \circ\}$, are represented by arrows that can shift between members of the multiplet, down $(\lambda-1 \leftarrow \lambda)$ and up $(\lambda \leftarrow \lambda+1)$; when acting on the lowest or the highest multiplet members respectively $(\lambda = \mp\frac{1}{2}\eta)$, they yield zero.

or half-integer. We thus speak of p, q as a spin-$\frac{1}{2}$ *doublet*, and of the three quadratic functions in (11.85) as an $s = 1$ *triplet*, etc. These multiplets transform under the finite-dimensional representations of the algebra $\mathsf{sp}(2, \mathsf{R})$ by $(2s + 1) \times (2s + 1)$ matrices (see Sect. 3.7). The theory of Lie group representations places much emphasis on vector spaces (finite- or infinite-dimensional) which are Hilbert spaces under some sesquilinear inner product, and which undergo *unitary* transformations; in the algebra [see (11.28)] this implies *skew-adjoint* representation matrices. There is a theorem which states that only compact groups can have finite-dimensional unitary representations [1, 22, 32]. The monomial multiplets introduced in (11.93)–(11.95) are finite-dimensional *non*-skew-adjoint representations of $\mathsf{sp}(2, \mathsf{R})$. With this connection in Part IV we shall use the well-known solid spherical harmonic functions $\mathcal{Y}^{\ell}_{m}(\vec{r})$ of $\vec{r} = (p^2, pq, q^2)$, to classify aberrations by their total (symplectic) spin ℓ and 'z'-projection m. •

We consider now the $\mathsf{sp}(4, \mathsf{R})$ multiplets (λ_1, λ_2) of monomials of degree $\eta \in \mathcal{Z}^+$ in four variables,

$$M^{\eta}(n_1, n_2; m_1, m_2) := p_1^{n_1} p_2^{n_2} q_1^{m_1} q_2^{m_2}, \qquad \begin{array}{c} 0 \le n_1, n_2, m_1, m_2 \le \eta, \\ n_1 + n_2 + m_1 + m_2 = \eta. \end{array} \tag{11.96}$$

On this basis, the Poisson operators of the weights $d_1 = \frac{1}{2}q_1p_1$ and $d_2 = \frac{1}{2}q_2p_2$ act as in (11.81)–(11.83), namely

$$\{d_1, \circ\}\, M^{\eta}(n_1, n_2; m_1, m_2) = \tfrac{1}{2}(n_1 - m_1)\, M^{\eta}(n_1, n_2; m_1, m_2), \tag{11.97}$$
$$\{d_2, \circ\}\, M^{\eta}(n_1, n_2; m_1, m_2) = \tfrac{1}{2}(n_2 - m_2)\, M^{\eta}(n_1, n_2; m_1, m_2), \tag{11.98}$$
$$\Rightarrow \quad M^{\eta}(n_1, n_2; m_1, m_2) \leftrightarrow \lambda_1 := \tfrac{1}{2}(n_1 - m_1), \ \lambda_2 := \tfrac{1}{2}(n_2 - m_2). \tag{11.99}$$

In Fig. 11.3 we show weight diagrams for multiplets of $\mathsf{sp}(4, \mathsf{R})$, noting that while λ_1 and λ_2 can be integers or half-integers, η is the integer degree of the monomial. For $\eta = 1$, there is the *fundamental* or '*quark*' multiplet.[13] For $\eta = 2$, the weight diagram of quadratic functions looks identical to (but is conceptually distinct from) the Cartan root diagram. For larger η's the $\mathsf{sp}(4, \mathsf{R})$

---

[13] We shall justify below that the name *quark* can be borrowed from elementary particle theory.

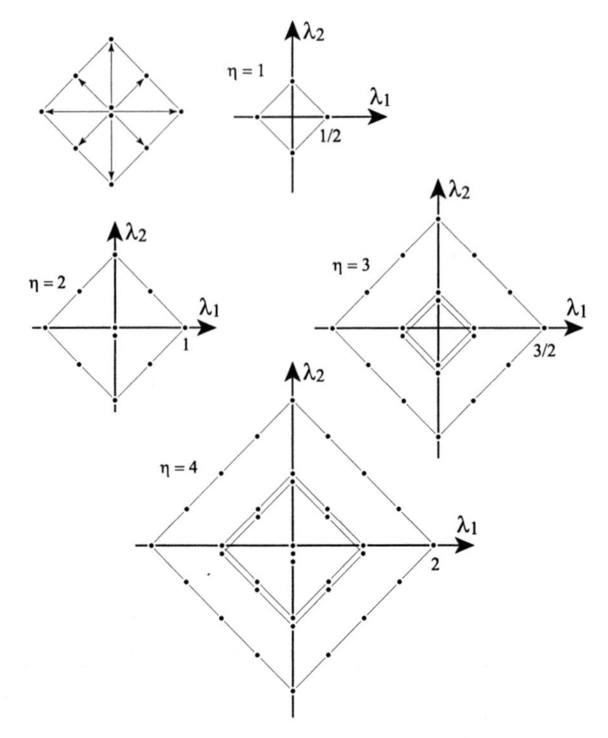

Fig. 11.3. Weight diagrams for $\mathsf{sp}(4, \mathsf{R})$ multiplets of degrees $\eta = 1, \ldots, 4$ and its root diagram (*center*). The two axes λ_1, λ_2 mark the eigenvalues of the monomials (*heavy dots*) under the two weight operators of $\mathsf{sp}(4, \mathsf{R})$. Starting with $\eta = 3$ there is degeneracy. The 8 arrows of the root diagram shift between the members of the multiplet.

multiplets remain arranged into larger rhombi, as shown in the figures. The roots of $\mathsf{sp}(4, \mathsf{R})$ (indicated by the eight arrows) shift between neighboring monomials in the multiplet, or yield zero when the arrows would cross the boundary of the multiplet, at $|\lambda_1| + |\lambda_2| = \frac{1}{2}\eta$.

Remark: Multiplets of aberrations. In Part IV we shall use the multiplets of phase space monomials $M^\eta(n_1, n_2; m_1, m_2)$ of degree η [in (11.96)] to generate optical *aberrations* of order $\eta - 1 \geq 2$ as nonlinear transformations of phase space,

$$\begin{aligned} &\{M(n_1, n_2; m_1, m_2), \circ\}\, f(\mathbf{p}, \mathbf{q}) \\ &= \left[m_1\, M(n_1, n_2; m_1 - 1, m_2) \frac{\partial}{\partial p_1} + m_2\, M(n_1, n_2; m_1, m_2 - 1) \frac{\partial}{\partial p_2} \right. \\ &\qquad \left. - n_1\, M(n_1 - 1, n_2; m_1, m_2) \frac{\partial}{\partial q_1} - n_2\, M(n_1, n_2 - 1; m_1, m_2) \frac{\partial}{\partial q_2} \right] f(\mathbf{p}, \mathbf{q}). \end{aligned} \tag{11.100}$$

The Poisson operators of the monomials, $\{M^\eta, \circ\}$, form multiplets under $\mathsf{sp}(4,\mathsf{R})$ because the eigenvalue equations (11.97)–(11.97) hold for the operators as they do for the monomials: $\{d_\ell, \circ\}\{M^\eta, \circ\} = \{\{d_\ell, M^\eta\}, \circ\}$. •

Remark: Degeneracy! As can be seen in Fig. 11.3, there is *degeneracy* in the $\mathsf{sp}(4,\mathsf{R})$ weight diagrams for monomials of degree η in (11.97)–(11.98), because more than one monomial can fall on the same point,

$$(\lambda_1, \lambda_2) \leftarrow \{\, p_1^{2\lambda_1}\, p_2^{2\lambda_2}\, (p_1q_1)^{\kappa-\nu}\, (p_2q_2)^{\nu}\,\}_{\nu=0}^{\kappa}, \quad \kappa := \tfrac{1}{2}\eta - \lambda_1 - \lambda_2. \quad (11.101)$$

These are all monomials which can be obtained from each other by replacing factors $p_1q_1 \leftrightarrow p_2q_2$. In Figs. 11.3, the outer rim of the multiplet is $\kappa = 0$ and corresponds to single monomial without degeneracy (i.e., degeneracy 1); successive concentric rhombi are numbered by $\kappa = 1, 2, \ldots$, ending with $\kappa = \frac{1}{2}\eta$ or $\frac{1}{2}(\eta - 1)$, according to whether the degree of homogeneity η is even or odd. When η is even, at the center of the weight diagram of the multiplet lie the $\frac{1}{2}\eta + 1$ monomials $(p_1q_1)^n (p_2q_2)^m$ with $n + m = \frac{1}{2}\eta$; when η is odd ('spinor' representations) there is no point at the origin, but the four points around it have the maximal degeneracy $\frac{1}{2}(\eta + 1)$. •

Remark: Resolution of degeneracy. To resolve the degeneracy in the weight diagram of a multiplet, we need some other atribute or label to distinguish between the phase space monomials, and generally between the abstact basis vectors of the representation space. The monomials of degree η are a particularly useful realization because their explicit form (11.101) shows that the degree of homogeneity in the coordinates p_1, q_1 is $\eta_1 := 2(\lambda_1 + \kappa - \nu)$ [and degree $\eta_2 := 2(\lambda_2 + \kappa - \nu)$ in p_2, q_2]. These $\kappa + 1$ degenerate states numbered by $0 \le \nu \le \kappa$ exhibit the range $\eta_1 \in \{2\lambda_1, 2\lambda_1 + 2, \ldots, 2(\lambda_1 + \kappa) = \eta - 2\lambda_2\}$. Each of these monomials therefore belongs to a distinct multiplet under the subalgebra $\mathsf{sp}(2,\mathsf{R})_1 \subset \mathsf{sp}(4,\mathsf{R})$ that involves only the first coordinate. This situation has been described for $\mathsf{sp}(2,\mathsf{R})$ in (11.93)–(11.95), and applies here replacing $\eta \mapsto \eta_1$. As shown in Fig. 11.4, the $(\kappa + 1)$-fold degeneracy at the point (λ_1, λ_2) of the $\mathsf{sp}(4,\mathsf{R})$ multiplet is thus resolved by asigning each monomial to a unique $\mathsf{sp}(2,\mathsf{R})$ *sub-multiplet* (with $\eta_1 + 1$ partners), of symplectic spin $s_1 = \frac{1}{2}\eta_1$. The weight diagram of the multiplet thus acquires an extra axis η_1 that resolves the degeneracy completely. •

Remark: Ordering a multiplet by weights. We establish an order relation between the weights of the members of $\mathsf{sp}(4,\mathsf{R})$ multiplets: (λ_1, λ_2) is said to be *higher* than (λ_1', λ_2') when $\lambda_1 > \lambda_1'$ or, when $\lambda_1 = \lambda_1'$, then $\lambda_2 > \lambda_2'$. This relation separates, out of 10 $\mathsf{sp}(4,\mathsf{R})$ generators, the 4 that raise it (i.e., p_1^2, p_1p_2, p_2^2, p_1q_2), from the 4 that lower it (i.e., q_1^2, q_1q_2, q_2^2, p_2q_1). The 'lexicographic' order of λ_j's can be extended straightforwardly to any number of them. •

Remark: Other representations and weight diagrams. It is not surprising that geometric optics cannot realize *all* multiplets of $\mathsf{sp}(4,\mathsf{R})$ with the ordinary phase space variables p_j and q_k. Multiplets of $\mathsf{sp}(4,\mathsf{R})$ exist which

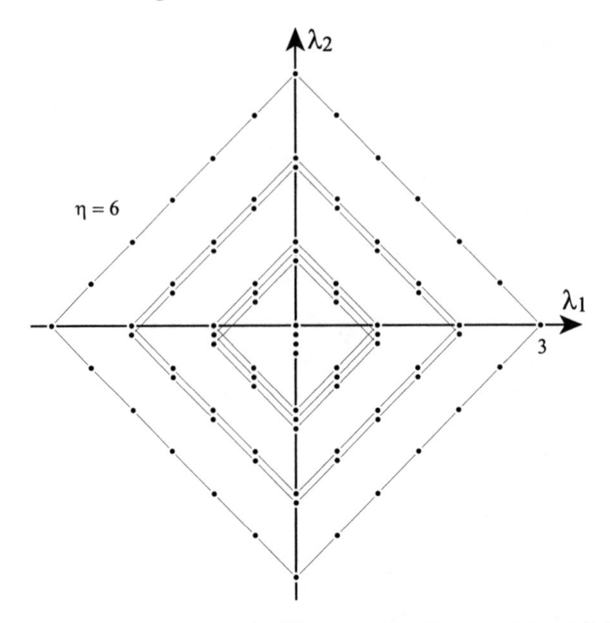

Fig. 11.4. Weight diagram of the $\mathsf{sp}(4, \mathsf{R})$-multiplet of monomials with homogeneity degree $\eta = 6$.

have weight diagrams of generally octagonal shape [56], whose sides alternate two unequal lengths. They are built with 'monomials' whose factors participate in Fermi statistics, or have definite intermediate permutational symmetry [100]. And it is *not* generally true that the subalgebra $\mathsf{sp}(2, \mathsf{R})_1 \subset \mathsf{sp}(4, \mathsf{R})$ resolves all degeneracies. The geometric optical (or classic mechanical) realization works with ordinary functions which commute as boson fields do; they thus yield only the *totally symmetric* representations of the algebra which are shown in the figures. The monomials are uniquely characterized by the degree of homogeneity η; this remark applies to all the multiplets under discussion. •

11.6 Roots and multiplets of the unitary algebras

The Cartan family of unitary algebras $A_{N-1} = \mathsf{su}(N)$ generates the Fourier transformations of geometric optics. This family has collected fame in physics, where it is extensively applied in nuclear and elementary particle models. Here we build the root diagrams and multiplets of the Lie algebras $\mathsf{su}(N)$ and $\mathsf{u}(N) = \mathsf{u}(1) \oplus \mathsf{su}(N)$, realized by monomial functions of the paraxial phase space coordinates. Under Poisson brackets, the N weight generator functions of $\mathsf{u}(N)$, $f_{k,k} = p_k^2 + q_k^2$, $k = 1, \ldots, N$ in (11.91) form a maximal set of weights which correspond to N orthogonal 1-dim harmonic guide/oscillator

Hamiltonians.[14] The rest of the generator functions of $\mathsf{u}(N)$ are the $l_{j,k}$'s and $f_{j,k}$'s for $1 \le j < k \le N$ [(11.74) and (11.74)]. The unitary algebras are well served by the complex *Bargmann* phase space coordinates[15]

$$z_k := \tfrac{1}{\sqrt{2}}(q_k + i p_k), \qquad a_k := \tfrac{1}{\sqrt{2}}(i q_k + p_k) = i z_k^*, \\ \{q_k, p_k\} = 1 \Leftrightarrow \{z_k, a_k\} = 1. \tag{11.102}$$

We note that the map $(p,q) \mapsto (a,z)$ is a *complex* canonical transformation of phase space [82] (see also [149, Chaps. 9 and 10]).

The N^2 quadratic monomials in the complex coordinates are

$$z_j\, a_k = \tfrac{1}{2}(q_j + i p_k)(i q_k + p_k) = \tfrac{1}{2} l_{j,k} + i \tfrac{1}{2} f_{j,k}. \tag{11.103}$$

The action of the N weight functions $z_j\, a_j = i\frac{1}{2} f_{j,j}$ on the monomials is

$$\{z_j a_j, \circ\}\, z_k a_\ell = (\delta_{j,\ell} - \delta_{j,k})\, z_k a_\ell, \tag{11.104}$$

and thus they are roots of the unitary algebras [cf. (11.97)–(11.98) for the symplectic algebras].

Remark: The case of $\mathsf{u}(2)$. When $j,k,\ell \in \{1,2\}$, the 4 generator functions of $\mathsf{u}(2)$ contain 2 weights, which asign to the monomials $z_k\, a_\ell$ the following root vectors

$$\begin{array}{ccccc} \text{GENERATORS} & & \text{ROOTS } (\lambda_1,\lambda_2) & & \mu := \tfrac{1}{2}(\lambda_1 - \lambda_2) \\ \boxed{\begin{array}{cc} z_1 a_1 & z_2 a_1 \\ z_1 a_2 & z_2 a_2 \end{array}} & \longleftrightarrow & \boxed{\begin{array}{cc} (0,0) & (+1,-1) \\ (-1,+1) & (0,0) \end{array}} & & \boxed{\begin{array}{cc} 0 & +1 \\ -1 & 0 \end{array}} \end{array} \tag{11.105}$$

Notice that the roots of $\mathsf{u}(2)$ all lie on the line $\lambda_1 + \lambda_2 = 0$, as shown in Fig. 11.5a. This reveals that $\mathsf{u}(2)$ has a central (i.e., Poisson-commuting) generator of $\mathsf{u}(1)$, $z_1 a_1 + z_2 a_2$, which may be taken out of the picture; this leaves $\frac{1}{2}(z_1 a_1 - z_2 a_2)$ as the single weight for $A_1 = \mathsf{su}(2)$, with eigenvalues $\mu \in \{-1, 0, +1\}$. •

Remark: The case of $\mathsf{u}(3)$. For $j,k,\ell \in \{1,2,3\}$ there are 9 generator functions $z_j a_k$ of $\mathsf{u}(3)$; the 3 weight functions and the 6 roots in the Cartan diagram are

$$\begin{array}{ccc} \text{GENERATORS} & & (\lambda_1,\lambda_2,\lambda_3) \\ \boxed{\begin{array}{ccc} z_1 a_1 & z_2 a_1 & z_3 a_1 \\ z_1 a_2 & z_2 a_2 & z_3 a_2 \\ z_1 a_3 & z_2 a_3 & z_3 a_3 \end{array}} & \leftrightarrow & \boxed{\begin{array}{ccc} (0,\,0,\,0) & (+1,-1,0) & (+1,0,-1) \\ (-1,+1,0) & (0,\,0,\,0) & (0,+1,-1) \\ (-1,0,+1) & (0,-1,+1) & (0,\,0,\,0) \end{array}} \end{array} \tag{11.106}$$

[14] We gloss over the factor $\frac{1}{2}$ that must be present in the true Hamiltonian function of a harmonic guide.

[15] The names of *annihilation* and *creation* are recognized and justified for operators corresponding to z_j and a_k, which annihilate and create energy quanta in a harmonic oscillator. *Lowering* and *raising* are also widely used in mathematics. See Bargmann's early work in Ref. [16] which provides a proper Hilbert space for the quantum mechanical version.

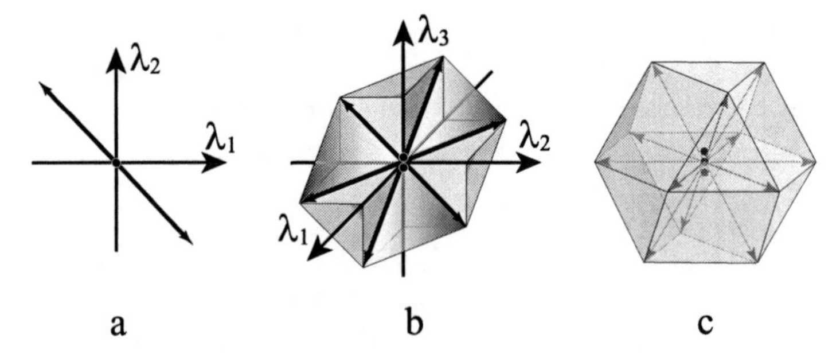

Fig. 11.5. Cartan root diagrams for the unitary algebras $A_N = \mathsf{su}(N{+}1)$. (**a**) $\mathsf{u}(2)$, (**b**) $\mathsf{u}(3)$, and (**c**) $\mathsf{su}(4)$. Since the figures lie in the subspaces $\sum_{i=1}^N \lambda_i = 0$, for $N = 4$ we show the 3-dim figure spanned by the generators.

Since $\lambda_1 + \lambda_2 + \lambda_3 = 0$ the Cartan roots shown in Fig. 11.5b are on a plane: that of $A_2 = \mathsf{su}(3)$. In the figure it is evident that $\mathsf{su}(3)$ contains three $\mathsf{su}(2)$ subalgebras on lines at 60° to each other. There is also a rather important and nontrivial subalgebra $\mathsf{so}(3) \subset \mathsf{su}(3)$ [103], whose generators are $l_{j,k} = z_j a_k + (z_j a_k)^* = z_j a_k - z_k a_j$, $1 \le j < k \le 3$, and which is not (directly) visible in the diagram. •

Remark: Traceless generators for $A_{N-1} = \mathsf{su}(N)$. The weight vectors of the $\mathsf{u}(N)$ roots lie on an $(N{-}1)$-dim plane within the N-dim $\vec{\lambda}$-space of their Cartan diagram, as can be seen from the trace function

$$f_{\rm tr} := \sum_{j=1}^{N} z_j a_j = |\mathbf{p}|^2 + |\mathbf{q}|^2, \qquad \{f_{\rm tr}, z_j a_k\} = 0 \Rightarrow \sum_{j=1}^{N} \lambda_j = 0. \tag{11.107}$$

[*Cf.* (11.77).] By subtracting this trace we obtain $N-1$ new weight operators whose eigenvalues are the axes for diagrams of the *special* unitary algebra $\mathsf{su}(N)$ – with one dimension less. The root diagram of $\mathsf{su}(2)$ is shown in Fig. 11.5a (which is identical to Fig. 11.2 on the axis $\mu \in \{1, 0, -1\}$). In this way the root diagram of $A_3 = \mathsf{su}(4)$ is shown in a 3-dim space in Fig. 11.5c, as be detailed below. •

Remark: Symmetry of the cylindrical guide. The Hamiltonian function of a 3-dim cylindrical optical guide (or a mechanical 2-dim harmonic oscillator), is $h = \frac{1}{2}(|\mathbf{p}|^2 + |\mathbf{q}|^2) = -i \sum_k z_k a_k$, and generates the $\mathsf{U}(1)$ group of central (inverse) Fourier transforms as the detector screen moves along the guide (see Sect. 10.3). This Hamiltonian Poisson-commutes with the $\mathsf{su}(2)$ generators of rotation (the *manifest* symmetry of a cylindrical guide) and, separately, of fractional Fourier transforms along the two axes (the *hidden* symmetry modulo the manifest one). In other words, a $\mathsf{U}(2)$-transformed input evolves along the guide into the $\mathsf{U}(2)$-transformed output. On phase space, the $\mathsf{u}(2)$ algebra generates the real ortho-symplectic matrices of Sect. 9.3. •

Remark: Unitary symmetry in nuclei. The old single-particle shell model of nuclei (see *e.g.*, Ref. [117, Chap. 7]), postulates that nucleons are subject to a 3-dim harmonic oscillator collective potential. *Mutatis mutandis*, the remark above leads to recognize $\mathsf{su}(3)$ as the symmetry algebra of this system [100], which corresponds to a 4-dim cylindrical guide of spherical section. In the quantum model, the hint that there is hidden symmetry lies in the degeneracy $d_n := \frac{1}{2}n(n+1)$ of the energy levels $n = 1, 2, \ldots$, which gives rise to the superior stability of closed-shell nuclei with the 'magic numbers' of 1, 4, 10, 20,... spin-pairs of protons or of neutrons. The generators of the $\mathsf{su}(3)$ symmetry algebra are the transition operators that intertwine states with the same energy. •

Remark: The Eightfold Way. Figure 11.5b must remind the reader of the Eightfold Way of Gell-Mann and Ne'eman for baryons and for mesons. The root diagram in Fig. 11.5b strictly corresponds to the mesons consisting of one quark and one antiquark.[16] Defining the *isospin* projection I_z, *strangeness* S, and *charge* Q as used in particle physics,

$$I_z := \tfrac{1}{2}(\lambda_2 - \lambda_1), \quad S := \lambda_1 + \lambda_2 + 1, \quad Q := \tfrac{1}{2}(\lambda_2 + \lambda_3 - \lambda_1), \tag{11.108}$$

the array (11.106) of $\mathsf{su}(3)$ traceless generators has the following correspondence with the basic baryon and meson octets:

$$\{I_z,\, S,\, Q\}$$

$$\boxed{\begin{array}{ccc} \{0,\,0,\,0\} & \{+1,+1,+1\} & \{\frac{1}{2},0,+1\} \\ \{-1,+1,-1\} & \{0,\,0,\,0\} & \{-\frac{1}{2},\,0,\,0\} \\ \{-\frac{1}{2},-2,-1\} & \{\frac{1}{2},-2,\,0\} & \{0,\,0,\,0\} \end{array}} \tag{11.109}$$

MESONS

$$\boxed{\begin{array}{ccc} \omega/\pi^0 & \pi^+ & K^+ \\ \pi^- & \omega/\pi^0 & K^0 \\ \overline{K}^- & \overline{K}^0 & \omega/\pi^0 \end{array}}$$

BARYONS

$$\boxed{\begin{array}{ccc} \Lambda/\Sigma^0 & \Sigma^+ & p \\ \Sigma^- & \Lambda/\Sigma^0 & n \\ \Xi^- & \Xi^0 & \Lambda/\Sigma^0 \end{array}}$$

The degeneracy in the Λ/Σ^0 and ω/π^0 eigenvalues is resolved by reduction along the subalgebra chain $\mathsf{u}(3) \supset \mathsf{su}(3) \supset \mathsf{su}(2)_{\text{isospin}}$. •

Remark: The case of $\mathsf{su}(4)$. In Fig. 11.5c we show the Cartan root diagram for $\mathsf{su}(4)$, which the reader will have no trouble in constructing with (11.104). It contains three $\mathsf{su}(3)$'s on three orthogonal planes, and a nontrivial $\mathsf{so}(4)$. Eugene Wigner and many other nuclear physicists have used $\mathsf{su}(4)$

[16] Baryons consist of three quarks in [2,1] permutational state, and its multiplet diagram *does* have the same form. There is of course a crucial physical difference between the quarks fields of the Standard Model and the a_j's here: products of the former obey Fermi statistics, while the present ones (being functions of phase space) commute, as if Bose statistics ruled.

to accomodate the spin and isospin states of nucleons into compound nuclei having observable external quantum numbers, in representations which respect the permutational symmetry of the two kinds of fermions, protons and neutrons. In contrast, classical uses of the unitary algebras result in the totally symmetric representations which are characteristic of boson systems.

•

We now build the (*totally symmetric*) **multiplets** of the special unitary Lie algebras $A_{N-1} = \mathsf{su}(N)$. The case of $\mathsf{su}(2)$ is already documented for $\mathsf{sp}(2,\mathsf{R})$ in Fig. 11.2; the two algebras are different,[17] but their finite multiplets follow the same pattern.

In the symplectic algebras $\mathsf{sp}(2N,\mathsf{R})$ we constantly referred to the $2N$-dim vector space of phase space coordinates because it transforms under the fundamental $2N \times 2N$ representation of the algebra; monomials with powers of these variables form the symplectic multiplets. The same rôle is played in the unitary algebras $\mathsf{su}(N)$ by the multiplet of components of the N-vector of the complex coordinates of phase space, $\{z_j\}_{j=1}^N$ or $\{a_j\}_{j=1}^N$ in (11.102). The functions that generate $\mathsf{u}(N)$ in (11.103) act on them through the fundamental representation of the algebra, as follows:

$$\{z_j a_k, \circ\}\, z_\ell = -\delta_{k,\ell}\, z_j, \qquad \{z_j a_k, \circ\}\, a_\ell = \delta_{j,\ell}\, a_k. \tag{11.110}$$

The weight diagram of this fundamental multiplet ($\eta = 1$) of $\mathsf{su}(3)$ is shown by the back triangle in Fig. 11.6.

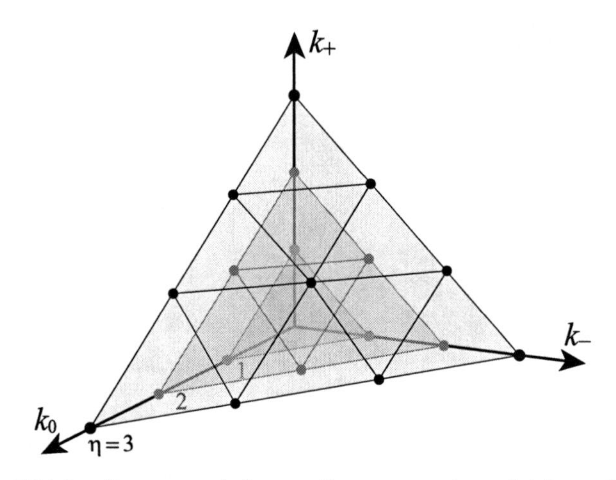

Fig. 11.6. Weight diagrams of the totally symmetric multiplets of $\mathsf{u}(3)$ lie on planes. Shown are the fundamental triplet $\eta = 1$, the $\eta = 2$ sextuplet of 3^d-order aberrations (Chap. 14), and the $\eta = 3$ decuplet of 5th-order aberrations.

---

[17] $\mathsf{su}(2)$ and $\mathsf{sp}(2,\mathsf{R})$ are related by the accidental homomorphism $\mathsf{su}(2) = \mathsf{usp}(1)$ and by the Weyl trick $\mathsf{usp}(1) \stackrel{\text{Weyl}}{\longrightarrow} \mathsf{sp}(2,\mathsf{R})$ in Sect. 11.3.

The (totally symmetric) multiplets of the unitary algebras $\mathsf{u}(N)$ are the monomials $z_1^{\eta_1} z_2^{\eta_2} \cdots z_N^{\eta_N}$ in the Bargmann coordinates of phase space (11.102). The weight diagrams can be found from

$$\{z_j a_j, \circ\} \prod_{k=1}^{N} z_k^{\eta_k} = -\eta_j \prod_{k=1}^{N} z_k^{\eta_k}, \qquad \sum_{k=1}^{N} \eta_k = \eta, \tag{11.111}$$

and are shown in Fig. 11.6 for $N = 3$ and $\eta = 2$ and 3. The form of these diagrams corresponds to those of the isotropic quantum oscillator states with $n = 1, 2, 3$ energy quanta, of elementary particles with 1, 2 (nonexistent), and 3 symmetric quarks (i.e., the decuplet of excited baryons which includes the Ω_-), and those of axis-symmetric aberrations of order 1, 3, and 5 (see Chap. 14). The shift generators of $\mathsf{u}(3)$ act on one monomial to yield another in the directions indicated by the roots of its Cartan diagram. The monomial is annihilated (mapped on 0) when the shift would land it outside the border of the multiplet.

Remark: Triality. The decuplet in Fig. 11.6 is famous because in the early 1960s it framed the prediction of the Ω^- particle and its properties: a totally symmetric three-quark state of spin $\frac{3}{2}$, isospin 0, and strangeness 2 [110] – and even its mass. Successive refinements of the theory have respected the basic 'three-ness' or *triality* of the original $\mathsf{su}(3)$, and the Standard Model keeps this philosophy. In Chap. 14 we shall use the $\mathsf{su}(3)$ triality to classify the functions that generate aberrations, which are monomials in the three axis-symmetric coordinates of phase space, [cf. Sect. 7.1, (7.3)],

$$M_{k_+,k_0,k_-} := \rho_+^{k_+} \rho_0^{k_0} \rho_-^{k_-}, \qquad \rho_+ := |\mathbf{p}|^2, \ \rho_0 := \mathbf{p}\cdot\mathbf{q}, \ \rho_- := |\mathbf{q}|^2. \tag{11.112}$$

Thus we shall classify aberrations by putting them in 1:1 correspondence with the states of the 3-dim quantum harmonic oscillator with k energy quanta.•

11.7 Roots of the orthogonal algebras

The orthogonal Lie algebras $\mathsf{so}(N)$ which generate rotations of R^N space, belong to the classical Cartan B- and D-families, according to whether N is odd or even. To find the maximal set of commuting weight generators we cannot use a simple subset of the weights of the unitary algebras, $f_{k,k}$ in (11.74), because they are *outside* the orthogonal subalgebras of generators $l_{j,k}$ (recall the discussion on page 214). We refer to Sect. 11.2, where $N \times N$ skew-symmetric matrices (11.29) realize the orthogonal algebras, and where we arranged the $\mathsf{so}(N)$ generators $\Lambda_{(\kappa,\lambda)}$ into the pattern (11.31), which applies to any realization.

The elements in the Lie algebra $\mathsf{so}(N)$ which generate rotations in planes κ–λ of a vector space $\vec{x} \in \mathsf{R}^N$, have a useful realization by self-adjoint differential operators acting on a Hilbert space of functions. It is (11.22) and

(11.29) multiplied by $-i$,[18]

$$J_{\kappa,\lambda} := -i\Lambda_{(\kappa,\lambda)} = -J_{\lambda,\kappa} = J^{\dagger}_{\kappa,\lambda} = -i(x_\kappa \partial_\lambda - x_\lambda \partial_\kappa). \tag{11.113}$$

They generate finite rotations of functions of $\vec{x} \in \mathsf{R}^N$ given by

$$\exp(i\theta J_{\kappa,\lambda})\, f(\vec{x}\,) = f(\vec{x}\,'), \quad \begin{array}{ccc} \begin{pmatrix} x'_\kappa \\ x'_\lambda \end{pmatrix} & = & \begin{pmatrix} \cos\theta & -\sin\theta \\ \sin\theta & \cos\theta \end{pmatrix} \begin{pmatrix} x_\kappa \\ x_\lambda \end{pmatrix}, \\ x'_\mu & = & x_\mu, \qquad \mu \neq \kappa, \lambda. \end{array} \tag{11.114}$$

Their Lie brackets (commutators) are found immediately from (11.33),

$$[J_{\mu,\nu}, J_{\kappa,\lambda}] = -i(\delta_{\nu,\kappa} J_{\mu,\lambda} + \delta_{\mu,\lambda} J_{\nu,\kappa} + \delta_{\kappa,\mu} J_{\lambda,\nu} + \delta_{\lambda,\nu} J_{\kappa,\mu}), \tag{11.115}$$

where again we appeal to the index mnemonics (11.37) with a $-i$ factor.

And now we choose the mutually commuting generators $J_{1,2}$, $J_{3,4}$, $J_{5,6}$, ... to be the weights in the generic $\mathsf{so}(N)$ pattern that we write explicitly for $\mathsf{so}(5)$ [*cf.* the pattern (11.31)]

$J_{1,2}$	$J_{1,3}$	$J_{1,4}$	$J_{1,5}$	...
	$J_{2,3}$	$J_{2,4}$	$J_{2,5}$	...
		$J_{3,4}$	$J_{3,5}$	...
			$J_{4,5}$	...

(11.116)

We shall relate the generators $J_{\mu,\nu}$ defined above, with the Cartan weight and roots of the algebra. To start, we examine the algebra $B_1 = \mathsf{so}(3)$ in the subspace of indices 1,2,3, which occupies the two first columns of the pattern (11.116). We have already chosen $J_{1,2}$ to be the weight, so we set conditions for the linear combinations $\alpha_\pm J_{1,3} + \beta_\pm J_{2,3}$ to have eigenvalues ± 1,

$$[J_{1,2},\, \alpha_\pm J_{1,3} + \beta_\pm J_{2,3}\,] = \pm(\alpha_\pm J_{1,3} + \beta_\pm J_{2,3}) \Rightarrow \beta_\pm = \pm i\alpha_\pm, \tag{11.117}$$

$$[\alpha_+(J_{1,3} + iJ_{2,3}),\, \alpha_-(J_{1,3} - iJ_{2,3})\,] = 2\alpha_+\alpha_-\, J_{1,2} \Rightarrow \alpha_+\,\alpha_- = \tfrac{1}{2}, \tag{11.118}$$

and we are free to choose $\alpha_\pm = \frac{1}{\sqrt{2}}$. Within $\mathsf{so}(N)$, an $\mathsf{so}(3)$ subalgebra can involve any three orthogonal axes μ, ν, κ, which will denote the generic $\mathsf{so}(3)$ *raising* ($\uparrow$) and *lowering* ($\downarrow$) generators:

[18] The $N \times N$ matrix representation $\mathbf{\Lambda}$ of $\mathsf{so}(N)$ is skew-symmetric and thus has pure imaginary eigenvalues; but these must be real to be physical observables. When we multiply the Λ's by $-i$, the J's form the angular momentum tensor of N-dim quantum mechanics, $\vec{J} = \vec{q}\wedge\vec{p}$. The $\mathsf{so}(3)$ Cartesian basis of quantum angular momentum theory is $\hat{J}_j \stackrel{\text{def}}{=} J_{k,\ell}$ (j, k, ℓ cyclic permutations of 1,2,3); its commutation relations are $[\hat{J}_j, \hat{J}_k] = i\hat{J}_\ell$.

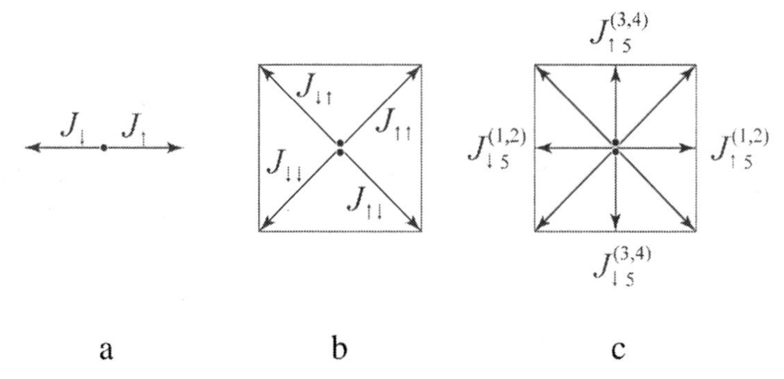

Fig. 11.7. Cartan diagrams for the orthogonal algebras of ranks 1 and 2. (**a**) $B_1 = \mathsf{so}(3)$, (**b**) $D_2 = \mathsf{so}(4)$, and (**c**) $B_2 = \mathsf{so}(5)$ – compare with the Cartan root diagram of $\mathsf{sp}(4,\mathsf{R})$ in Fig. 11.1b. The root vectors are represented by arrows while the weights $(J_{1,2},\ J_{3,4})$ are dots at the center.

$$J^{(\mu,\nu)}_{\updownarrow,\kappa} := \tfrac{1}{\sqrt{2}}(J_{\mu,\kappa} \pm iJ_{\nu,\kappa}) \Longrightarrow \begin{cases} [J_{\mu,\nu}, J^{(\mu,\nu)}_{\updownarrow,\kappa}] = \pm J^{(\mu,\nu)}_{\updownarrow,\kappa}, \\ [J^{(\mu,\nu)}_{\uparrow,\kappa}, J^{(\mu,\nu)}_{\downarrow,\kappa}] = J_{\mu,\nu}. \end{cases} \tag{11.119}$$

In Fig. 11.7a we show the 1-dim Cartan diagram for $B_1 = \mathsf{so}(3)$ (which is identical with that of $C_1 = \mathsf{usp}(1) \overset{\text{Weyl}}{\longrightarrow} \mathsf{sp}(2,\mathsf{R})$ in Fig. 11.1a) and that of $A_1 = \mathsf{su}(2)$ in Fig. 11.5a. The roots have Euclidean length 1.

Remark: The isomorphism $\mathsf{su}(2) = \mathsf{so}(3)$. This is the classic connection between the generators of spin and those of angular momentum. Since we know the relation between the matrices of the groups from (9.56)–(9.57), it is most straightforward to investigate elements near to the identity [as in (11.73)], namely $\mathbf{U} \approx \mathbf{1} + \varepsilon\,\mathbf{u}$, with complex parameters α, β constrained by $|\alpha|^2 + |\beta|^2 = 1$, and approximated by $\alpha \approx 1 + \varepsilon\,(a^0 + ia^1)$ and $\beta \approx \varepsilon(b^0 + ib^1)$ for $\varepsilon^2 \approx 0$. From $\det \mathbf{U} = 1 \Rightarrow \operatorname{tr}\mathbf{u} = 0$, it follows that

$$\begin{aligned} \mathbf{u} &= \begin{pmatrix} ia^0 & b^0 + ib^1 \\ -b^0 + ib^1 & -ia^0 \end{pmatrix} \in \mathsf{su}(2) \quad \leftrightarrow \\ \mathbf{D}^{\mathrm{so}(3)}(\mathbf{u}) &= 2\begin{pmatrix} 0 & a^1 & -b^0 \\ -a^1 & 0 & b^1 \\ b^0 & b^1 & 0 \end{pmatrix} \in \mathsf{so}(3). \end{aligned} \tag{11.120}$$

We note the factor of 2 in front of the $\mathsf{so}(3)$ matrix; this tells us that when a (fixed) $\mathsf{su}(2)$ element generates a closed loop in the spin group, the corresponding $\mathsf{so}(3)$ element will have completed its loop twice in the rotation group. Yet, the distinction between the two *algebras* is a subtle one which is most clearly seen in their *realizations*. Multiplets of $\mathsf{su}(2)$ [see (11.111) for $z_1^{\eta-\eta_2} z_2^{\eta_2}$, $\eta = 0, 1, 2, \ldots$] can have an even or odd number of members, while the multiplets of $\mathsf{so}(3)$ (built with powers of x_1, x_2, x_3) can only have odd numbers, i.e., integer spin. •

The generators of the Lie algebra $D_2 = \mathsf{so}(4)$ occupy the first three columns in the pattern (11.116); it has rank 2 and hence there are 2 weight generators: $J_{1,2}$ and $J_{3,4}$. The eigenvectors of the first are $J^{(1,2)}_{\updownarrow,3}$ and $J^{(1,2)}_{\updownarrow,4}$ in (11.119); to find the simultaneous eigenvectors of the second, we use the same raising/lowering technique as above, but now for the weight $J_{3,4}$, and the linear combinations $\gamma J_{\rho,3} + \delta J_{\rho,4}$ for roots. We thus write the $\mathsf{so}(4)$ complement to (11.117)–(11.118) as

$$[J_{3,4},\, \gamma_\pm J_{\rho,3}+\delta_\pm J_{\rho,4}] = \pm(\gamma_\pm J_{\rho,3}+\delta_\pm J_{\rho,4}) \Rightarrow \delta_\pm = \pm i\gamma_\pm, \quad (11.121)$$

$$[\gamma_+ (J_{\rho,3}+iJ_{\rho,4}),\, \gamma_- (J_{\rho,3}-iJ_{\rho,4})] = 2\gamma_+\gamma_-\, J_{3,4} \Rightarrow \gamma_+\,\gamma_- = \tfrac{1}{2}. \quad (11.122)$$

Again our choice of phase will be $\gamma_\pm = \frac{1}{\sqrt{2}}$.

We note that the 'row' index ρ in (11.122) can stand for 1,2 (generically μ, ν), and for any linear combination of generators, such as $\uparrow, \downarrow$. We thus build the 4 shift generators belonging to the subalgebra $\mathsf{so}(4) \subset \mathsf{so}(N)$ which generates rotations in the 4-dim subspace of indices $\mu, \nu, \kappa, \lambda$, and label them

$$J^{\binom{\mu,\nu}{\kappa,\lambda}}_{\uparrow,\uparrow} := \frac{1}{\sqrt{2}}(J^{(\mu,\nu)}_{\uparrow,\kappa} + iJ^{(\mu,\nu)}_{\uparrow,\lambda}) = \tfrac{1}{2}(J_{\mu,\kappa} - J_{\nu,\lambda} + iJ_{\nu,\kappa} + iJ_{\mu,\lambda}), \quad (11.123)$$

$$J^{\binom{\mu,\nu}{\kappa,\lambda}}_{\uparrow,\downarrow} := \frac{1}{\sqrt{2}}(J^{(\mu,\nu)}_{\uparrow,\kappa} - iJ^{(\mu,\nu)}_{\uparrow,\lambda}) = \tfrac{1}{2}(J_{\mu,\kappa} + J_{\nu,\lambda} + iJ_{\nu,\kappa} - iJ_{\mu,\lambda}), \quad (11.124)$$

$$J^{\binom{\mu,\nu}{\kappa,\lambda}}_{\downarrow,\uparrow} := \frac{1}{\sqrt{2}}(J^{(\mu,\nu)}_{\downarrow,\kappa} + iJ^{(\mu,\nu)}_{\downarrow,\lambda}) = \tfrac{1}{2}(J_{\mu,\kappa} + J_{\nu,\lambda} - iJ_{\nu,\kappa} + iJ_{\mu,\lambda}), \quad (11.125)$$

$$J^{\binom{\mu,\nu}{\kappa,\lambda}}_{\downarrow,\downarrow} := \frac{1}{\sqrt{2}}(J^{(\mu,\nu)}_{\downarrow,\kappa} - iJ^{(\mu,\nu)}_{\downarrow,\lambda}) = \tfrac{1}{2}(J_{\mu,\kappa} - J_{\nu,\lambda} - iJ_{\nu,\kappa} - iJ_{\mu,\lambda}). \quad (11.126)$$

For the $\mathsf{so}(4)$ algebra in the first three columns of the pattern (11.116), we may omit the systematic (but awkward) super-index $\binom{\mu,\nu}{\kappa,\lambda} = \binom{1,2}{3,4}$, to write the $\mathsf{so}(4)$ commutation relations in the Cartan basis as

$$\begin{aligned} [J_{1,2}, J_{\uparrow,\updownarrow}] &= +J_{\uparrow,\updownarrow}, \qquad & [J_{3,4}, J_{\updownarrow,\uparrow}] &= +J_{\updownarrow,\uparrow}, \\ [J_{1,2}, J_{\downarrow,\updownarrow}] &= -J_{\downarrow,\updownarrow}, \qquad & [J_{3,4}, J_{\updownarrow,\downarrow}] &= -J_{\updownarrow,\downarrow}, \end{aligned} \quad (11.127)$$

$$\begin{aligned} [J_{\uparrow,\uparrow}, J_{\downarrow,\downarrow}] &= J_{1,2} + J_{3,4}, \qquad & [J_{\uparrow,\uparrow}, J_{\downarrow,\uparrow}] &= 0 = [J_{\uparrow,\uparrow}, J_{\uparrow,\downarrow}], \\ [J_{\uparrow,\downarrow}, J_{\downarrow,\uparrow}] &= J_{1,2} - J_{3,4}, \qquad & [J_{\downarrow,\downarrow}, J_{\downarrow,\uparrow}] &= 0 = [J_{\downarrow,\downarrow}, J_{\uparrow,\downarrow}]. \end{aligned} \quad (11.128)$$

In this way the 6 generators of plane rotations in 4 dimensions that form the Lie algebra $D_2 = \mathsf{so}(4)$, are organized into two weights with indices 1-2 and 3-4, and four roots, $J_{\updownarrow,\updownarrow}$ at $(\lambda_{1\text{-}2}, \lambda_{3\text{-}4}) = (\pm 1, \pm 1)$, as shown in the Cartan diagram of Fig. 11.7b. The lengths of these roots is $\sqrt{2}$.

Remark: The accident $\mathsf{so}(4) = \mathsf{su}(2) \oplus \mathsf{su}(2)$. The Cartan diagram of $\mathsf{so}(4)$ in Fig. 11.7b indicates at a glance that the algebra decomposes into a direct sum of two rank-1 subalgebras. It is an *accidental* (and unique) feature among the classical algebras. Two mutually commuting sets of $\mathsf{so}(3) = \mathsf{su}(2)$

generators, $\vec{J}^a$ and $\vec{J}^b$, each arranged into the familiar pattern (11.116), provide the generators $J_{\mu,\kappa}$ of $\mathsf{so}(4)$ as

$$\begin{array}{|c|c|}\hline J^a_3 & -J^a_2 \\ \hline \multicolumn{1}{c|}{} & J^a_1 \\ \cline{2-2}\end{array} \oplus \begin{array}{|c|c|}\hline J^b_3 & -J^b_2 \\ \hline \multicolumn{1}{c|}{} & J^b_1 \\ \cline{2-2}\end{array} = \begin{array}{|c|c|c|}\hline J^a_3 + J^b_3 & -J^a_2 - J^b_2 & J^a_1 - J^b_1 \\ \hline \multicolumn{1}{c|}{} & J^a_1 + J^b_1 & J^a_2 - J^b_2 \\ \cline{2-3} \multicolumn{1}{c}{} & \multicolumn{1}{c|}{} & J^a_3 - J^b_3 \\ \cline{3-3}\end{array} \tag{11.129}$$

The fact that the summand algebras are $\mathsf{su}(2)$ rather than $\mathsf{so}(3)$ is indicated by their weights in (11.129): $J^a_3 = \frac{1}{2}(J_{1,2} + J_{3,4})$, $J^b_3 = \frac{1}{2}(J_{1,2} - J_{3,4})$. When the weights in $\mathsf{so}(4)$ have integer values, those in the decomposition can be either integer or half-integer. This isomorphism is widely used in quantum angular momentum theory [21] to couple angular momenta. •

Remark: Generators of noncompact forms $\mathsf{so}(N_+, N_-)$. We indicate the coordinates that will be subject to the Weyl trick by bars as in (11.44), so

$$x_\kappa \overset{\text{Weyl}}{\longrightarrow} -ix_{\bar{\kappa}}, \quad \partial_\kappa \overset{\text{Weyl}}{\longrightarrow} i\partial_{\bar{\kappa}}, \quad J_{\mu,\kappa} \overset{\text{Weyl}}{\longrightarrow} iJ_{\mu,\bar{\kappa}}. \tag{11.130}$$

In this way we transmute the rotation generators (11.113) into boost generators in the same plane,

$$J_{\mu,\bar{\kappa}} = J_{\bar{\kappa},\mu} = J^\dagger_{\mu,\bar{\kappa}} = -i(x_\mu \partial_{\bar{\kappa}} + x_{\bar{\kappa}} \partial_\mu). \tag{11.131}$$

Their exponentiated action on functions of $\vec{x} \in \mathsf{R}^N$, with $\zeta \in \mathsf{R}$, is then

$$\exp(i\zeta J_{\mu,\bar{\kappa}})\, f(\vec{x}\,) = f(\vec{x}\,'), \qquad \begin{pmatrix} x'_\mu \\ x'_{\bar{\kappa}} \end{pmatrix} = \begin{pmatrix} \cosh\zeta & \sinh\zeta \\ \sinh\zeta & \cosh\zeta \end{pmatrix} \begin{pmatrix} x_\mu \\ x_{\bar{\kappa}} \end{pmatrix}. \tag{11.132}$$

[Cf. (11.114).] Generators with two barred indices, $J_{\bar{\kappa},\bar{\lambda}}$, are left invariant by the Weyl trick; together with the generators $J_{\mu,\nu}$, they will form the maximal compact subalgebra $\mathsf{so}(N_+) \oplus \mathsf{so}(N_-) \subset \mathsf{so}(N_+, N_-)$. The Lie brackets of $\mathsf{so}(N_+, N_-)$ are (11.115) after replacing the Kronecker $\delta_{\mu,\nu}$'s by the components of the metric tensor $g_{\mu,\nu} = \tau_\mu \delta_{\mu,\nu}$,

$$[J_{\mu,\nu}, J_{\kappa,\lambda}] = -i(g_{\nu,\kappa} J_{\mu,\lambda} + g_{\mu,\lambda} J_{\nu,\kappa} + g_{\kappa,\mu} J_{\lambda,\nu} + g_{\lambda,\nu} J_{\kappa,\mu}), \tag{11.133}$$

with positive signature τ_μ for the unbarred coordinates and negative for the barred ones. •

Remark: The isomorphism $\mathsf{su}(1,1) = \mathsf{so}(2,1)$. As we did before in (11.120) with the compact versions of these algebras, we find from (9.55) the announced isomorphism between their fundamental representations of $\mathsf{su}(1,1)$ and $\mathsf{so}(2,1)$. From $\mathbf{V} \approx 1 + \varepsilon \mathbf{v}$, $\varepsilon^2 \approx 0$, and the restrictions of trace as before, it follows that

$$\mathbf{v} = \begin{pmatrix} ia^0 & b^0 + ib^1 \\ b^0 - ib^1 & -ia^0 \end{pmatrix} \in \mathsf{su}(1,1) \quad \leftrightarrow$$
$$\mathbf{D}^{\mathrm{SO}(2,1)}(\mathbf{v}) = 2 \begin{pmatrix} 0 & a^1 & b^1 \\ -a^1 & 0 & b^0 \\ b^1 & b^0 & 0 \end{pmatrix} \in \mathsf{so}(2,1). \tag{11.134}$$

As in (11.120), we again note the factor 2 in front of the last matrix; this suggests the 2:1 covering in the generated Lie groups. •

Remark: The isomorphism $\mathsf{sp}(2,\mathsf{R}) = \mathsf{sl}(2,\mathsf{R}) = \mathsf{so}(2,1)$. It will be most useful to establish a relation between the symplectic and 3-dim Lorentz matrices. As in (11.120) and (11.134), from (9.53) with $\mathbf{M} \approx 1 + \varepsilon\mathbf{m}$, $\varepsilon^2 \approx 0$, we obtain

$$\mathbf{m} = \begin{pmatrix} a & b \\ c & -a \end{pmatrix} \in \mathsf{sp}(2,\mathsf{R}) = \mathsf{sl}(2,\mathsf{R}) \quad \leftrightarrow$$
$$\mathbf{D}^{\mathrm{so}(2,1)}(\mathbf{m}) = 2 \begin{pmatrix} 0 & c-b & -a \\ b-c & 0 & b+c \\ -a & b+c & 0 \end{pmatrix} \in \mathsf{so}(2,1). \tag{11.135}$$

Or, using the parameters of $\mathsf{so}(2,1)$,

$$\frac{1}{2}\begin{pmatrix} r_{1,3} & r_{2,3} - r_{1,2} \\ r_{2,3} + r_{1,2} & -r_{1,3} \end{pmatrix} \quad \leftrightarrow \quad \begin{pmatrix} 0 & r_{1,2} & -r_{1,3} \\ -r_{1,2} & 0 & r_{2,3} \\ -r_{1,3} & r_{2,3} & 0 \end{pmatrix}. \tag{11.136}$$

We use the same Cartan diagram to depict a classical algebra and its noncompact forms obtained through the Weyl trick. An accidental isomorphism or decomposition may *or may not* have a noncompact sibling relation. •

Remark: The D_2 accidents. The 6-dim Lie algebra $D_2 = \mathsf{so}(4)$ and its noncompact forms suffer the following accidental isomorphisms:

$\mathsf{so}(4) = \mathsf{su}(2) \oplus \mathsf{su}(2)$,	6 compact generators,		
$\mathsf{so}(3,1) = \mathsf{sl}(2,\mathsf{C})$,	3	"	3 noncompact,
$\mathsf{so}(2,2) = \mathsf{su}(1,1) \oplus \mathsf{su}(1,1)$,	2	"	4 "

The first was examined above; the other two now follow. •

Remark: The Lorentz accident $\mathsf{so}(3,1) = \mathsf{sl}(2,\mathsf{C})$. The Lorentz algebra $\mathsf{so}(3,1)$ is obtained from $\mathsf{so}(4)$ with the Weyl trick applied on the coordinate $4 \mapsto \bar{4}$. There are 3 compact rotation generators (including the weight $J_{1,2}$), and 3 noncompact boosts (including the weight $J_{3,\bar{4}}$). The decomposition of $\mathsf{so}(3,1)$ into two summands is impossible, because among its subalgebras only $\mathsf{so}(2,1)$ can provide boosts, so there would be 2 or 4, but not *three*. Searching beyond the classical algebras, among the linear algebras of Sect. 11.1 we find that 2×2 complex matrices form a Lie algebra with the same structure. The isomorphism is easily proposed and checked using the representation (11.12) by Pauli matrices:

$$\begin{array}{rlll} \text{rotations:} & J_{1,2} \leftrightarrow \frac{1}{2}\,\boldsymbol{\sigma}_3, & J_{2,3} \leftrightarrow \frac{1}{2}\,\boldsymbol{\sigma}_1, & J_{1,3} \leftrightarrow -\frac{1}{2}\,\boldsymbol{\sigma}_2, \\ \text{boosts:} & J_{3,\bar{4}} \leftrightarrow \frac{1}{2} i\boldsymbol{\sigma}_3, & J_{1,\bar{4}} \leftrightarrow \frac{1}{2} i\boldsymbol{\sigma}_1, & J_{2,\bar{4}} \leftrightarrow -\frac{1}{2} i\boldsymbol{\sigma}_2. \end{array} \tag{11.137}$$

The compact generators of $\mathsf{so}(3) \subset \mathsf{so}(3,1)$ correspond to the self-adjoint matrices in $\mathsf{su}(2) \subset \mathsf{sl}(2,\mathsf{C})$, while the noncompact ones are i times the previous, and correspond to skew-adjoint matrices. This isomorphism underlies the use of the dotted and undotted spinor indices in relativity theory. •

Remark: The accident $\mathsf{so}(2,2) = \mathsf{su}(1,1) \oplus \mathsf{su}(1,1)$. The Weyl trick applied on the coordinates 3 and 4 of $\mathsf{so}(4)$ yields $\mathsf{so}(2,2)$, so the 2 weights $J_{1,2}$ and $J_{\bar{3},\bar{4}}$ are compact, while the 4 shifts are noncompact. Their sums and differences in the pattern (11.129) show that the Weyl trick $\mathsf{so}(4) \overset{\text{Weyl}}{\longrightarrow} \mathsf{so}(2,2)$ also transmutes $\mathsf{su}(2)_a \oplus \mathsf{su}(2)_b \overset{\text{Weyl}}{\longrightarrow} \mathsf{su}(1,1)_a \oplus \mathsf{su}(1,1)_b$. •

Now follows the Lie algebra $B_2 = \mathsf{so}(5)$. To the 6 generators of $\mathsf{so}(4)$ in the pattern (11.116), we add 4 generators in a new column ι, to obtain the 10 generators of $\mathsf{so}(5)$. The new roots for the Cartan diagram are the raising and lowering combinations for elements in the column ι, obtained as in (11.119), namely

$$\begin{array}{ll} J^{(\mu,\nu)}_{\updownarrow,\iota} = \frac{1}{\sqrt{2}}(J_{\mu,\iota} \pm iJ_{\nu,\iota}), & J^{(\kappa,\lambda)}_{\updownarrow,\iota} = \frac{1}{\sqrt{2}}(J_{\kappa,\iota} \pm iJ_{\lambda,\iota}), \\ (\lambda_\mu, \lambda_\nu) = (\pm 1, 0), & (\lambda_\mu, \lambda_\nu) = (0, \pm 1). \end{array} \tag{11.138}$$

The rank of $\mathsf{so}(5)$ is 2, and the weight generators are $J_{\mu,\nu}$ and $J_{\nu,\kappa}$ – the same as in $\mathsf{so}(4)$. The Cartan diagram of $\mathsf{so}(5)$ is shown in Fig. 11.7c: it has the 4 roots of length $\sqrt{2}$ belonging to the $\mathsf{so}(4)$ subalgebra, and the 4 roots of length 1 that we newly added. As is evident in the diagram, the commutator of two shifts in the latter set is one in the former:

$$\begin{array}{ll} [J^{(\mu,\nu)}_{\uparrow,\iota}, J^{(\kappa,\lambda)}_{\uparrow,\iota}] = J^{\binom{\mu,\nu}{\kappa,\lambda}}_{\uparrow,\uparrow}, & [J^{(\mu,\nu)}_{\uparrow,\iota}, J^{(\kappa,\lambda)}_{\downarrow,\iota}] = +J^{\binom{\mu,\nu}{\kappa,\lambda}}_{\uparrow,\downarrow}, \\ [J^{(\mu,\nu)}_{\downarrow,\iota}, J^{(\kappa,\lambda)}_{\downarrow,\iota}] = J^{\binom{\mu,\nu}{\kappa,\lambda}}_{\downarrow,\downarrow}, & [J^{(\mu,\nu)}_{\downarrow,\iota}, J^{(\kappa,\lambda)}_{\uparrow,\iota}] = -J^{\binom{\mu,\nu}{\kappa,\lambda}}_{\downarrow,\uparrow}. \end{array} \tag{11.139}$$

The Cartan diagrams of $B_N = \mathsf{so}(2N+1)$ have the same rank and dimension as those of $D_N = \mathsf{so}(2N)$.

Remark: The accident $B_2 = C_2$ and its noncompact forms. The Cartan diagram of $B_2 = \mathsf{so}(5)$ in Fig. 11.7c is a rotated copy of that of $C_2 = \mathsf{usp}(2)$ in Fig. 11.1b. This uncovers another accident between 10-dim compact Cartan algebras, and between their noncompact forms, which are

$$\begin{array}{lll} \mathsf{so}(5) = \mathsf{usp}(2), & 10 \text{ compact generators}, & \\ \mathsf{so}(4,1) = \mathsf{usp}(1,1), & 6 \quad " & 4 \text{ noncompact}, \\ \mathsf{so}(3,2) = \mathsf{sp}(4,\mathsf{R}), & 4 \quad " & 6 \quad " \end{array}$$

To establish an equality we must find a 1:1 correspondence between the two sets of generators which is preserved under Lie brackets. There is a (harmless) ambiguity in comparing Cartan diagrams, because they posess a finite symmetry group – called the *Weyl* group of the algebra, which entails a limited freedom to elect the orientation of the coordinate axes through linear combinations of the weights. Below we shall use this strategy to analyze in detail the important isomorphism which involves the real symplectic algebra of paraxial optics. The algebras $\mathsf{usp}(2)$ and $\mathsf{usp}(1,1)$ have yet to be proven useful for optical models. •

To analyze the isomorphism $\mathsf{so}(3,2) = \mathsf{sp}(4,\mathsf{R})$ between the algebra of '3+2' relativity and the symplectic algebra of paraxial geometric optics (that will be our main tool in chapter 12), we recall the Cartan bases of weights and shifts for the compact form $\mathsf{so}(5)$ from (11.127)–(11.128), and of $\mathsf{sp}(4,\mathsf{R})$ from (11.81)–(11.83); the Cartesian axes for the 2-dim screen will be denoted x and y. The Weyl trick on the axes $4 \mapsto \bar{4}$ and $5 \mapsto \bar{5}$ of $\mathsf{so}(5)$ in the pattern (11.116), separates the 3×2 rectangle of noncompact generators; the remaining 4 generators will form the maximal compact subalgebra $\mathsf{so}(3) \oplus \mathsf{so}(2) \subset \mathsf{so}(3,2)$. This must be isomorphic to the maximal compact subalgebra $\mathsf{u}(2) \subset \mathsf{sp}(4,\mathsf{R})$ in (11.74). We thus establish the following correspondence between the operators in the first and functions of phase space in the second:

$$\mathsf{so}(3,2) \supset \mathsf{so}(3) \oplus \mathsf{so}(2) = \mathsf{u}(2) = \mathsf{su}(2) \oplus \mathsf{u}(1) \subset \mathsf{sp}(4,\mathsf{R})$$

$$J_{1,2} = -i(x_1\partial_2 - x_2\partial_1) \leftrightarrow j_{1,2} := \tfrac{1}{2} l_{1,2} = \tfrac{1}{2}(q_x p_y - q_y p_x), \tag{11.140}$$

$$J_{2,3} = -i(x_2\partial_3 - x_3\partial_2) \leftrightarrow j_{2,3} := \tfrac{1}{2} f_{1,2} = \tfrac{1}{2}(p_x p_y + q_x q_y), \tag{11.141}$$

$$J_{1,3} = -i(x_1\partial_3 - x_3\partial_1) \leftrightarrow j_{1,3} := \tfrac{1}{4}(f_{2,2} - f_{1,1}) = \tfrac{1}{4}(-p_x^2 + p_y^2 - q_x^2 + q_y^2) \tag{11.142}$$

$$J_{\bar{4},\bar{5}} = -i(x_{\bar{4}}\partial_{\bar{5}} - x_{\bar{5}}\partial_{\bar{4}}) \leftrightarrow j_{\bar{4},\bar{5}} := \tfrac{1}{4}(f_{1,1} + f_{2,2}) = \tfrac{1}{4}(|\mathbf{p}|^2 + |\mathbf{q}|^2). \tag{11.143}$$

The self-adjoint operators $J_{\mu,\nu}$ satisfy (11.133), *e.g.*, $[J_{1,2}, J_{1,3}] = iJ_{2,3} = -iJ_{3,2}$ etc., while the corresponding generator functions on phase space $j_{\mu,\nu}(\mathbf{p},\mathbf{q})$ close into the same algebra under Poisson brackets, *e.g.* $\{j_{1,2}, j_{1,3}\} = j_{2,3} = -j_{3,2}$, etc. The association between the two Lie brackets is $[J, J'] = -iJ'' \leftrightarrow \{j, j'\} = j''$, so

$$\{j_{\mu,\nu}, j_{\kappa,\lambda}\} = g_{\nu,\kappa} j_{\mu,\lambda} + g_{\mu,\lambda} j_{\nu,\kappa} + g_{\kappa,\mu} j_{\lambda,\nu} + g_{\lambda,\nu} j_{\kappa,\mu}, \tag{11.144}$$

where as before $g_{\mu,\nu}$ is the metric.

Remark: Angular momentum and harmonic guide Hamiltonian. The generator functions of $\mathsf{so}(3,2)$ in (11.140)–(11.143) correspond to sums of monomials of phase space which generate the $\mathsf{U}(2)$-Fourier transforms. Above, the physical quantities corresponding to the compact weight generators of $\mathsf{so}(3,2)$ are the angular momentum $2j_{1,2}$, and the harmonic guide Hamiltonian $h = 2j_{\bar{4},\bar{5}}$. The other two compact generators are the separated plane guide Hamiltonians $h_x := j_{3,1} + j_{\bar{4},\bar{5}}$ and $h_y := -j_{3,1} + j_{\bar{4},\bar{5}}$. And note that the weight generators that we used for $\mathsf{sp}(4,\mathsf{R})$, $d_k = \frac{1}{2} q_k p_k$, $k = 1, 2$, in (11.81)–(11.83), are *not* among the compact generators listed above. •

Since angular momentum $2j_{1,2}$ commutes with the noncompact generators $j_{3,\bar{4}}$ and $j_{3,\bar{5}}$, these two functions must be scalars. We have met such realization before [see Sect. 7.1, (7.3), (7.19) and (11.131)], so we can complement the 4 matched generators above with 2 more:

$$\mathsf{so}(3,2) \supset \mathsf{so}(2,1) = \mathsf{u}(1,1) \subset \mathsf{sp}(4,\mathsf{R})$$

$$J_{3,\bar{4}} = -i(x_3\partial_{\bar{4}} + x_{\bar{4}}\partial_3) \leftrightarrow j_{3,\bar{4}} := \tfrac{1}{2}\mathbf{p}\cdot\mathbf{q} \tag{11.145}$$

$$J_{3,\bar{5}} = -i(x_3\partial_{\bar{5}} + x_{\bar{5}}\partial_3) \leftrightarrow j_{3,\bar{5}} := \tfrac{1}{4}(|\mathbf{p}|^2 - |\mathbf{q}|^2). \tag{11.146}$$

So $j_{3,\bar{4}}$ is the isotropic scaling generator function, and $j_{3,\bar{5}}$ is the repulsive guide Hamiltonian; together with the harmonic guide Hamiltonian $2j_{\bar{4},\bar{5}}$ they form the noncompact $\mathsf{so}(2,1)$ 'isotropic' subalgebra in the subspace of coordinates $3, \bar{4}, \bar{5}$, as is evident in the pattern (11.116).

Finally, from the commutators between (11.140)–(11.143) and (11.145)–(11.146), *e.g.* $\{j_{1,2}, j_{3,\bar{4}}\} = j_{1,\bar{4}}$, etc., we determine the remaining 4 noncompact generators:

$$J_{1,\bar{4}} = -i(x_1\partial_{\bar{4}} + x_{\bar{4}}\partial_1) \leftrightarrow j_{1,\bar{4}} := \tfrac{1}{4}(p_x^2 - p_y^2 - q_x^2 + q_y^2), \tag{11.147}$$

$$J_{1,\bar{5}} = -i(x_1\partial_{\bar{5}} + x_{\bar{5}}\partial_1) \leftrightarrow j_{1,\bar{5}} := \tfrac{1}{2}(p_x q_x - p_y q_y), \tag{11.148}$$

$$J_{2,\bar{4}} = -i(x_2\partial_{\bar{4}} + x_{\bar{4}}\partial_2) \leftrightarrow j_{2,\bar{4}} := \tfrac{1}{2}(-p_x p_y + q_x q_y), \tag{11.149}$$

$$J_{2,\bar{5}} = -i(x_2\partial_{\bar{5}} + x_{\bar{5}}\partial_2) \leftrightarrow j_{2,\bar{5}} := \tfrac{1}{2}(p_x q_y + p_y q_x). \tag{11.150}$$

Thus we display explicitly the isomorphism between the differential operators on functions of R^5 under commutation, and quadratic functions of R^4 phase space under Poisson brackets.

The isomorphism $\mathsf{so}(3,2) = \mathsf{sp}(4,\mathsf{R})$ can be written equivalently in matrix form. The generators $J_{\mu,\nu}$ of $\mathsf{so}(3,2)$ act linearly on the 'mathematical' space $(x_1, x_2, x_3, x_{\bar{4}}, x_{\bar{5}}) \in \mathsf{R}^5$, and are realized by 5×5 matrices $\mathbf{r}$ [see (11.113)–(11.114) and (11.131)–(11.132)], while the generators $\{j_{\mu,\nu}, \circ\}$ of $\mathsf{sp}(4,\mathsf{R})$ are realized by 4×4 matrices $\mathbf{m}(\mathbf{r})$ acting on paraxial optical phase space $(p_x, p_y, q_x, q_y) \in \mathsf{R}^4$ in (9.20). The two matrix realizations are related through

$$\mathbf{r} = \left(\begin{array}{ccc|cc} 0 & r_{1,2} & r_{1,3} & r_{1,\bar{4}} & r_{1,\bar{5}} \\ -r_{1,2} & 0 & r_{2,3} & r_{2,\bar{4}} & r_{2,\bar{5}} \\ -r_{1,3} & -r_{2,3} & 0 & r_{3,\bar{4}} & r_{3,\bar{5}} \\ \hline r_{1,\bar{4}} & r_{2,\bar{4}} & r_{3,\bar{4}} & 0 & r_{\bar{4},\bar{5}} \\ r_{1,5} & r_{2,5} & r_{3,5} & -r_{\bar{4},\bar{5}} & 0 \end{array}\right) \qquad \leftrightarrow \tag{11.151}$$

$$\mathbf{m}(\mathbf{r}) = \frac{1}{2}\left(\begin{array}{cc|cc} r_{1,\bar{5}} + r_{3,\bar{4}} & r_{1,2} + r_{2,\bar{5}} & -r_{1,3} - r_{1,\bar{4}} - r_{3,\bar{5}} + r_{\bar{4},\bar{5}} & r_{2,3} + r_{2,\bar{4}} \\ -r_{1,2} + r_{2,\bar{5}} & -r_{1,\bar{5}} + r_{3,\bar{4}} & r_{2,3} + r_{2,\bar{4}} & r_{1,3} + r_{1,\bar{4}} - r_{3,\bar{5}} + r_{\bar{4},\bar{5}} \\ \hline r_{1,3} - r_{1,\bar{4}} - r_{3,\bar{5}} - r_{\bar{4},\bar{5}} & -r_{2,3} + r_{2,\bar{4}} & -r_{1,\bar{5}} - r_{3,\bar{4}} & r_{1,2} - r_{2,\bar{5}} \\ -r_{2,3} + r_{2,\bar{4}} & -r_{1,3} + r_{1,\bar{4}} - r_{3,\bar{5}} - r_{\bar{4},\bar{5}} & -r_{1,2} - r_{2,\bar{5}} & r_{1,\bar{5}} - r_{3,\bar{4}} \end{array}\right). \tag{11.152}$$

Remark: The group homomorphism $\mathsf{SO}(3,2) \overset{1:2}{=} \mathsf{Sp}(4,\mathsf{R})$. When two Lie algebras are isomorphic, there is a 1:1 correspondence between their elements which is preserved under Lie brackets; such is the case for $\mathbf{r}$ and $\mathbf{m}(\mathbf{r})$ in (11.151)–(11.152). When we exponentiate the Lie algebras to Lie groups, isomorphism among the former assures only a homomorphism among the latter. It is hard to exponentiate full matrices with symbolic entries that are larger than 2×2, but we can track the one-parameter subgroups, noting that all $\mathsf{so}(3,2)$ compact parameters, $r_{1,2}, r_{1,3}, r_{2,3}$ and $r_{\bar{4},\bar{5}}$, have a factor of $\frac{1}{2}$ in the $\mathsf{sp}(4,\mathsf{R})$ matrices, so $\mathsf{SO}(3,2)$ is covered by $\mathsf{Sp}(4,\mathsf{R})$ twice. [This is exactly as $\mathsf{SO}(3)$ is covered twice by $\mathsf{SU}(2)$ in (11.120), and $\mathsf{SO}(2,1)$ by $\mathsf{SU}(1,1)$ in (11.134).] •

Remark: Lorentz subalgebras $\mathsf{so}(3,1) \subset \mathsf{sp}(4,\mathsf{R})$. Thinking of the group $\mathsf{Sp}(4,\mathsf{R})$ of paraxial optical transformations in terms of $\mathsf{SO}(3,2)$ allows the easy recognition of some interesting subgroups, their generators, and the nature of the group action. When we cross out the column $\bar{5}$ in the $\mathsf{so}(3,2)$ pattern (11.116) of generators (11.140)–(11.146), we obtain the pattern and generators of a Lorentz subalgebra $\mathsf{so}(3,1) \subset \mathsf{Sp}(4,\mathsf{R})$. The corresponding 'Lorentz' transformations of optical phase space will consist of $\mathsf{SU}(2)$-Fourier transformations (i.e., *excluding* the central $\mathsf{U}(1)$ transforms generated by $j_{\bar{4},\bar{5}}$), of *magnifications* (also called *scalings* or *squeezings*, generated by the noncompact $j_{3,\bar{4}}$), and of all their products. This 'Lorentz' $\mathsf{SO}(3,1)$ group is distinct from the Lorentz group of space-time relativity, but they are mathematically identical of course [79]. Another $\mathsf{SO}(3,1)$ subgroup of phase space transformations appears when we cross out the row and column $\bar{4}$ from the $\mathsf{so}(3,2)$ algebra pattern, which consists of the same Fourier subgroup, plus hyperbolic expanders (10.12) or a repulsive guide evolution generated by $j_{3,\bar{5}}$, and of all their products. Similarly, $\mathsf{SO}(2,2)$ subgroups of phase space transformations can be identified crossing out indices 1, 2 or 3 in the $\mathsf{so}(3,2)$ pattern (11.116). •

We proceed with the Cartan diagrams of the orthogonal algebras of rank three: $D_3 = \mathsf{so}(6)$ and $B_3 = \mathsf{so}(7)$. This will illustrate the alternation of B- and D-algebras when the Cartesian pattern (11.116) of generators grows. The weight generators of $\mathsf{so}(6)$ are $J_{1,2}$, $J_{3,4}$, and $J_{5,6}$; they define the axes $\lambda_1, \lambda_2, \lambda_3$ of the 3-dim Cartan diagram. The algebra has now 15 generators, and the 12 shift operators can be grouped into three 2×2 blocks that we label by $\binom{1,2}{3,4}$, $\binom{1,2}{5,6}$, and $\binom{3,4}{5,6}$ following (11.123)–(11.126). The Cartan diagram of $\mathsf{so}(6)$ consists of three $\mathsf{so}(4)$ diagrams on orthogonal planes, as shown in Fig. 11.8a; all roots have the same length $\sqrt{2}$. Now compare its Cartan diagram with that for $A_3 = \mathsf{su}(4)$ in Fig. 11.5c. They are identical, save for a rotation of the axes. This is again an accident, and the isomorphism extends to other matching noncompact forms that we remark below.

Remark: The accident $D_3 = A_3$ and its noncompact forms. The orthogonal and unitary algebras of rank 3 have vector dimension 15, and suffer four isomorphism accidents:

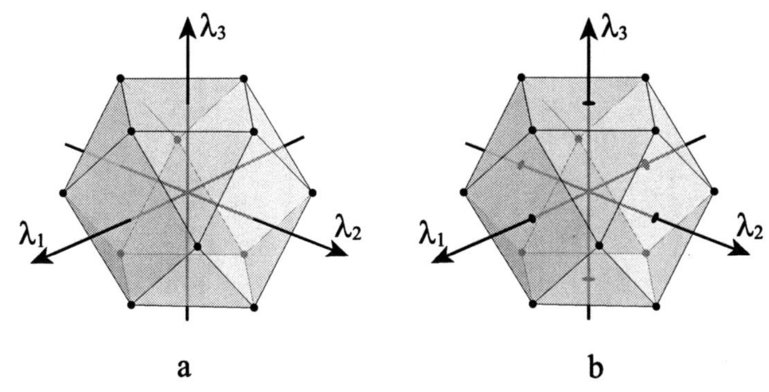

Fig. 11.8. Cartan root diagrams for the orthogonal algebras of rank 3. (**a**) $D_3 = \mathsf{so}(6)$; the axes show the eigenvalues of $J_{1,2}$, $J_{3,4}$, $J_{5,6}$ (note that this is a rotated version of the diagram for $\mathsf{su}(4)$ in Fig. 11.5c). (**b**) $B_3 = \mathsf{so}(7)$; the extra generators in column 7 occupy the midpoints of the square faces (cf. the relation between the $\mathsf{so}(4)$ and $\mathsf{so}(5)$ diagrams in Figs. 11.7b and c).

$\mathsf{so}(6) = \mathsf{su}(4)$,	15 compact generators,		
$\mathsf{so}(5,1) = \mathsf{sl}(2,\mathsf{Q})$,	10 "	5 noncompact,	
$\mathsf{so}(4,2) = \mathsf{su}(2,2)$,	7 "	8 "	
$\mathsf{so}(3,3) = \mathsf{sl}(4,\mathsf{R})$,	6 "	9 "	

The identification of the generators in the noncompact forms can be done again by identifying the maximal compact subalgebras. Of interest to us is $\mathsf{so}(4,2) = \mathsf{su}(2,2)$, because the conformal algebra $\mathsf{so}(4,2)$ was used in Chap. 6 for the Maxwell fish-eye. The conformal algebra has the maximal compact subalgebra $\mathsf{so}(4) \oplus \mathsf{so}(2)$, while that of $\mathsf{su}(2,2)$ is $\mathsf{su}(2) \oplus \mathsf{su}(2) \oplus \mathsf{u}(1)$[19] We shall not detail these isomorphisms further, but we refer the reader to their very complete treatement in Ref. [56]. •

The Cartan root diagram of the Lie algebra $B_3 = \mathsf{so}(7)$ is obtained by adding 6 new generators in the seventh column of the pattern (11.116). These are combined to form 3 pairs of shift operators $J^{(\mu,\nu)}_{\updownarrow,7}$ in (11.119), with upper-index pairs (1,2), (3,4), and (5,6), and appear in the Cartan diagram as 3 extra pairs of arrows on the $\mathsf{so}(6)$ diagram. As in other B-algebras, the 'last column' leads to 'short' roots of length 1. The $\mathsf{so}(7)$ Cartan root diagram is shown in Fig. 11.8b. There are now 21 generators, 3 weights and 18 shifts. Notwithstanding that $C_3 = \mathsf{sp}(6,\mathsf{R})$ also has 21 generators, its Cartan diagram in Fig. 11.1c proves at a glance that two the algebras are distinct. There are no further accidental isomorphisms in orthogonal algebras of this or higher rank.

[19] To count the number of unitary generators properly, start with the 16 generators of $\mathsf{u}(2,2)$, whose maximal compact subalgebra is $\mathsf{u}(2) \oplus \mathsf{u}(2)$. Each $\mathsf{u}(2)$ decomposes as $\mathsf{u}(2) \supset \mathsf{su}(2) \oplus \mathsf{u}(1)$, *but* the two $\mathsf{u}(1)$'s are restricted by the tracelessness condition to a single one. So the two compact algebras have vector dimension 5 and are again isomorphic.

12 Hamiltonian orbits

The trajectories of light rays in guides,[1] when plotted on the planes (p_i, q_j) of paraxial phase space R^{2D}, can be closed (ellipses) or open (straight lines or hyperbola branches). [For the case of $D = 1$-dim screens see Figs. 3.2 and 9.1]. Using linear canonical transformations $\mathbf{M} \in \mathsf{Sp}(2D, \mathsf{R})$ we can rotate, squeeze, or skew phase space; but open trajectories will never close, nor *vice versa*. Elliptic, straight, and hyperbolic trajectories form equivalence classes that will remain disjoint sets. These trajectories are generated by the Hamiltonian function of the guide medium, $H(\mathbf{m}; \mathbf{p}, \mathbf{q})$, through

$$\begin{pmatrix} \mathbf{p}(z) \\ \mathbf{q}(z) \end{pmatrix} = \exp(\, z\, \{H, \circ\}) \begin{pmatrix} \mathbf{p}(0) \\ \mathbf{q}(0) \end{pmatrix}, \qquad z \in \mathsf{R}, \tag{12.1}$$

and are characterized by the Hamiltonian matrix $\mathbf{m} \in \mathsf{sp}(2D, \mathsf{R})$ [see (11.56) and (11.71)]. We now classify paraxial optical systems into equivalence classes, called Hamiltonian *orbits*, that we define as follows:

$$\text{orbit}\,(\mathbf{m}) := \{\alpha \mathbf{M}\, \mathbf{m}\, \mathbf{M}^{-1} \in \mathsf{sp}(2D, \mathsf{R}) \ \mid \mathbf{M} \in \mathsf{Sp}(2D, \mathsf{R}),\ 0 \neq \alpha \in \mathsf{R}\}. \tag{12.2}$$

Two Hamiltonian systems $H(\mathbf{m})$ and $H(\mathbf{m}')$ are said to be equivalent when their matrices belong to the same orbit. The ray trajectories in the two systems, $\begin{pmatrix} \mathbf{p}(z) \\ \mathbf{q}(z) \end{pmatrix}$ and $\begin{pmatrix} \mathbf{p}'(z') \\ \mathbf{q}'(z') \end{pmatrix}$, can then be brought one onto the other by means of the canonical transformation $\mathbf{M}$ in (12.2), and the scale factor $z' = \alpha z$ for their evolution parameters. In this chapter we classify the orbits of paraxial optical systems in 2 and 3 dimensions ($D = 1, 2$). In both cases we shall apply the accidental isomorphisms between the symplectic and pseudo-orthogonal algebras, $\mathsf{sp}(2, \mathsf{R}) = \mathsf{so}(2, 1)$ and $\mathsf{sp}(4, \mathsf{R}) = \mathsf{so}(3, 2)$ respectively (see pages 242 and 244–245).

12.1 Orbits in $\mathsf{sp}(2, \mathsf{R})$ for plane systems

We first examine the orbit structure of the set of plane paraxial optical systems, with $D = 1$-dim screens, and thus develop some concepts for the $D = 2$

[1] We recall that media which are invariant under translations along the optical axis z are generically called *guides*.

astigmatic case that looms ahead. We use the representation of quadratic functions of phase space by Hamiltonian matrices, written for generic dimension in (11.71). For the 1-dim case of our immediate interest, this is

$$H(\mathbf{m}; p, q) = \tfrac{1}{2}c\,p^2 - a\,qp - \tfrac{1}{2}b\,q^2 \quad \leftrightarrow \quad \mathbf{m} = \begin{pmatrix} a & b \\ c & -a \end{pmatrix}, \tag{12.3}$$

It is natural to inquire first about the eigenvalues of this Hamiltonian matrix, because they are left invariant by the similarity transformation in (12.2), and can be used thus to characterize orbits. The generic 2×2 Hamiltonian matrix (12.3) has two eigenvalues, $\lambda^{\pm}$, that are solutions of

$$\det(\mathbf{m} - \lambda\mathbf{1}) = \begin{vmatrix} a-\lambda & b \\ c & -a-\lambda \end{vmatrix} = 0, \quad \text{i.e.,} \tag{12.4}$$

$$\lambda^2 + \Delta = 0, \qquad \Delta := \det \mathbf{m} = -(a^2 + bc), \tag{12.5}$$

$$\Rightarrow \lambda^{\pm} = \pm\sqrt{-\Delta} = \pm\sqrt{a^2 + bc}. \tag{12.6}$$

In the eigenvalue equation (12.5), the term linear in λ is absent because the trace of real Hamiltonian matrices (12.3), equal to the sum of its eigenvalues, is zero; their product is $\Delta \in \mathsf{R}$, and so they are complex conjugate, symmetric pairs.

The free scale parameter $\alpha \neq 0$ in (12.2) multiplies the eigenvalues $\lambda^{\pm}$ by α, and Δ by α^2, so for $\lambda \neq 0$ we can normalize $|\lambda| = 1$. The *sign* of Δ thus classifies Hamiltonians into the three promised orbits, whose standard representatives we choose to have $c = 1$, namely:

$$\begin{array}{ll} \text{H: HARMONIC (elliptic) orbit,} & \text{trajectories: ellipses,} \\ \Delta = 1, \quad \lambda^{\pm} = \pm i, & \text{for } H := H\begin{pmatrix} 0 & -1 \\ 1 & 0 \end{pmatrix} = \tfrac{1}{2}p^2 + \tfrac{1}{2}q^2, \\ \text{F: FREE (parabolic) orbit,} & \text{trajectories: straight lines,} \\ \Delta = 0, \quad \lambda^{\pm} = 0, & \text{for } F := H\begin{pmatrix} 0 & 0 \\ 1 & 0 \end{pmatrix} = \tfrac{1}{2}p^2, \\ \text{R: REPULSIVE (hyperbolic) orbit,} & \text{trajectories: hyperbolas,} \\ \Delta = -1, \lambda^{\pm} = \pm 1, & \text{for } R := H\begin{pmatrix} 0 & 1 \\ 1 & 0 \end{pmatrix} = \tfrac{1}{2}p^2 - \tfrac{1}{2}q^2. \end{array} \tag{12.7}$$

Strictly speaking, there is also a fourth, trivial Hamiltonian orbit of $H = 0$, which we disregard. To determine the orbit of a given Hamiltonian $H(\mathbf{m})$, it suffices to find the sign of $\Delta(\mathbf{m})$. In Fig. 12.1 we show the parameter space $(a, b, c) \in \mathsf{R}^3$ of the $\mathsf{sp}(2, \mathsf{R})$ algebra of paraxial optical Hamiltonians $\mathbf{m} = \begin{pmatrix} a & b \\ c & -a \end{pmatrix}$ given by (12.3), and the conic surfaces of equivalence classes under the group.

Remark: Transformation to the orbit representative An arbitrary paraxial Hamiltonian represented by $\mathbf{m} \in \mathsf{sp}(2, \mathsf{R})$, with $\sigma := \operatorname{sign} \det \mathbf{m} \in \{-1, 0, +1\}$, can be brought to one of the three orbit representatives $H(\mathbf{m}_\circ)$

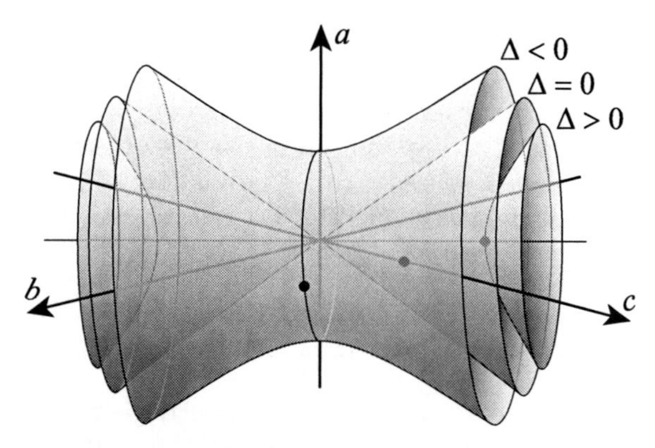

Fig. 12.1. The *abc*-manifold R^3 of the algebra $\mathsf{sp}(2,\mathsf{R})$ is divided into the three Hamiltonian orbits. The open interior of the cone ($\Delta > 0$) is the elliptic orbit that contains harmonic guides [whose standard representative we chose to be $(a,b,c) = (0,-1,1)$]; the surface of the cone ($\Delta = 0$) is the parabolic orbit of free propagation [representative $(0,0,1)$]; and the open exterior ($\Delta < 0$) is the hyperbolic orbit of repulsive guides [representative $(0,1,1)$]. Similarity transformations with $\mathsf{Sp}(2,\mathsf{R})$ move the Hamiltonians within the surface of each hyperboloid $\Delta =$ constant; scaling by $0 \neq \alpha \in \mathsf{R}$ identifies all hyperboloids within each orbit. The trivial orbit of the null Hamiltonian is at the apex.

listed in (12.7), through a similarity transformation by a matrix $\mathbf{M} = \begin{pmatrix} A & B \\ C & D \end{pmatrix} \in \mathsf{Sp}(2,\mathsf{R})$ satisfying $\mathbf{m} = \alpha \mathbf{M}\, \mathbf{m}_\circ\, \mathbf{M}^{-1}$, or

$$\begin{pmatrix} a & b \\ c & -a \end{pmatrix} \begin{pmatrix} A & B \\ C & D \end{pmatrix} = \alpha \begin{pmatrix} A & B \\ C & D \end{pmatrix} \begin{pmatrix} 0 & -\sigma \\ 1 & 0 \end{pmatrix}, \tag{12.8}$$

with $\alpha = \pm\sqrt{|\det \mathbf{m}|}$. There is a one-parameter family of solutions $\mathbf{M}(\chi)$, since any right-hand factor $\exp(\chi \mathbf{m}_\circ)$ will commute with $\mathbf{m}_\circ$. This freedom allows us to choose an $\mathbf{M}$ with $C = 0$, or $B = 0$, or with some convenient value for one of its parameters (except at some boundaries). Two such 'anti-diagonalizing' matrices are:

$$\mathbf{M}_{(C=0)} = \begin{pmatrix} \sqrt{\alpha/c} & a/\sqrt{\alpha c} \\ 0 & \sqrt{c/\alpha} \end{pmatrix}, \tag{12.9}$$

$$\mathbf{M}_{(B=0)} = \begin{pmatrix} \sqrt{-\sigma b/\alpha} & 0 \\ \sigma a/\sqrt{-\sigma \alpha b} & \sqrt{-\alpha/\sigma b} \end{pmatrix}. \tag{12.10}$$

The matrix (12.9) is an imager that reduces Hamiltonians with a free propagation term ($c \neq 0$) to the harmonic, free or repulsive representatives; the matrix (12.10) is useful to reduce imager Hamiltonians ($c = 0$). •

12.2 Trajectories in $\mathsf{Sp}(2,\mathsf{R})$

The classification of ray trajectories follows from the division of the Hamiltonians $H(\mathbf{m})$ into orbits, and we can choose one convenient 'optical' representative $H(\mathbf{m}_circ)$ for each distinct orbit. The similarity transformation in the definition (12.2) implies the same orbit classification for the one-parameter subgroups which they generate:

$$\begin{aligned} \mathbf{m} = \alpha \mathbf{M}\, \mathbf{m}_\circ\, \mathbf{M}^{-1} \in \mathsf{sp}(2,\mathsf{R}) \quad &\Leftrightarrow \\ \exp(z\mathbf{m}) = \mathbf{M} \exp(\alpha z \mathbf{m}_\circ)\, \mathbf{M}^{-1} \in \mathsf{Sp}(2,\mathsf{R}). & \end{aligned} \tag{12.11}$$

This in turn classifies the phase space ray trajectories in guides, and establishes their equivalence to the ray trajectory of the orbit representative $\mathbf{m}_\circ$,

$$\begin{pmatrix} \mathbf{p}(z) \\ \mathbf{q}(z) \end{pmatrix} = e^{z\,\mathbf{m}} \begin{pmatrix} \mathbf{p}(0) \\ \mathbf{q}(0) \end{pmatrix} = \mathbf{M} \begin{pmatrix} \mathbf{p}_\circ(\alpha z) \\ \mathbf{q}_\circ(\alpha z) \end{pmatrix}, \quad \begin{pmatrix} \mathbf{p}_\circ(\alpha z) \\ \mathbf{q}_\circ(\alpha z) \end{pmatrix} := e^{\alpha z \mathbf{m}_\circ} \begin{pmatrix} \mathbf{p}_\circ(0) \\ \mathbf{q}_\circ(0) \end{pmatrix}. \tag{12.12}$$

The 2×2 orbit representative matrices in (12.7) are easily exponentiated [see (11.3)–(11.6)]. For generic matrices $\mathbf{m} \in \mathsf{sp}(2,\mathsf{R})$ we use a method that will also serve below for the 4×4 case. By the Cayley-Hamilton theorem, the eigenvalue equation (12.5) is satisfied by the matrices themselves (replacing λ by $\mathbf{m}$, see [55, Sect. II.10.4]), namely $\mathbf{m}^2 = -\Delta \mathbf{1}$, $\Delta = \det \mathbf{m}$. The terms of the exponential matrix series (11.1) can then be grouped into even and odd subseries, as

$$\begin{aligned} \exp(z\mathbf{m}) &= \sum_{n=0}^{\infty} \frac{z^{2n}}{(2n)!} \mathbf{m}^{2n} + \sum_{n=0}^{\infty} \frac{z^{2n+1}}{(2n+1)!} \mathbf{m}^{2n+1} \qquad (12.13) \\ &= \mathbf{1} \sum_{n=0}^{\infty} \frac{(-1)^n z^{2n}}{(2n)!} \Delta^n + \mathbf{m} \sum_{n=0}^{\infty} \frac{(-1)^n z^{2n+1}}{(2n+1)!} \Delta^n \\ &= \begin{cases} \mathbf{1} \cos(z\sqrt{\Delta}) + \mathbf{m}\, \frac{1}{\sqrt{\Delta}} \sin(z\sqrt{\Delta}), & \Delta > 0, \\ \mathbf{1} + z\,\mathbf{m}, & \Delta = 0, \\ \mathbf{1} \cosh(z\sqrt{-\Delta}) + \mathbf{m}\, \frac{1}{\sqrt{-\Delta}} \sinh(z\sqrt{-\Delta}), & \Delta < 0, \end{cases} \qquad (12.14) \\ &=: \begin{pmatrix} \mathrm{C}(z\omega) + a\omega^{-1}\, \mathrm{S}(z\omega) & b\omega^{-1}\, \mathrm{S}(z\omega) \\ c\omega^{-1}\, \mathrm{S}(z\omega) & \mathrm{C}(z\omega) - a\omega^{-1}\, \mathrm{S}(z\omega) \end{pmatrix}, \qquad (12.15) \end{aligned}$$

with $\omega := \pm\sqrt{|\Delta|}$. According to the orbit of the generator Hamiltonian, the trajectories in the group $\mathsf{Sp}(2,\mathsf{R})$ are divided into

$$\begin{array}{llll} \text{ELLIPTIC:} & \Delta > 0, & \mathrm{C}(z\omega) = \cos z\omega, & \mathrm{S}(z)\omega = \sin z\omega, \\ \text{PARABOLIC:} & \Delta = 0, & \mathrm{C}(z\omega) = 1, & \omega^{-1}\mathrm{S}(z\omega) = z, \\ \text{HYPERBOLIC:} & \Delta < 0, & \mathrm{C}(z)\omega = \cosh z\omega, & \mathrm{S}(z)\omega = \sinh z\omega. \end{array} \tag{12.16}$$

Remark: Integral of the exponential series. With steps analogous to (12.13)–(12.15) we can sum the series of $\int_0^z dz' \exp(z'\mathbf{m})$, namely

$$\frac{\exp(z\mathbf{m}) - 1}{\mathbf{m}} = \mathbf{1}\frac{\sin(z\sqrt{\Delta})}{\sqrt{\Delta}} + \mathbf{m}\,\frac{1-\cos(z\sqrt{\Delta})}{\Delta}, \tag{12.17}$$

for $\Delta = \det \mathbf{m} > 0$. Hyperbolic functions appear when $\Delta < 0$; and when $\Delta = 0$, the series truncates to $z\mathbf{1} + \frac{1}{2}z^2\mathbf{m}$. •

Remark: Non-exponential region. Not *every* element (optical system) in Sp(2, R) belongs to one of the orbits of one-parameter subgroups in (12.16) because, beside harmonic, repulsive, and free systems, Sp(2, R) also has the *conatus* of a fourth region: *non-exponential* systems [see (11.2)], which are *not* realizable through evolution in guides. •

In Fig. 12.2 we show one-parameter subgroups i.e., *trajectories*, in the group space ABC of Sp(2, R) (cf. Fig. 9.4a whose axes have been rotated for visibility), generated by the paraxial optical Hamiltonians $H = \frac{1}{2}cp^2 - \frac{1}{2}bq^2$ in the algebra sp(2, R), that are in the $a = 0$ plane of the previous figure. At the group unit $(A, B, C) = (1, 0, 0)$, these trajectories are tangent to the vectors $(0, b, c)$ of $H\begin{pmatrix} 0 & b \\ c & 0 \end{pmatrix}$ in Fig. 12.1,[2] and lie on the intersection of the

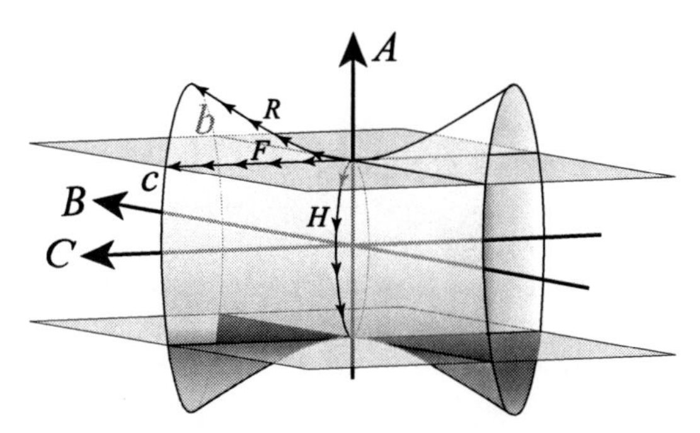

Fig. 12.2. The ABC manifold of the group Sp(2, R) generated by optical sp(2, R)-Hamiltonians $H = \frac{1}{2}cp^2 - \frac{1}{2}bq^2$ in the horizontal plane $a = 0$ of the previous figure; this plane is shown tangent at the group unit $(A, B, C) = (1, 0, 0)$. The hyperboloid is a 2-dim submanifold of the full group; the trajectories generated in the group are one-parameter subgroups, indicated by $\rightarrow\!\!\!\rightarrow\!\!\!\rightarrow$. Harmonic guide trajectories H girdle the waist, with a pure Fourier circle or by plane ellipses that pass also through the minus-unit $(-1, 0, 0)$. Homogeneous medium trajectories F are straight lines. Repulsive guide trajectories R are branches of plane hyperbolas. On this hyperboloid, the shaded region below $A = -1$ is non-exponential; its elements cannot be reached through Hamiltonian evolution (except for the minus-unit).

[2] Distinguish carefully between the coordinates of the algebra, a, b, c, and those of the group, A, B, C, as used here.

plane $B/b = C/c$ with the one-sheeted equilateral hyperboloid of the matrices (12.15); they are plane conics.

A $\mathsf{U}(1)$-circle generated by the $b = -c$ harmonic Hamiltonian covers twice the $\mathsf{SO}(2)$-circle of fractional Fourier transforms (see the discussion on the metaplectic sign in Sect. 9.4). Other planes containing the a-axis define trajectories which are ellipses when the guides are harmonic ($bc < 0$ for $\sigma = +1$), or hyperbola branches connected to the unit for repulsive guides ($bc > 0$ for $\sigma = -1$), or straight lines $(1,\, 0,\, C \in \mathsf{R})$ for homogeneous media ($a = 0 = b$). All compact subgroups (ellipses) contain the two central (i.e., commuting) group elements: the unit $(1,0,0) \leftrightarrow \mathbf{1}$ and the minus-unit $(-1,0,0) \leftrightarrow -\mathbf{1}$. Except for this last point, the $A \le -1$ region of the group is non-exponential, and *cannot* be reached by one-parameter subgroup trajectories.

Remark: Traces distinguish between $\mathsf{Sp}(2,\mathsf{R})$ orbits. The trace of $\mathsf{Sp}(2,\mathsf{R})$ matrices, $\tau := \operatorname{tr}\mathbf{M} = A + D$, is the only invariant under similarity transformations (12.11) beside the determinant; its range readily distinguishes between orbits since, according to (12.15)–(12.16) for $\mathbf{M} \neq \mathbf{1}$, these are

$$\begin{aligned} \text{HYPERBOLIC: } & \tau = 2\cosh z\omega \in (2,\infty), \\ \text{PARABOLIC: } & \tau = 2, \\ \text{ELLIPTIC: } & \tau = 2\cos z\omega \in [-2,2), \\ \text{NON-EXPONENTIAL: } & \tau \in (-\infty,-2). \end{aligned} \tag{12.18}$$

Non-exponential matrices can be thus recognized for having traces in the complement of the range of exponential matrices. The former can be obtained multiplying the latter by $-\mathbf{1} \in \mathsf{Sp}(2,\mathsf{R})$. •

12.3 $\mathsf{sp}(4,\mathsf{R})$ Hamiltonians and eigenvalues

We investigate here the orbit structure of the 10-dim Lie algebra $\mathsf{sp}(4,\mathsf{R})$ of Hamiltonians that generate $\mathsf{Sp}(4,\mathsf{R})$ paraxial astigmatic systems. First we follow the route of the previous section to find the eigenvalues; they will provide a sufficient (but not necessary) criterion to distinguish orbits. We start again associating quadratic functions of phase space $(\mathbf{p},\mathbf{q}) \in \mathsf{R}^4$, generically called *Hamiltonians*, with 4×4 Hamiltonian matrices $\mathbf{m} \in \mathsf{sp}(4,\mathsf{R})$. This correspondence was established in (11.71) for the case $D = 2$; indicating the screen axes by x and y, it is

$$\left.\begin{aligned} & H(\mathbf{m};\mathbf{p},\mathbf{q}) \\ & = \tfrac{1}{2}c_x p_x^2 + \tfrac{1}{2}c_y p_y^2 + c_\times p_x p_y \\ & - \tfrac{1}{2}b_x q_x^2 - \tfrac{1}{2}b_y q_y^2 - b_\times q_x q_y \\ & - a_x q_x p_x - a_y q_y p_y \\ & - a_{xy} q_x p_y - a_{yx} q_y p_x \end{aligned}\right\} \leftrightarrow \mathbf{m} = \left(\begin{array}{cc|cc} a_x & a_{xy} & b_x & b_\times \\ a_{yx} & a_y & b_\times & b_y \\ \hline c_x & c_\times & -a_x & -a_{yx} \\ c_\times & c_y & -a_{xy} & -a_y \end{array}\right). \tag{12.19}$$

Four-by-four Hamiltonian matrices are bulky; their eigenvalue equation $\det(\mathbf{m} - \lambda\mathbf{1}) = 0$ is of fourth degree in λ, but *can* be manipulated by hand. The matrices are real, so if λ is an eigenvalue, then also its complex conjugate λ^* is an eigenvalue. Next, from the definition of Hamiltonian matrices in (11.56) with the symplectic metric matrix $\boldsymbol{\Omega} := \begin{pmatrix} \mathbf{0} & -\mathbf{1} \\ \mathbf{1} & \mathbf{0} \end{pmatrix}$, it follows that $\boldsymbol{\Omega}^{-1}\mathbf{m}\,\boldsymbol{\Omega} = -\mathbf{m}^\top$, and from this,

$$\det\left(\boldsymbol{\Omega}^{-1}(\mathbf{m} - \lambda\mathbf{1})\boldsymbol{\Omega}\right) = \det(-\mathbf{m}^\top - \lambda\mathbf{1}) = \det(\mathbf{m} + \lambda\mathbf{1}), \tag{12.20}$$

so if λ is an eigenvalue, then so is $-\lambda$. The eigenvalues of $\mathsf{sp}(4, \mathsf{R})$ matrices thus come in symmetric, complex conjugate quartets; null eigenvalues must be either double or four-fold. Hamiltonian matrices are traceless [cf. (11.19)] and the sum of their four eigenvalues is zero. It also follows that the fourth-degree eigenvalue equation for $\mathsf{sp}(4, \mathsf{R})$ Hamiltonian matrices has no terms of odd degree; so it is biquadratic:

$$\det(\mathbf{m} - \lambda\mathbf{1}) = \lambda^4 + \Gamma\lambda^2 + \Delta = 0, \tag{12.21}$$

with $\Delta = \det \mathbf{m}$ and

$$\Gamma = \begin{vmatrix} a_x & b_x \\ c_x & -a_x \end{vmatrix} + 2\begin{vmatrix} a_{xy} & b_\times \\ c_\times & -a_{yx} \end{vmatrix} + \begin{vmatrix} a_y & b_y \\ c_y & -a_y \end{vmatrix}. \tag{12.22}$$

The four eigenvalues of $\mathbf{m} \in \mathsf{sp}(4, \mathsf{R})$ are the solutions to (12.21), namely

$$\lambda^\sigma := \pm\sqrt{-\tfrac{1}{2}\Gamma + \sigma\sqrt{(\tfrac{1}{2}\Gamma)^2 - \Delta}}, \qquad \sigma \in \{+, -\}. \tag{12.23}$$

The eigenvalues are thus $\pm\lambda^+$ and $\pm\lambda^-$.

Remark: Determinant of $\mathsf{sp}(4, \mathsf{R})$ matrices. While the determinant of generic 4×4 matrices has $4! = 24$ summands, that of Hamiltonian matrices has only 17 terms. We group these into 'non-cross' and 'cross' summands according to the absence or presence of the coefficients $b_\times, c_\times$ in (12.19), as follows:

$$\Delta = \Delta_{\substack{\text{non}\\\text{cross}}} + \Delta_{\text{cross}}, \tag{12.24}$$

$$\begin{aligned}\Delta_{\substack{\text{non}\\\text{cross}}} &= \begin{vmatrix} a_x & b_x \\ c_x & -a_x \end{vmatrix}\begin{vmatrix} a_y & b_y \\ c_y & -a_y \end{vmatrix} \\ &\quad + a_{xy}^2 a_{yx}^2 + a_{xy}^2 b_y c_x + a_{yx}^2 b_x c_y - 2a_{xy}a_{yx}a_x a_y,\end{aligned} \tag{12.25}$$

$$\begin{aligned}\Delta_{\text{cross}} &= b_\times^2 c_\times^2 - b_\times^2 c_x c_y - c_\times^2 b_x b_y + 2b_\times c_\times(a_x a_y + a_{xy}a_{yx}) \\ &\quad - 2b_\times(a_x a_{yx} c_y + a_y a_{xy} c_x) - 2c_\times(a_x a_{xy} b_y + a_y a_{yx} b_x).\end{aligned} \tag{12.26}$$

In x–y-separable optical systems, $b_\times = 0 = c_\times$, and so $\Delta_{\text{cross}} = 0$. •

Eigenvalues of Hamiltonian matrices come in quartets and are determined by the two real parameters (Γ, Δ) that we represent on the plane of Fig. 12.3. The subsets with two coincident eigenvalues correspond to boundaries that divide this plane into open regions, while four coincident eigenvalues occur only at the origin. The eigenvalue patterns in the complex λ-plane which characterize nonequivalent orbits of paraxial optical guide systems are:

BOUNDARY	REGION	BOUNDARY
$\Gamma<0,\ \Delta=\frac{1}{4}\Gamma^2$	$\Gamma\in\mathsf{R},\ \Delta>\frac{1}{4}\Gamma^2$	$\Gamma>0,\ \Delta=\frac{1}{4}\Gamma^2$
REGION	POINT	REGION
$\Gamma<0,\ 0\leq\Delta\leq\frac{1}{4}\Gamma^2$	$\Gamma=0,\ \Delta=0$	$\Gamma>0,\ 0<\Delta<\frac{1}{4}\Gamma^2$
BOUNDARY	REGION	BOUNDARY
$\Gamma<0,\ \Delta=0$	$\Gamma\in\mathsf{R},\ \Delta<0$	$\Gamma>0,\ \Delta=0$

(12.27)

The scale factor $\alpha \neq 0$ in the definition of Hamiltonian orbits (12.2) makes the points (Γ, Δ) and $(\alpha^2\Gamma, \alpha^4\Delta)$ equivalent. Hamiltonian orbits in $\mathsf{sp}(4,\mathsf{R})$ correspond thus to the convertical half-parabolas that foliate the (Γ, Δ) plane of Fig. 12.3, except for the origin, where lies the four-fold zero eigenvalue. We can intercept these parabolas with a circle $\mathcal{S}^1$, whose distinct points will thus determine distinct eigenvalue patterns up to scale, and distinguish between orbits of $\mathsf{sp}(4,\mathsf{R})$ Hamiltonians. And there is $\{0\}$, the four-fold zero pattern at the origin.

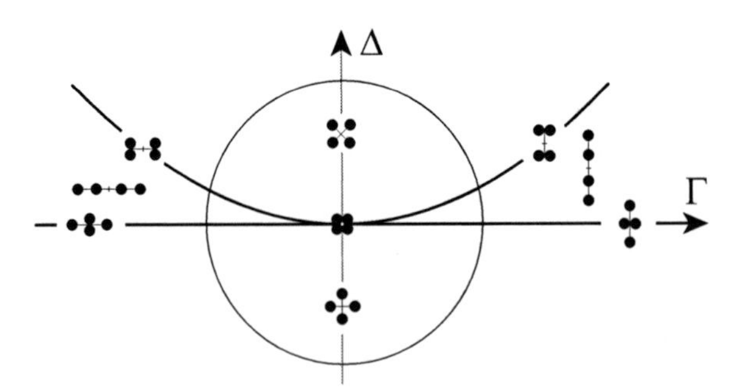

Fig. 12.3. The (Γ, Δ) plane of eigenvalues of $\mathsf{sp}(4,\mathsf{R})$ Hamiltonians and their patterns. There are four regions in the plane, separated by four boundaries (two half-lines, two half-parabolas), and one osculation point between these boundaries. When the system is scaled by $\alpha \neq 0$, the points of the plane slide on the convertical parabolas $(\alpha^2\Gamma, \alpha^4\Delta)$ representing Hamiltonian orbits. These are projected on the points on the circle – plus the center.

But is $\mathcal{S}^1 + \{0\}$ the manifold of $\mathsf{sp}(4,\mathsf{R})$ Hamiltonian orbits? The answer is *no*, because while in the $D = 1$ case the similarity transformation is performed with general matrices $\mathbf{M} \in \mathsf{Sp}(2,\mathsf{R}) = \mathsf{Sl}(2,\mathsf{R})$ and the eigenvalues indeed provide the complete orbit classification, in the $D = 2$ case of $\mathsf{sp}(4,\mathsf{R})$, Hamiltonian orbits are equivalence classes under similarity by matrices which are not general, but only symplectic: $\mathsf{Sp}(4,\mathsf{R}) \subset \mathsf{Sl}(4,\mathsf{R})$ is a proper subgroup. Classes of Hamiltonians which are equivalent under $\mathsf{Sl}(4,\mathsf{R})$ may break up into separate subclasses when the equivalence is restricted to $\mathsf{Sp}(4,\mathsf{R})$. So we have a sufficiency criterion: two distinct points in $\mathcal{S}^1 + \{0\}$ correspond to two distinct $\mathsf{sp}(4,\mathsf{R})$ orbits; but two symplectic-inequivalent Hamiltonians do not necessarily fall on distinct points in Fig. 12.3.

Example: Manifold of separable Hamiltonians. Paraxial optical systems built with cylindrical lenses aligned with the x- and y-directions of the screen, are termed $(x$–$y)$-*separable*. Separable guides are generated by Hamiltonians of the generic form

$$G_\theta(\mathbf{p},\mathbf{q}) := \alpha(G(p_x,q_x)\cos\theta + G'(p_y,q_y)\sin\theta), \quad 0 \neq \alpha \in \mathsf{R}, \quad G, G' \in \{H, F, R\}, \text{ as in (12.7).} \tag{12.28}$$

Since $\pm\alpha$ makes G_θ and $G_{\theta+\pi}$ equivalent, the range of θ restricts to $-\frac{1}{2}\pi < \theta \leq \frac{1}{2}\pi$. •

Remark: Reduction by Fourier transformation. The central Fourier transformation $\mathcal{F}$, which is isotropic and separable, leaves the harmonic generators H invariant but changes the sign of the repulsive ones, $\mathcal{F} : R = -R$. Thus, when one of the summands of (12.28) is R, then the range of θ is reduced to $0 \leq \theta \leq \frac{1}{2}\pi$. If the two summands are R, the range is further reduced to $0 \leq \theta \leq \frac{1}{4}\pi$ with the Fourier transforms $\mathcal{F}_x$ and $\mathcal{F}_y$. •

Remark: Reduction by pure magnifiers. A separable pure magnifier in the y-direction, $\exp(\tau\{I_y, \circ\})$ generated by $p_y q_y$, will multiply a free-propagation summand F_y in (12.28) by $e^{2\tau} > 0$, while leaving $G(p_x, q_x)$ invariant. Hence, in the cases involving free Hamiltonians, the ratio of the x- and y-summands is equivalent to $1 : \sigma$, $\sigma \in \{+1, 0, -1\}$; this index substitutes the continuous parameter θ to characterize those inequivalent orbits. •

Remark: Reduction by rotation. A rotation by $\frac{1}{2}\pi$ around the optical axis of a separable guide exchanges the summands $G \leftrightarrow G'$ in (12.28), rendering the two systems equivalent, and so the 9 cases G–G' fold into 6. In the 3 cases $G = G'$, this implies the equivalence of G_θ and $G_{\frac{1}{2}\pi-\theta}$ and halves the range to $-\frac{1}{4}\pi \leq \theta \leq \frac{1}{4}\pi$. Note that rotations are *not* separable transformations (except for angles 0 and π). •

Example: Orbits of separable Hamiltonians. Separable Hamiltonians contain no cross terms $b_\times, c_\times$, so their $\mathsf{sp}(4,\mathsf{R})$ representation matrices (12.19) display a checkerboard pattern of zeros, and belong to the proper subalgebra $\mathsf{sp}(2,\mathsf{R})_x \oplus \mathsf{sp}(2,\mathsf{R})_y \subset \mathsf{sp}(4,\mathsf{R})$ of vector dimension $3 + 3 = 6$. Their coordinates in the eigenvalue plane (Γ, Δ) [see (12.5), (12.22) and (12.24)–(12.26)] are

$$\begin{aligned} \Gamma_\theta &= \Delta_x + \Delta_y = \alpha^2(\sigma_x \cos^2\theta + \sigma_y \sin^2\theta), \\ \Delta_\theta &= \quad \Delta_x\, \Delta_y \quad = \tfrac{1}{4}\alpha^4\, \sigma_x\sigma_y \sin^2 2\theta, \end{aligned} \tag{12.29}$$

where $\Delta_x := -(a_x^2 + b_x c_x)$ has the sign $\sigma_x = \operatorname{sign}\Delta_x$; this can be $+1$ (harmonic case H), 0 (free case F) or -1 (repulsive case R) – and correspondingly for y. Separable Hamiltonians thus fall on the following regions of the eigenvalue plane:

$$\begin{array}{lll} \text{H-H}_\theta\text{: } \Gamma = \alpha^2 > 0, & 0 \le \Delta \le \tfrac{1}{4}\Gamma^2, \\ \text{H-F}_\theta\text{: } \Gamma = \alpha^2\cos^2\theta > 0, & \Delta = 0, \\ \text{H-R}_\theta\text{: } \Gamma = \tfrac{1}{2}\alpha^2 \cos 2\theta \in \mathsf{R}, & \Delta \le 0, \\ \text{F-F }\ \text{: } \Gamma = 0, & \Delta = 0, \\ \text{R-F}_\theta\text{: } \Gamma = -\alpha^2\cos^2\theta < 0, & \Delta = 0, \\ \text{R-R}_\theta\text{: } \Gamma = -\alpha^2 < 0, & 0 \le \Delta \le \tfrac{1}{4}\Gamma^2, \end{array} \tag{12.30}$$

In the eigenvalue plane of Fig. 12.3, separable Hamiltonians occupy the three regions which are under and include the parabolic boundary. The region above the parabola corresponds to patterns of four complex eigenvalues, , and harbors systems which are not separable. Also, we may suspect that the F-F orbit at the center of the (Γ, Δ) plane, cohabits the eigenvalue with other nonseparable orbits. •

Remark: Eigenvalue quartets for $\mathsf{Sp}(2D, \mathsf{R})$. For generic dimension $2D \times 2D$, the eigenvalues of Hamiltonian matrices also come in symmetric and complex conjugate quartets $\{\lambda_j, -\lambda_j, \lambda_j^*, -\lambda_j^*\}$, $j = 1, 2, \ldots, \frac{1}{2}D$ for D even; and there will be one more symmetric pair for D odd [85]. A general reference on the symplectic eigen-problem is [49]. •

12.4 Hamiltonians in the $\mathsf{so}(3, 2)$ basis

The orbit analysis of astigmatic optical Hamiltonians is facilitated by the fortunate accident $\mathsf{sp}(4, \mathsf{R}) = \mathsf{so}(3, 2)$. This is analogous to the acceident $\mathsf{sp}(2, \mathsf{R}) = \mathsf{so}(2, 1)$ that we used in Sect. 12.1 to visualize the manifold of Hamiltonians and their generated trajectories in the group as intersections of planes and conic surfaces in 3 dimensions. Now we shall have conics in 5 dimensions and a 10-parameter Lie algebra [see Sect. 11.7, pages 244–245]. One major advantage of using the pseudo-orthogonal algebras and groups is that we can argue transformations and orbit equivalences of Hamiltonians through rotations and boosts[3] of vectors in successive planes, thus avoiding

[3] We are using the word *boost* borrowed from special relativity: a hyperbolic linear transformation between a spacelike and a timelike coordinate.

the explicit use of large matrices. On this basis we shall report from [78] the resolution of the orbit degeneracies that exist on the (Γ, Δ) eigenvalue plane of the previous section. (No more 'accidents' of this kind occur in higher dimensions.)

Remark: Notation for $D = 1$ plane systems. The elements of $\mathsf{so}(2,1) = \mathsf{sp}(2,\mathsf{R})$ can be realized as differential operators on an abstract space of 3-vectors $\vec{v} = (v_1, v_2, v_{\bar{3}})^\top \in \mathsf{R}^3$, where we indicate by bars the coordinates of negative metric, here $(+1,+1,-1)$ [see (11.22)–(11.25), (11.113) and (11.130).] The orbit representatives chosen in (12.7) then correspond to the following operators, Hamiltonians and guide systems:

$$\begin{array}{llll} \text{Elliptic:} & J_{1,2} = -i(v_1\partial_2 - v_2\partial_1) & \leftrightarrow \tfrac{1}{2}H := \tfrac{1}{4}(p^2+q^2) & \text{harmonic,} \\ \text{Hyperbolic:} & \left\{ \begin{array}{l} J_{1,\bar{3}} = -i(v_1\partial_{\bar{3}} + v_{\bar{3}}\partial_1) \\ J_{2,\bar{3}} = -i(v_2\partial_{\bar{3}} + v_{\bar{3}}\partial_2) \end{array} \right. & \begin{array}{l} \leftrightarrow \tfrac{1}{2}R := \tfrac{1}{4}(p^2-q^2) \\ \leftrightarrow \tfrac{1}{2}I := \tfrac{1}{2}pq \end{array} & \begin{array}{l} \text{repulsive,} \\ \text{magnifier,} \end{array} \\ \text{Parabolic:} & \left\{ \begin{array}{l} J_{1,2} + J_{1,\bar{3}} \\ J_{1,2} - J_{1,\bar{3}} \end{array} \right. & \begin{array}{l} \leftrightarrow F := \tfrac{1}{2}p^2 \\ \leftrightarrow L := \tfrac{1}{2}q^2 \end{array} & \begin{array}{l} \text{free,} \\ \text{lens.} \end{array} \end{array} \tag{12.31}$$

Although the magnifier is not a 'true' optical guide generator, it will also be of notational use below. •

With the previous notation for the x- and y-axes of $D = 2$-dim screens, we can write the basis for $\mathsf{so}(3,2)$ given by the ten $j_{\mu,\nu}$'s, $\mu, \nu \in \{1, 2, 3, \bar{4}, \bar{5}\}$, in (11.140)–(11.150) and identify the Hamiltonians as follows:

HARMONIC: Fourier generators:

$$j_{1,2} = \tfrac{1}{2}(q_x p_y - q_y p_x), \qquad \tfrac{1}{2}\text{-angular momentum,} \tag{12.32}$$

$$j_{1,3} = \tfrac{1}{4}(-p_x^2 + p_y^2 - q_x^2 + q_y^2), \quad \text{counter-harmonic } \tfrac{1}{2}(-H_x + H_y), \tag{12.33}$$

$$j_{2,3} = \tfrac{1}{2}(p_x p_y + q_x q_y), \qquad \text{cross-harmonic,} \tag{12.34}$$

$$j_{\bar{4},\bar{5}} = \tfrac{1}{4}(|\mathbf{p}|^2 + |\mathbf{q}|^2), \qquad \text{isotropic harmonic } \tfrac{1}{2}(H_x + H_y). \tag{12.35}$$

REPULSIVE generators:

$$j_{1,\bar{4}} = \tfrac{1}{4}(p_x^2 - p_y^2 - q_x^2 + q_y^2), \qquad \text{counter-repulsive } \tfrac{1}{2}(R_x - R_y), \tag{12.36}$$

$$j_{2,\bar{4}} = \tfrac{1}{2}(-p_x p_y + q_x q_y), \qquad -\text{cross-repulsive,} \tag{12.37}$$

$$j_{3,\bar{5}} = \tfrac{1}{4}(|\mathbf{p}|^2 - |\mathbf{q}|^2), \qquad \text{isotropic repulsive } \tfrac{1}{2}(R_x + R_y). \tag{12.38}$$

MAGNIFIER generators:

$$j_{1,\bar{5}} = \tfrac{1}{2}(p_x q_x - p_y q_y), \qquad \text{counter-magnifier } \tfrac{1}{2}(I_x - I_y), \tag{12.39}$$

$$j_{2,\bar{5}} = \tfrac{1}{2}(p_x q_y + p_y q_x), \qquad \text{cross-magnifier,} \tag{12.40}$$

$$j_{3,\bar{4}} = \tfrac{1}{2}\mathbf{p}\cdot\mathbf{q}, \qquad \text{isotropic magnifier } \tfrac{1}{2}(I_x + I_y). \tag{12.41}$$

The 4 harmonic Hamiltonians are in the compact subalgebra $\mathsf{u}(2) = \mathsf{so}(2) \oplus \mathsf{so}(3) \subset \mathsf{so}(3,2)$ that generates $\mathsf{U}(2)$-Fourier transforms (recall Sect. 10.3).

The 3 repulsive and 3 magnifier Hamiltonians are the remaining generators of 6 boosts. Free propagation Hamiltonians and lens generators are obtained as sums and differences of a harmonic and a repulsive generator:

FREE displacement generators:

$$j_{1,-} := j_{1,3} - j_{1,\bar{4}} = -\tfrac{1}{2}(p_x^2 - p_y^2), \quad \text{counter-free } -F_x + F_y, \tag{12.42}$$

$$j_{2,-} := j_{2,3} - j_{2,\bar{4}} = p_x p_y, \quad \text{cross-free}, \tag{12.43}$$

$$j_{\bar{5},-} := j_{\bar{5},3} - j_{\bar{5},\bar{4}} = \tfrac{1}{2}(p_x^2 + p_y^2), \quad \text{isotropic free } F_x + F_y. \tag{12.44}$$

Remark: Euclidean subalgebras of $\mathsf{so}(3,2)$. The 3 free Hamiltonians $j_{k,-}$ in (12.42)–(12.44) mutually Poisson-commute. They are the generators of translations in the (pseudo-) Euclidean subalgebra $\mathsf{iso}(2,1)_- \subset \mathsf{so}(3,2)$, together with the generator of rotation $j_{1,2}$, and of boosts $j_{1,\bar{5}}$ and $j_{2,\bar{5}}$. Lens generators ($\sim q_i q_j$) are the Fourier transforms ($\mathbf{p} \mapsto \mathbf{q}$, $\mathbf{q} \mapsto -\mathbf{p}$) of the generators of free displacements $j_{\alpha,-}$, namely $j_{\alpha,+} := j_{\alpha,3} + j_{\alpha,\bar{4}}$, $\alpha \in \{1, 2, \bar{5}\}$. Together with the same $j_{1,2}$, $j_{1,\bar{5}}$, and $j_{2,\bar{5}}$, the three lens generators form an $\mathsf{iso}(2,1)_+$ subalgebra which is Fourier-equivalent to the previous one. •

The generic $\mathsf{so}(3,2)$ Hamiltonian is a linear combination of the 10 previous generators in the $\mathsf{so}(3,2)$ basis,

$$H(\mathbf{r}; \mathbf{p}, \mathbf{q}) = \sum_{1 \le m < n \le 3} r_{m,n}\, j_{m,n} + \sum_{m=1,2,3} \sum_{\bar{n}=\bar{4},\bar{5}} r_{m,\bar{n}}\, j_{m,\bar{n}} + r_{\bar{4},\bar{5}}\, j_{\bar{4},\bar{5}} \tag{12.45}$$

[*cf.* the monomial form in (11.71)]. The 10 coefficients $\mathbf{r} := \|r_{\alpha,\beta}\|$ can be organized as the entries of a 5×5 matrix (11.151). Since this is an infinitesimal pseudo-orthogonal matrix, it will be sufficient to display only its elements above its diagonal. We use again the pattern of generators in (11.116), here adapted to distinguish between the compact rotation and the noncompact boost generators, as follows:

$$H(\mathbf{r}; \mathbf{p}, \mathbf{q}) \leftrightarrow \mathcal{H}(\mathbf{r}) := \begin{array}{cc|cc} r_{1,2} & r_{1,3} & r_{1,\bar{4}} & r_{1,\bar{5}} \\ & r_{2,3} & r_{2,\bar{4}} & r_{2,\bar{5}} \\ & & r_{3,\bar{4}} & r_{3,\bar{5}} \\ & & & r_{\bar{4},\bar{5}} \end{array} \quad \text{are Hamiltonian parameters of:} \tag{12.46}$$

angular momentum	counter harmonic	counter repulsive	counter magnifier
	cross harmonic	cross repulsive	cross magnifier
		isotropic magnifier	isotropic repulsive
			isotropic harmonic

(12.47)

The relation between the $\mathsf{sp}(4,\mathsf{R})$ and $\mathsf{so}(3,2)$ bases of Hamiltonian coefficients, (12.19) and (12.45) respectively, is given by (11.151)–(11.152) in the previous chapter.

Example: Manifold of U(2)-Fourier generators. The region of the (Γ, Δ) eigenvalue plane which is covered by the Hamiltonians $H(\mathbf{r}_{\rm F}; \mathbf{p}, \mathbf{q})$ contained in the first and last summands in (12.45), which generate the maximal compact subgroup $\mathsf{U}(2) \subset \mathsf{Sp}(4, \mathsf{R})$ of Fourier transforms [using (12.46), (11.151)–(11.152), and (12.22)–(12.26)], is

$$\mathcal{H}(\mathbf{r}_{\rm F}) = \begin{array}{|cc|cc|} \hline r_{1,2} & r_{1,3} & 0 & 0 \\ & r_{2,3} & 0 & 0 \\ & & 0 & 0 \\ \hline & & & r_{\bar{4},\bar{5}} \\ \hline \end{array} \leftrightarrow H_{\rm F}(\mathbf{p}, \mathbf{q}) = \sum_{1 \le m < n \le 3} r_{m,n}\, j_{m,n} + r_{\bar{4},\bar{5}}\, j_{\bar{4},\bar{5}},$$

$$\leftrightarrow \mathbf{m}(\mathbf{r}_{\rm F}) = \frac{1}{2} \left(\begin{array}{cc|cc} 0 & r_{1,2} & -r_{1,3} + r_{\bar{4},\bar{5}} & r_{2,3} \\ -r_{1,2} & 0 & r_{2,3} & r_{1,3} + r_{\bar{4},\bar{5}} \\ \hline r_{1,3} - r_{\bar{4},\bar{5}} & -r_{2,3} & 0 & r_{1,2} \\ -r_{2,3} & -r_{1,3} - r_{\bar{4},\bar{5}} & -r_{1,2} & 0 \end{array} \right)$$

$$\Rightarrow \begin{cases} \Gamma = \frac{1}{2}\left(r_{1,2}^2 + r_{1,3}^2 + r_{2,3}^2 + r_{\bar{4},\bar{5}}^2\right) > 0, \\ \Delta = \frac{1}{16}\left(r_{1,2}^2 + r_{1,3}^2 + r_{2,3}^2 - r_{\bar{4},\bar{5}}^2\right)^2 \Rightarrow 0 \le \Delta \le \frac{1}{4}\Gamma^2. \end{cases} \tag{12.48}$$

In Fig. 12.3, these Hamiltonians occupy the region of eigenvalue patterns , including its two boundaries, and . •

Example: Orbits of U(2)-Fourier generators. The three Hamiltonians that generate SU(2)-Fourier transforms, $j_{1,2}$, $j_{1,3}$, $j_{2,3}$, can be rotated amongst themselves, so can always choose $j_{1,3} = \frac{1}{2}(-H_x + H_y)$ as representative; on the other hand, the isotropic harmonic Hamiltonian $j_{\bar{4},\bar{5}} = \frac{1}{2}(H_x + H_y)$ that generates the central U(1) fractional Fourier transforms, remains invariant. The orbit representatives for the compact Fourier subalgebra $\mathsf{u}(1) \oplus \mathsf{su}(2) = \mathsf{u}(2) \subset \mathsf{sp}(4, \mathsf{R})$ are

$$\begin{gathered} H(\alpha_x{:}\alpha_y; \mathbf{p}, \mathbf{q}) = \alpha_x H_x + \alpha_y H_y = (-\alpha_x + \alpha_y) j_{1,3} + (\alpha_x + \alpha_y) j_{\bar{4},\bar{5}}, \\ \Gamma = \alpha_x^2 + \alpha_y^2 > 0, \quad \Delta = \alpha_x^2 \alpha_y^2 \ge 0, \quad \tfrac{1}{4}\Gamma^2 - \Delta \ge 0, \end{gathered} \tag{12.49}$$

for every ratio $\alpha_x : \alpha_y$, and where we can normalize to $\Gamma = 1$. A two-fold degeneracy between $H(\alpha_x{:}\alpha_y)$ and $H(\alpha_y{:}\alpha_x)$ is present at every point (Γ, Δ) of this region. As we shall see below, the boundaries of this region have even higher degeneracy. •

Example: Isotropic systems with/out angular momentum. Guide systems which are invariant under rotations around the optical axis are generated by the three isotropic Hamiltonians $j_{3,\bar{4}}$, $j_{3,\bar{5}}$, $j_{\bar{4},\bar{5}}$ in the pattern (12.47), and by angular momentum $j_{1,2}$. These close under Poisson brackets into the 'isotropic' Lie subalgebra $\mathsf{so}(2)_{1,2} \oplus \mathsf{so}(2, 1)_{\rm iso} \subset \mathsf{sp}(4, \mathsf{R})$. For these systems,

$$\mathcal{H}(\mathbf{r}_{\rm iso}) = \begin{array}{|cc|cc|} \hline r_{1,2} & 0 & 0 & 0 \\ & 0 & 0 & 0 \\ & & r_{3,\bar{4}} & r_{3,\bar{5}} \\ \hline & & & r_{\bar{4},\bar{5}} \\ \hline \end{array} \leftrightarrow H_{\rm iso}(\mathbf{p}, \mathbf{q}) = r_{1,2} j_{1,2} + \sum_{3 \le m < n \le \bar{5}} r_{m,n}\, j_{m,n},$$

$$\leftrightarrow \mathbf{m}(\mathbf{r}_{\rm iso}) = \frac{1}{2} \left(\begin{array}{cc|cc} r_{3,\bar{4}} & r_{1,2} & -r_{3,\bar{5}} + r_{\bar{4},\bar{5}} & 0 \\ -r_{1,2} & r_{3,\bar{4}} & 0 & -r_{3,\bar{5}} + r_{\bar{4},\bar{5}} \\ \hline -r_{3,\bar{5}} - r_{\bar{4},\bar{5}} & 0 & -r_{3,\bar{4}} & r_{1,2} \\ 0 & -r_{3,\bar{5}} - r_{\bar{4},\bar{5}} & -r_{1,2} & -r_{3,\bar{4}} \end{array} \right)$$

$$\Rightarrow \begin{cases} \Gamma = \frac{1}{2}(r_{1,2}^2 - r_{3,\bar{4}}^2 - r_{3,\bar{5}}^2 + r_{\bar{4},\bar{5}}^2) \in \mathsf{R}, \\ \Delta = \frac{1}{16}(r_{1,2}^2 + r_{3,\bar{4}}^2 + r_{3,\bar{5}}^2 - r_{\bar{4},\bar{5}}^2)^2 \\ \quad = \frac{1}{4}\Gamma^2 + \frac{1}{4}r_{1,2}^2(r_{3,\bar{4}}^2 + r_{3,\bar{5}}^2 - r_{\bar{4},\bar{5}}^2) \geq 0. \end{cases} \tag{12.50}$$

Isotropic systems *without* angular momentum ($r_{1,2} = 0$) thus occupy the parabolic boundary $\Delta = \frac{1}{4}\Gamma^2$ of Fig. 12.3, and include the three eigenvalue patterns , and . When angular momentum is present (as in particle guides with a coaxial magnetic field), the system will above the $\Delta = 0$ line, and below, on, or above the parabola $\Delta = \frac{1}{4}\Gamma^2$, according to whether $r_{3,\bar{4}}^2 + r_{3,\bar{5}}^2 - r_{\bar{4},\bar{5}}^2$ is less, equal, or greater than zero, corresponding to harmonic, free, or repulsive guides. •

12.5 Equivalence under Fourier transformers and magnifiers

To classify the set of linear optical Hamiltonians $H(\mathbf{r}; p, q)$, $\mathbf{r} \in \mathsf{so}(3,2)$, into inequivalent orbits, and to find appropriate representatives $\mathbf{r}_\circ$ for each orbit, our strategy will be to transform the $\mathsf{so}(3,2)$ patterns of coefficients through similarity, $\mathbf{r} \mapsto \mathbf{T}\,\mathbf{r}\,\mathbf{T}^{-1}$, with $\mathbf{T} \in \mathsf{SO}(3,2)$ being produced by succesive rotations and boosts. We ought to arrive thus at equivalent but *reduced* patterns that will contain as many zeros as possible, and where inequivalence should be manifest [see (12.45)–(12.47).]

Similarity transformations by a rotation or boost in the (α,β)-plane of R^5, are exponential Lie-Poisson operators that we denote by

$$T^{(\alpha,\beta)}(\tau) := \exp(\tau\,\{j_{\alpha,\beta}, \circ\}) \in \mathsf{SO}(3,2). \tag{12.51}$$

These produce linear combinations of the coefficients $r_{\rho,\kappa}$ in the Hamiltonians $H(\mathbf{r};\mathbf{p},\mathbf{q}) = \sum_{\rho,\kappa} r_{\rho,\kappa}\, j_{\rho,\kappa}(\mathbf{p},\mathbf{q})$ through matrices $\mathbf{T}^{(\alpha,\beta)}(\tau)$, which act on the rows ρ (from the left) or on the columns κ (from the right). This action can be subsumed, for any two commuting one-parameter subgroups ($\alpha,\beta,\gamma,\delta$ all distinct), by writing only the affected elements in the 2×2 sub-pattern,

$$T^{(\alpha,\beta)}(\tau)\,T^{(\gamma,\delta)}(\tau') : \boxed{\begin{matrix} r_{\alpha,\gamma} & r_{\alpha,\delta} \\ r_{\beta,\gamma} & r_{\beta,\delta} \end{matrix}} = \mathbf{T}^{(\alpha,\beta)}(\tau) \boxed{\begin{matrix} r_{\alpha,\gamma} & r_{\alpha,\delta} \\ r_{\beta,\gamma} & r_{\beta,\delta} \end{matrix}}\, \mathbf{T}^{(\gamma,\delta)}(\tau')^{\top}. \tag{12.52}$$

The transforming submatrices contain trigonometric or hyperbolic functions of $\tau \in \mathcal{S}^1$ or $\tau \in \mathsf{R}$ respectively,

$$\mathbf{T}^{(\alpha,\beta)}(\tau) = \begin{cases} \begin{pmatrix} \cos\tau & -\sin\tau \\ \sin\tau & \cos\tau \end{pmatrix}, & \text{when } \alpha \in \{1,2,3\} \ni \beta, \\ & \text{or } \alpha \in \{\bar{4},\bar{5}\} \ni \beta; \\ \begin{pmatrix} \cosh\tau & \sinh\tau \\ \sinh\tau & \cosh\tau \end{pmatrix}, & \text{when } \alpha \notin \{1,2,3\} \ni \beta, \\ & \text{or } \alpha \in \{\bar{4},\bar{5}\} \not\ni \beta. \end{cases} \tag{12.53}$$

The values of τ and/or τ' that are needed to bring one or two of the coefficients in this sub-pattern to zero are obtained directly from (12.52).

We **rotate** the Hamiltonian $\mathcal{H}(\mathbf{r})$ by $\mathsf{U}(2)$-Fourier transforms. The effect of these rotations in the planes $(1,2)$, $(2,3)$ and $(\bar{4},\bar{5})$ of R^5 on the coefficients of the pattern (12.46), is to linearly combine them as follows:

$$\begin{array}{|c|cc|}\hline \bullet\, r_{1,2}\ \updownarrow & \updownarrow & \updownarrow \\ \hline & r_{3,\bar{4}} & r_{3,\bar{5}} \\ & & r_{\bar{4},\bar{5}} \\ \hline\end{array}\,, \qquad \begin{array}{|c|cc|}\hline \longleftrightarrow & r_{1,\bar{4}} & r_{1,\bar{5}} \\ \bullet\, r_{2,3} & \updownarrow & \updownarrow \\ & & r_{\bar{4},\bar{5}} \\ \hline\end{array}\,, \qquad \begin{array}{|cc|c|}\hline r_{1,2} & r_{1,3} & \longleftrightarrow \\ & r_{2,3} & \longleftrightarrow \\ & & \longleftrightarrow \\ & & \bullet\, r_{\bar{4},\bar{5}} \\ \hline\end{array} \qquad (12.54)$$

where we indicate the generator of the rotation with $\bullet$, connect by arrows the coefficient sites that are linearly combined, and leave in place explicitly those which are invariant. With 4 successive rotations one can annul 4 coefficients in the Hamiltonian pattern (12.46) and reduce it to a form with at most 6 remaining parameters,

$$\mathsf{U}(2) : \mathcal{H}(\mathbf{r}) \mapsto \mathcal{H}_{\mathrm{U}} = \begin{array}{|cc|cc|}\hline m & r & s & u \\ & 0 & 0 & 0 \\ & & 0 & v \\ & & & w \\ \hline\end{array}\,. \qquad (12.55)$$

The conclusion is that any quadratic Hamiltonian is $\mathsf{U}(2)$-Fourier equivalent to a Hamiltonian containing an angular momentum term with coefficient m, counter-(harmonic, repulsive, magnifier) terms with $(r,\, s,\, u)$, and isotropic-(repulsive, harmonic) terms with (v, w). Except for angular momentum $m = r_{1,2}$, the coefficients $r_{2,\kappa}$ of all *cross* terms are thus eliminated.

We **boost** Hamiltonians $\mathcal{H}(\mathbf{r})$ in R^5 with the noncompact generators of $\mathsf{SO}(3,2)$. An isotropic magnifier in the plane $(3,\bar{4})$, generated by $j_{3,\bar{4}}$ in (12.41), produces the following linear combination between their parameters with the hyperbolic matrix in (12.53),

$$T^{(3,\bar{4})}(\tau) : \begin{array}{|cc|cc|}\hline r_{1,2} & r_{1,3} & r_{1,\bar{4}} & r_{1,\bar{5}} \\ & r_{2,3} & r_{2,\bar{4}} & r_{2,\bar{5}} \\ & & r_{3,\bar{4}} & r_{3,\bar{5}} \\ & & & r_{\bar{4},\bar{5}} \\ \hline\end{array} \quad = \quad \begin{array}{|c|ccc|}\hline r_{1,2} & \longleftrightarrow & & r_{1,\bar{5}} \\ & \longleftrightarrow & & r_{2,\bar{5}} \\ & \bullet\, r_{3,\bar{4}} & & \updownarrow \\ \hline\end{array}. \qquad (12.56)$$

When we apply this transformation to produce one more zero in the Hamiltonian coefficient pattern, as we know from special relativity, we find that there are three nonequivalent cases: timelike, spacelike, and lightlike, for each of the three pairs of coefficients that are bound by arrows in (12.56).

Applying (12.56) on the previous Fourier-reduced Hamiltonian $\mathcal{H}_{\mathrm{U}}$, whose coefficient pattern is (12.55), the relativistic trichotomy divides all $\mathsf{so}(3,2)$ Hamiltonians into three coarse equivalence classes which, modulo further rotation and precisions, we indicate by

S: SEPARABLE   L: LORENTZIAN   E: EUCLIDEAN

$$\mathcal{H}_{\mathrm{S}} = \begin{array}{|cc|cc|}\hline 0 & r & s & u \\ & 0 & 0 & 0 \\ & & 0 & 0 \\ & & & w \\ \hline\end{array} \qquad \mathcal{H}_{\mathrm{L}} = \begin{array}{|cc|cc|}\hline m & r & 0 & u \\ & 0 & 0 & 0 \\ & & 0 & v \\ & & & 0 \\ \hline\end{array} \qquad \mathcal{H}_{\mathrm{E}} = \begin{array}{|cc|cc|}\hline m & \pm s & s & u \\ & 0 & 0 & 0 \\ & & 0 & 1 \\ & & & \pm 1 \\ \hline\end{array}$$

$$\in \mathsf{so}(2,2) \qquad\qquad \in \mathsf{so}(3,1) \qquad\qquad \in \mathsf{iso}(2,1) \qquad (12.57)$$

In this way the tensor $r_{\alpha,\beta}$ of Hamiltonian coefficients is brought to lie entirely in one of three subspaces $\mathsf{R}^4 \subset \mathsf{R}^5$ where the eliminated coordinate is, respectively, the spacelike 2, the timelike $\bar{4}$, and the lightlike $3 \pm \bar{4}$. Correspondingly, the Hamiltonian will lie in one of the three distinct subalgebras of $\mathsf{so}(3,2)$. The problem of classifying the Hamiltonian orbits in $\mathsf{so}(3,2)$ is thus reduced to the classification of orbits in the separable algebra $\mathsf{so}(2,2) = \mathsf{so}(2,1) \oplus \mathsf{so}(2,1)$, in the Lorentz algebra $\mathsf{so}(3,1)$, and in the pseudo-Euclidean algebra $\mathsf{iso}(2,1)$. This is analogous to the $D = 1$ case seen in Sect. 12.1, where the division is made according to the three nonequivalent subalgebras of $\mathsf{so}(2,1)$, namely H (harmonic, $\mathsf{so}(2)$), R (repulsive, $\mathsf{so}(1,1)$) and F (free, $\mathsf{iso}(1) = \mathsf{T}_1$). There, this was the full and irreducible classification of orbits; in the $D = 2$ case the process continues within the subalgebras.

12.6 Separable, Lorentzian and Euclidean Hamiltonians

The orbit analyses of $\mathsf{so}(2,2)$, $\mathsf{so}(3,1)$, and $\mathsf{iso}(2,1)$ follow the strategy outlined in the previous section: through similarity transformations, rotations and boosts (and in the Euclidean case also translations), one of a pair of coefficients in the patterns (12.57) is brought to zero (or in the Euclidean case, the two are brought to $\pm$-equality). At each application of a boost or magnifier, the process forks into three branches as in (12.57). When no further reduction is possible, we are left with either a parametrized continuum of Hamiltonian orbits or, if all parameters can be scaled or transformed away, with an isolated orbit. Then, $\mathsf{Sp}(4,\mathsf{R})$ similarity transformations outside each of the generated subgroups are used to reduce the remaining parameters and identify end-branches. We point to [78] for the details of this process in our optical context; at the end of this section we remark on the literature in the framework of mechanics.

Separable Hamiltonians are sums of one function of the x-coordinates of phase space plus another function of the y-coordinates. In the paraxial régime they are quadratic in the coordinates of phase space and have the generic form (12.28)–(12.30), where the range of the linear combination angle $\theta \in \mathcal{S}^1$ was reduced, by separable Fourier transformation, magnification and rotation, to 3 continua of orbits – called *strata* – and 6 isolated orbits. With the previous notation, the orbit structure of separable Hamiltonians is as follows:[4]

[4] The angle θ we use here to parametrize the strata is *distinct* from the angle on the circle.

SEPARABLE		HAMILTONIANS	PARAMETER RANGE
H-H$_\theta$	stratum:	$H_x \cos\theta + H_y \sin\theta,$	$-\frac{1}{4}\pi < \theta \le \frac{1}{4}\pi,$
H-R$_\theta$	stratum:	$H_x \cos\theta + R_y \sin\theta,$	$0 < \theta < \frac{1}{2}\pi,$
R-R$_\theta$	stratum:	$R_x \cos\theta + R_y \sin\theta,$	$0 \le \theta \le \frac{1}{4}\pi,$
H-F$_\sigma$	orbits:	$H_x + \sigma F_y,$	$\sigma \in \{-1, +1\},$
R-F	orbit:	$R_x + F_y,$	
F-F$_\sigma$	orbits:	$F_x + \sigma F_y,$	$\sigma \in \{-1, 0, +1\}.$

(12.58)

We identify these strata and orbits with regions and boundaries of the eigenvalue plane in Fig. 12.4, and with segments and points of the unit circle – and its center.

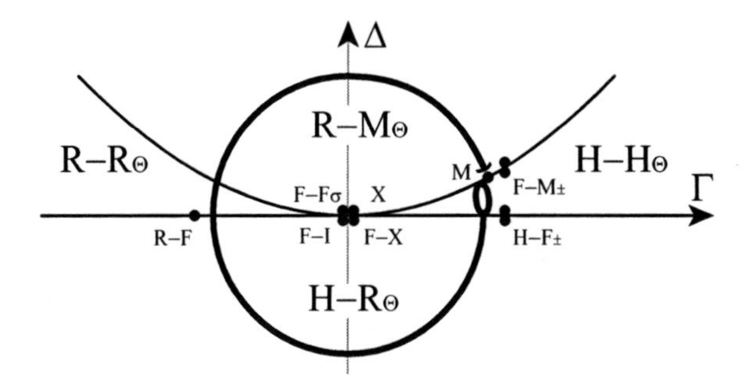

Fig. 12.4. Strata and orbits of paraxial optical Hamiltonians in the (Γ, Δ) plane, projected on the circle and center and identified where degeneracy occurs. There are four strata: Separable H-H$_\theta$ (doubly degenerate), H-R$_\theta$, and R-R$_\theta$, and Lorentzian M-R$_\theta$. On their boundaries are superposed 6 isolated orbits: separable H-F$_\pm$ and R-F, Lorentzian M, and Euclidean F-M$_\pm$. At the center there coexist 6 orbits: separable F-F$_{0,\pm}$, Lorentzian X, and Euclidean F-I and F-X.

Remark: Eigenvalue degeneracies among separable Hamiltonians. The 2-fold degeneracy $\theta \leftrightarrow -\theta$ of eigenvalues in the H-H$_\theta$ region is now resolved through separating the parameter range $-\frac{1}{4}\pi < \theta \le \frac{1}{4}\pi$ into the open range $-\frac{1}{4}\pi < \theta < 0$ for Hamiltonians $c\,H_x + s\,H_x$ where c and s have different signs, and the closed range $0 \le \theta \le \frac{1}{4}\pi$ where they have the same sign or one is zero at the region boundary. Yet the boundaries of this region are not free of degeneracy, the isolated pair of orbits $H_x \pm F_y$ is degenerate with the Hamiltonian H_x of the H-H$_\theta$ stratum, at the positive-Γ axis, which is thus 3-fold degenerate. There is also a 2-fold degeneracy on negative-Γ axis: the Hamiltonians R_x (boundary of the H-R$_\theta$ stratum) and $R_x + F_y$ fall on the same eigenvalue point but are distinct orbits. And the three isolated orbits of free Hamiltonians F-F$_\sigma$ cohabitate the center. •

Lorentzian Hamiltonians are the second outcome in the reduction of the pattern $\mathcal{H}_{\mathrm{U}}$ in (12.56), when the generators have been rotated wholy into the subspace with indices $(1, 2, 3, \bar{5})$. They are elements of a Lorentz algebra $\mathsf{so}(3,1)$ comprising generators of $\mathsf{SU}(2)$-Fourier transforms including angular momentum, counter- and cross-magnifiers, and isotropic repulsive terms. The Lorentzian Hamiltonians of $\mathsf{sp}(4,\mathsf{R}) = \mathsf{so}(3,2)$ are divided into 3 distinct equivalence sets: 1 open stratum [recall the example (12.50) of the isotropic repulsive case with angular momentum] and 2 isolated orbits, whose representatives are chosen as follows [78]:

LORENTZIAN		HAMILTONIANS	PARAMETER RANGE	
M-R$_\theta$	stratum:	$M\cos\theta + R\sin\theta,$	$0 < \theta < \frac{1}{2}\pi,$	(12.59)
M	orbit:	M (angular momentum)		
X	orbit:	$q_x p_y.$		

We identify this stratum and orbits in Fig. 12.4; Lorentzian Hamiltonians occupy the upper region together with its right half-parabola boundary (the left half-parabola is R – a separable Hamiltonian).

Remark: Angular momentum and the Lorentzian stratum. Angular momentum M could in principle be considered as the $\theta = 0$ boundary of the M-R$_\theta$ stratum. But M is compact, so it generates bounded trajectories (cylindrical helices in the guide), while all other Hamiltonians in the stratum are noncompact (exponential spirals), so we choose to list M as an isolated orbit. The example (12.50) comprised angular momentum plus isotropic harmonic, repulsive and magnifier Hamiltonians; the harmonic case M-H$_\theta$ actually belongs to the separable Hamiltonian orbit because [as can be seen in the pattern (12.47) with a $\frac{1}{2}\pi$ rotation in the (2,3) plane] we can rotate angular momentum $j_{1,2}$ to the counter-harmonic $j_{1,3}$; and an isotropic-plus a counter-harmonic Hamiltonian is an anisotropic, separable harmonic Hamiltonian, belonging to the H-H$_\theta$ stratum in (12.58). •

Remark: The double-cross X-orbit. A lightlike case in the reduction process has yielded one quaint orbit, called X, whose generator $p_x q_y = j_{2,\bar{5}} - j_{1,2}+$ (or $p_y q_x$) will 'double-cross' phase space through $p \leftrightarrow q$ and $x \leftrightarrow y$. Its eigenvalues are a fourfold zero, $\oplus$, so X falls on the origin of the (Γ, Δ) plane of Fig. 12.3. We do not see it realized in any optical device. •

Euclidean Hamiltonians are the third branch of the $\mathsf{so}(3,2)$ Hamiltonian orbits in (12.57). The orbit analysis of the Euclidean algebra $\mathsf{iso}(2,1)$, with 3 generators of translations, 1 of rotation and 2 of boosts is a rather arduous exercise using light-cone coordiantes that we shall omit (see [78]). It provides one new pair of significant Hamiltonian orbits and two 'non-optical' ones which are degenerate at zero:

EUCLIDEAN		HAMILTONIANS	
F-M$_\pm$	orbits:	$F \pm M = \frac{1}{2}\lvert\mathbf{p}\rvert^2 \pm (q_x p_y - q_y p_x),$	(12.60)
F-I	orbit:	$p_x p_y + \frac{1}{2}(p_x q_x - p_y q_y),$	
F-X	orbit:	$\frac{1}{2}(p_x + p_y)^2 + p_y q_x.$	

These complete the orbit classification of $\mathsf{sp}(4, \mathsf{R}) = \mathsf{so}(3, 2)$ paraxial optical guide systems and are identified in Fig. 12.4.

Remark: Eigenvalues of Euclidean Hamiltonians. On the (Γ, Δ) eigenvalue plane, the twin Euclidean Hamiltonians fall on the right parabola while the two isolated orbits are at the center of Fig. 12.3 [see (12.22)–(12.26)]. They are degenerate with the separable and Lorentzian Hamiltonian orbits that also occupy this boundary and point. •

Remark: Literature on Hamiltonian orbits. The classification of quantum mechanical Hamiltonians into classes with a physically relevant representative was addressed by Moshinsky and Winternitz in [104] for the case $\mathsf{sp}(4, \mathsf{R})$ of two space dimensions. Nuclear physics models require the quantum numbers provided by a maximal set of commuting operators, so these authors classified the maximal abelian subalgebras, later to be extended to the physically relevant case of $\mathsf{sp}(6, \mathsf{R})$. The generic case for $\mathsf{sp}(2D, \mathsf{R})$ was tackled shortly afterwards in [114]. The Lie algebra $\mathsf{so}(3, 2)$ had merited the attention of Dirac [39], since it serves as model for field theories with a fundamental length [52, 115] and is known under the name of *anti-de Sitter* algebra. (*The* de-Sitter algebra is $\mathsf{so}(4, 1)$.) The maximal abelian subalgebras of all de Sitter-type algebras can be found in [113]. In the context of wave optics, and with the purpose of classifying anisotropic Gaussian Schell-model beams (twisted and untwisted) through their second-order moments, Simon, Sudarshan and Mukunda enumerated the $\mathsf{so}(3, 2)$ orbits in [128] The present treatment follows that of [78]. •

12.7 Evolution along $\mathsf{sp}(4, \mathsf{R})$-guides

We started this chapter on Hamiltonian orbits by following the evolution of ray trajectories along $\mathsf{sp}(2, \mathsf{R})$ guides, i.e., along z-independent but otherwise general paraxial optical systems. Here we find the guide transformations $\mathbf{M}(z) = \exp z\mathbf{m} \in \mathsf{Sp}(4, \mathsf{R})$ for any Hamiltonian matrix $\mathbf{m} \in \mathsf{sp}(4, \mathsf{R})$. The exponential series of 4×4 Hamiltonian matrices can be summed along the lines of the 2×2 case in (12.13)–(12.15). It is generally not an easy proposition to exponentiate matrices larger than 2×2 (see [98, 59]), but in the symplectic case, the Cayley-Hamilton theorem on the biquadratic eigenvalue equation (12.21) implies that

$$\mathbf{m}^4 = -\Gamma \mathbf{m}^2 - \Delta \mathbf{1}, \tag{12.61}$$

where Γ and Δ are given by (12.22) and (12.24)–(12.26).

Example: Powers of $\mathbf{m} \in \mathsf{sp}(4, \mathsf{R})$. The powers $\mathbf{m}^n$ of $\mathbf{m} \in \mathsf{sp}(4, \mathsf{R})$ are expressible as the sum of the two lowest powers of $\mathbf{m}$ with the same parity, and the expressions come in quartets with a recurrence relation that is found from (12.61):

$$\begin{aligned} \mathbf{m}^{4n} &= \alpha_n \mathbf{m}^2 + \beta_n \mathbf{1}, & \alpha_{n+1} &= (\Gamma^2 - \Delta)\alpha_n - \Gamma\beta_n, \\ \mathbf{m}^{4n+1} &= \alpha_n \mathbf{m}^3 + \beta_n \mathbf{m}, & \beta_{n+1} &= \Gamma\Delta\alpha_n - \Delta\beta_n; \\ \mathbf{m}^{4n+2} &= (-\Gamma\alpha_n + \beta_n)\mathbf{m}^2 - \Delta\alpha_n \mathbf{1}, & \alpha_0 &= 0, \quad \alpha_1 = -\Gamma, \\ \mathbf{m}^{4n+3} &= (-\Gamma\alpha_n + \beta_n)\mathbf{m}^3 - \Delta\alpha_n \mathbf{m}; & \beta_0 &= 1, \quad \beta_1 = -\Delta. \end{aligned} \tag{12.62}$$

But because α's and β's are interwined, it is more convenient to search for the eigen-solutions of $(\mathbf{m}^2+v\mathbf{1})\mathbf{m}^4 = \mu(\mathbf{m}^2+v\mathbf{1})$ and thus find closed expressions for the eigenvalues μ and coefficients v, – see [164]. •

Remark: Step-4 sub-series of the exponential function. Few series with step 4 can be summed analytically; but the exponential series does yield:

$$\begin{aligned} \sum \frac{x^{4n}}{(4n)!} &= \tfrac{1}{2}(\cosh x + \cos x), & \sum \frac{x^{4n+1}}{(4n+1)!} &= \tfrac{1}{2}(\sinh x + \sin x), \\ \sum \frac{x^{4n+2}}{(4n+2)!} &= \tfrac{1}{2}(\cosh x - \cos x), & \sum \frac{x^{4n+3}}{(4n+3)!} &= \tfrac{1}{2}(\sinh x - \sin x), \end{aligned} \tag{12.63}$$

where all sums $\sum$ are over $n|_0^\infty$. •

Using the techniques outlined above one finds that the exponential series of a 4×4 Hamiltonian matrix $\mathbf{m}$ is

$$\exp z\mathbf{m} = E_0(z)\,\mathbf{1} + E_1(z)\,\mathbf{m} + E_2(z)\,\mathbf{m}^2 + E_3(z)\,\mathbf{m}^3, \tag{12.64}$$

$$\begin{aligned} E_0(z) &= \tfrac{1}{2}(\cosh z\lambda^+ + \cosh z\lambda^-) + \frac{\Gamma}{2\sqrt{\Gamma^2-4\Delta}}(\cosh z\lambda^+ - \cosh z\lambda^-) \\ &= -\frac{1}{2\sqrt{\Gamma^2-4\Delta}}(\lambda^- \cosh z\lambda^+ - \lambda^+ \cosh z\lambda^-), \end{aligned} \tag{12.65}$$

$$\begin{aligned} E_1(z) &= \tfrac{1}{2}\left(\frac{\sinh z\lambda^+}{\lambda^+} + \frac{\sinh z\lambda^-}{\lambda^-}\right) + \frac{\Gamma}{2\sqrt{\Gamma^2-4\Delta}}\left(\frac{\sinh z\lambda^+}{\lambda^+} - \frac{\sinh z\lambda^-}{\lambda^-}\right) \\ &= -\frac{1}{2\sqrt{\Gamma^2-4\Delta}}\left(\frac{\lambda^-}{\lambda^+}\sinh z\lambda^+ + \frac{\lambda^+}{\lambda^-}\sinh z\lambda^-\right), \end{aligned} \tag{12.66}$$

$$E_2(z) = \frac{1}{\sqrt{\Gamma^2-4\Delta}}(\cosh z\lambda^+ - \cosh z\lambda^-), \tag{12.67}$$

$$E_3(z) = \frac{1}{\sqrt{\Gamma^2-4\Delta}}\left(\frac{\sinh z\lambda^+}{\lambda^+} - \frac{\sinh z\lambda^-}{\lambda^-}\right). \tag{12.68}$$

[Cf. (12.13)–(12.16).]

Remark: Check on (12.65)–(12.68). To check that the functions $E_k(z)$ given above indeed are the coefficients of (12.64), we verify that the initial conditions match, $E_k(0) = \delta_{k,0}$, and that $de^{z\mathbf{m}}/dz = \mathbf{m}\,e^{z\mathbf{m}}$ is satisfied:

$$\frac{d}{dz}\exp z\mathbf{m} = E_0'(z)\,\mathbf{1} + E_1'(z)\,\mathbf{m} + E_2'(z)\,\mathbf{m}^2 + E_3'(z)\,\mathbf{m}^3, \tag{12.69}$$

$$\mathbf{m}\exp z\mathbf{m} = -\Delta E_3\,\mathbf{1} + E_0\,\mathbf{m} + (E_1 - \Gamma E_3)\,\mathbf{m}^2 + E_2\,\mathbf{m}^3, \tag{12.70}$$

where primes indicate derivatives. This requires proving $E_1' = E_0$ and $E_3' = E_2$, which are immediate, then $E_0' = -\Delta E_3$, which follows from the identity $\frac{1}{2}(\sqrt{\Gamma^2 - 4\Delta} \pm \Gamma)(\lambda^\pm)^2 = \mp\Delta$, and finally $E_2' = E_1 - \Gamma E_3$, is a consequence of $\frac{1}{2}(\sqrt{\Gamma^2 - 4\Delta} \mp \Gamma)/\lambda^\pm = \pm\lambda^\pm$. •

Remark: Exponential series for inhomogeneous matrices. Inhomogeneous Lie algebras exponentiate to their inhomogeneous Lie groups following the matrix realization where the Hamiltonian matrix occupies the first block and translations the last column [see (5.41) and (12.83)–(12.86) below]. The matrix series which needs to be summed is

$$\mathbf{m}^{-1}(\exp(z\mathbf{m}) - \mathbf{1}) = \sum_{k=0}^{3} G_k(z)\mathbf{m}^k, \quad \begin{array}{ll} G_0 = -\Delta E_3, & G_1 = E_0, \\ G_2 = E_1 - \Gamma E_3, & G_3 = E_2. \end{array} \tag{12.71}$$

•

The coefficients $E_k(z)$ in (12.65)–(12.68) are well defined when the common denominator $\sqrt{\Gamma^2 - 4\Delta}$ is not zero, i.e., everywhere on the (Γ, Δ) eigenvalue plane of $\mathsf{Sp}(4, \mathsf{R})$ Hamiltonians in Fig. 12.3, except on their $\Delta = \frac{1}{4}\Gamma$ boundary parabola. There, two nonzero eigenvalues (12.23) coalesce: $\lambda^+ = \lambda^-$ (real for $\Gamma < 0$ or pure imaginary for $\Gamma > 0$). In particular on that parabola lie all isotropic systems (12.50), for which the 2×2-case must be regained. Also, at the center of the eigenvalue plane we may expect surprises. When two eigenvalues $\lambda^\pm$ approach each other, their point on the (Γ, Δ) plane approaches the parabola by

$$\varepsilon := (\lambda^+)^2 - (\lambda^-)^2 = \sqrt{\Gamma^2 - 4\Delta} \to 0, \quad \varepsilon^2 \approx 0, \tag{12.72}$$

$$(\lambda^\pm)^2 = -\tfrac{1}{2}\Gamma \pm \tfrac{1}{2}\varepsilon \Rightarrow \lambda^\pm \approx \lambda \pm \varepsilon/4\lambda, \quad \lambda := \sqrt{-\tfrac{1}{2}\Gamma}, \tag{12.73}$$

$$\Rightarrow \quad \cosh z\lambda^\pm \approx \cosh z\lambda \pm \tfrac{1}{4} z\varepsilon \frac{\sinh z\lambda}{\lambda}, \tag{12.74}$$

$$\frac{\sinh z\lambda^\pm}{\lambda^\pm} \approx \frac{\sinh z\lambda}{\lambda} \pm \frac{\varepsilon}{4\lambda^2}\left(z \cosh z\lambda - \frac{\sinh z\lambda}{\lambda}\right). \tag{12.75}$$

(Note that ε can be complex.) Now, to find the coefficients $E_k(z)$ of $\exp z\mathbf{m}$ when $\mathbf{m}$ has double eigenvalues $\pm\lambda$, we replace (12.74)–(12.75) in (12.66)–(12.68), and cancel denominators in ε. Then, we let $\varepsilon \to 0$,

$$E_0(z) = \cosh z\lambda - \tfrac{1}{2}\lambda^2 z \frac{\sinh z\lambda}{\lambda}, \tag{12.76}$$

$$E_1(z) = \frac{\sinh z\lambda}{\lambda} - \frac{1}{2}\left(z \cosh z\lambda - \frac{\sinh z\lambda}{\lambda}\right), \tag{12.77}$$

$$E_2(z) = \tfrac{1}{2} z \frac{\sinh z\lambda}{\lambda}, \tag{12.78}$$

$$E_3(z) = \frac{1}{2\lambda^2}\left(z \cosh z\lambda - \frac{\sinh z\lambda}{\lambda}\right), \tag{12.79}$$

where $\lambda^4 = \frac{1}{4}\Gamma^2 = \Delta$.

Remark: Step-2 matrices. The Hamiltonian matrices $\mathbf{m}$ corresponding to any one of the $\mathsf{so}(3, 2)$ generators $j_{\alpha,\beta}$ in (12.32)–(12.41), can be found from (11.152). They lie on the boundary parabola $\Delta = \frac{1}{4}\Gamma^2$ of double eigenvalues. They are also special in that $\mathbf{m}^2 = \lambda^2\mathbf{1}$, as are also the Hamiltonians of isotropic guides. The four summands of the expression for $\exp z\mathbf{m}$ in (12.64) reduce to only two. In the 2×2 case, the determinant Δ appears as $\mathbf{m}^2 = -\Delta\mathbf{1}$; in the 4×4 case it is $\mathbf{m}^2 = \pm\sqrt{|\Delta|}\mathbf{1}$. We sum the coefficients of $\mathbf{1}$ and $\mathbf{m}^2$, and those of $\mathbf{m}$ and $\mathbf{m}^3$, to find that

$$(E_0 + \lambda^2 E_2)(z) = \cosh z\lambda, \qquad (E_1 + \lambda^2 E_3)(z) = \frac{\sinh z\lambda}{\lambda}. \tag{12.80}$$

Thus we recover the two summands of the 2×2 case. •

Hamiltonian systems with double eigenvalues can be slid down the boundary parabola of the (Γ, Δ) plane of Fig. 12.3, to approach the highly orbit-degenerate origin. For $\lambda^4 = \Delta = \frac{1}{4}\Gamma^2 \to 0$, expanding (12.76)–(12.79) in powers of λ, the constant terms yield the series in (12.64) with

$$E_0(z) = 1, \quad E_1(z) = z, \quad E_2(z) = \tfrac{1}{2}z^2, \quad E_3(z) = \tfrac{1}{6}z^3. \tag{12.81}$$

If the Hamiltonian matrix $\mathbf{m}$ is step-1 nilpotent, i.e., $\mathbf{m}^2 = \mathbf{0}$ as in free propagation, the sum (12.64) reduces to $\exp z\mathbf{m} = \mathbf{1} + z\mathbf{m}$.

Remark: Traces of $\mathsf{Sp}(4, \mathsf{R})$ matrices. The value of the trace $\tau(z) := \operatorname{tr} \mathbf{M}(z)$, $\mathbf{M}(z) = \exp z\mathbf{m} \in \mathsf{Sp}(4, \mathsf{R})$, is invariant under similarity transformations. In the 2×2 case (12.18) it uniquely determines the subgroup orbit generated by the Hamiltonian $\mathbf{m} \in \mathsf{sp}(2, \mathsf{R})$, and in particular serves to recognize non-exponential paraxial systems for having traces $\tau < -2$. In the 4×4 case, $\mathbf{m}$ has conjugate and symmetric eigenvalue pairs $\pm\lambda^1, \pm\lambda^2$ (or a complex quartet $\pm\lambda^{\mathrm{R}} \pm i\lambda^{\mathrm{I}}$) in various patterns [see (12.27)]. We assume that $\mathbf{m} \in \mathsf{sp}(4, \mathsf{R})$ has been brought to a form with eigenvalues on the diagonal by means of a $\mathsf{GL}(4, \mathsf{R})$ transformation, so $\tau(z) = 2\cosh z\lambda^1 + 2\cosh z\lambda^2$ (or $\tau(z) = 4\cosh z\lambda^{\mathrm{R}} \cos z\lambda^{\mathrm{I}}$). According to the Hamiltonian stratum of its generator, the trace of the symplectic guide-evolution matrices $\mathbf{M}(z) \neq \mathbf{1}$ $(z \neq 0)$, will be:

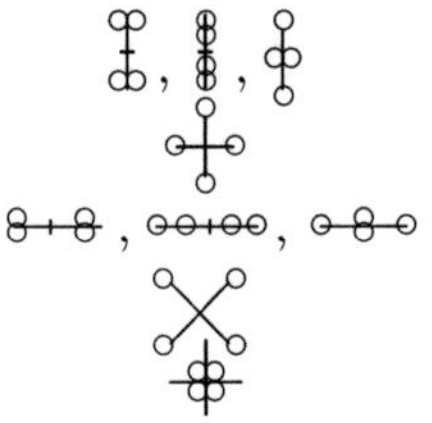

$$\begin{array}{ll} \text{H-H}_\theta: & \tau(z) \in [-4, 4), \\ \text{H-R}_\theta: & \tau(z) \in (0, \infty), \\ \text{R-R}_\theta: & \tau(z) \in (4, \infty), \\ \text{M-R}_\theta: & \tau(z) \in \mathsf{R}, \\ \text{center}: & \tau(z) = 4. \end{array} \tag{12.82}$$

Hence, there is no characteristic interval of τ that would allow us to recognize the non-exponential matrices in $\mathsf{Sp}(4, \mathsf{R})$, as there was in $\mathsf{Sp}(2, \mathsf{R})$. •

12.8 Inhomogeneous Hamiltonians

Most of our concern has been with the canonical transformations generated by strictly quadratic paraxial Hamiltonians and the $\mathsf{Sp}(4, \mathsf{R})$-matrix methods associated to this model. With matrices larger by one dimension, we can also represent systems whose Hamiltonians have linear as well as quadratic anisotropic terms,

$$H(\mathbf{e}(\mathbf{w}, \mathbf{m}); \mathbf{p}, \mathbf{q}) = \tfrac{1}{2}\mathbf{p}^\top \mathbf{c}\mathbf{p} - \mathbf{q}^\top \mathbf{a}\mathbf{p} - \tfrac{1}{2}\mathbf{q}^\top \mathbf{b}\mathbf{q} + \mathbf{v} \cdot \mathbf{p} - \mathbf{u} \cdot \mathbf{q} \tag{12.83}$$

$$\leftrightarrow \quad \mathbf{e}(\mathbf{w}, \mathbf{m}) := \begin{pmatrix} \mathbf{m} & \mathbf{w} \\ \mathbf{0}^\top & 0 \end{pmatrix}, \quad \mathbf{m} = \begin{pmatrix} \mathbf{a} & \mathbf{b} \\ \mathbf{c} & -\mathbf{a}^\top \end{pmatrix}, \quad \mathbf{w} = \begin{pmatrix} \mathbf{u} \\ \mathbf{v} \end{pmatrix}, \tag{12.84}$$

where $\mathbf{u}$ and $\mathbf{v}$ are 2-vectors, $\mathbf{m} \in \mathsf{sp}(4, \mathsf{R})$, and $\mathbf{e}(\mathbf{w}, \mathbf{m}) \in \mathsf{isp}(4, \mathsf{R}) = \mathsf{t}_4 \oplus \mathsf{sp}(4, \mathsf{R})$ are 5×5 *inhomogeneous* Hamiltonian matrices.

To generate a representation of the inhomogeneous symplectic group $\mathsf{ISp}(4, \mathsf{R}) = \mathsf{T}_4 \otimes \mathsf{Sp}(4, \mathsf{R})$, we use the notation of Sect. 5.5 and sum the exponential series of (12.84) to

$$\mathbf{E}(\mathbf{W}(z), \mathbf{M}(z)) := \exp(z\mathbf{e}(\mathbf{w}, \mathbf{m})) \tag{12.85}$$

$$\begin{pmatrix} \mathbf{M}(z) & \mathbf{W}(z) \\ \mathbf{0}^\top & 1 \end{pmatrix} = \begin{pmatrix} \exp(z\,\mathbf{m}) & \dfrac{\exp(z\,\mathbf{m})-1)}{z\,\mathbf{m}}\mathbf{w} \\ \mathbf{0}^\top & 1 \end{pmatrix}, \tag{12.86}$$

where the series summed in (12.71) appears. [For 2×2 matrices we have the explicit results in (12.13)–(12.16) and (12.17), and for 4×4 ones those in (12.64)–(12.68) and (12.71).] This represents and parametrizes the elements of the group of inhomogeneous linear canonical transformations of paraxial phase space R^{2D}, now including translations $\mathcal{T}$ by $\mathbf{W}(z) = \begin{pmatrix} \mathbf{U}(z) \\ \mathbf{V}(z) \end{pmatrix}$,

$$\mathcal{E}(\mathbf{E}) = \mathcal{E}(\mathbf{W}, \mathbf{M}) = \mathcal{T}(\mathbf{W})\,\mathcal{M}(\mathbf{M}) \leftrightarrow \mathcal{E}(\mathbf{W}, \mathbf{M}) = \mathcal{E}(\mathbf{W}, \mathbf{1})\,\mathcal{E}(\mathbf{0}, \mathbf{M}). \tag{12.87}$$

Translations of paraxial optical momentum are produced by prisms; displacements of the optical center (or misalignments between elements) translate positions. (These translations need not be 'small' in the paraxial approximation.) As we noted in Sect. 5.5, and again in (9.16), the action of the optical system $\mathcal{E}(\mathbf{W}, \mathbf{M})$ proceeds through the *inverse* of its associated matrix.

Example: Media of linear + quadratic index. In the paraxial régime, the refractive index of a guide medium

$$n(\mathbf{q}) = \overline{n} - \nu(\mathbf{q}), \quad \nu(\mathbf{q}) = n_0 + \mathbf{n}_1 \cdot \mathbf{q} + \tfrac{1}{2}\mathbf{q}^\top \widehat{\mathbf{n}}_2\,\mathbf{q}, \tag{12.88}$$

where $|\nu|^2 \approx 0$, $\mathbf{n}_1$ is a 2-vector and $\widehat{\mathbf{n}}_2$ a symmetric 2×2 matrix, has a Hamiltonian represented by

$$h(\mathbf{p}, \mathbf{q}) = \frac{1}{2\overline{n}}\mathbf{p}^2 + \nu(\mathbf{q}) \;\leftrightarrow\; \begin{pmatrix} \mathbf{0} & -\widehat{\mathbf{n}}_2 & -\mathbf{n}_1 \\ \mathbf{1}/\overline{n} & \mathbf{0} & \mathbf{0} \\ \mathbf{0}^\top & \mathbf{0}^\top & 0 \end{pmatrix}. \tag{12.89}$$

We note that $n_0(z)$ in (12.88) is absent from (12.89). •

Having spent some effort in classifying the orbits of inequivalent $\mathsf{sp}(4, \mathsf{R})$ Hamiltonians, we can extend the main results to $\mathbf{e}(\mathbf{w}, \mathbf{m}) \in \mathsf{isp}(4, \mathsf{R})$ under the adjoint action of the group elements $\mathbf{E}(\mathbf{W}, \mathbf{M}) \in \mathsf{ISp}(4, \mathsf{R})$ through the similarity transformation

$$\begin{pmatrix} \mathbf{M} & \mathbf{W} \\ \mathbf{0} & 1 \end{pmatrix}\begin{pmatrix} \mathbf{m} & \mathbf{w} \\ \mathbf{0} & 0 \end{pmatrix}\begin{pmatrix} \mathbf{M} & \mathbf{W} \\ \mathbf{0} & 1 \end{pmatrix}^{-1} = \begin{pmatrix} \mathbf{M}\,\mathbf{m}\,\mathbf{M}^{-1} & \mathbf{M}(\mathbf{w} - \mathbf{m}\,\mathbf{M}^{-1}\mathbf{W}) \\ \mathbf{0} & 0 \end{pmatrix}. \tag{12.90}$$

The orbit structure of the $\mathsf{sp}(4,\mathsf{R})$ subalgebra of $\mathsf{isp}(4,\mathsf{R})$ was solved in Sect. 12.6, so we can assume that the 1-1 element of (12.90) has been brought to its orbit representative $\mathbf{m}_\circ = \mathbf{M}\,\mathbf{m}\,\mathbf{M}^{-1}$, by some $\mathbf{M} \in \mathsf{Sp}(4,\mathsf{R})$. For each $\mathbf{m}_\circ$ we must now determine whether there exists a translation $\mathbf{W}$ by which one can set to $\mathbf{0}$ the 1-2 element of that matrix, *or not.* When $\Delta := \det \mathbf{m} = \det \mathbf{m}_\circ \neq 0$, a unique solution exists, $\mathbf{W} = \mathbf{m}_\circ^{-1}\mathbf{M}\,\mathbf{w}$, that brings the translation part to zero; the orbit representative will then be the bare $\mathbf{m}_\circ$ found previously. But when $\Delta = \det \mathbf{m} = 0$, no such solution may exist, and $\mathbf{e}(\mathbf{w},\mathbf{m})$ will lie in an $\mathsf{isp}(4,\mathsf{R})$ orbit inequivalent to any previous $\mathsf{sp}(4,\mathsf{R})$ orbit. One has then to examine, for each $\Delta = 0$ orbit in the (Γ,Δ) plane of eigenvalues (Figs. 12.3 and 12.4), the vector components of $\mathbf{M}\,\mathbf{w} - \mathbf{m}_\circ \mathbf{W}$ that *cannot* be made to vanish through translations $\mathbf{W}$. Among those orbits, our main interest lies in the 'physical' free-propagation orbits.

Example: Media of linear refractive index. In the previous example (12.88)–(12.89), the linear term with the 2-vector $\mathbf{n}_1$ can be eliminated from the Hamiltonian by a translation of the optical center of the screen $\mathbf{W} = \binom{\mathbf{0}}{\mathbf{V}}$, such that $\mathbf{n}_1 = -\widehat{\mathbf{n}}_2\mathbf{V}$, *provided* that $\det \widehat{\mathbf{n}}_2 \neq 0$. The free propagation of rays in a generally anisotropic medium [replacing the term $\frac{1}{2n}\mathbf{p}^2$ in (12.89) by $\frac{1}{2}\mathbf{p}^\top\mathbf{c}\mathbf{p}$ from (12.83)] with a fixed linear dependence of the refractive index on position $\sim \mathbf{n}_1\!\cdot\!\mathbf{q}$ (and $\widehat{\mathbf{n}}_2 = \mathbf{0}$) belongs to a distinct orbit of inhomogeneous Hamiltonians. These can be readily exponentiated using three isomorphic realizations:

$$\begin{array}{c} \exp\left(-z\left\{\tfrac{1}{2}\mathbf{p}^\top\mathbf{c}\mathbf{p} + \mathbf{n}_1\mathbf{q}, \circ\right\}\right) \\ \mathcal{E}\left(\begin{pmatrix} z\mathbf{n}_1 \\ -\frac{1}{2}z^2\mathbf{n}_1 \end{pmatrix}, \begin{pmatrix} 1 & 0 \\ -z\mathbf{c} & 1 \end{pmatrix}\right) \end{array} \begin{array}{c} \leftrightarrow \\ \leftrightarrow \end{array} \begin{pmatrix} \mathbf{1} & \mathbf{0} & z\mathbf{n}_1 \\ -z\mathbf{c} & \mathbf{1} & -\frac{1}{2}z^2\mathbf{c}\,\mathbf{n}_1 \\ \mathbf{0}^\top & \mathbf{0}^\top & 1 \end{pmatrix}, \tag{12.91}$$

$$\mathcal{E} : f(\mathbf{p},\mathbf{q}) \mapsto f(\mathbf{p} - z\mathbf{n}_1,\ \mathbf{q} + z\mathbf{c}\mathbf{p} - \tfrac{1}{2}z^2\mathbf{c}\,\mathbf{n}_1). \tag{12.92}$$

I.e., rays fall in the direction of $\mathbf{n}_1$, in a plane of 3-dim optical space $(\mathbf{q}(z), z)$, in a plane of 4-dim phase space $(\mathbf{p}(z), \mathbf{q}(z))$, and in a plane of the 14-dim $\mathsf{ISp}(4,\mathsf{R})$ group space, where only $\mathbf{C}(z) = z\mathbf{c}$ and $(\mathbf{U}(z), \mathbf{V}(z))$ depend on z.•

The Hamiltonian orbit representatives of the $\mathsf{Sp}(4,\mathsf{R})$ part in the previous example (12.91), are one of the three separable Hamiltonians $F_x + \sigma F_y$, $\sigma \in \{-1, 0, +1\}$, as given in (12.58), which lie at the center of the eigenvalue plane in Fig. 12.4. For $\sigma = +1$ the representative free system is isotropic, so $\mathbf{n}_1$ (the gradient of the refractive index) can be always rotated in a chosen direction; an appropriate representative for the $\mathsf{ISp}(4,\mathsf{R})$-orbit of the previous example is $h_\circ(\mathbf{p},\mathbf{q}) = \frac{1}{2}\mathbf{p}^2 + q_x$. The two other free cases (with $\sigma = 0$ or -1) are not isotropic, so for every direction of $\mathbf{n}_1$ on the circle there is a distinct orbit, and the collection of them forms two new $\mathsf{ISp}(4,\mathsf{R})$ strata. Other $\Delta = 0$ inhomogeneous orbits need not be analyzed further.

C Canonical Fourier optics

The paraxial régime of geometric optics has essentially the same Hamiltonian and phase space structures as the wave theory called *Fourier* optics, that we characterize as *canonical*. We follow the approach of appendix B, where we first examine models on the spaces of cosets of the invariance group of phase space; in the paraxial régime, this is the noncommutative Heisenberg–Weyl group of phase space translations. We determine the action of paraxial optical systems on the Hilbert space of monochromatic wavefields through a group of unitary integral transforms. The generators of this group form an algebra of *second*-order differential operators, whose action is distinct from the *Berührungstransformationen* (*contact* transformations) originally studied by S.M. Lie, who worked with algebras of *first*-order differential operators only.

Moshinsky and Quesne introduced the group $\mathsf{Sp}(2D, \mathsf{R})$ of integral transforms, which they called *canonical* within the context of quantum oscillator mechanics [101]; the results we recount here derive from [149, Chaps. 9–10] and other papers. The relevance of canonical transforms for wave models of light was unrecognized for some time [109, 31]; they are efficacious for our purposes since, as in geometric optics, the composition of elements into systems is reduced to multiplication of matrices – except for the so-called *metaplectic sign*. To treat this oft-misunderstood feature we have developed the necessary tools in the previous chapters of this part.

C.1 The Royal Road to Fourier optics

To walk the Royal Road from classical to quantum mechanics we require a complex Hilbert space of wavefunctions and a correspondence between phase space quantities and essentially self-adjoint operators. This mathematical baggage is also sufficient to pass from geometric paraxial optics to the model of monochromatic, scalar wave optics in the paraxial régime. As before, there is little extra cost to consider D dimensions.

The wavefields $f(\mathbf{q})$ belong to the Hilbert space $\mathcal{L}^2(\mathsf{R}^D)$ of Lebesgue square-integrable functions on the standard screen. Their absolute square, $|f(\mathbf{q})|^2 < \infty$, is the local *intensity* of light, and their norm $\|f\|^2 := (f, f)_{\mathcal{L}^2(\mathsf{R}^D)} < \infty$ is proportional to the total energy in the field. Unlike quantum mechanics, where $|f(\mathbf{q})|^2$ is a probability density and the norm of the wavefunction must be unity, the wavefields of optics need not be normalized; distribution theory further allows for plane waves and Dirac δ's. Lossless linear optical systems form a group $\mathsf{Sp}(2,\mathsf{R})$ of unitary transformations of $\mathcal{L}^2(\mathsf{R}^D)$; systems with loss [109, 82] can be described moreover as a complex extension of this group.

The direct and inverse Fourier integral transforms, and their Parseval identity are

$$(\mathcal{F}:f)(\mathbf{r}) = \widetilde{f}(\mathbf{r}) := (2\pi)^{-\frac{1}{2}D}\int_{\mathsf{R}^D} d\mathbf{r}'\, e^{-i\mathbf{r}\cdot\mathbf{r}'} f(\mathbf{r}'), \tag{C.1}$$

$$(\mathcal{F}^{-1}:\widetilde{f})(\mathbf{r}') = f(\mathbf{r}') = (2\pi)^{-\frac{1}{2}D}\int_{\mathsf{R}^D} d\mathbf{r}\, e^{+i\mathbf{r}'\cdot\mathbf{r}} \widetilde{f}(\mathbf{r}), \tag{C.2}$$

$$(f,g) := \int_{\mathsf{R}^D} d\mathbf{r}\, f(\mathbf{r})^* g(\mathbf{r}) = \int_{\mathsf{R}^D} d\mathbf{r}'\, \widetilde{f}(\mathbf{r}')^* \widetilde{g}(\mathbf{r}'). \tag{C.3}$$

It is well known that the square of the Fourier transform, $\mathcal{F}^2 = \mathcal{I}$, is the inversion $\mathcal{I} : f(\mathbf{r}) = f(-\mathbf{r})$, $\mathcal{F}^3 = \mathcal{F}^{-1}$, and $\mathcal{F}^4 = 1$ is the unit operator. The vectors $\mathbf{r}$, $\mathbf{r}' \in \mathsf{R}^D$ in the previous formulas stand for the dimensionless optical momentum $\mathbf{p}$ and the position $\mathbf{q}/\lambda\!\!\!^{-}$, counted in units of the *reduced wavelength* of the field, $\lambda\!\!\!^{-} := \lambda/2\pi$. The exponent of the Fourier kernel is thus $-i$ times $\mathbf{r}\cdot\mathbf{r}' = \mathbf{p}\cdot\mathbf{q}/\lambda\!\!\!^{-}$, so $\lambda\!\!\!^{-}$ takes the rôle of the reduced Heisenberg constant in quantum mechanics, $\hbar = h/2\pi$. Alternatively, the exponent can be written as $\mathbf{r}\cdot\mathbf{r}' = \mathbf{k}\cdot\mathbf{q}$, with the *wavenumber* vector $\mathbf{k} := \mathbf{p}/\lambda\!\!\!^{-}$. While $\hbar$ is fixed by Nature, $\lambda\!\!\!^{-}$ can take any asigned nonzero value.

We assume that $f(\mathbf{q})$ can be sensed in amplitude and phase on a D-dim screen in a $(D{+}1)$-dim optical space. A plane wave $\sim e^{i\mathbf{p}\cdot\mathbf{q}/\lambda\!\!\!^{-}}$ is defined to have *momentum* $\mathbf{p} \in \mathsf{R}^D$. As in geometric paraxial optics, the model of canonical Fourier optics relaxes the bound $|\mathbf{p}| \le n$ of the refractive index (see Chap. 1) and extends the support of $\widetilde{f}(\mathbf{p})$ to the whole R^D plane. This paraxial momentum space is subject to rigid translation by prisms (see Sect. 12.8), in the same way that the screen is translated in configuration space. Again we shall refer to the manifold $(\mathbf{p},\mathbf{q}) \in \mathsf{R}^{2D}$ as *phase space*, and introduce its translation operators:

$$\mathcal{T}_\mathbf{u} : f(\mathbf{q}) := f(\mathbf{q} + \lambda\!\!\!^{-}\mathbf{u}), \qquad \widetilde{\mathcal{T}}_\mathbf{v} : f(\mathbf{q}) := \exp(i\mathbf{v}\cdot\mathbf{q})\, f(\mathbf{q}), \tag{C.4}$$

where $\mathbf{u}$ is dimensionless while $\mathbf{v}$ has units of $\lambda\!\!\!^{-\,-1}$. These operators are unitary and Fourier conjugates of each other: $\widetilde{\mathcal{T}}_{\mathbf{u}/\lambda\!\!\!^{-}} = \mathcal{F}\,\mathcal{T}_\mathbf{u}\,\mathcal{F}^{-1}$. Evidently, $\mathcal{T}_{\mathbf{u}_1}\mathcal{T}_{\mathbf{u}_2} = \mathcal{T}_{\mathbf{u}_1+\mathbf{u}_2}$ and similarly for the $\widetilde{\mathcal{T}}_\mathbf{v}$'s. They do not commute; instead,

$$\begin{aligned} \mathcal{T}_\mathbf{u}\,\widetilde{\mathcal{T}}_\mathbf{v} : f(\mathbf{q}) &= \mathcal{T}_\mathbf{u} : e^{i\mathbf{v}\cdot\mathbf{q}} f(\mathbf{q}) = e^{i\mathbf{v}\cdot(\mathbf{q}+\lambda\!\!\!^{-}\mathbf{u})} f(\mathbf{q} + \lambda\!\!\!^{-}\mathbf{u}), \\ \widetilde{\mathcal{T}}_\mathbf{v}\,\mathcal{T}_\mathbf{u} : f(\mathbf{q}) &= \widetilde{\mathcal{T}}_\mathbf{v} : f(\mathbf{q} + \lambda\!\!\!^{-}\mathbf{u}) = e^{+i\mathbf{v}\cdot\mathbf{q}} f(\mathbf{q} + \lambda\!\!\!^{-}\mathbf{u}), \end{aligned} \tag{C.5}$$

$$\Rightarrow \quad \mathcal{T}_\mathbf{u}\,\widetilde{\mathcal{T}}_\mathbf{v} = e^{i\lambda\!\!\!^{-}\,\mathbf{v}\cdot\mathbf{u}}\,\widetilde{\mathcal{T}}_\mathbf{v}\,\mathcal{T}_\mathbf{u}. \tag{C.6}$$

The last is called the *Weyl* group commutator; it introduces the extra operator of *phase*,

$$\Phi_w : f(\mathbf{q}) = e^{i\lambda w}\, f(\mathbf{q}), \qquad \lambda w \in \mathsf{R} \bmod 2\pi. \tag{C.7}$$

Products of $\mathcal{T}$'s, $\widetilde{\mathcal{T}}$'s, and Φ's, form the $(2D+1)$-parameter *Heisenberg-Weyl* group W_D.

Phase space translations and phases have generators which are realized on the screen $\mathbf{q} \in \mathsf{R}^D$ as

$$\mathcal{T}_{\mathbf{u}} = \exp(i\,\mathbf{u}\cdot\widehat{\mathbf{P}}) \quad \Rightarrow \quad \widehat{\mathbf{P}} : f(\mathbf{q}) = -i\lambda\frac{\partial}{\partial\mathbf{q}} f(\mathbf{q}), \tag{C.8}$$

$$\widetilde{\mathcal{T}}_{\mathbf{v}} = \exp(i\,\mathbf{v}\cdot\widehat{\mathbf{Q}}) \quad \Rightarrow \quad \widehat{\mathbf{Q}} : f(\mathbf{q}) = \mathbf{q}\, f(\mathbf{q}), \tag{C.9}$$

$$\Phi_w = \exp(\,i\,w\,\widehat{\Lambda}\,) \quad \Rightarrow \quad \widehat{\Lambda} : f(\mathbf{q}) = \lambda\, f(\mathbf{q}), \tag{C.10}$$

that we may call the *Fourier–Schrödinger* realization.[1] These generators form a basis for the $(2D+1)$-dim Lie algebra w_D, presented by

$$[\widehat{Q}_j, \widehat{P}_k] = i\delta_{j,k}\widehat{\Lambda}, \qquad \begin{array}{ll} [\widehat{Q}_j, \widehat{\Lambda}\,] = 0, & [\widehat{P}_j, \widehat{\Lambda}\,] = 0, \\ [\widehat{Q}_j, \widehat{Q}_k] = 0, & [\widehat{P}_j, \widehat{P}_k] = 0, \end{array} \tag{C.11}$$

and known as the *Heisenberg* commutation relations; w_D is called the Heisenberg-Weyl Lie algebra.

In the following remarks we continue the method of appendix B to define optical models on spaces of group cosets. We recall that the gist of that method was to propose subgroups which define the symmetry of the fundamental object in each model, such as lines in the geometric model, planes in the polychromatic wave model, etc. The Heisenberg-Weyl group is simpler than the Euclidean group, so the models are fewer.

Remark: The Heisenberg–Weyl group and algebra. The abstract Heisenberg-Weyl Lie group has elements $\omega(\mathbf{u}, \mathbf{v}, w)$ whose multiplication rule given by[2]

$$\omega(\mathbf{u}, \mathbf{v}, w) := \widetilde{\mathcal{T}}_{\mathbf{v}}\, \mathcal{T}_{\mathbf{u}}\, \Phi_w = \mathcal{T}_{\mathbf{u}}\, \widetilde{\mathcal{T}}_{\mathbf{v}}\, \Phi_{w-\mathbf{v}\cdot\mathbf{u}}, \quad \mathbf{u}, \lambda\mathbf{v} \in \mathsf{R}^D,\ \lambda w \in \mathsf{R}, \tag{C.12}$$

$$\omega(\mathbf{u},\, \mathbf{v},\, w)\; \omega(\mathbf{u}', \mathbf{v}', w') = \omega(\mathbf{u}+\mathbf{u}',\, \mathbf{v}+\mathbf{v}',\, w+w'+\mathbf{u}\cdot\mathbf{v}'). \tag{C.13}$$

The unit is $\omega(\mathbf{0}, \mathbf{0}, 0)$ and the inverse $\omega(\mathbf{u}, \mathbf{v}, w)^{-1} = \omega(-\mathbf{u}, -\mathbf{v}, -w+\mathbf{u}\cdot\mathbf{v})$. Note carefully that the parameter w need not be cyclic modulo λ, as it is in the Fourier–Schrödinger realization (C.7). In the Hilbert space of functions on the group, $F(\omega) = F(\mathbf{u}, \mathbf{v}, w)$, with the right- and left-invariant Haar

[1] In quantum mechanics this is called the Schrödinger realization, with $\hbar$ in place of λ. In Fourier optics we hesitate to call it simply the *Fourier* realization.

[2] Several closely related parametrizations of W_D are possible; the widely used *polar* one is $\omega_{\rm pol}(\mathbf{x}, \mathbf{y}, z) := \exp i(\mathbf{x}\cdot\widehat{\mathbf{P}} + \mathbf{y}\cdot\widehat{\mathbf{Q}} + z\widehat{\Lambda})$.

measure $d\omega = d\mathbf{u}\,d\mathbf{v}\,dw$ (see page 141 *et seq.*), the group action from the right, $\omega' : F(\omega) = F(\omega\,\omega')$, is generated by the following differential operators:

$$\widehat{\mathbf{P}}_{\mathrm{W}} = -i\frac{\partial}{\partial \mathbf{u}}, \quad \widehat{\mathbf{Q}}_{\mathrm{W}} = -i\Big(\frac{\partial}{\partial \mathbf{v}} + \mathbf{u}\frac{\partial}{\partial w}\Big), \quad \widehat{\Lambda}_{\mathrm{W}} = -i\frac{\partial}{\partial w}. \tag{C.14}$$

•

Remark: The polychromatic model. There are three relevant and nonequivalent one-parameter subgroups in W_D, namely $\mathsf{T} = \{\mathcal{T}_{\mathbf{u}}\}_{\mathbf{u}\in\mathsf{R}^D}$, $\widetilde{\mathsf{T}} = \{\widetilde{\mathcal{T}}_{\mathbf{v}}\}_{\lambda\mathbf{v}\in\mathsf{R}^D}$, and $\Phi = \{\Phi_w\}_{\lambda w\in\mathsf{R}}$. When the functions $F(\omega)$ depend only on the coset representatives of $\widetilde{\mathsf{T}}\backslash\mathsf{W}_D$; i.e., when $F(\mathbf{u}, w) \in \mathcal{L}^2(\mathsf{R}^{D+1})$ is independent of $\mathbf{v}$, then the realization of the Lie algebra w_D in (C.14) reduces to

$$\widehat{\mathbf{P}}_{\mathrm{C}} = -i\frac{\partial}{\partial \mathbf{u}}, \quad \widehat{\mathbf{Q}}_{\mathrm{C}} = -i\mathbf{u}\frac{\partial}{\partial w}, \quad \widehat{\Lambda}_{\mathrm{C}} = -i\frac{\partial}{\partial w}. \tag{C.15}$$

A Fourier expansion in $w \in \mathsf{R}$ displays a *polychromatic* wavefield $F(\mathbf{u}, w)$ as an integral of monochromatic components $F^{\lambda}(\mathbf{u})$,

$$\begin{aligned} F(\mathbf{u}, w) &= (2\pi)^{-\frac{1}{2}D} \int_{\mathsf{R}} d\lambda\, e^{+i\lambda w}\, F^{\lambda}(\mathbf{u}), \\ F^{\lambda}(\mathbf{u}) &:= (2\pi)^{-\frac{1}{2}D} \int_{\mathsf{R}} dw\, e^{-i\lambda w}\, F(\mathbf{u}, w). \end{aligned} \tag{C.16}$$

This divides $\mathcal{L}^2(\mathsf{R}^{D+1})$ into a direct integral of eigenspaces of $\widehat{\Lambda}$ distinguished by $\lambda \in \mathsf{R}$. On each eigenspace there is a distinct Fourier–Schrödinger realization of the algebra, (C.8)–(C.10), acting on wavefields $f(\mathbf{q}) = F^{\lambda}(\mathbf{q}/\lambda)$ of position coordinate $\mathbf{q} = \lambda\mathbf{u}$. •

Remark: The geometric model. The case $\lambda = 0$ should be excluded from the preceding remark. We need not regard it as the 'geometric limit' of canonical Fourier optics, but return to functions on the W_D group manifold (C.12) which are independent of the coordinate w of Φ, and consider functions $F(\mathbf{u}, \mathbf{v})$ of the space of cosets $\Phi\backslash\mathsf{W}_D$. The generators (C.15) reduce then to $\widehat{\mathbf{P}}_{\mathrm{G}} = -i\partial/\partial\mathbf{u}$, $\widehat{\mathbf{Q}}_{\mathrm{G}} = -i\partial/\partial\mathbf{v}$, while $\widehat{\Lambda}_{\mathrm{G}}$ is the zero operator; this collapses w_D to the algebra of translations i_{2D}. Of course, the Poisson operators of the geometric model, $\{\mathbf{p}, \circ\}$ and $\{\mathbf{q}, \circ\}$, commute. •

Remark: Fourier optics approximates Maxwell optics. It has been shown by Bacry and Cadilhac [11] that the Maxwell equations for electromagnetic fields, in an approximation where the refractive index of the medium varies slowly over the distance of a wavelength, and where light propagates close to the direction of an optical axis (the paraxial approximation), leads to fields that satisfy a Schrödinger-type equation. This equation can be obtained following the Royal Road from geometric/classical to the wave/quantum models [see (1.13)–(1.17)]. When the refractive index/potential is quadratic, this is the canonical Fourier model treated here. The authors also examine how the metaplectic phase arises. •

C.2 Linear canonical transforms

There occurs a remarkable accident: the group of automorphisms of the the Heisenberg-Weyl algebra $\mathsf{w}_D = \text{span}\{\widehat{\mathbf{P}}, \widehat{\mathbf{Q}}, \widehat{\Lambda}\}$ is (*much*) larger than the group W_D itself. In fact, it is the semidirect product (see Sect. 5.5) of the Heisenberg–Weyl and symplectic groups, called the *Weyl-symplectic* group, denoted by $\mathsf{WSp}(2D, \mathsf{R}) = \mathsf{W}_D \circledast \mathsf{Sp}(2D, \mathsf{R})$. This means that the primed operators in

$$\begin{pmatrix} \widehat{\mathbf{P}}' \\ \widehat{\mathbf{Q}}' \\ \widehat{\Lambda}' \end{pmatrix} = \begin{pmatrix} \mathbf{a} & \mathbf{b} & \mathbf{u} \\ \mathbf{c} & \mathbf{d} & \mathbf{v} \\ \mathbf{0}^\top & \mathbf{0}^\top & 1 \end{pmatrix}^{-1} \begin{pmatrix} \widehat{\mathbf{P}} \\ \widehat{\mathbf{Q}} \\ \widehat{\Lambda} \end{pmatrix}, \tag{C.17}$$

obey the same commutation relations (C.11) as their unprimed counterparts, provided that $\mathbf{M} := \left(\begin{smallmatrix} \mathbf{a} & \mathbf{b} \\ \mathbf{c} & \mathbf{d} \end{smallmatrix}\right)$ is symplectic. This linear transformation, which is canonical in the phase space of the geometric paraxial model, shall be now extended to the Fourier optical model as a unitary integral transform of the Hilbert space of monochromatic wavefields [101].

We present canonical transforms first for the $D = 1$ case, with no translations and $\lambda\!\!\!^{-} = 1$, because formulas are simpler and contain all their essential features. Consider an operator indicated $\mathcal{C}_{\mathrm{M}} = \mathcal{C}(\mathbf{M})$, $\mathbf{M} = \left(\begin{smallmatrix} a & b \\ c & d \end{smallmatrix}\right) \in \mathsf{Sp}(2, \mathsf{R})$, whose adjoint action on the algebraic basis of $\mathsf{w}_1 = \text{span}\{\widehat{P}, \widehat{Q}, \widehat{1}\}$ [as in (C.17)],[3] is linear,

$$\begin{pmatrix} \widehat{P}' \\ \widehat{Q}' \end{pmatrix} = \mathcal{C}_{\mathrm{M}} \begin{pmatrix} \widehat{P} \\ \widehat{Q} \end{pmatrix} \mathcal{C}_{\mathrm{M}}^{-1} = \mathbf{M}^{-1} \begin{pmatrix} \widehat{P} \\ \widehat{Q} \end{pmatrix} = \begin{pmatrix} d\widehat{P} - b\widehat{Q} \\ -c\widehat{P} + a\widehat{Q} \end{pmatrix}, \tag{C.18}$$

and naturally $\mathcal{C}_{\mathrm{M}} \widehat{\Lambda}\, \mathcal{C}_{\mathrm{M}}^{-1} = \widehat{\Lambda}$. It follows that

$$\mathcal{C}(\mathbf{M}_1)\mathcal{C}(\mathbf{M}_2) = \pm\mathcal{C}(\mathbf{M}_1\mathbf{M}_2), \quad \mathcal{C}(\mathbf{1}) = \pm\mathit{1}, \quad \mathcal{C}(\mathbf{M}^{-1}) = \pm\mathcal{C}(\mathbf{M})^{-1}, \tag{C.19}$$

and thus the $\mathcal{C}$'s realize the group $\mathsf{Sp}(2, \mathsf{R})$ – up to a sign. The sign is ambiguous because the adjoint action of $\mathsf{Sp}(2, \mathsf{R})$ on the w_1 generators is *quadratic*, i.e., $\mathcal{C}_{\mathrm{M}} \cdots \mathcal{C}_{\mathrm{M}}^{-1}$, so either sign satisfies the composition, unit and inverse. For each matrix $\mathbf{M} \in \mathsf{Sp}(2, \mathsf{R})$ there are thus *two* operators, $\pm\mathcal{C}(\mathbf{M})$. When acting linearly on wavefields however, the sign of $\mathcal{C}(\mathbf{M})$ must be defined consistently. Of course, the unit operator *1* and the inverse $\mathcal{C}^{-1}$ of an operator $\mathcal{C}$ are unique, but the labelling of $\mathcal{C}$'s by 2×2 matrices is not. This is referred to as the '*metaplectic* sign problem,' and reflects the fact that the set of $\mathcal{C}(\mathbf{M})$'s actually realizes the *two-fold cover* of $\mathsf{Sp}(2, \mathsf{R})$, i.e., the metaplectic group $\mathsf{Mp}(2, \mathsf{R})$ seen in Chap. 9. The group $\mathsf{Mp}(2, \mathsf{R})$ has **no** finite-dimensional *faithful* matrix representation. In the course of this appendix we shall see how

---

[3] Please note that throughout this volume we use $\left(\begin{smallmatrix} P \\ Q \end{smallmatrix}\right)$, and *not* $\left(\begin{smallmatrix} Q \\ P \end{smallmatrix}\right)$, as many authors – including the present one – have used. In all formulas, the two conventions are connected by $a \leftrightarrow d$ and $b \leftrightarrow c$.

to control this sign; for the time being it will be sufficient to stay within a finite neighborhood of the group unit *1*.

In the Fourier–Schrödinger realization of w_1, (C.8)–(C.10), the action (C.18) on the operators entails a corresponding action on the wavefields, $\mathcal{C}_\mathrm{M} : f = f_\mathrm{M}$, such that

$$\begin{aligned}\mathcal{C}_\mathrm{M} : \widehat{P} f &= \mathcal{C}_\mathrm{M}\, \widehat{P}\, \mathcal{C}_\mathrm{M}^{-1}\, \mathcal{C}_\mathrm{M} : f = (d\widehat{P} - b\widehat{Q}) f_\mathrm{M}, \\ &\Rightarrow \Big(\mathcal{C}_\mathrm{M} : (-i\tfrac{d}{dq}) f\Big)(q) = \quad d(-i\tfrac{d}{dq} f_\mathrm{M}(q)) - b\, q\, f_\mathrm{M}(q); \qquad \text{(C.20)} \\ \mathcal{C}_\mathrm{M} : \widehat{Q} f &= \mathcal{C}_\mathrm{M}\, \widehat{Q}\, \mathcal{C}_\mathrm{M}^{-1}\, \mathcal{C}_\mathrm{M} : f = (-c\widehat{P} + a\widehat{Q}) f_\mathrm{M}, \\ &\Rightarrow \Big(\mathcal{C}_\mathrm{M} : q\, f\Big)(q) = -c(-i\tfrac{d}{dq} f_\mathrm{M}(q)) + a\, q\, f_\mathrm{M}(q). \qquad \text{(C.21)}\end{aligned}$$

For $\mathbf{M} = \left(\begin{smallmatrix} 0 & -1 \\ 1 & 0 \end{smallmatrix}\right)$, these are properties of the Fourier transform, so we conclude that the action of $\mathcal{C}_\mathrm{M}$ will be generally an *integral* transform,

$$f_\mathrm{M}(q) := (\mathcal{C}_\mathrm{M} : f)(q) = \int_\mathsf{R} dq'\, C_\mathrm{M}(q, q')\, f(q'), \tag{C.22}$$

with a kernel $C_\mathrm{M}(q, q')$ to be found below. Applying (C.20)–(C.21) to (C.22) and integrating by parts, the kernel becomes the solution of the coupled pair of differential equations

$$-i\frac{\partial}{\partial q'} C_\mathrm{M}(q, q') = \left(b\, q + id\frac{\partial}{\partial q}\right) C_\mathrm{M}(q, q'), \tag{C.23}$$

$$q'\, C_\mathrm{M}(q, q') = \left(a\, q + ic\frac{\partial}{\partial q}\right) C_\mathrm{M}(q, q'). \tag{C.24}$$

The transform kernel is thus a complex Gaussian,

$$C_\mathrm{M}(q, q') = K_\mathrm{M} \exp i\frac{aq^2 - 2qq' + dq'^2}{2c} = C_{\mathrm{M}^{-1}}(q', q)^*, \tag{C.25}$$

where K_M will be shown to be the normalization constant [101]

$$K_\mathrm{M} := \frac{1}{\sqrt{2\pi\, ic}} := \frac{e^{-i\frac{1}{4}\pi} \exp(-i\frac{1}{2}\arg c)}{\sqrt{2\pi\, |c|}}, \qquad \begin{matrix} c \neq 0 \\ -\pi \leq \arg c \leq 0. \end{matrix} \tag{C.26}$$

From the relation between the extreme members of (C.25) follows that the transformation is unitary in $\mathcal{L}^2(\mathsf{R})$, namely $(f_\mathrm{M}, g) = (f, g_{\mathrm{M}^{-1}})$. When $\mathbf{M}$ is in a finite neighborhood of unity, where $a, d > 0$, the absolute value of the kernel decreases for large q and q' provided c is in the lower complex half-plane. In the normalization constant (C.26) thus, $c > 0$ has $\arg c = 0$, while $c < 0$ is understood to have $\arg c = -\pi$. In particular, acting on a centered Dirac δ,

$$(\mathcal{C}_\mathrm{M} : \delta)(q) = C_\mathrm{M}(q, 0) = \exp(iaq^2/2c)/\sqrt{2\pi ic}. \tag{C.27}$$

Imager systems are characterized by matrices $\mathbf{M} = \left(\begin{smallmatrix} a & b \\ 0 & a^{-1} \end{smallmatrix}\right)$ [see page 180], where the integral kernel in (C.25)–(C.26) is singular. In geometric optics, imager systems $\mathcal{M}(\mathbf{M})$ reproduce the object up to scale; in the corresponding wave model, it should be a transformation $\mathcal{C}(\mathbf{M}) : f(q) \mapsto f_{\mathrm{M}}(q)$ which is *not* an integral as (C.22) but, taking the lead from the geometric model,

$$\mathcal{M}\begin{pmatrix} a & b \\ 0 & a^{-1} \end{pmatrix} : \begin{pmatrix} p \\ q \end{pmatrix} = \begin{pmatrix} a & b \\ 0 & a^{-1} \end{pmatrix}^{-1} \begin{pmatrix} p \\ q \end{pmatrix} = \begin{pmatrix} a^{-1}p - bq \\ aq \end{pmatrix}, \tag{C.28}$$

$$\mathcal{C}\begin{pmatrix} a & b \\ 0 & a^{-1} \end{pmatrix} : f(q) = \sqrt{a}\, \exp(-i\tfrac{1}{2}ab\, q^2)\, f(aq), \quad a > 0. \tag{C.29}$$

To prove that this form is correct, we return to (C.20)–(C.21) and substitute $f_{\mathrm{M}}(q) \sim e^{i\beta q^2} f(\alpha q)$, to find $\alpha = a$, $\beta = -\frac{1}{2}\alpha b$. The normalization constant $\sqrt{a}$ makes the transformation unitary in $\mathcal{L}^2(\mathsf{R})$, so $(f_{\mathrm{M}}, g_{\mathrm{M}}) = (f, g)$. For the moment we eschew the range $a < 0$ of inverting imagers, which belongs to the non-exponential region of $\mathsf{Sp}(2, \mathsf{R})$ (see pages 254 *et seq.*), pending the full clarification of the phase. These contact (*non-integral*) canonical transforms should be the limit of the integral transforms (C.22) when $c \to 0$ from the lower complex half-plane. To check this directly, and thereby find the proper phase of the integral kernel in (C.26), we must integrate complex Gaussians carefully, as we remark below.

Remark: The integral of complex Gaussians. We determine the integral

$$I(r, s) := \int_{\mathsf{R}} dq\, \exp i(r^2 q^2 + sq) = \exp\Big(-i\frac{s^2}{4r^2}\Big) \int_{\mathsf{R}} dq\, \exp i(rq + s/2r)^2, \tag{C.30}$$

for r and s complex, by reducing it to Euler's formula, $\int_{\mathsf{R}} dz\, e^{-z^2} = \sqrt{\pi}$. To this end we change variables to

$$z := e^{-i\frac{1}{4}\pi}(rq + s/2r), \quad q \in \mathsf{R} \Rightarrow z \in \begin{cases} \text{line at angle } -\frac{1}{4}\pi + \arg r, \\ \text{passing through } e^{-i\frac{1}{4}\pi}\, s/2r. \end{cases} \tag{C.31}$$

The Gaussian e^{-z^2} is analytic in the finite complex-z plane; on the real axis it has the real bell form while on the imaginary axis it blows up; elsewhere it oscillates with frequency $\sim 2\,\mathrm{Re}\,z\,\mathrm{Im}\,z$. Its amplitude decreases in the open regions $\arg z \in (-\frac{1}{4}\pi, \frac{1}{4}\pi)$ and $\arg z \in (\frac{3}{4}\pi, \frac{5}{4}\pi)$, and increases in the open complement. The integration contour can be thus shifted to pass through the origin, and rotated to the real axis $(-\infty, \infty)$ when r is in the first complex quadrant (including the boundaries), or $(\infty, -\infty)$ when in the third (including boundaries). Hence,

$$I(r, s) = \sigma(r)\, \frac{e^{i\frac{1}{4}\pi}\sqrt{\pi}}{r} \exp -i\frac{s^2}{4r^2}, \tag{C.32}$$

$$\sigma(r) := \begin{cases} +1 \text{ when } \arg r \in [0, \frac{1}{2}\pi], \\ -1 \text{ when } \arg r \in [-\pi, -\frac{1}{2}\pi]. \end{cases} \tag{C.33}$$

When real, $r = \sigma(r)\, |r|$, but not otherwise. •

Remark: The limit $c \to 0$. To verify that the normalization constant (C.26) appears in the limit $c \to 0$ of contact transformations (C.29), we compute

$$\begin{aligned}\int_{\mathsf{R}} dq'\, C\Big(\begin{smallmatrix} a & b \\ c & d \end{smallmatrix}\Big)(q,q') &= K_{\mathrm{M}} \exp\Big(i\frac{aq^2}{2c}\Big)\, I\Big(\sqrt{d/2c}, -q/c\Big) \\ &= K_{\mathrm{M}}\, \sigma(\sqrt{d/c})\, e^{i\frac{1}{4}\pi} \sqrt{\frac{2\pi c}{d}} \exp i\frac{bq^2}{2d} = \frac{\sigma(\sqrt{d/c})}{\sqrt{d}} \exp i\frac{bq^2}{2d}.\end{aligned} \tag{C.34}$$

The sign $\sigma(r)$ is given by (C.33); its value depends only on the argument of d/c, so we are allowed to let c approach 0, while d is prevented from being null by the symplectic condition $ad - bc = 1$. When the system becomes the non-inverting imager (C.29) with $d = a^{-1} > 0$ and $-\pi \le \arg c \le 0$ (so $0 \le \arg c^{-\frac{1}{2}} \le \frac{1}{2}\pi$), then $\sigma = +1$, and

$$\lim_{c \to 0} \int_{\mathsf{R}} dq'\, C\Big(\begin{smallmatrix} a & b \\ c & d \end{smallmatrix}\Big)(q,q') = \sqrt{a}\, \exp i\tfrac{1}{2}abq^2. \tag{C.35}$$

The integrand (C.25)–(C.26) becomes infinitely oscillating, but has a stationary point when the numerator of the exponent vanishes: $aq^2 - 2qq' + a^{-1}q'^2 = 0$, i.e., for $q' = aq$. The Gaussian kernel thus provides a sequence of functions weakly converging to the Dirac δ,

$$C\Big(\begin{smallmatrix} a & b \\ 0 & a^{-1} \end{smallmatrix}\Big)(q,q') = \delta(q' - aq)\, \sqrt{a}\, \exp i\tfrac{1}{2}abq^2, \tag{C.36}$$

which agrees with (C.29). •

The integral we solved in (C.30)–(C.32) is also instrumental to verify that the product of two canonical transforms $\mathcal{C}(\mathbf{M}_1)$ and $\mathcal{C}(\mathbf{M}_2)$ is $\mathcal{C}(\mathbf{M}_3)$ with $\mathbf{M}_3 = \mathbf{M}_1\, \mathbf{M}_2$, as ascertained in (C.19). For these three matrices it holds that

$$\int_{\mathsf{R}} dq'\, C_{\mathrm{M}_1}(q,q')\, C_{\mathrm{M}_2}(q',q'') = \sigma\Big(\sqrt{\frac{c_3}{c_1 c_2}}\Big)\, C_{\mathrm{M}_3}(q,q''), \tag{C.37}$$

where σ is given by (C.33). We should be careful to note however, that this does not 'solve' the metaplectic sign problem; because of the presence of the square roots, even $\sqrt{1} = \pm 1$. We must refer to the Iwasawa decomposition of $\mathbf{M}$ in (9.69), into a fractional Fourier transformer and a noninverting imager (C.36); it is the Fourier factor which bears the onus of the integral and the double cover.

Example: Free propagation. Free propagation in the paraxial régime of geometric optics is the matrix transformation produced by $\mathcal{M}\big(\begin{smallmatrix} 1 & 0 \\ -z & 1 \end{smallmatrix}\big)$ [see (9.3) and (9.18)] on the vector $\big(\begin{smallmatrix} \mathbf{p} \\ \mathbf{q} \end{smallmatrix}\big)$ of phase space. In canonical Fourier optics, the propagation of wavefields is represented by a canonical integral transform of kernel

$$C\Big(\begin{smallmatrix} 1 & 0 \\ -z & 1 \end{smallmatrix}\Big)(q,q') = \frac{e^{i\frac{1}{4}\pi}}{\sqrt{2\pi\, z}} \exp i\frac{-(q-q')^2}{2z}, \qquad z > 0. \tag{C.38}$$

This is known as the Fresnel, or imaginary Gauss-Weierstrass transform; it still has to be multiplied by $\exp iz\lambda\!\!\!^{-}$ to match the phase of the oscillating wavefield over the distance from the standard screen. •

Example: Propagation in a harmonic waveguide. In geometric optics, propagation by $z > 0$ along a harmonic waveguide [of refractive index $n(q) = n_o - \frac{1}{2}k^2q^2$, see (3.38)], is an inverse fractional Fourier transform. [*Cf.* (3.19)–(3.20) and (10.8)–(10.9) with (3.38)]. The $\mathsf{Sp}(2,\mathsf{R})$ matrix should have a 2–1 element $c(z)$ starting with the same sign, $c(z) \approx -kz$, as free propagation (C.38). The corresponding canonical transform kernel, for $\tau := kz$ in the range $0 < \tau < \pi$, is

$$C\begin{pmatrix} \cos\tau & \sin\tau \\ -\sin\tau & \cos\tau \end{pmatrix}(q,q') = \frac{e^{+i\frac{1}{4}\pi}}{\sqrt{2\pi\,\sin\tau}} \exp i\frac{-(q^2-q'^2)\cos\tau + 2qq'}{2\sin\tau}. \tag{C.39}$$

This gives a handle on the metaplectic sign, because when $\tau \to 0^+$ the kernel limits to the Dirac $\delta(q-q')$, and when $\tau \to \pi^-$ to $i\,\delta(q+q')$. By straightforward composition, the kernel for $\tau \to 2\pi^-$ is $-\delta(q-q')$, and for $\tau \to 3\pi^-$ it is $-i\delta(q+q')$. Only for $\tau \to 4\pi^-$ the kernel returns to the unit $\delta(q-q')$. Again we should multiply the wavefield by $\exp iz\lambda\!\!\!^{-}$ for distance z from the screen. •

Example: Diffusion. The real Gauss-Weierstrass integral transform describes heat diffusion for time $t > 0$ with a Gaussian integral kernel of width $w = t$. This is the canonical transform (C.38) for imaginary displacement $z = it$. The 2–1 matrix element is in the lower complex half-plane, and the *complex* canonical transform kernel is real, *viz.*,

$$C\begin{pmatrix} 1 & 0 \\ -it & 1 \end{pmatrix}(q,q') = \frac{1}{\sqrt{2\pi\,t}} \exp -\frac{(q-q')^2}{2t} =: G_t(q-q'). \tag{C.40}$$

This transform conserves 'total heat' $\int_{\mathsf{R}} dq\, G_t(q) = 1$, but *not* the quadratic norm $\int_{\mathsf{R}} dq\, G_t(q)^2 = 1/\sqrt{4\pi t}$. •

Example: Canonical transformation of Gaussian beams. Complex canonical transforms can be consistently used in conjunction with real ones. According to (C.27), a centered Gaussian wavefield of width $w > 0$ is the complex canonical transform of a centered Dirac $\delta(q)$,

$$G_w(q) = \left[\mathcal{C}\begin{pmatrix} 1 & 0 \\ -iw & 1 \end{pmatrix} : \delta\right](q) = C\begin{pmatrix} 1 & 0 \\ -iw & 1 \end{pmatrix}(q,0). \tag{C.41}$$

From the composition of transform kernels (C.37), it follows that the $\mathcal{C}\left(\begin{smallmatrix} a & b \\ c & d \end{smallmatrix}\right)$-transform of a centered Gaussian wavefield can be found as (cf. [81]),

$$\begin{aligned}\left[\mathcal{C}\begin{pmatrix} a & b \\ c & d \end{pmatrix} : G_w\right](q) &= \left[\mathcal{C}\begin{pmatrix} a & b \\ c & d \end{pmatrix}\mathcal{C}\begin{pmatrix} 1 & 0 \\ -iw & 1 \end{pmatrix} : \delta\right](q) = \left[\mathcal{C}\begin{pmatrix} a-iwb & b \\ c-iwd & d \end{pmatrix} : \delta\right](q) \\ &= \left[\mathcal{C}\begin{pmatrix} 1 & 0 \\ -iw_{\mathrm{M}} & 1 \end{pmatrix}\mathcal{C}\begin{pmatrix} a-iwb & b \\ 0 & (a-iwb)^{-1} \end{pmatrix} : \delta\right](q) \end{aligned} \tag{C.42}$$

$$= \frac{1}{\sqrt{a - ibw}} \Big[\mathcal{C} \begin{pmatrix} 1 & 0 \\ -iw_{\rm M} & 1 \end{pmatrix} : \delta \Big](q) = \frac{1}{\sqrt{a - ibw}} G w_{\rm M}(q)$$

$$\text{with} \quad w_{\rm M} := (wd + ic)/(a - iwb), \tag{C.43}$$

•

Remark: Complex canonical transforms. In (C.40) we gave an example where the parameters of the canonical transform are allowed into the complex plane. Another particular case of interest is the *Bargmann-Segal* transform [16, 124],

$$C \frac{1}{\sqrt{2}} \begin{pmatrix} 1 & -i \\ -i & 1 \end{pmatrix} (q, q') = \frac{1}{\sqrt{\pi\sqrt{2}}} \exp[-\tfrac{1}{2}(q^2 + q'^2) + \tfrac{1}{\sqrt{2}} qq']. \tag{C.44}$$

According to (C.18), this maps the Fourier–Schrödinger operators of position and momentum, $\widehat{P}$ and $\widehat{Q}$, to $\widehat{P}' = i\widehat{A}^\dagger := i\frac{1}{\sqrt{2}}(\widehat{Q} - i\widehat{P})$ and $\widehat{Q}' = \widehat{A} := \frac{1}{\sqrt{2}}(\widehat{Q} + i\widehat{P})$, i.e., the $i\times$creation and the annihilation operators for the quantum oscillator. The set of complex matrices $\mathbf{M} = \begin{pmatrix} a & b \\ c & d \end{pmatrix} \in \mathsf{Sp}(2, \mathsf{C})$ [mantaining the symplectic condition $ad - bc = 1$ and keeping c/d in the lower complex half-plane – see (C.33) and (C.34)] form a *semi*-group (*Halb*-gruppe), denoted $\mathsf{HSp}(2, \mathsf{C})$ [82]. This contains the diffusion and Bargmann-Segal transforms, (C.40) and (C.44), and the one-parameter subgroups leading to them – but *not* their inverses. Nevertheless, unitarity can be upheld between $\mathcal{L}^2(\mathsf{R})$ and *Bargmann-Segal* Hilbert spaces of analytic functions in $z := \frac{1}{\sqrt{2}}(q + ip)$ [see (11.102)], whose inner product has the Gaussian measure $\sim \exp -|z|^2$. See [147] and [149, Sect. 9.2]. •

C.3 Hyperdifferential forms

The group of canonical integral transforms (C.22) with kernel (C.25)–(C.26), including the contact transformations (C.29), is generated by a realization of the Lie algebra $\mathsf{sp}(2, \mathsf{R})$ by up-to-*second*-order differential operators, whose exponential series we call *hyperdifferential*. The one-parameter subgroups of interest in Sect. 10.1 are generated by the following:

$$\widehat{J}^{\rm H} := \tfrac{1}{2}(\widehat{P}^2 + \widehat{Q}^2), \qquad \exp i\tau\widehat{J}^{\rm H} = \mathcal{C}\begin{pmatrix} \cos\tau & \sin\tau \\ -\sin\tau & \cos\tau \end{pmatrix}, \tag{C.45}$$

$$\widehat{J}^{\rm F} := \tfrac{1}{2}\widehat{P}^2, \qquad \exp i\tau\widehat{J}^{\rm F} = \mathcal{C}\begin{pmatrix} 1 & 0 \\ -\tau & 1 \end{pmatrix}, \tag{C.46}$$

$$\widehat{J}^{\rm L} := \tfrac{1}{2}\widehat{Q}^2, \qquad \exp i\tau\widehat{J}^{\rm L} = \mathcal{C}\begin{pmatrix} 1 & \tau \\ 0 & 1 \end{pmatrix}, \tag{C.47}$$

$$\widehat{J}^{\rm R} := \tfrac{1}{2}(\widehat{P}^2 - \widehat{Q}^2), \qquad \exp i\tau\widehat{J}^{\rm R} = \mathcal{C}\begin{pmatrix} \cosh\tau & -\sinh\tau \\ -\sinh\tau & \cosh\tau \end{pmatrix}, \tag{C.48}$$

$$\widehat{J}^{\rm I} := \tfrac{1}{2}(\widehat{Q}\widehat{P} + \widehat{P}\widehat{Q}), \qquad \exp i\tau\widehat{J}^{\rm I} = \mathcal{C}\begin{pmatrix} e^\tau & 0 \\ 0 & e^{-\tau} \end{pmatrix}. \tag{C.49}$$

Example: Baker–Campbell–Hausdorff relations. The correspondence between matrices and exponentials of $\mathsf{sp}(2,\mathsf{R})$ generators listed in (C.45)–(C.49) allows us to find Baker-Campbell-Hausdorff relations between hyperdifferential operators. In particular, harmonic evolution by $|\tau| < \frac{1}{2}\pi$ can be decomposed as

$$\begin{pmatrix} \cos\tau & \sin\tau \\ -\sin\tau & \cos\tau \end{pmatrix} = \begin{pmatrix} 1 & 0 \\ -\tan\tau & 1 \end{pmatrix} \begin{pmatrix} \cos\tau & 0 \\ 0 & \sec\tau \end{pmatrix} \begin{pmatrix} 1 & \tan\tau \\ 0 & 1 \end{pmatrix}, \quad (C.50)$$

$$\begin{aligned} \Rightarrow \exp[i\tau\tfrac{1}{2}(\widehat{P}^2 + \widehat{Q}^2)] &= \exp[i(\tan\tau)\tfrac{1}{2}\widehat{P}^2] \\ &\quad \times \exp[i(\ln\cos\tau)\tfrac{1}{2}(\widehat{P}\widehat{Q} + \widehat{Q}\widehat{P})] \\ &\quad \times \exp[i(\tan\tau)\tfrac{1}{2}\widehat{Q}^2]. \end{aligned} \quad (C.51)$$

•

The harmonic guide (C.45) is the representative of the compact elliptic orbit in $\mathsf{Sp}(2,\mathsf{R})$ (see Chap. 12). Free propagation (C.46) and lenses (C.47) are related by the Fourier transform and belong to the parabolic orbit. Finally, the square root of the Fourier transform intertwines the repulsive guide (C.48) with pure positive magnifiers (C.49), which belong to the hyperbolic orbit. Their spectra and eigenfunctions are well known from quantum mechanics:

$$\widehat{J}^{\mathrm{H}}\,\Psi^{\mathrm{H}}_n = (n+\tfrac{1}{2})\Psi^{\mathrm{H}}, \qquad \Psi^{\mathrm{H}}_n(q) = e^{-\frac{1}{2}q^2}\,H_n(q)\Big/\sqrt{2^n n!\sqrt{\pi}}, \quad \text{spectrum: } n \in \{0,1,2,\ldots\}; \quad (C.52)$$

$$\begin{aligned} \widehat{J}^{\mathrm{F}}\,\Psi^{\mathrm{F}}_r &= \tfrac{1}{2}r^2\Psi^{\mathrm{F}}_p, & \Psi^{\mathrm{F}}_r(q) &= e^{-irq}/\sqrt{2\pi}, \\ \widehat{J}^{\mathrm{L}}\,\Psi^{\mathrm{L}}_r &= \tfrac{1}{2}r^2\Psi^{\mathrm{L}}_r, & \Psi^{\mathrm{L}}_r(q) &= \delta(q-r), \\ && &\text{spectrum: } r \in \mathsf{R}; \end{aligned} \quad (C.53)$$

$$\begin{aligned} \widehat{J}^{\mathrm{R}}\,\Psi^{\mathrm{R}}_{s,\varepsilon} &= s\Psi^{\mathrm{R}}_{s,\varepsilon}, \quad \Psi^{\mathrm{R}}_{s,\varepsilon}(q) = \frac{\exp[\frac{1}{4}i\pi(\frac{1}{2}-is)]}{2^{3/4}\pi}\Gamma(\tfrac{1}{2}-is)\,D_{-(\frac{1}{2}-is)}(\varepsilon e^{i\frac{3}{4}\pi}\sqrt{2}\,q), \\ \widehat{J}^{\mathrm{I}}\,\Psi^{\mathrm{I}}_{s,\varepsilon} &= s\Psi^{\mathrm{I}}_{s,\varepsilon}, \quad \Psi^{\mathrm{I}}_{s,\varepsilon}(q) = \frac{1}{\sqrt{2\pi}}\,q_\varepsilon^{-(\frac{1}{2}-is)}, \quad q_\varepsilon := \begin{cases} |q|, & \varepsilon q>0, \\ 0, & \varepsilon q\le 0, \end{cases} \\ &\text{spectrum: } s \in \mathsf{R}, \quad \varepsilon \in \{+,-\}; \end{aligned} \quad (C.54)$$

where $H_n(x)$ are the Hermite polynomials, $D_s(z)$ are the parabolic cylinder functions [48, Vol. 2, p. 119], and q_ε are the positive- and negative-cut power functions.

The wavefields (C.52)–(C.54) are eigenfunctions of the generators of the canonical transforms (C.45)–(C.49), with eigenvalues $e^{i\tau\mu}$, where τ is the subgroup parameter and μ is any of their eigenvalues [n, r or s in (C.54)]. To find canonical transforms of these eigenfields, we can factorize the corresponding $\mathsf{sp}(2,\mathsf{R})$ matrices in a way similar to (C.50), as a product of a contact imaging transformation (C.29) times the corresponding guide evolution,

$$\begin{aligned} \left[\mathcal{C}\begin{pmatrix} a & b \\ c & d \end{pmatrix} : \Psi_\mu\right](q) &= \left[\mathcal{C}\begin{pmatrix} \alpha & \beta \\ 0 & \alpha^{-1} \end{pmatrix} e^{i\tau\widehat{J}} : \Psi_\mu\right](q) = e^{i\mu\tau}\left[\mathcal{C}\begin{pmatrix} \alpha & \beta \\ 0 & \alpha^{-1} \end{pmatrix}\Psi_\mu\right](q) \\ &= e^{i\mu\tau}\sqrt{\alpha}\,e^{-i\frac{1}{2}\alpha\beta q^2}\,\Psi_\mu(\alpha q). \end{aligned} \quad (C.55)$$

Example: Canonical transforms of Hermite functions. For the harmonic guide normal modes $\Psi_n^{\mathrm{H}}(q)$ in (C.52), the above decomposition of a $\mathcal{C}\left(\begin{smallmatrix} a & b \\ c & d \end{smallmatrix}\right)$ system into geometric and guide factors is

$$\begin{gathered}\begin{pmatrix} a & b \\ c & d \end{pmatrix} = \begin{pmatrix} \alpha & \beta \\ 0 & \alpha^{-1} \end{pmatrix} \begin{pmatrix} \cos\tau & \sin\tau \\ -\sin\tau & \cos\tau \end{pmatrix}, \\ \alpha = \frac{1}{\sqrt{c^2+d^2}}, \quad \beta = \frac{ac+bd}{\sqrt{c^2+d^2}}, \quad \tan\tau = -\frac{c}{d}, \quad e^{2i\tau} = \frac{d-ic}{d+ic}. \end{gathered} \tag{C.56}$$

The transformation of the Hermite functions is thus found to be

$$\begin{aligned} \left[\mathcal{C}\left(\begin{smallmatrix} a & b \\ c & d \end{smallmatrix}\right) : \Psi_n^{\mathrm{H}}\right](q) &= e^{i(n+\frac{1}{2})\tau} \left[\mathcal{C}\left(\begin{smallmatrix} \alpha & \beta \\ 0 & \alpha^{-1} \end{smallmatrix}\right) \Psi_n^{\mathrm{H}}\right](q) \\ &= \sqrt{\frac{(d-ic)^n}{(d+ic)^{n+1}}} \frac{1}{\sqrt{2^n n! \sqrt{\pi}}} \exp\left(-\frac{a+ib}{d-ic}\frac{q^2}{2}\right) H_n\left(\frac{q}{\sqrt{c^2+d^2}}\right). \end{aligned} \tag{C.57}$$

In particular, under free propagation $\mathcal{C}\left(\begin{smallmatrix} 1 & 0 \\ -z & 1 \end{smallmatrix}\right)$, the argument of the Hermite polynomial is $q/\sqrt{1+z^2}$. Hence, its set of zeros, $\{x_{n,m}\}_{m=0}^{n}$ $(H_n(x_{n,m}) = 0)$, traces out a family concentric and convertical hyperbolas[4] $q^2/x_{n,m}^2 - z^2 = 1$ in the optical q–z plane. On the other hand, under diffusion $z = it$, the set of zeros traces corresponding ellipses. Decompositions analogous to (C.56) hold for eigenfunctions in the other two Hamiltonian orbits with their corresponding spectra; see [149, Tables 10.1 and 10.2]. •

The eigenfunction sets (C.52)–(C.54) are orthogonal on $q \in \mathsf{R}$ as well as *complete*; they allow a Dirichlet-type limit to a Dirac δ on the screen,[5]

$$\left.\begin{array}{r} \sum_{n=0}^{\infty} \Psi_n^{\mathrm{H}}(q)\, \Psi_n^{\mathrm{H}}(q') \\[2ex] \int_{\mathsf{R}} dr\, \Psi_r^{\mathrm{F}}(q)^*\, \Psi_r^{\mathrm{F}}(q') \\[2ex] \sum_{\varepsilon=\pm} \int_{\mathsf{R}} ds\, \Psi_{s,\varepsilon}^{\mathrm{R}}(q)^*\, \Psi_{s,\varepsilon}^{\mathrm{R}}(q') \end{array}\right\} = \delta(q-q'). \tag{C.58}$$

A canonical trasformation $\mathcal{C}\left(\begin{smallmatrix} a & b \\ c & d \end{smallmatrix}\right)$ in the coordinate q' thus reproduces the kernel (C.25)–(C.26) as a bilinear generating function of the kernel,

$$\left.\begin{array}{r} \sum_{n=0}^{\infty} \Psi_n^{\mathrm{H}}(q) \left[\mathcal{C}\left(\begin{smallmatrix} a & b \\ c & d \end{smallmatrix}\right) : \Psi_n^{\mathrm{H}}\right](q') \\[2ex] \int_{\mathsf{R}} dr\, \Psi_r^{\mathrm{F}}(q)^* \left[\mathcal{C}\left(\begin{smallmatrix} a & b \\ c & d \end{smallmatrix}\right) : \Psi_r^{\mathrm{F}}\right](q') \\[2ex] \sum_{\varepsilon=\pm} \int_{\mathsf{R}} ds\, \Psi_{s,\varepsilon}^{\mathrm{R}}(q)^* \left[\mathcal{C}\left(\begin{smallmatrix} a & b \\ c & d \end{smallmatrix}\right) : \Psi_{s,\varepsilon}^{\mathrm{R}}\right](q') \end{array}\right\} = C\left(\begin{smallmatrix} a & b \\ c & d \end{smallmatrix}\right)(q, q'). \tag{C.59}$$

In particular, the set of eigenfields of a guide $\exp(iz\widehat{J}) = \mathcal{C}\left(\begin{smallmatrix} a(z) & b(z) \\ c(z) & d(z) \end{smallmatrix}\right)$ of eigenvalues $\{\mu\}$ (in each of the three orbits, $\mu \in \mathcal{Z}_0^+ + \frac{1}{2}$, $\mu \in \mathsf{R}$, or $\mu \in \mathsf{R} \oplus \mathsf{R}$),

[4] The family of hyperbolas has the same *second* ('invisible') vertices.

[5] The Hermite functions $\Psi_n^{\mathrm{H}}(q)$ are real, so they need not be complex conjugated.

are only multiplied by phases $e^{i\tau\mu}$. Thus we are provided with one-parameter subgroups of 'fractional' transforms with kernel (C.59).

Example: Generator function for harmonic waveguides. The set of Hermite functions yields the evolution kernel for a harmonic waveguide given by (C.39) and (C.59), as

$$\begin{aligned} C\begin{pmatrix} \cos\tau & \sin\tau \\ -\sin\tau & \cos\tau \end{pmatrix}(q,q') &= \sum_{n=0}^{\infty} \Psi_n^{\rm H}(q)\, e^{i(n+\frac{1}{2})\tau} \Psi_n^{\rm H}(q') \\ &= \frac{e^{+i\frac{1}{4}\pi}}{\sqrt{2\pi \sin\tau}} \exp i \frac{-(q^2-q'^2)\cos\tau + 2q\,q'}{2\sin\tau}. \end{aligned} \tag{C.60}$$

We draw attention in particular to the values $\tau = +\frac{1}{2}\pi$ and $\tau = -\frac{1}{2}\pi$:

$$C\begin{pmatrix} 0 & 1 \\ -1 & 0 \end{pmatrix}(q,q') = e^{+i\frac{1}{4}\pi} \frac{e^{+iq\,q'}}{\sqrt{2\pi}}, \quad C\begin{pmatrix} 0 & -1 \\ 1 & 0 \end{pmatrix}(q,q') = e^{-i\frac{1}{4}\pi} \frac{e^{-iq\,q'}}{\sqrt{2\pi}}. \tag{C.61}$$

The right-hand sides are the kernels of the inverse and the direct Fourier integral transforms (C.1)–(C.2) respectively, times $e^{\pm i\frac{1}{4}\pi}$. •

Remark: Airy functions. The Hamiltonian operators (C.45), (C.46), and (C.48) are representatives of $\mathsf{sp}(2,\mathsf{R})$ orbits. When we include translations of phase space, the algebra grows to the inhomogeneous $\mathsf{isp}(2,\mathsf{R})$ [see Sect. 12.8, (12.91)], and there appears a new orbit whose representative Hamiltonian can be added to the list:

$$\widehat{J}^{\rm C} := \tfrac{1}{2}\widehat{P}^2 + \widehat{Q}, \qquad \exp i\tau\widehat{J}^{\rm C} = \mathcal{C}\begin{pmatrix} 1 & 0 & \tau \\ -\tau & 1 & -\frac{1}{2}\tau^2 \\ 0 & 0 & 1 \end{pmatrix}, \tag{C.62}$$

where we use the 3×3 matrix representation of $\mathsf{isp}(2,\mathsf{R})$. This Hamiltonian corresponds to an optical medium with unit refractive index gradient, $n = n_0 - q$, or a linear potential (free fall) in mechanics. Its spectrum and $\mathcal{L}^2(\mathsf{R})$-normalized eigenfunctions are:

$$\widehat{J}^{\rm C}\,\Psi_\nu^{\rm H} = \nu\,\Psi_\nu^{\rm H}, \qquad \Psi_\nu^{\rm H}(q) = 2^{\frac{1}{3}}\,\mathrm{Ai}\,(2^{\frac{1}{3}}(q-\nu)), \quad \nu\in\mathsf{R}, \tag{C.63}$$

where $\mathrm{Ai}(x)$ are the Airy functions, which should be added to the list (C.52)–(C.54). With the measure $\int_{\mathsf{R}} d\nu$, they participate in completeness relations (C.58) and provide another bilinear generating function of the kind (C.59) for the canonical transform kernel. Their transformation under various evolutions and connection with separation of variables has been examined in [149, Chap. 10]. •

We now have the background and examples to state the relation between the evolution in an optical harmonic waveguide $\mathcal{C}\begin{pmatrix} \cos\tau & \sin\tau \\ -\sin\tau & \cos\tau \end{pmatrix}(q,q')$, and the ordinary Fourier transform $\mathcal{F}$ in (C.1) —actually its inverse (C.2). The $\mathcal{L}^2(\mathsf{R})$ basis of Hermite functions $\Psi_n^{\rm H}(q)$ has the well-known property

$$\mathcal{F}: \Psi_n^{\rm H}(q) = (-i)^n \Psi_n^{\rm H}(q). \tag{C.64}$$

It is suggestive to use this property to define the kernel $F^a(q,q')$ of a *fractional* Fourier transform $\mathcal{F}^a$ through interpreting the phase in (C.64) to be $(-i)^n := e^{-i\frac{1}{2}\pi n}$ as done by Condon [34][6] and Namias [105]. This determines the kernel of $\mathcal{F}^a$ by the generating function

$$F^a(q,q') := \sum_{n=0}^{\infty} \Psi_n^{\mathrm{H}}(q)\, e^{-i\frac{1}{2}\pi n a}\, \Psi_n^{\mathrm{H}}(q'). \tag{C.65}$$

Under waveguide evolution on the other hand, the Hermite functions display an *extra* phase $e^{-i\frac{1}{4}\pi}$ [see (C.61)], due to the shift by $\frac{1}{2}$ between the energy spectrum (C.52) and the mode number set $n \in \mathcal{Z}_0^+$. Comparison of (C.65) with the waveguide evolution in (C.60) shows that

$$F^a(q,q') = e^{i\frac{1}{4}\pi a}\, C\begin{pmatrix} \cos\frac{1}{2}\pi a & \sin\frac{1}{2}\pi a \\ -\sin\frac{1}{2}\pi a & \cos\frac{1}{2}\pi a \end{pmatrix}(q,q'), \tag{C.66}$$

$$\mathcal{F}^a = e^{i\frac{1}{4}\pi a} \exp\left(-i\pi\, a\, \tfrac{1}{4}\,(\widehat{P}^2 + \widehat{Q}^2)\right). \tag{C.67}$$

where $e^{i\frac{1}{4}\pi a}$ is the metaplectic phase.

C.4 *D*-dim and radial canonical transforms

The case of two (or D) screen dimensions follows very closely the $D = 1$ case seen above. Linear transformations (C.17) preserve the Heisenberg–Weyl Lie algebra (C.11) when the $2D \times 2D$ matrix $\mathbf{M} = \left(\begin{smallmatrix} \mathbf{a} & \mathbf{b} \\ \mathbf{c} & \mathbf{d} \end{smallmatrix}\right)$ is symplectic, as can be verified straightforwardly. The corresponding transformation of the wavefields is the D-dim version of (C.22),

$$f_{\mathrm{M}}(\mathbf{q}) := (\mathcal{C}_{\mathrm{M}} : f)(\mathbf{q}) = \int_{\mathsf{R}^D} d\mathbf{q}'\, C_{\mathrm{M}}(\mathbf{q},\mathbf{q}')\, f(\mathbf{q}'). \tag{C.68}$$

Equations analogous to (C.18)–(C.24) follow for the components of $\mathbf{q} \in \mathsf{R}^D$, and for $\det \mathbf{c} \neq 0$ the kernel is

$$C\begin{pmatrix} \mathbf{a} & \mathbf{b} \\ \mathbf{c} & \mathbf{d} \end{pmatrix}(\mathbf{q},\mathbf{q}') = K_{\mathrm{M}} \exp i\tfrac{1}{2}(\mathbf{q}^\top \mathbf{c}^{-1}\mathbf{a}\,\mathbf{q} - 2\mathbf{q}^\top \mathbf{c}^{-1}\mathbf{q}' + \mathbf{q}'\mathbf{d}\mathbf{c}^{-1}\mathbf{q}'), \tag{C.69}$$

where the multiplication order of the submatrices is now important. When the submatrix $\mathbf{c}$ is diagonalizable, the normalization constant K_{M} is the product of the constants (C.26) for each screen dimension, so

$$K_{\mathrm{M}} := \frac{1}{\sqrt{(2\pi i)^D \det \mathbf{c}}} := \frac{e^{-i\frac{1}{4}D\pi} \exp(-i\frac{1}{2}\arg\det\mathbf{c})}{\sqrt{(2\pi)^D\,|\det\mathbf{c}|}}, \qquad \begin{array}{c} \det\mathbf{c} \neq 0, \\ -\pi \leq \arg\det\mathbf{c} \leq 0. \end{array} \tag{C.70}$$

[6] Note that this author uses the kernel e^{ipq} instead of the more common e^{-ipq} that we use here.

When $\mathbf{c} = \mathbf{0}$, the D-dim analogue of (C.29) for positive definite $\mathbf{a}$ is

$$\Big[\mathcal{C}\begin{pmatrix} \mathbf{a} & \mathbf{b} \\ \mathbf{0} & \mathbf{a}^{\top -1}\end{pmatrix} : f\Big](\mathbf{q}) = \sqrt{\det \mathbf{a}}\, \exp(-i\tfrac{1}{2}\mathbf{q}^\top \mathbf{a}\,\mathbf{b}^\top \mathbf{q})\, f(\mathbf{a}^\top q). \qquad \text{(C.71)}$$

The most notable example of the $\mathbf{c} = \mathbf{0}$ case are the lens transformations $\mathcal{C}\begin{pmatrix} \mathbf{1} & \mathbf{b} \\ \mathbf{0} & \mathbf{1}\end{pmatrix}$, which multiply the wavefield by the quadratic phase $\exp(-i\frac{1}{2}\mathbf{q}^\top\, \mathbf{b}\,\mathbf{q})$. We shall not detail further the boundary cases when $\det \mathbf{c} = 0$ but $\mathbf{c} \neq \mathbf{0}$; the limit (C.35) applies and the canonical transform kernel will have one or more δ's.

Radial canonical transforms were introduced in [102] for D-dim canonical transforms produced by systems which have rotational symmetry around the evolution axis. Wavefields are generally *not* axis-symmetric of course, but we can use their multipole expansion to write, in the $D = 2$ case,

$$f(\mathbf{q}) = \sum_{m \in \mathcal{Z}} f_m(r)\, \frac{e^{im\theta}}{\sqrt{2\pi}}, \qquad \begin{array}{ll} q_x = r\cos\theta, & r \in [0, \infty), \\ q_y = r\sin\theta, & \theta \in \mathcal{S}^1. \end{array} \qquad \text{(C.72)}$$

Since $\mathcal{C}\begin{pmatrix} a1 & b1 \\ c1 & d1\end{pmatrix}$ commutes with rotations of the screen, its action will not mix the summands in (C.72), but only transform the radial factors $f_m(r)$ through

$$(\mathcal{C}_{\mathrm{M}} : f_m)(r) = \int_0^\infty dr'\, C^{(m)}_{\mathrm{M}}(r, r')\, f_m(r'), \quad \mathbf{M} = \begin{pmatrix} a & b \\ c & d\end{pmatrix} \in \mathsf{Sp}(2, \mathsf{R}) \quad \text{(C.73)}$$

[cf. (C.22) and (C.68)], with the integral kernel

$$C^{(m)}_{\mathrm{M}}(r, r') = e^{-i\frac{1}{2}\pi(|m|+1)}\, \frac{\sqrt{r\,r'}}{c}\, J_{|m|}\Big(\frac{r\,r'}{c}\Big)\, \exp i\frac{a\,r^2 + d\,r'^2}{2c}, \quad c \neq 0 \quad \text{(C.74)}$$

[cf. (C.25)–(C.26) and (C.69)–(C.26)]. Again, when $c \to 0$, the limit is the contact transformation

$$\Big[\mathcal{C}\begin{pmatrix} a & b \\ 0 & a^{-1}\end{pmatrix} : f_m\Big](r) = \sqrt{a}\, \exp(-i\tfrac{1}{2}ab\,r^2)\, f_m(a\,r), \quad a > 0 \qquad \text{(C.75)}$$

[cf. (C.29) and (C.71)].

It should be observed that the metaplectic sign disappears in radial canonical transforms obtained from 2-dim ones: there is no square root on c in the kernel (C.74). Radial canonical transforms thus represent $\mathsf{Sp}(2, \mathsf{R})$ faithfully when m is the integer angular momentum of the field component. However, when the index m is continued to real values $\mu \in \mathsf{R}$, then radial canonical transforms generally represent the covering group $\overline{\mathsf{Sp}(2, \mathsf{R})}$. In particular for $\mu = \pm\frac{1}{2}$, one recovers the $D = 1$ case of the previous sections.

Remark: Hyperbolic canonical transforms. When we use hyperbolic coordinates for the screen, and expand the wavefield in a continuous integral of eigenfunctions of a Hamiltonian belonging to the hyperbolic orbit [whose spectrum is given by (C.54)], we obtain *hyperbolic* canonical transforms [150]. The energy is then a continuous number $s \in \mathsf{R}$, and there is also a sign/parity index ε because two charts of hyperbolic coordinates are needed to cover the screen. The reduced integral kernel [cf. (C.74)] has 2×2 components $(\varepsilon, \varepsilon')$ containing the usual oscillating Gaussians, but now instead of Bessel, one has Hankel and Macdonald functions of imaginary index is, whose argument is evaluated on the ε-side of their branch cut. This integral kernel realizes all $\mathsf{Sp}(2,\mathsf{R})$ unitary and irreducible representations known as the *continuous* series, indicated by Bargmann as $C^{\varepsilon}_{s^2-1}$ [15]. (Radial transforms realize the Bargmann *discrete* series $D^{\text{sign}\,s}_{\frac{1}{2}(1+|s|)}$.) Finally, there is an *exceptional* representation series in the interval $0 < |s| < 1$ built with a *sui generis* Hilbert space measure. All previous properties hold, except for the complex extension of the group parameters. •

Remark: The screen and two other bases. All the above canonical integral kernels have for arguments the positions of two points on the screen [see (C.25), (C.69) or the radii in (C.74)]. Generally, the kernels are matrix elements of $\mathcal{C}_{\text{M}} := \mathcal{C}(\mathbf{M}) \in \mathsf{Sp}(2,\mathsf{R})$ between two eigenfunction sets $\Psi^{\text{X}}_{\mu}(q)$, namely

$$C^{\text{X}}_{\text{M}}(\mu,\mu') := \left(\Psi^{\text{X}}_{\mu},\, \mathcal{C}_{\text{M}} : \Psi^{\text{X}}_{\mu'}\right) =: D^{\text{X}}_{\mu,\mu'}(\mathbf{M}), \qquad \text{X} \in \{\text{H, F, L, R or I}\}. \tag{C.76}$$

Above, we have chosen the position eigenbasis of the parabolic orbit generator $\widehat{J}^{\text{L}} = \frac{1}{2}\widehat{Q}^2$, i.e., the set of $\delta(q-\mu)$'s, with $\mu \in \mathsf{R}$ determined by the squared position and parity. But any other basis X among the three generator orbits (C.52)–(C.54) can be chosen, together with its proper Parseval-conjugate measure $d\mu$ under which the set is complete [see (C.3) and (C.58)]. In the elliptic orbit, the rows and columns of the matrix $D^{\text{X}}_{\mu,\mu'}(\mathbf{M})$ are discrete, lower-bound, and labelled by the mode number. Finally, there is the hyperbolic basis, which provides a 2×2 matrix-integral representation, whose cut-power basis can be selected for its scaling properties. All these kernels/matrices —as well as those built between two different bases – plus their analytic properties, can be found in [20]. •

Part IV

Hamilton-Lie aberrations

Introduction

Beyond the paraxial régime of geometric optics lies the *metaxial régime.* In addition to the linear transformations of phase space that occupied part III, in the metaxial model we consider polynomial maps of the phase space coordinates, obtained from Taylor expansions truncated to a common degree A —the *aberration order.* An example appeared in Sect. 4.4 for the symmetric root transformation with $A = 5$ and screen dimension $D = 1$; in a remark in Sect. 5.7 we expanded the relativistic coma aberration with $A = 3$ and generic D.

In Chap. 13, the first of this part, we shall classify the $D = 1$ polynomials in (p, q) —and the aberrations they generate— by their degree and other 'quantum' numbers of Lie-theoretical significance. We call them *Hamilton-Lie* aberrations because they transform as members of finite multiplets under the paraxial Lie algebra $\mathsf{sp}(2, \mathsf{R})$, and are canonical (*up to the aberration order*). The generators of linear maps and aberrations, truncated to a finite order, form a Lie algebra that exponentiates to a corresponding Lie group of deformations of phase space. Elements of this group of aberrations realize the free spaces and refracting surfaces of optical setups, and their multiplication corresponds 1:1 with the ordered left-to-right concatenation of the aberrating systems. Although they could be represented in principle by (huge) matrices, we eschew such realization because it is unwieldy; polynomials of phase space serve to keep track of the brood more economically.

We should point out that most of the optical aberration literature follows the classification strategy introduced by Ludwig Seidel [125], which is based on the Hamilton point-point characteristic function and was organized for applications by H.A. Buchdahl [25, 26] in what we would regard as the *Lagrangian* formulation of optics (see Sect. A.5). The Seidel classification of aberrations regards *Bildfehler* (*image errors*) in the ray positions at input and output screens of an apparatus with pupils inside. The Hamilton-Lie classification on the other hand performs on the scenario of phase space with groups of transformations. But, since both schemes essentially coincide in the lowest aberration order, we use the traditional nomenclature of spherical aberration, coma, astigmatism, curvature of field, distorsion, ..., as far as is convenient.

In Chap. 14 we inquire on the aberrations in $D = 2$-dim *axis-symmetric* optical systems, and find them to be in 1:1 correspondence with the eigenstates of the 3-dim quantum harmonic oscillator, in Cartesian and in spherical basis. The former begets the basis of *monomial* aberrations, while the latter characterizes them by *symplectic spin* ('angular momentum') and weight ('z-projection'). We find the *symplectic-harmonic* basis to be very convenient; it is essentially that of solid spherical harmonics in the three coordinates of reduced phase space (see Sect. 7.4). The group multiplication is given explicitly for the group of seventh-order aberrations, as well as the parametrized group elements that correspond to free displacement and refracting surfaces of polynomial shape. Elements of this group will be composed in the same order as the corresponding building blocks on the optical table, and *de*composed into a paraxial factor and an ordered product of *pure* aberrations of increasing order.

Hamiltonians generating aberrations were used first in the theory of perturbations for celestial mechanics by G. Hori [69] and A. Deprit [37]. This method was applied early to optical aberrations by Alex J. Dragt [41, 42], provided with a Lie algebraic structure [43], and later developed extensively with his collaborators [44, 45] for the (failed) Superconducting Supercollider project. Although many commercial programs use efficient ray-tracing methods that practically obviate the need of any theory in designing optical instruments, the strategy of higher-order perturbations does arise in many other physical contexts; a cogent structure of polynomial approximations to nonlinear maps of phase space is kinder to understanding than brute-force computation.

Hamilton-Lie aberrations quantify the *departure* of a system from linearity, and the coefficients are thus independent of the imaging or non-imaging purpose of the apparatus —but they do respect its symmetries. In this vein we examine three fractional Fourier transformer arrangements in Chap. 15, whose paraxial approximation was seen in previous chapters, and develop some new tactics to reduce their aberrations through introducing higher-order polynomial corrections to their refracting and reflecting surfaces.

13 Polynomials and aberrations in one dimension

The concepts and notation that we developed for linear transformations of phase space under the paraxial régime in Part III, will be expanded to non-linear transformations – *aberrations* – under a régime that we refer to as *metaxial*, up to an aberration *order*. Here we introduce the key features of this model of Hamilton–Lie aberrations for the case $D = 1$ of plane optics.

13.1 Monomials in multiplets

Analytic functions of phase space $A(p, q)$ will be associated as before to Lie-Poisson operators [see *e.g.* (3.25)]

$$\widehat{A} := \{A(p, q), \circ\} = \frac{\partial A(p, q)}{\partial q}\frac{\partial}{\partial p} - \frac{\partial A(p, q)}{\partial p}\frac{\partial}{\partial q}. \tag{13.1}$$

These functions form a linear vector space $\mathcal{A}$, and so do the operators. Their Lie exponentials, $\exp \alpha\widehat{A}$, generate one-parameter groups of transformations of phase space which are *canonical* [(3.31)–(3.32)], i.e., they preserve Poisson brackets between the functions $A, B \in \mathcal{A}$. We introduce a grading on this space of functions by homogeneous polynomial degree $2k$ in the phase space coordinates, with integer or half-integer $k \geq 0$. Thus we decompose $\mathcal{A}$ into a direct sum of finite-dimensional subspaces $\mathcal{A}_k$, each of which is invariant under the action of linear transformations $\mathbf{M} \in \mathsf{Sp}(2, \mathsf{R})$. Vector bases of the subspaces form the *multiplets* of $\mathsf{sp}(2, \mathsf{R})$ seen in Sect. 11.5, (11.92).

Our prefered basis for the $D = 1$ case are the monomials of degree $2k$,

$$M_{k,m} = M_{k,m}(p, q) := p^{k+m} q^{k-m} \quad \begin{cases} \text{rank} \quad k \in \{0, \frac{1}{2}, 1, \frac{3}{2}, 2, \ldots\}, \\ \text{weight } m \in \{k, k-1, \ldots, -k\}. \end{cases} \tag{13.2}$$

There are $2k + 1$ partners in a multiplet of rank k, distinguished by their weight m, i.e., by their eigenvalue under the operator $\frac{1}{2}\widehat{M_{1,0}} = \{\frac{1}{2}pq, \circ\}$, which counts the powers of p, subtracts the powers of q, and divides by 2 [weight in an $\mathsf{sp}(2, \mathsf{R})$ multiplet was defined in (11.94)]. Analytic functions $A \in \mathcal{A}$ thus decompose as

$$A(p, q) = \sum_{2k=0}^{\infty} A_k(p, q), \quad A_k(p, q) := \sum_{m=-k}^{k} A_{k,m}\, M_{k,m}(p, q) \in \mathcal{A}_k, \tag{13.3}$$

where we sum over all integers $2k \geq 0$, and $\mathbf{A} := \{A_{k,m}\}$ are the linear combination coefficients of $A(p,q)$. The basis vectors, $\{M_{k,m}\}_{m=-k}^{k} \in \mathcal{A}_k$ in (13.2), will transform among themselves under the action of linear systems $\mathcal{M}(\mathbf{M})$, $\mathbf{M} = \begin{pmatrix} a & b \\ c & d \end{pmatrix} \in \mathsf{Sp}(2,\mathsf{R})$, whose action, from (9.15)–(9.17), is

$$\mathcal{M}\begin{pmatrix} a & b \\ c & d \end{pmatrix} : \begin{pmatrix} p \\ q \end{pmatrix} = \begin{pmatrix} a & b \\ c & d \end{pmatrix}^{-1} \begin{pmatrix} p \\ q \end{pmatrix} = \begin{pmatrix} d & -b \\ -c & a \end{pmatrix} \begin{pmatrix} p \\ q \end{pmatrix} \quad \Rightarrow \tag{13.4}$$

$$\begin{aligned}\mathcal{M}\begin{pmatrix} a & b \\ c & d \end{pmatrix} : M_{k,m}(p,q) &= (dp - bq)^{k+m}(-cp + aq)^{k-m} \\ &= \sum_{n_1=0}^{k+m} \binom{k+m}{n_1} d^{k-m-n_2}(-b)^{n_1} \\ &\quad \times \sum_{n_2=0}^{k-m} \binom{k-m}{n_2} a^{n_2}(-c)^{k-m-n_2} p^{2k-n_1-n_2} q^{n_1+n_2} \\ &= \sum_{m'=-k}^{k} D_{m,m'}^{k}\begin{pmatrix} a & b \\ c & d \end{pmatrix}^{-1} M_{k,m'}(p,q).\end{aligned} \tag{13.5}$$

In the last step we introduced $k - m' := n_1 + n_2$ to organize the coefficients into a $(2k+1) \times (2k+1)$ matrix acting on the column vector of monomials $M_{k,m'}(p,q)$. Throughout, k, m and m' are either all integer, or all half-integer. With $n_2 =: k - \mu$ and $\mathbf{M}$ in place of $\mathbf{M}^{-1}$, the matrix elements are

$$D_{m,m'}^{k}\begin{pmatrix} a & b \\ c & d \end{pmatrix} := \sum_{\mu} \binom{k+m}{\mu-m'} \binom{k-m}{k-\mu} a^{k-\mu+m+m'} b^{\mu-m'} c^{\mu-m} d^{k-\mu}, \tag{13.6}$$

where the binomials restrict the range of μ to $-k \leq m \leq \mu \leq k$ and $-k \leq m' \leq \mu \leq k + m + m'$, and so the exponents of a, b, c, d are nonnegative. By construction, the matrices $\mathbf{D}^k(\mathbf{M}) = \|D_{m',m}^{k}(\mathbf{M})\|$ *represent* the group, i.e., they satisfy

$$\mathbf{D}^k(\mathbf{M}_1)\,\mathbf{D}^k(\mathbf{M}_2) = \mathbf{D}^k(\mathbf{M}_1\mathbf{M}_2), \quad \mathbf{D}^k(\mathbf{1}) = \mathbf{1}, \quad \mathbf{D}^k(\mathbf{M}^{-1}) = \mathbf{D}^k(\mathbf{M})^{-1}, \tag{13.7}$$

and are irreducible representations of $\mathsf{Sp}(2,\mathsf{R})$.[1] We note that $\mathbf{D}^k(-\mathbf{M}) = (-1)^{2k}\mathbf{D}(\mathbf{M})$, so the matrices represent $\mathsf{Sp}(2,\mathsf{R})$ *faithfully* only when k is half-integer; when k is integer, they represent faithfully $\mathsf{SO}(2,1)$ [cf. (9.52)–(9.53) with rearranged rows and columns.]

Example: Representation of the Fourier transform. The Fourier transform $\mathcal{F} := \mathcal{M}(\mathbf{F})$, $\mathbf{F} := \begin{pmatrix} 0 & -1 \\ 1 & 0 \end{pmatrix}$, acts on phase space as $\mathcal{F} : \begin{pmatrix} p \\ q \end{pmatrix} =$

[1] Do not confuse these matrices $D_{m,m'}^{k}(\mathbf{M})$, with their homographous Wigner matrices $D_{m,m'}^{j}(\phi,\theta,\psi)$, the unitary irreducible representations of the spin group $\mathsf{SU}(2)$ in Euler angles.

$\mathbf{F}^{-1}\binom{p}{q} = \binom{q}{-p}$ [see (3.21)], and is represented in the monomial basis by an antidiagonal matrix with unit elements of alternating signs,

$$D^k_{m,m'}(\mathbf{F}^{-1}) = \delta_{m+m',0}(-1)^{k-m} \Rightarrow \mathcal{F}: M_{k,m} = (-1)^{k-m} M_{k,-m}. \quad (13.8)$$

Aberrations of the image q and of the ray momenta p will be thus related by Fourier conjugation. •

Remark: The Fourier aberration basis. The monomial basis for $\mathcal{A}_k$ will privilege imaging systems because it involves the position and momentum coordinates of rays. If our purpose were to describe aberrations in waveguides or optical fibers, the relevant coordinates could be the complex combinations

$$z := \tfrac{1}{\sqrt{2}}(q + ip), \qquad a := i\tfrac{1}{\sqrt{2}}(q - ip) = iz^*, \quad (13.9)$$

given in (11.102). These are obtained from (p, q) through the canonical Bargmann–Segal transform in (C.44). The resulting basis for aberrations — constructed with monomials $a^{k+m}(iz)^{k-m}$ exactly as above— is such that the coefficients are only multiplied by phases under displacements in a paraxial harmonic guide; metaxial departures from harmonicity are then appropriately described in this basis; see Refs. [152, 153]. •

Example: Representation of reflections. The representation property (13.7) holds in fact for any matrix $\mathbf{M} \in \mathsf{GL}(2, \mathsf{C})$. In the paraxial régime, the effect of reflections on phase space was seen in Sect. 10.4, to be given by a nonsymplectic matrix $\mathbf{K} := \begin{pmatrix} -1 & 0 \\ 0 & 1 \end{pmatrix} = \mathbf{K}^{-1}$, which reverses the direction of optical momentum (10.38). In the basis of monomials of rank k, reflection is diagonal,

$$D^k_{m,m'}(\mathbf{K}) = \delta_{m,m'}(-1)^{k+m} \Rightarrow \mathcal{M}(\mathbf{K}): M_{k,m} = (-1)^{k+m} M_{k,m}, \quad (13.10)$$

as confirmed by the explicit form of $M_{k,m}(p,q)$ in (13.2). Comparing with the Fourier transform (13.8), note that for k integer, $(-1)^{k+m} = (-1)^{k-m}$, while for k half-integer, $(-1)^{k+m} = -(-1)^{k-m}$. •

Example: Displacements and imagers. Free displacements along the optical axis and imager systems are represented by lower- and upper-triangular D^k-matrices respectively, in which the sum (13.6) reduces to a single term,

$$D^k_{m,m'}\begin{pmatrix} 1 & 0 \\ z & 1 \end{pmatrix} = \binom{k-m}{k-m'} z^{m'-m} \quad \begin{cases} -k \le m \le m' \le k, \\ 0 \text{ otherwise}, \end{cases} \quad (13.11)$$

$$D^k_{m,m'}\begin{pmatrix} a & b \\ 0 & a^{-1} \end{pmatrix} = \binom{k+m}{m-m'} (ab)^m (a/b)^{m'} \begin{cases} -k \le m' \le m \le k, \\ 0 \text{ otherwise}. \end{cases} \quad (13.12)$$

•

Example: The first five ranks. In rank $k = 0$ there is only $D^0_{0,0}(\mathbf{M}) = 1$. It will be useful to have the next four cases $D^k_{m,m'}$ explicitly, recalling that the rows m are numbered decreasingly, k to $-k$, from top to bottom, and the columns m' from left to right. Thus,

$$\mathbf{D}^{\frac{1}{2}}\begin{pmatrix} a & b \\ c & d \end{pmatrix} = \begin{pmatrix} D_{\frac{1}{2},\frac{1}{2}} & D_{\frac{1}{2},-\frac{1}{2}} \\ D_{-\frac{1}{2},\frac{1}{2}} & D_{-\frac{1}{2},-\frac{1}{2}} \end{pmatrix} = \begin{pmatrix} a & b \\ c & d \end{pmatrix}, \tag{13.13}$$

$$\mathbf{D}^{1}\begin{pmatrix} a & b \\ c & d \end{pmatrix} = \begin{pmatrix} D_{1,1} & D_{1,0} & D_{1,-1} \\ D_{0,1} & D_{0,0} & D_{0,-1} \\ D_{-1,1} & D_{-1,0} & D_{-1,-1} \end{pmatrix} = \begin{pmatrix} a^2 & 2ab & b^2 \\ ac & ad+bc & bd \\ c^2 & 2cd & d^2 \end{pmatrix}, \tag{13.14}$$

$$\mathbf{D}^{\frac{3}{2}}\begin{pmatrix} a & b \\ c & d \end{pmatrix} = \begin{pmatrix} a^3 & 3a^2b & 3ab^2 & b^3 \\ a^2c & a(ad+2bc) & b(2ad+bc) & b^2d \\ ac^2 & c(2ad+bc) & d(ad+2bc) & bd^2 \\ c^3 & 3c^2d & 3cd^2 & d^3 \end{pmatrix}, \tag{13.15}$$

$$\mathbf{D}^{2}\begin{pmatrix} a & b \\ c & d \end{pmatrix} = \begin{pmatrix} a^4 & 4a^3b & 6a^2b^2 & 4ab^3 & b^4 \\ a^3c & a^2(ad+3bc) & 3ab(ad+bc) & b^2(3ad+bc) & b^3d \\ a^2c^2 & 2ac(ad+bc) & a^2d^2+4abcd+b^2c^2 & 2bc(ad+bc) & b^2d^2 \\ ac^3 & c^2(3ad+bc) & 3cd(ad+bc) & d^2(ad+3bc) & bd^3 \\ c^4 & 4c^3d & 6c^2d^2 & 4cd^3 & d^4 \end{pmatrix}. \tag{13.16}$$

The mth row is the expansion of $(ax+b)^{k+m}(cx+d)^{k-m}$ with the coefficients of $x^{k+m'}$ collected in column m'. Finally, we note that for all 2×2 matrices $\mathbf{M}$,

$$\det \mathbf{D}^k(\mathbf{M}) = (\det \mathbf{M})^{k(2k+1)}. \tag{13.17}$$

•

13.2 Rank-K aberration algebras

We now regard the phase space monomials, $M_{k,m}(p,q)$ in (13.2) as generators of a Lie algebra $\mathcal{A}$. Direct computation yields the generic Poisson bracket,

$$\{M_{k,m}, M_{k',m'}\} = 2(km' - k'm)\, M_{k+k'-1,\, m+m'}. \tag{13.18}$$

This operation generally mixes ranks because the Poisson brackets relate $\{\mathcal{A}_k, \mathcal{A}_{k'}\} = \mathcal{A}_{k+k'-1}$ among the vector subspaces of homogeneous polynomials $\mathcal{A}_k$ [recall (5.50)].

There is however the distinguished subalgebra $\mathsf{sp}(2,\mathsf{R}) = \mathcal{A}_1 \subset \mathcal{A}$, with the vector basis

$$M_{1,1}(p,q) = p^2, \quad M_{1,0}(p,q) = pq, \quad M_{1,-1}(p,q) = q^2, \tag{13.19}$$

which generates the linear part of the transformations of phase space. The Lie brackets (commutation relations) between the Poisson operators $\widehat{M}_{1,m} := \{M_{1,m}, \circ\}$ present this algebra by

$$[\widehat{M}_{1,m}, \widehat{M}_{1,m'}] = 2(m' - m)\widehat{M}_{1,m+m'}. \tag{13.20}$$

The spaces of functions $\mathcal{A}_k$ serve as homogeneous spaces for the action of the Lie algebra $\mathcal{A}_1$,

$$\text{raising:}\quad [\widehat{M}_{1,1}\,, \widehat{M}_{k,m}] = 2(m-k)\,\widehat{M}_{k,m+1}, \tag{13.21}$$

$$\text{weight:}\quad [\widehat{M}_{1,0}\,, \widehat{M}_{k,m}] = 2m\,\widehat{M}_{k,m}, \tag{13.22}$$

$$\text{lowering:}\quad [\widehat{M}_{1,-1}, \widehat{M}_{k,m}] = 2(m+k)\,\widehat{M}_{k,m-1}. \tag{13.23}$$

For ranks $k, k' > 1$, repeated Poisson brackets will give rise to ever higher monomials, since $k + k' - 1 > k, k'$.

Remark: Finite and infinite-dimensional subalgebras. Beside $\mathcal{A}_1$, finite-dimensional Lie subalgebras of $\mathcal{A}$ also occur for $k, k' \in \{0, \frac{1}{2}\}$, with the basic bracket $\{q, p\} = 1$ and $\{1, \circ\} = 0$; also, when $k, k' \in \{0, \frac{1}{2}, 1\}$, we have $\mathsf{isp}(2, \mathsf{R})$. In Sect. 3.6 we saw also the 'spherical aberration' and comatic transformations generated by functions $f(p)$ and $q\,g(p)$ respectively, and their Fourier conjugates. Although there is little problem in exponentiating any single generator $\widehat{A} \in \mathcal{A}$ to a one-parameter Lie group of canonical transformations, whose action on phase space can be found as a summable or formal Lie series, the composition of two or more generally baloons into groups with an infinite number of parameters (called *functional* or *pseudo*-groups because of their pathologies) which can hardly serve us in this part. •

The grading by rank k allows the consistent truncation of the infinite-dimensional Lie algebra $\mathcal{A}$ generated by all analytic functions of phase space, to Lie algebras with a manageably finite number of parameters. This works in the same way as the polynomial approximation to analytic functions by the first K terms in their Taylor series, with the additional structure stemming from their Lie (Poisson) brackets.

Definition: *The* **rank-K aberration algebra** *of* $\mathsf{sp}(2, \mathsf{R})$ *(of aberration* **order** $a_K = 2K - 1$*), denoted* $\mathsf{a}_K\mathsf{sp}(2, \mathsf{R})$*, is the Lie algebra spanned by the generators* $\widehat{M}_{k,m}$*,* $1 \le k \le K$ *integer or half-integer as in (13.2), and characterized by the commutation relations*

$$[\widehat{M}_{k,m}, \widehat{M}_{k',m'}] = \begin{cases} 2(km' - k'm)\,\widehat{M}_{k+k'-1,\,m+m'}, & 1 \le k, k', k+k'-1 \le K, \\ 0, & \textit{otherwise}. \end{cases} \tag{13.24}$$

The three generators $\{\widehat{M}_{1,m}\}_{m=-1}^{1}$ *form the* **paraxial** *subalgebra* $\mathsf{sp}(2, \mathsf{R})$*, while their vector complement,* $\mathsf{a}_K := \oplus_{k>1}^{K} \mathcal{A}_k$*, is the* **pure aberration** *subalgebra. In this basis, the coordinates* $\mathbf{A} := \{A_{k,m}\}_{k>1}^{K}$ *of an element* $\widehat{A} \in \mathsf{a}_K$ *are its* **Hamilton–Lie** *aberration coefficients.*

Remark: Why we exclude translations. We discarded the rank-$\frac{1}{2}$ pair (p, q) because their Poisson operator will *lower* the rank of polynomials, precluding the consistent truncation to *finite* aberration algebras. As a consequence, we are also discarding translations and paraxial tilts of the screen, which are generated by $M_{\frac{1}{2},\frac{1}{2}} = p$ and $M_{\frac{1}{2},-\frac{1}{2}} = q$. But we can always fix the origins of the object and image phase spaces by choosing a *design ray* and screens that are perpendicular to it, around which we can perform the aberration expansion [42]. •

The generators $\widehat{M}_{k,m}$ of $\mathsf{a}_K\mathsf{sp}(2,\mathsf{R})$ can be realized by the Poisson operators of the monomials (13.2), *modulo* ranks higher than K, i.e., replacing $\widehat{M}_{k',m}$ by zero whenever $k' > K$. The functions $M_{k,m}(p,q) \in \mathcal{A}$ and the generators $\widehat{M}_{k,m} \in \mathsf{a}_K$ will be loosely called *aberrations*, – **symmetric** when k is integer (so $M_{k,m}(-p,-q) = M_{k,m}(p,q)$), and **skew-symmetric** when k is half-integer (so $M_{k,m}(-p,-q) = -M_{k,m}(p,q)$). When only symmetric aberrations are present, they form a subalgebra $\mathsf{a}_K^{\mathsf{s}}\mathsf{sp}(2,\mathsf{R}) \subset \mathsf{a}_K\mathsf{sp}(2,\mathsf{R})$. We recall (3.66): the commutator of the Poisson operators of two functions of phase space $A, B \in \mathcal{A}$ is the Poisson operator of their Poisson bracket, $[\{A,\circ\},\{B,\circ\}] = \{\{A,B\},\circ\}$, so all Lie algebraic computations with aberrations will only involve Poisson brackets between functions.

The aberration algebras have the structure of a semidirect sum [see Sect. 5.4, (5.47) *et seq.*], between the paraxial and the aberration parts,

$$\mathsf{a}_K\mathsf{sp}(2,\mathsf{R}) = \mathsf{a}_K \oplus\!\!\!\!\!\!\;\raisebox{0.3ex}{$\scriptstyle\supset$}\, \mathsf{sp}(2,\mathsf{R}), \tag{13.25}$$

where $\mathsf{sp}(2,\mathsf{R})$ is the radical subalgebra and a_K is the invariant subalgebra. Unlike the Euclidean algebras of Sect. 5.4, where the invariant subalgebra of translations is abelian, aberrations generally do not commute; they form a *nilpotent* subalgebra because repeated Poisson brackets will always end in 0 modulo ranks higher than K. In the cases $\mathsf{a}_{\frac{3}{2}}\mathsf{sp}(2,\mathsf{R}) = \mathsf{i}_4\mathsf{sp}(2,\mathsf{R})$ and $\mathsf{a}_2^{\mathsf{s}}\mathsf{sp}(2,\mathsf{R}) = \mathsf{i}_5\mathsf{sp}(2,\mathsf{R})$ all aberrations do commute.

Remark: The number of $D = 1$ aberrations. Beside the 3 parameters of the radical subalgebra $\mathsf{sp}(2,\mathsf{R})$, the pure aberration subalgebra a_K has $n_k := 2k+1$ parameters at each rank $1 < k \leq K$ (aberration order $a_k = 2k-1$), which accumulate to its total vector dimension, as counted in the table below:

rank	a_k	n_k	dim a_K	rank	a_k	n_k	dim $\mathsf{a}_K^{\mathsf{s}}$
$k = \frac{3}{2}$	2	4	4	$k = 2$	3	5	5
2	3	5	9	3	5	7	12
$\frac{5}{2}$	4	6	15	4	7	9	21
3	5	7	22	5	9	11	32
k	$2k-1$	$2k+1$	$2K^2+3K-5$	integer k	$2k-1$	$2k+1$	K^2+2K-3

•

13.3 Rank-K aberration groups

The Lie exponential of the rank-K aberration algebra (13.25) is the rank-K aberration *group*. This group is a semidirect product

$$\mathsf{A}_K\mathsf{Sp}(2,\mathsf{R}) = \mathsf{A}_K \otimes\!\!\!\!\!\!\;\raisebox{0.3ex}{$\scriptstyle\supset$}\, \mathsf{Sp}(2,\mathsf{R}), \tag{13.26}$$

where the *radical* subgroup is $\mathsf{Sp}(2, \mathsf{R})$, and the invariant subgroup is the Lie exponential of the generators of pure aberrations, denoted A_K. It is important to define an adequate parametrization of the aberration groups for the purposes of computational simplicity, and of applicability to optical systems which are not necessarily imaging apparata.

Definition: *The* **factored-product parametrization** *of an element of the rank-K aberration group $\mathsf{A}_K\mathsf{Sp}(2, \mathsf{R})$ is given by the* **Hamilton–Lie** *aberration coefficients $A_{k,m} \in \mathsf{R}$ of polynomial functions of phase space $A_k(p, q) \in \mathcal{A}_k$ [see (13.3)], and has its generators $\widehat{A}_k = \{A_k, \circ\}$ ordered by rank as*

$$\mathcal{G}_K(\mathbf{A}; \mathbf{M}) := \mathcal{G}(\mathbf{A}_K, \ldots, \mathbf{A}_2, \mathbf{A}_{3/2}; \mathbf{1})\, \mathcal{G}(\mathbf{0}; \mathbf{M}(\mathbf{A}_1)), \text{ where} \tag{13.27}$$

$$\mathcal{G}(\mathbf{A}; \mathbf{1}) := \exp \widehat{A}_K \times \cdots \times \exp \widehat{A}_2 \times \exp \widehat{A}_{3/2}, \tag{13.28}$$

$$\widehat{A}_k := \textstyle\sum_{m=-k}^{k} A_{k,m}\, \widehat{M}_{k,m}, \qquad 1 \leq k \leq K, \tag{13.29}$$

$$\mathcal{G}(\mathbf{0}, \mathbf{M}) := \exp \widehat{A}_1 = \mathcal{M}\left(\exp \begin{pmatrix} -A_{1,0} & -2A_{1,-1} \\ 2A_{1,1} & A_{1,0} \end{pmatrix}\right). \tag{13.30}$$

The right factor of (13.27), $\mathcal{G}(\mathbf{0}, \mathbf{M}) \in \mathsf{Sp}(2, \mathsf{R})$, is the linear map generated by the rank-1 generators $\widehat{A}_1$ through (12.3) and (12.13)–(12.15), while the left factor $\mathcal{G}(\mathbf{A}, \mathbf{1})$, is the invariant subgroup of pure aberrations. This is the factorization we used for Euclidean translations and rotations in (5.41)–(5.45). Clearly, the unit of the aberration groups is $\mathcal{G}_K(\mathbf{0}; \mathbf{1})$.

The philosophy of factoring aberrations to the left and linear transformations to the right, is that the input 'object' phase space is *first* aberrated, and *then* linearly transformed. This gives Hamilton–Lie aberrations a meaning independent of the final magnification, paraxial skewing, or fractional Fourier transform of the 'image' phase space of rays. Also –in contrast with the Seidel treatment of aberrations [25] –, the coefficients are independent of any pupils in the optical apparatus. The paraxial model is the foundation of the metaxial model with its 'flat' R^{2D} phase space. Hamilton–Lie aberrations describe the *non*linearities of phase space maps as perturbation expansions in Hamiltonian systems.

The multiplication of two rank-K aberration group elements has the overall structure of semidirect products [see (5.48)],

$$\mathcal{G}_K(\mathbf{A}; \mathbf{M})\, \mathcal{G}_K(\mathbf{B}; \mathbf{N}) = \mathcal{G}_K(\mathbf{A}; \mathbf{1})\, \mathcal{M}(\mathbf{M})\, \mathcal{G}_K(\mathbf{B}; \mathbf{1})\, \mathcal{M}(\mathbf{N}) \tag{13.31}$$

$$= \mathcal{G}_K(\mathbf{A}; \mathbf{1})\, \mathcal{G}_K(\mathcal{M}(\mathbf{M}) : \mathbf{B};\ \mathbf{1})\, \mathcal{M}(\mathbf{M}\mathbf{N}) \tag{13.32}$$

$$= \mathcal{G}_K(\mathbf{A} \sharp \mathcal{M}(\mathbf{M}) : \mathbf{B};\ \mathbf{M}\mathbf{N}), \tag{13.33}$$

where the first line contains the factored-product form; in the second we use (13.5) to apply $\mathcal{M}(\mathbf{M})$ on the exponent operators $\widehat{B}_k$, i.e. on the functions $B_k(p, q)$ and thus on its aberration coefficients $B_{k,m}$ [cf. (13.28)–(13.29)], and also on the rightmost factor $\mathcal{M}(\mathbf{N})$. With the coefficients of (13.6), this action is

$$\mathcal{M}(\mathbf{M}) : B_k(p,q) = \sum_{m=-k}^{k} \sum_{m'=-k}^{k} B_{k,m}\, D^k_{m,m'}(\mathbf{M}^{-1})\, M_{k,m'}(p,q), \tag{13.34}$$

$$\mathcal{M}(\mathbf{M}) : B_{k,m} = \sum_{m'=-k}^{k} B_{k,m'}\, D^k_{m,m'}(\mathbf{M}^{-1}), \tag{13.35}$$

$$\mathcal{M}(\mathbf{M}) : \mathbf{B}_k = \mathbf{D}^k(\mathbf{M}^{-1})^\top\, \mathbf{B}_k. \tag{13.36}$$

When we organize the aberration coefficients into column subvectors $\mathbf{B}_k = (B_{k,m})$ within the column vector $\mathbf{B} = (\mathbf{B}_k)$, this action is block-diagonal,

$$\mathcal{M}(\mathbf{M}) : \begin{pmatrix} \mathbf{B}_{3/2} \\ \vdots \\ \mathbf{B}_K \end{pmatrix} = \begin{pmatrix} \mathbf{D}^{3/2}(\mathbf{M}^{-1})^\top & & \mathbf{0} \\ & \ddots & \\ \mathbf{0} & & \mathbf{D}^K(\mathbf{M}^{-1})^\top \end{pmatrix} \begin{pmatrix} \mathbf{B}_{3/2} \\ \vdots \\ \mathbf{B}_K \end{pmatrix}. \tag{13.37}$$

The crux of the aberration group multiplication remains the third line of the product in (13.33), namely the composition of two pure aberrations indicated by

$$\mathcal{G}_K(\mathbf{A};\mathbf{1})\,\mathcal{G}_K(\mathbf{B};\mathbf{1}) = \mathcal{G}_K(\mathbf{A}\,\sharp\,\mathbf{B};\mathbf{1}). \tag{13.38}$$

The $\sharp$ (*gato*) composition rule can be found through the interchange of non-commuting exponents with Baker-Campbell-Hausdorff relations [96] or, more manageably, using the Poisson brackets between two phase space functions, as we shall do below. Ultimately, we search for explicit algebraic expressions between the Hamilton–Lie factored-product aberration coefficients of $\mathbf{C} = \mathbf{A}\,\sharp\mathbf{B}$ in terms of those of $\mathbf{A}$ and $\mathbf{B}$. We shall find by hand the *gato* product in $\mathsf{A}_2\mathsf{Sp}(2,\mathsf{R})$, which contains aberrations of ranks $k = \frac{3}{2}$ and 2 (aberration orders 2 and 3). In each rank we deal with functions A_k, $B_{k'}$, their Poisson operators and brackets, and indicate by '$\approx$' every time we disregard summands of rank $k > 2$ (monomials of degree higher than 3). Thus, in the factored product order, an element of the pure aberration group is written as a sum of Poisson operators

$$\begin{aligned} \mathcal{G}_2(\mathbf{A};\mathbf{1}) &:= \exp \widehat{A}_2 \exp \widehat{A}_{\frac{3}{2}} \\ &\approx (1+\widehat{A}_2)\,(1+\widehat{A}_{\frac{3}{2}} + \tfrac{1}{2!}\widehat{A}^2_{\frac{3}{2}}) \\ &\approx 1 + \{A_{\frac{3}{2}}, \circ\} + \{A_2, \circ\} + \tfrac{1}{2!}\{A_{\frac{3}{2}}, \{A_{\frac{3}{2}}, \circ\}\}. \end{aligned} \tag{13.39}$$

The ordered group product of two such aberrations is then

$$\begin{aligned} \mathcal{G}_2(\mathbf{A};\mathbf{1})\,\mathcal{G}_2(\mathbf{B};\mathbf{1}) &\approx (1+\widehat{A}_{\frac{3}{2}}+\widehat{A}_2+\tfrac{1}{2!}\widehat{A}^2_{\frac{3}{2}})(1+\widehat{B}_{\frac{3}{2}}+\widehat{B}_2+\tfrac{1}{2!}\widehat{B}^2_{\frac{3}{2}}) \\ &\approx 1 + \widehat{A}_{\frac{3}{2}} + \widehat{B}_{\frac{3}{2}} + \widehat{A}_2 + \widehat{B}_2 + \tfrac{1}{2!}\widehat{A}^2_{\frac{3}{2}} + \tfrac{1}{2!}\widehat{B}^2_{\frac{3}{2}} + \widehat{A}_{\frac{3}{2}}\widehat{B}_{\frac{3}{2}} \\ &= 1+\widehat{C}_{\frac{3}{2}}+\widehat{C}_2+\tfrac{1}{2!}\widehat{C}^2_{\frac{3}{2}} \approx \mathcal{G}_2(\mathbf{C};\mathbf{1}). \end{aligned} \tag{13.40}$$

Equating terms with the same rank in the last two lines, we find the composition of the aberration functions to third order. To order four they are:

$$
\begin{aligned}
C_{\frac{3}{2}}(p,q) &= A_{\frac{3}{2}}(p,q) + B_{\frac{3}{2}}(p,q), && (13.41)\\
C_2(p,q) &= A_2(p,q) + B_2(p,q) + \tfrac{1}{2}\{A_{\frac{3}{2}}, B_{\frac{3}{2}}\}(p,q), && (13.42)\\
C_{\frac{5}{2}}(p,q) &= A_{\frac{5}{2}}(p,q) + B_{\frac{5}{2}}(p,q) + \{A_{\frac{3}{2}}, B_2\}(p,q) && (13.43)\\
&\quad + \tfrac{1}{3}\{A_{\frac{3}{2}}, \{A_{\frac{3}{2}}, B_{\frac{3}{2}}\}\}(p,q) - \tfrac{1}{6}\{\{A_{\frac{3}{2}}, B_{\frac{3}{2}}\}, B_{\frac{3}{2}}\}(p,q).
\end{aligned}
$$

To obtain (13.42) we used (13.41) to subtract the term $\frac{1}{2}\widehat{C}^2_{\frac{3}{2}}$ from (13.40); then we used the Jacobi identity $\widehat{A}\widehat{B} - \widehat{B}\widehat{A} = \widehat{\{A, B\}}$ to form the new functions. For (13.43) and higher ranks, symbolic machine computation becomes the most practical tool. [Equations (13.41)–(13.43) also describe the *symmetric* aberration group $\mathsf{A}^{\mathsf{s}}_4\mathsf{Sp}(2,\mathsf{R})$, upon the replacement of ranks given by $\frac{3}{2} \mapsto 2$, $2 \mapsto 3$, $\frac{5}{2} \mapsto 4$, i.e. of aberration orders $2 \mapsto 3$, $3 \mapsto 5$, $4 \mapsto 7$.]

Remark: Associativity of $\sharp$. The *gato* composition (13.38) is associative,

$$
(\mathbf{A} \sharp \mathbf{B}) \sharp \mathbf{C} = \mathbf{A} \sharp (\mathbf{B} \sharp \mathbf{C}) =: \mathbf{A} \sharp \mathbf{B} \sharp \mathbf{C}, \qquad (13.44)
$$

since all are multiplications in a group. This can be verified from (13.40)–(13.43). •

Remark: Composition of aberration coefficients. The algebraic expressions for the Hamilton–Lie factored-product aberration coefficients of $\mathbf{C} = \mathbf{A} \sharp \mathbf{B}$, explicitly giving $C_{k,m}$ in terms of $A_{k',m'}$ and $B_{k'',m''}$, can be found from the composition formulas (13.41)–(13.43), expanding all functions $A_k(p,q)$, etc., in the monomial basis with (13.29). In this way, from (13.41) follows

$$
C_{\frac{3}{2},m} = A_{\frac{3}{2},m} + B_{\frac{3}{2},m}, \qquad m \in \{\tfrac{3}{2}, \tfrac{1}{2}, -\tfrac{1}{2}, -\tfrac{3}{2}\}. \qquad (13.45)
$$

The lowest-rank aberration coefficients always sum. From (13.43) we find

$$
\begin{aligned}
C_{2,2} &= A_{2,2} + B_{2,2} + \tfrac{3}{2}A_{\frac{3}{2},\frac{1}{2}}B_{\frac{3}{2},\frac{3}{2}} - \tfrac{3}{2}A_{\frac{3}{2},\frac{3}{2}}B_{\frac{3}{2},\frac{1}{2}},\\
C_{2,1} &= A_{2,1} + B_{2,1} + 3A_{\frac{3}{2},-\frac{1}{2}}B_{\frac{3}{2},\frac{3}{2}} - 3A_{\frac{3}{2},\frac{3}{2}}B_{\frac{3}{2},-\frac{1}{2}},\\
C_{2,0} &= A_{2,0} + B_{2,0} + \tfrac{9}{2}A_{\frac{3}{2},-\frac{3}{2}}B_{\frac{3}{2},\frac{3}{2}} + \tfrac{3}{2}A_{\frac{3}{2},-\frac{1}{2}}B_{\frac{3}{2},\frac{1}{2}}\\
&\qquad - \tfrac{3}{2}A_{\frac{3}{2},\frac{1}{2}}B_{\frac{3}{2},-\frac{1}{2}} - \tfrac{9}{2}A_{\frac{3}{2},\frac{3}{2}}B_{\frac{3}{2},-\frac{3}{2}}, \qquad (13.46)\\
C_{2,-1} &= A_{2,-1} + B_{2,-1} + 3A_{\frac{3}{2},-\frac{3}{2}}B_{\frac{3}{2},\frac{1}{2}} - 3A_{\frac{3}{2},\frac{1}{2}}B_{\frac{3}{2}-\frac{3}{2}},\\
C_{2,-2} &= A_{2,-2} + B_{2,-2} + \tfrac{3}{2}A_{\frac{3}{2},-\frac{3}{2}}B_{\frac{3}{2},-\frac{1}{2}} - \tfrac{3}{2}A_{\frac{3}{2},-\frac{1}{2}}B_{\frac{3}{2},-\frac{3}{2}}.
\end{aligned}
$$

In Chap. 14 we shall display the composition rules for the aberration coefficients of ranks 2, 3, 4 (aberration orders 3, 5, 7) in $D = 2$ dimensions. •

With the *gato* product determined, we find the explicit *inverse* of any rank-$\frac{5}{2}$ pure aberration group element,

$$\mathcal{G}(\mathbf{A}^{\mathrm{I}};\mathbf{1}) := \mathcal{G}(\mathbf{A};\mathbf{1})^{-1} \Rightarrow \tag{13.47}$$
$$\mathbf{A}^{\mathrm{I}} \sharp \mathbf{A} = \mathbf{0} = \mathbf{A} \sharp \mathbf{A}^{\mathrm{I}} \ \Rightarrow$$
$$A^{\mathrm{I}}_{\frac{3}{2}} = -A_{\frac{3}{2}}, \qquad \text{aberration order 2,} \Rightarrow \tag{13.48}$$
$$A^{\mathrm{I}}_{2} = -A_{2}, \qquad \text{order 3,} \Rightarrow \tag{13.49}$$
$$A^{\mathrm{I}}_{\frac{5}{2}} = -A_{\frac{5}{2}} + \{A_{\frac{3}{2}}, A_2\}, \qquad \text{order 4.} \tag{13.50}$$

From here we find the inverse of generic linear+aberration transformations of phase space,

$$\begin{aligned}\mathcal{G}(\mathbf{A};\mathbf{M})^{-1} &= \mathcal{G}(\mathbf{0};\mathbf{M})^{-1}\,\mathcal{G}(\mathbf{A};\mathbf{1})^{-1}\\ &= \mathcal{G}(\mathbf{0};\mathbf{M}^{-1})\,\mathcal{G}(\mathbf{A}^{\mathrm{I}};\mathbf{1})\,\mathcal{G}(\mathbf{0};\mathbf{M})\,\mathcal{G}(\mathbf{0};\mathbf{M}^{-1})\\ &= \mathcal{G}(\mathbf{D}(\mathbf{M}^{-1})\,\mathbf{A}^{\mathrm{I}};\mathbf{M}^{-1}).\end{aligned} \tag{13.51}$$

Remark: Inverting aberrations? In the paraxial régime, all transformations of optical phase space can be realized by lenses and positive free displacements along the optical axisi, as we saw in Sect. 10.5, with minimal lens arrangements of up to three thin lenses and two displacements. Can the same be done with rank-$\frac{5}{2}$ aberrating $D = 1$ systems say, composed of up-to-fifth degree refracting interfaces and free displacements? This question is open for optical systems, but if computer simulations can aberrate signals, they should also be able to correct aberrated ones. •

Remark: Associativity in $\mathsf{A}_K\mathsf{Sp}(2,\mathsf{R})$. From (13.44) and the semidirect product form of the aberration groups, the property of associativity for the triple product of $\mathcal{G}(\mathbf{A};\mathbf{M}_{\mathrm{A}})\,\mathcal{G}(\mathbf{B};\mathbf{M}_{\mathrm{B}})\,\mathcal{G}(\mathbf{C};\mathbf{M}_{\mathrm{C}})$ is

$$\begin{aligned}\mathbf{A}\sharp\mathcal{M}(\mathbf{M}_{\mathrm{A}}) : [\mathbf{B}\sharp\mathcal{M}(\mathbf{M}_{\mathrm{B}}) : \mathbf{C}] &= [\mathbf{A}\sharp\mathcal{M}(\mathbf{M}_{\mathrm{A}}) : \mathbf{B}]\sharp\mathcal{M}(\mathbf{M}_{\mathrm{A}}\mathbf{M}_{\mathrm{B}}) : \mathbf{C}\\ &= \mathbf{A}\sharp[\mathcal{M}(\mathbf{M}_{\mathrm{A}}) : \mathbf{B}]\sharp[\mathcal{M}(\mathbf{M}_{\mathrm{A}}\mathbf{M}_{\mathrm{B}}) : \mathbf{C}].\end{aligned} \tag{13.52}$$

•

13.4 Aberrations of phase space

The rank-K aberration groups $\mathsf{A}_K\mathsf{Sp}(2,\mathsf{R})$ are meant to deform the phase space of geometric optical rays of the paraxial model. In the factored-product parametrization (13.27), the action of the generic element is

$$\begin{aligned}\mathcal{G}(\mathbf{A};\mathbf{M}) : \begin{pmatrix} p \\ q \end{pmatrix} &= \mathcal{G}(\mathbf{A};\mathbf{1})\,\mathcal{G}(\mathbf{0};\mathbf{M}) : \begin{pmatrix} p \\ q \end{pmatrix}\\ &= \mathcal{G}(\mathbf{A};\mathbf{1}) : \mathbf{M}^{-1}\begin{pmatrix} p \\ q \end{pmatrix} = \mathbf{M}^{-1}\begin{pmatrix} \mathcal{G}(\mathbf{A};\mathbf{1}) : p \\ \mathcal{G}(\mathbf{A};\mathbf{1}) : q \end{pmatrix}.\end{aligned} \tag{13.53}$$

Thus there remains the task to determine the action of *pure* aberrations on the phase space coordinates. We do this in the metaxial régime up to some rank K i.e. aberration order $a_K = 2K - 1$. The realization of the aberration group with Lie exponentials of Poisson operators, $\widehat{A}_k = \{A_k, \circ\}$ as in (13.28), guarantees that the transformation will be *canonical* (see Sect. 3.5). This is important because it means that the description which Lie methods provide to approximate aberrating optical systems with a finite number of parameters, does so along the *sub*manifold which is canonical, within the manifold of *all* possible transformations of R^{2D}. Hence the number of Hamilton–Lie aberration parameters is minimal to start with. Further reductions can be achieved when we discover further symmetries in the paraxial part of the system.

The explicit formulas for the aberrated phase space coordinates $p_{\mathrm{A}}(p,q) := \mathcal{G}_K(\mathbf{A};\mathbf{1}) : p$ and $q_{\mathrm{A}}(p,q) := \mathcal{G}_K(\mathbf{A};\mathbf{1}) : q$ in (13.53), still involve infinite series. Denoting $r = p$ or q, we can collect the summands according to their homogeneous degree $2k$ (rank k) as follows

$$r_{\mathrm{A}}(p,q) := \mathcal{G}_K(\mathbf{A};\mathbf{1}) : r \tag{13.54}$$

$$\begin{aligned} &= \exp \widehat{A}_K \times \cdots \times \exp \widehat{A}_{\frac{5}{2}} \exp \widehat{A}_2 \exp \widehat{A}_{\frac{3}{2}} : r \\ &= r \qquad \text{original monomial of degree 1} \\ &\quad + \{A_{\frac{3}{2}}, r\} \qquad \text{degree 2} \\ &\quad + \{A_2, r\} + \tfrac{1}{2!}\{A_{\frac{3}{2}}, \{A_{\frac{3}{2}}, r\}\} \qquad \text{degree 3} \\ &\quad + \{A_{\frac{5}{2}}, r\} + \{A_2, \{A_{\frac{3}{2}}, r\}\} \\ &\qquad\quad + \tfrac{1}{3!}\{A_{\frac{3}{2}}, \{A_{\frac{3}{2}}, \{A_{\frac{3}{2}}, r\}\}\} \qquad \text{degree 4} \\ &\quad + \cdots \qquad \text{degrees } 5, \ldots, 2K-2 \\ &\quad + \{A_K, r\} + \cdots \qquad \text{degree } 2K-1 = a_K \\ &\quad + \cdots \qquad \text{higher degrees.} \end{aligned} \tag{13.55}$$

We see that in each degree up to the aberration order $a_K = 2K - 1$, the first summand is $\{A_k, r\}$ while the remaining summands involve $\{A_{k'}, \circ\}$'s of smaller rank $k' < k$; beyond a_K there are no new aberration parameters. The truncation strategy of Hamilton–Lie aberrations is served by disregarding all higher degrees of the phase space coordinates, and denoting the finite number of remaining terms by

$$r_{\mathrm{A}}^{[a_K]} := r_{\mathrm{A}} - \text{terms of degree higher than } a_K. \tag{13.56}$$

When we do so, the map $(p,q) \mapsto (p_{\mathrm{A}}^{[a_K]}, q_{\mathrm{A}}^{[a_K]})$ is no longer strictly canonical, but this property holds *up to the aberration order*, *viz.*,

$$\{q_{\mathrm{A}}^{[a_K]}, p_{\mathrm{A}}^{[a_K]}\} = 1 + \text{terms of degree higher than } a_K. \tag{13.57}$$

[Again, as we pointed out below (13.43), this result holds also for the symmetric aberration group $\mathsf{A}_4^{\mathrm{s}}\mathsf{Sp}(2,\mathsf{R})$ replacing the ranks $\frac{3}{2} \mapsto 2$, $2 \mapsto 3$, $\frac{5}{2} \mapsto 4$, i.e. aberration orders $2 \mapsto 3$, $3 \mapsto 5$, $4 \mapsto 7$.]

Remark: Third-order aberrations. Let us get acquainted with the individual features of each aberration, by name and countenance. In the $D = 1$ case we can draw the deformations of phase space in plane figures. For the quintuplet of pure symmetric third-order aberrations, where the aberrated coordinate in (13.54) consists of a single term beyond the identity, and is a polynomial of degree 3 in (p, q). The designations are borrowed from the Seidel nomenclature – except for $\widehat{M}_{2,-2}$, which does not appear in his scheme because it does not change ray positions at the screen; thus it did not receive his attention. With the notation $\mathcal{G}_2(\alpha M_{2,m}) := \mathcal{G}_2(\mathbf{A}_2; \mathbf{1})$ for $A_2 = \alpha M_{2,m} = \alpha p^{2+m} q^{2-m}$, they are:

$$\begin{aligned}
\mathcal{G}_2(\alpha M_{2,2}) : \begin{pmatrix} p \\ q \end{pmatrix} &= \begin{pmatrix} p \\ q - 4\alpha\, p^3 \end{pmatrix} && \text{spherical aberration,} \\
\mathcal{G}_2(\alpha M_{2,1}) : \begin{pmatrix} p \\ q \end{pmatrix} &= \begin{pmatrix} p + \alpha\, p^3 \\ q - 3\alpha\, p^2 q \end{pmatrix} && \text{coma,} \\
\mathcal{G}_2(\alpha M_{2,0}) : \begin{pmatrix} p \\ q \end{pmatrix} &= \begin{pmatrix} p + 2\alpha\, p^2 q \\ q - 2\alpha\, pq^2 \end{pmatrix} && \text{astigmatism / curvature of field,} \\
\mathcal{G}_2(\alpha M_{2,-1}) : \begin{pmatrix} p \\ q \end{pmatrix} &= \begin{pmatrix} p + 3\alpha\, pq^2 \\ q - \alpha\, q^3 \end{pmatrix} && \text{distorsion,} \\
\mathcal{G}_2(\alpha M_{2,-2}) : \begin{pmatrix} p \\ q \end{pmatrix} &= \begin{pmatrix} p + 4\alpha\, q^3 \\ q \end{pmatrix} && \text{pocus.}
\end{aligned} \tag{13.58}$$

•

Remark: The faces of aberrations. In Fig. 13.1, we show the maps of phase space due to the aberration group elements expanded up to the first power of their generators only, i.e.,

$$\begin{aligned}
\mathcal{G}_k(\alpha M_{k,m}) &= 1 + \alpha\{M_{k,m}, \circ\}, \\
\mathcal{G}_k(\alpha M_{k,m}) : \begin{pmatrix} p \\ q \end{pmatrix} &= \begin{pmatrix} p + \alpha(k-m)\, p^{k+m} q^{k-m-1} \\ q - \alpha(k+m)\, p^{k+m-1} q^{k-m} \end{pmatrix},
\end{aligned} \tag{13.59}$$

for $k = 1, \frac{3}{2}, 2, \frac{5}{2}, 3$ (aberration orders $1, 2, 3, 4, 5$) and $m \in \{k, k-1, \ldots, -k\}$. The three cases $k = 1$, $m \in \{1, 0, -1\}$ belong to the paraxial part and are linear. Then follow the phase space deformations due to the elements of $\mathsf{A}^{\mathrm{s}}_k\mathsf{Sp}(2,\mathsf{R})$. Compare the skew-symmetric aberrations of orders 2 and 4 (ranks $\frac{3}{2}$ and $\frac{5}{2}$) with the symmetric ones of order 3 (rank 2) of (13.58). In particular, compare the linear shear of Fig. 9.1 with the spherical aberrations $m = k$ (note the change of sign), the pure magnifier in Fig. 10.1 with the astigmatisms $m = 0$, and the thin lens action of Fig. 9.2 with the pocuses $m = -K$. •

Remark: Aberration carried to higher orders. In the two previous remarks we examined one-parameter subgroups of the rank-2 aberration group, truncated after the first term, namely $1 + \alpha\{M_{2,m}, \circ\}$. In $\mathsf{A}_K\mathsf{Sp}(2,\mathsf{R})$,

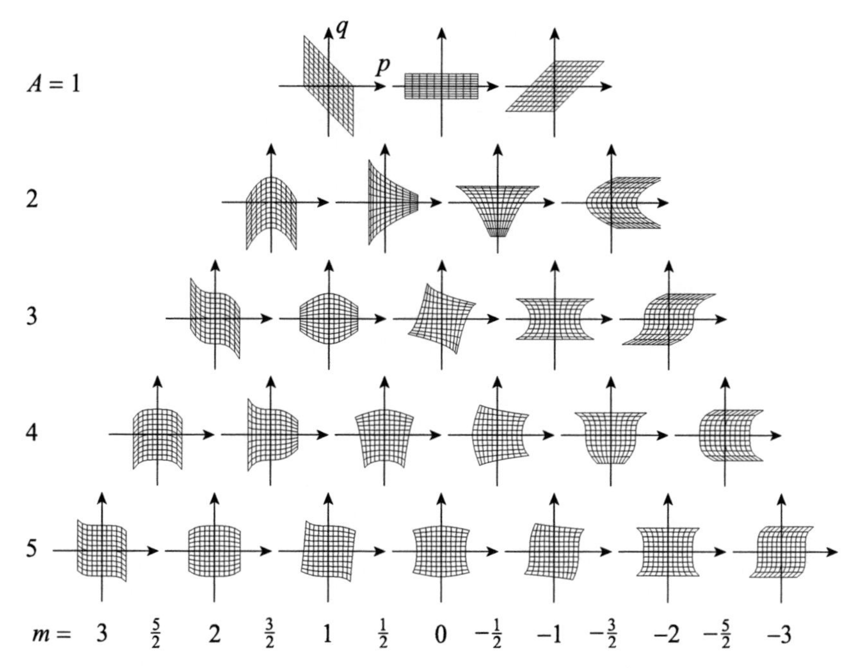

Fig. 13.1. Linear transformations and aberrations of R^2 phase space due to $\mathcal{G}_k(\alpha M_{k,m})$ approximated to its first term as in (13.59). The rows show aberration orders $1, 2, 3, 4, 5$ (ranks $k = 1, \frac{3}{2}, 2, \frac{5}{2}, 3$); within each row the maps are classified by $m \in \{k, k-1, \ldots, -k\}$ from left to right.

$K > 2$, the approximating series of a rank-2 aberration carries higher terms in (13.55). Some series may be computed in closed form; for instance, with the notation of (13.58), astigmatism yields

$$\mathcal{G}_K(\alpha M_{2,0}) : \begin{pmatrix} p \\ q \end{pmatrix} = \sum_{2n+1 \leq K} \frac{\alpha^n}{n!} 2^n \begin{pmatrix} p^{n+1} q^n \\ (-1)^n p^n q^{n+1} \end{pmatrix} \approx \begin{pmatrix} e^{2\alpha pq} p \\ e^{-2\alpha pq} q \end{pmatrix}. \quad (13.60)$$

Figure 13.2 show successive approximations to the exact result. •

Example: Free displacement in 5th aberration order. The Hamiltonian function that generates free displacement in a homogeneous optical medium of unit refractive index is given by (2.6), and has been seen and used often before. This example is trivial in the sense that Lie series are Taylor series where all operators commute, but sheds light on our approximation. We truncate the Taylor expansion of the Hamiltonian (2.11) to aberration order $a_K = 5$ (i.e., rank $K = 3$) keeping its terms up to degree $2K = 6$,

$$h_{\mathrm{F}}^{[6]}(p, q) := -1 + \tfrac{1}{2}p^2 + \tfrac{1}{8}p^4 + \tfrac{1}{16}p^6, \quad \text{i.e., of the form} \quad (13.61)$$

$$\widehat{h}^{[6]} = h_{1,1}\widehat{M}_{1,1} + h_{2,2}\widehat{M}_{2,2} + h_{3,3}\widehat{M}_{3,3}, \quad (13.62)$$

$$\exp(-z\widehat{h}^{[6]}) = \exp(-zh_{3,3}\widehat{M}_{3,3})\exp(-zh_{2,2}\widehat{M}_{2,2})\,\mathcal{M}\begin{pmatrix} 1 & 0 \\ -zh_{1,1} & 1 \end{pmatrix}$$

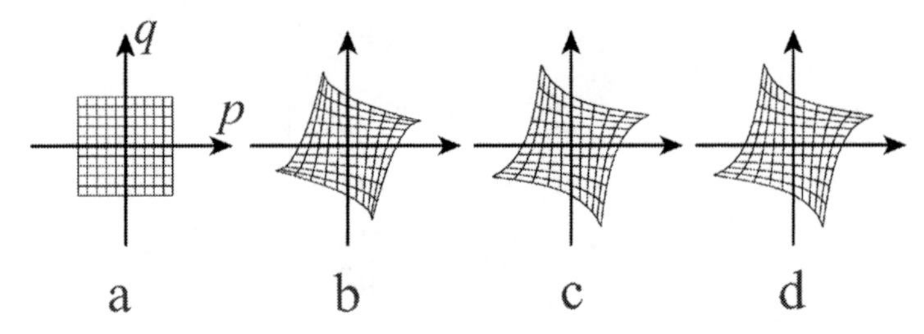

Fig. 13.2. Third-order astigmatism on phase space, generated by $M_{2,0} = p^2q^2$ to rank K. (**a**) Undeformed space, (**b**) deformed to aberration order $A = 3$ (rank $K = 2$), (**c**) to order $A = 5$ (rank $K = 3$), and (**d**) to order $A = 7$ (rank $K = 4$).

$$\approx (1 - zh_{3,3}\{p^6, \circ\})(1 - zh_{2,2}\{p^4, \circ\})\mathcal{M}\begin{pmatrix} 1 & 0 \\ -zh_{1,1} & 1 \end{pmatrix}, \quad (13.63)$$

$$\begin{pmatrix} p^{[5]}(z) \\ q^{[5]}(z) \end{pmatrix} = \begin{pmatrix} 1 & 0 \\ zh_{1,1} & 1 \end{pmatrix} \begin{pmatrix} p \\ q + 4zh_{2,2}p^3 + 6zh_{3,3}p^5 \end{pmatrix}. \quad (13.64)$$

The $h_{k,k}$ are *spherical aberration* coefficients of order $2k-1$. Free displacement is invariant under reflection across the optical axis, $(p, q) \leftrightarrow (-p, -q)$, so the series of its generating Hamiltonian only contains terms of even degree in (p, q), i.e., of integer rank and odd aberration order. All spherical aberration coefficients are also invariant under reflection across the screen $p \leftrightarrow -p$ [see (13.10) and recall Sect. 4.6]. •

Example: Aberration coefficients of the root transformation. The root transformation $\mathcal{R}_{n;\zeta}$ maps the phase space of rays in a medium of refractive index n, from the standard screen $z = 0$ to the phase space on the symmetric line $z = \zeta(q) = \zeta_2 q^2 + \zeta_4 q^4 + \zeta_6 q^6 + \cdots$. (As we saw in Chap. 4, two such factors yield the refracting and reflecting surface transformations.) In Sect. 4.4 we expanded the root transformation as an aberration series to fifth order in the coordinates of $D = 1$ optical phase space (p, q). In this example we show how to find its aberration coefficients. We refer to (4.25) and (4.26) for the full expressions. Those equations can be put in the form of (13.53), as

$$\begin{pmatrix} \bar{p}^{[5]} \\ \bar{q}^{[5]} \end{pmatrix} \approx \mathcal{R}_{n;\zeta} : \begin{pmatrix} p \\ q \end{pmatrix} = \begin{pmatrix} p + 2n\zeta_2 q + \cdots \\ q + \cdots \end{pmatrix}$$
$$= \begin{pmatrix} 1 & -2n\zeta_2 \\ 0 & 1 \end{pmatrix}^{-1} \begin{pmatrix} \mathcal{G}(\mathbf{R}; \mathbf{1}) : p \\ \mathcal{G}(\mathbf{R}; \mathbf{1}) : q \end{pmatrix}, \quad (13.65)$$

where '$+\cdots$' indicates summands that we know for degrees 3 and 5. The root transformation is an element of the aberration group, $\mathcal{G}(\mathbf{R}; \mathbf{M}_{\mathrm{R}}) := \mathcal{R}_{n;\zeta}$, whose paraxial part is therefore

$$\mathbf{M}_{\mathrm{R}} := \mathbf{M}(\mathbf{R}_1) = \begin{pmatrix} 1 & -2n\zeta_2 \\ 0 & 1 \end{pmatrix}, \quad \mathbf{R}_1 = \begin{pmatrix} 0 \\ 0 \\ n\zeta_2 \end{pmatrix}. \quad (13.66)$$

[Cf. Sect. 9.1, (9.4)–(9.5) and (9.6)–(9.7).] To find the fifth-order aberration parameters $\mathbf{R}$ of the root transformation, from (13.65) we can extract

$$\begin{aligned}
\mathcal{G}_3(\mathbf{R};\mathbf{1}) : p &= \bar{p}^{[5]} - 2n\zeta_2 q^{[5]} && (13.67)\\
&= p - \frac{\zeta_2}{n}p^2q + 4n\zeta_4\, q^3 && \text{1}^{\text{st}} \text{ and } 3^{\text{rd}} \text{ orders}\\
&\quad - \frac{\zeta_2}{4n^3}p^4q - \frac{\zeta_2^2}{n^2}p^3q^2 - \frac{2\zeta_4}{n}p^2q^3 + 12\zeta_2\zeta_4\, p\, q^4 + 6n\zeta_6\, q^5, && 5^{\text{th}} \text{ order}\\
\mathcal{G}_3(\mathbf{R};\mathbf{1}) : q &= \bar{q}^{[5]} && (13.68)\\
&= q + \frac{\zeta_2}{n}pq^2 && \text{1}^{\text{st}} \text{ and } 3^{\text{rd}} \text{ orders}\\
&\quad + \frac{\zeta_2}{2n^3}p^3q^2 + \frac{2\zeta_2^2}{n^2}\, p^2\, q^3 + \frac{\zeta_4}{n}p\, q^4. && 5^{\text{th}} \text{ order}
\end{aligned}$$

Since only odd aberration orders appear, the only coefficients present in $\mathcal{G}_3(\mathbf{R};\mathbf{1})$ will be of integer ranks 2 and 3. We return to the second line of the expansion (13.54)–(13.55) [with the noted symmetric replacements $k = \frac{3}{2} \mapsto 2$ and $2 \mapsto 3$], and see that the 3rd order terms will determine the rank-2 aberration polynomial of the root transformation, as follows:

$$R_2(p,q) := R_{2,2}p^4 + R_{2,1}p^3q + R_{2,0}p^2q^2 + R_{2,-1}pq^3 + R_{2,-2}q^4 \quad (13.69)$$

$$\begin{aligned}
\Rightarrow \{R_2, p\} &= R_{2,1}p^3 + 2R_{2,0}p^2q + 3R_{2,-1}pq^2 + 4R_{2,-2}q^3 && (13.70)\\
&= -\frac{\zeta_2}{n}p^2q + 4n\zeta_4\, q^3, \quad \text{from (13.67), and}
\end{aligned}$$

$$\{R_2, q\} = -4R_{2,2}p^3 - 3R_{2,1}p^2q - 2R_{2,0}pq^2 - R_{2,-1}q^3 = \frac{\zeta_2}{n}pq^2, \quad (13.71)$$

from (13.68). Each equation determines 4 of the 5 coefficients: (13.70) leaves out $R_{2,2}$ while (13.71) leaves $R_{2,-2}$ out; together, they are an overdetermined set which must be consistent if the input transformation is indeed canonical. This serves as a tight check on the computation. We have thus found the third-order aberration coefficients of the root transformation:

$$\mathbf{R}_2 := (R_{2,2}, R_{2,1}, R_{2,0}, R_{2,-1}, R_{2,-2})^\top = \Big\{0,\ 0,\ -\frac{\zeta_2}{2n},\ 0,\ n\zeta_4\Big\}. \quad (13.72)$$

Finally, the fifth-order aberration coefficients $R_{3,m}$ are found from the terms of degree 5 in the third line of (13.54)–(13.55) [with $\frac{3}{2} \mapsto 2$, $2 \mapsto 3$, $\frac{5}{2} \mapsto 4$, this is $\{R_3, r\} + \frac{1}{2!}\{R_2, \{R_2, r\}\}$; thus we must compute the second summand with the now-known $R_2(p,q)$ and subtract it]. The result is

$$\begin{aligned}
\mathbf{R}_3 &:= (R_{3,3}, R_{3,2}, R_{3,1}, R_{3,0}, R_{3,-1}, R_{3,-2}, R_{3,-3})^\top\\
&= \Big\{0,\ 0,\ -\frac{\zeta_2}{8n^3},\ -\frac{\zeta_2^2}{2n^2},\ -\frac{\zeta_4}{2n},\ 2\zeta_2\zeta_4,\ n\zeta_6\Big\}.
\end{aligned} \quad (13.73)$$

We note that the first two coefficients, those of spherical aberration and coma, are zero for any integer rank $k \geq 1$. •

The algorithm of the previous example sketches constructively the proof of the *Dragt–Finn theorem* [40]. This states that, given a canonical transformation of phase space leaving the origin invariant (i.e., subject to a Taylor expansion starting with degree 1), there exists a factored-product form of an aberration group element which approximates this transformation to any desired degree/rank/order. When using aberration series though, there are problems in determining the radius, rate, and rank for practical convergence. The example also highlights that the computation of aberrations for whole systems, such as a refracting surface which is the composition of a direct and inverse root transformation [see (4.13) and (13.40)–(13.46)], or of a thick lens made of two surfaces and a free flight, or of an optical system composed of several lenses, can become a major enterprise when one attempts to perform it by hand. On the other hand, when aided by symbolic computer programs, the Hamilton–Lie theory of aberrations allows the composition of previously-computed aberrating optical elements (from left to right, as they are placed on the table) with free parameters. This can serve efficiently in magnetic optics [44] to find the Poincaré phase space map of a particle after 2^N turns in a tokamak or particle accelerator, or to find the stable fields of rays between two warped mirrors in a resonator once one turn or bounce has been computed carefully. In the following chapter, we shall remark on existing computer systems for Hamilton–Lie aberrations.

14 Axis-symmetric aberrations

The definition and principal properties of the rank-K aberration algebras (and groups) in $D = 2$ dimensions can be extrapolated from the $D = 1$ case seen in the previous chapter. They are the semidirect sum (and product) of a paraxial part and an aberration part with a correspondingly larger dimension, generated by the monomials of phase space $(\mathbf{p}, \mathbf{q}) \in \mathsf{R}^4$ written in Sect. 11.5, (11.96). In this chapter we develop the theory of $D = 2$-dim aberrations in Hamiltonian systems which are *axis-symmetric*, i.e., invariant under the group of rotations around the optical axis, and reflections across planes containing this axis (invariance under reflections excludes magnetic systems). Our interest lies mostly in understanding the *structure* of this set of aberrations. We also give the full parametric expressions for the coefficients of common optical elements, and for the product rule to concatenate them into optical systems.

14.1 Axis-symmetric aberrations in the Cartesian basis

We set up a basis of monomials in the quadratic functions of the phase space coordinates $\mathbf{p} = \begin{pmatrix} p_x \\ p_y \end{pmatrix}$, $\mathbf{q} = \begin{pmatrix} q_x \\ q_y \end{pmatrix}$ on the standard screen (see Sect. 7.1) that are invariant under the group $\mathsf{O}(2)$ of rotations around the optical axis, $\mathbf{R} = \begin{pmatrix} \cos\theta & -\sin\theta \\ \sin\theta & \cos\theta \end{pmatrix}$, and reflections $\begin{pmatrix} -1 & 0 \\ 0 & 1 \end{pmatrix}$ across lines through the optical center. The monomials are

$$M_{k_+,k_0,k_-}(\mathbf{p},\mathbf{q}) := (|\mathbf{p}|^2)^{k_+}(\mathbf{p}\cdot\mathbf{q})^{k_0}(|\mathbf{q}|^2)^{k_-}, \tag{14.1}$$

with integer $k_\sigma \geq 0$, $\sigma \in \{+, 0, -\}$ [cf. (7.3) and (11.112)]. These are analogous to the monomials in the previous chapter, (13.2), but now in 3 variables, and also a basis for a vector space $\mathcal{A}$ of denumerably-infinite dimension. We exclude $\mathbf{p}\times\mathbf{q} = p_x q_y - p_y q_x$ because it changes sign under reflection $(p_x, p_y; q_x, q_y) \leftrightarrow (-p_x, p_y; -q_x, q_y)$.

We use the *reduced* phase space coordinates $(|\mathbf{p}|^2, \mathbf{p}\cdot\mathbf{q}, |\mathbf{q}|^2)$ (cf. Sect. 7.3) to build the monomials, whose *rank* will be their total homogeneous degree; their *weight* will be their eigenvalue under the operator $\{\frac{1}{2}\mathbf{p}\cdot\mathbf{q}, \circ\}$, which counts powers of $|\mathbf{p}|^2$ and subtracts the powers of $|\mathbf{q}|^2$, i.e.,

$$M_{k_+,k_0,k_-}(\mathbf{p},\mathbf{q}) \quad \text{has} \begin{cases} \text{rank} \quad k := k_+ + k_0 + k_- \in \{1,2,3,\ldots\}, \\ \text{weight } m := k_+ - k_- \in \{k, k-1, \ldots, -k\}. \end{cases} \tag{14.2}$$

As before, these monomials are associated to corresponding Poisson operators, indicated $\widehat{M}_{k_+,k_0,k_-} := \{M_{k_+,k_0,k_-}, \circ\}$, whose commutator yields the Lie algebraic structure of the space $\mathcal{A}$. The paraxial part of axis-symmetric systems is generated by the 3 monomials of rank 1,

$$M_{1,0,0} = |\mathbf{p}|^2, \quad M_{0,1,0} = \mathbf{p}\cdot\mathbf{q}, \quad M_{0,0,1} = |\mathbf{q}|^2, \tag{14.3}$$

whose Poisson operators close into the algebra $\mathsf{sp}(2,\mathsf{R})$ that generates linear transformations of $\mathcal{A}$. Aberrations will be generated by monomials of higher ranks. Again we display the structure of $\mathcal{A}$ as the direct sum of finite-dimensional subspaces of definite rank $\mathcal{A}_k$, where each $\mathcal{A}_k$ is invariant under the action of linear transformations $\mathbf{M} \in \mathsf{Sp}(2,\mathsf{R})$, and has dimension $d_k := \sum_{n=1}^{k+1} n = \frac{1}{2}(k+1)(k+2)$. For ranks $k = 2,3,4\ldots$, there will be thus $d_k = 6, 10, 15, \ldots$ aberrations of orders $a_k := (2k-1) = 3, 5, 7, \ldots$. The Fourier transform relates monomials through

$$\mathcal{F} : M_{k_+,k_0,k_-}(\mathbf{p},\mathbf{q}) = M_{k_+,k_0,k_-}(\mathbf{q},-\mathbf{p}) = (-1)^{k_0} M_{k_-,k_0,k_+}(\mathbf{p},\mathbf{q}). \tag{14.4}$$

Example: Monomials of rank 2. There are 6 monomials of rank 2, with indices $\{(2,0,0),\ (1,1,0),\ (1,0,1),\ (0,2,0),\ (0,1,1),\ (0,0,2)\}$ – ordered lexicographically. This sextuplet is indicated in Fig. 14.1 by points of integer coordinates in the first octant of an R^3 space of axes (k_+, k_0, k_-), on the plane $k = k_+ + k_0 + k_- = 2$. This figure should be compared to Fig. 11.6,

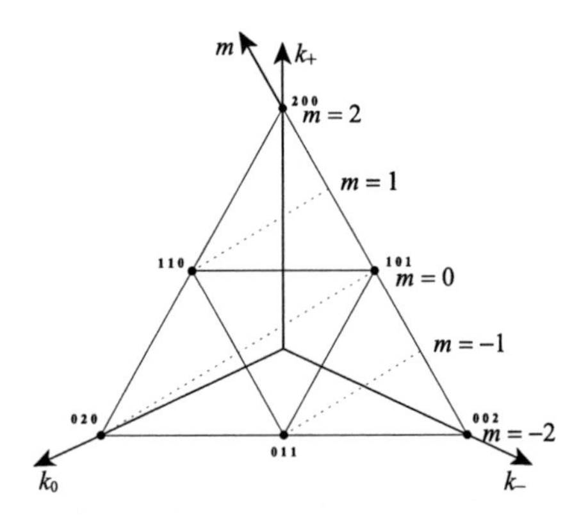

Fig. 14.1. Sextuplet of rank-2 monomials in the basis M_{k_+,k_0,k_-} of Cartesian coordinates (k_+, k_0, k_-) lie on integer points of the plane $k = k_+ + k_0 + k_- = 2$. Each point represents one distinct third-order aberration. The axis of weights, $m = k_+ - k_-$, is also shown. The names associated to this multiplet are: spherical aberration (2,0,0), coma (1,1,0), curvature of field (1,0,1), astigmatism (0,2,0), distorsion (0,1,1), and pocus (0,0,2). The last is not counted in the traditional Seidel scheme.

which represents the weight diagram of symmetric 2-quark representation of su(3) [cf. (11.112)]. Although the resemblance is formal, it associates 1:1 the monomials of phase space M_{k_+,k_0,k_-} to the energy eigenstates of the 3-dim quantum harmonic oscillator with (k_+, k_0, k_-) energy quanta along three orthogonal Cartesian axes. This quantum system is well known to separate also in spherical coordinates, with eigenstates classified by energy, orbital angular momentum and its third-axis projection as (k, j, m); the same linear combinations will serve below to classify optical aberrations by rank, symplectic spin, and weight. •

Remark: On labelling monomials. In contradistinction to the $D = 1$ case of the previous chapter, the two numbers of rank and weight do not suffice to characterize uniquely the set of monomials $M_{k_+,k_0,k_-}(\mathbf{p}, \mathbf{q})$ in (14.1). For example, $M_{1,0,1}$ (curvature of field) and $M_{0,2,0}$ (astigmatism) have the same rank and weight, so we must use their power indices k_σ to distinguish them. But note that 'k_σ-counting' operators, such as would be $|\mathbf{p}|^2\, \partial/\partial|\mathbf{p}|^2 = \frac{1}{2}\mathbf{p}\cdot\partial/\partial\mathbf{p}$, are *not* the Poisson operators of any function, so these labels are extraneous to the Lie algebra generated by the monomials. The axis-symmetric rank-k spaces $\mathcal{A}_k$ are thus further *reducible* under the action of the paraxial part. In the following sections we shall resolve this degeneracy introducing the symplectic spin label. •

The spaces of functions $\mathcal{A}_k$ serve as homogeneous spaces for the Lie algebra sp(2, R) of $\mathcal{A}_1$ generated by the monomials (14.3); but instead of the simple form of (13.21)–(13.23), the action of the raising, weight, and lowering generators is now

$$[\widehat{M}_{1,0,1}, \widehat{M}_{k_+,k_0,k_-}] = -2k_0\widehat{M}_{k_++1,k_0-1,k_-} - 4k_-\widehat{M}_{k_+,k_0+1,k_--1}, \tag{14.5}$$

$$[\widehat{M}_{0,1,0}, \widehat{M}_{k_+,k_0,k_-}] = 2(k_+ - k_-)\widehat{M}_{k_+,k_0,k_-}, \tag{14.6}$$

$$[\widehat{M}_{0,0,1}, \widehat{M}_{k_+,k_0,k_-}] = 4k_+\widehat{M}_{k_+-1,k_0+1,k_-} + 2k_0\widehat{M}_{k_+,k_0-1,k_-+1}, \tag{14.7}$$

respectively. The generic commutator between monomials that presents the Lie-algebraic structure of the space $\mathcal{A}$ is thus

$$\begin{aligned}[\widehat{M}_{k_+,k_0,k_-}, \widehat{M}_{k'_+,k'_0,k'_-}] &= 4(k_-k'_+ - k_+k'_-)\,\widehat{M}_{k_++k'_+-1,k_0+k'_0+1,k_-+k'_--1}\\ &+ 2\Big(k_0(k'_+ - k'_-) - (k_+ - k_-)k'_0\Big)\,\widehat{M}_{k_++k'_+,k_0+k'_0-1,k_-+k'_-}\,.\end{aligned} \tag{14.8}$$

This can be compared with the $D = 1$ case in (13.18), and seen to exhibit also the graded composition of ranks, namely $\{\mathcal{A}_k, \mathcal{A}_{k'}\} = \mathcal{A}_{k+k'-1}$, which allows us again to consistently truncate terms to a given rank K as in (13.24), through replacing all monomials of higher rank by zero. We thus define the *rank-K axis-symmetric aberration algebra* denoted $\mathsf{a}_K^{\mathsf{AS}}\mathsf{sp}(2, \mathsf{R})$. As in the $D = 1$ case (13.25), it is a semidirect sum of the radical algebra sp(2, R) generated by (14.3), with an invariant and nilpotent subalgebra of aberrations $\mathsf{a}_K^{\mathsf{AS}}$, whose dimension is now larger: for ranks $K = 2, 3, 4$ (aberration orders $a_K = 3, 5, 7$) there are now 6, $6 + 10 = 16$, and $6 + 10 + 15 = 31$ aberrations.

Example: Highest and lower members of the $k = 2$ sextuplet. The highest-weight member of a multiplet of rank k is $M_{k,0,0} = (|\mathbf{p}|^2)^k$, with weight $m = k$. Applying (14.7) repeatedly, we produce all other members down to $m = -k$; one further lowering yields 0. For $k = 2$ this process, schematically modulo constants, is

$$\begin{array}{ccccccccc} & & & & m=0 & & & & \\ m=2 & & m=1 & \nearrow & (1,0,1) & \searrow & m=-1 & & m=-2 \\ (2,0,0) & \rightarrow & (1,1,0) & & |\mathbf{p}|^2|\mathbf{q}|^2 & & (0,1,1) & \rightarrow & (0,0,2)\,. \\ (|\mathbf{p}|^2)^2 & & |\mathbf{p}|^2\mathbf{p}\cdot\mathbf{q} & \searrow & (0,2,0) & \nearrow & \mathbf{p}\cdot\mathbf{q}\,|\mathbf{q}|^2 & & (|\mathbf{q}|^2)^2 \\ & & & & (\mathbf{p}\cdot\mathbf{q})^2 & & & & \end{array} \tag{14.9}$$

The two highest and the two lowest m-multiplet members are the only ones of their weight (i.e., they are nondegenerate), but then there is a branching into the two weight-degenerate monomials $m = 0$; see Fig. 14.1. For higher ranks k, the degeneracy pattern is: $m = \pm k$ and $\pm(k-1)$ are single, $m = \pm(k-2)$ and $\pm(k-3)$ are doubly degenerate, $m = \pm(k-4)$ and $\pm(k-5)$ are triply degenerate, etc.; the mid-member of the multiplet, $m = 0$, is k-fold degenerate. •

As in the previous chapter, the Poisson operators of the Cartesian monomials (14.2) are applied on the phase space coordinates $(\mathbf{p}, \mathbf{q}) \in \mathsf{R}^4$, which can be asigned to rank $\frac{1}{2}$, and written in matrix form as

$$\begin{aligned} \widehat{M}_{k_+,k_0,k_-}\begin{pmatrix}\mathbf{p}\\ \mathbf{q}\end{pmatrix} &= \begin{pmatrix}\partial/\partial\mathbf{q}\\ -\partial/\partial\mathbf{p}\end{pmatrix} M_{k_+,k_0,k_-}(\mathbf{p},\mathbf{q}) \\ &= \begin{pmatrix} k_0\, M_{k_+,k_0-1,k_-}(\mathbf{p},\mathbf{q}) & k_-\, M_{k_+,k_0,k_--1}(\mathbf{p},\mathbf{q}) \\ -k_+\, M_{k_+-1,k_0,k_-}(\mathbf{p},\mathbf{q}) & -k_0\, M_{k_+,k_0-1,k_-}(\mathbf{p},\mathbf{q}) \end{pmatrix}\begin{pmatrix}\mathbf{p}\\ \mathbf{q}\end{pmatrix}. \end{aligned} \tag{14.10}$$

This is a nonlinear action bringing a rank-$\frac{1}{2}$ vector to one of rank $k - \frac{1}{2}$.

When we exponentiate the Poisson operators to generate one-parameter groups of phase space transformations as in the previous chapter, we decide on an aberration order (the degree of homogeneity $a_K = 2K - 1$) beyond which we truncate the series. We thus define the *rank-K axis-symmetric aberration group* $\mathsf{A}_K^{\mathsf{AS}}\mathsf{Sp}(2,\mathsf{R})$, whose elements we also denote by $\mathcal{G}_K(\mathbf{A};\mathbf{M})$ in complete analogy with (13.26), and the *factored product parametrization* in (13.27)–(13.30), except that all ranks are now integer and the aberration polynomials [cf. (13.29)] now sum over the monomials (14.1), i.e.

$$\mathcal{G}_K(\mathbf{A};\mathbf{M}) := \mathcal{G}(\mathbf{A}_K, \ldots, \mathbf{A}_4, \mathbf{A}_3, \mathbf{A}_2; \mathbf{1})\, \mathcal{G}(\mathbf{0}; \mathbf{M}(\mathbf{A}_1)), \tag{14.11}$$

$$\mathcal{G}(\mathbf{A};\mathbf{1}) := \exp \widehat{A}_K \times \cdots \times \exp \widehat{A}_4 \times \exp \widehat{A}_3 \times \exp \widehat{A}_2, \tag{14.12}$$

$$\widehat{A}_k := \textstyle\sum_{k_++k_0+k_-=k} A_{k_+,k_0,k_-}\, M_{k_+,k_0,k_-}(\mathbf{p},\mathbf{q}), \tag{14.13}$$

$$\mathcal{G}(\mathbf{0},\mathbf{M}) := \exp \widehat{A}_1 = \mathcal{M}\left(\exp\begin{pmatrix} -A_{1,1,0} & -2A_{1,1,-1} \\ 2A_{1,1,1} & A_{1,1,0}\end{pmatrix}\right). \tag{14.14}$$

The coefficients $\{A_{k_+,k_0,k_-}\}$ for $2 \le k \le K$ will be called the Hamilton–Lie *Cartesian* aberration coefficients of the system.

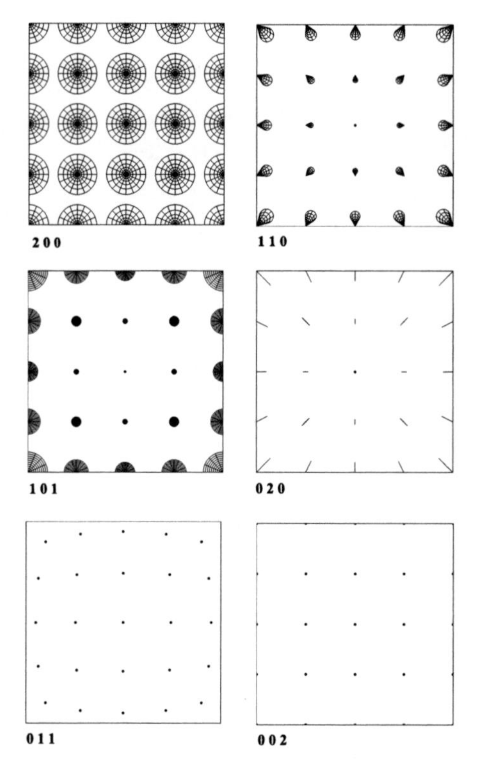

Fig. 14.2. Spots diagrams of the 6 aberrations of order 3 in the Cartesian basis. (2,0,0): spherical aberration is field-independent i.e., spot size does not depend on position, (1,1,0): coma has linear field dependence, (1,0,1): curvature of field, (0,2,0): astigmatism, (0,1,1): distorsion, and (0,0,2): pocus.

Remark: The spots of $K = 2$ aberrations. As we did for the 5 third-order aberrations of the $D = 1$ case in Fig. 13.1, here in Fig. 14.2 we show the visible face of the 6 $D = 2$ axis-symmetric aberrations given by their spots diagrams (cf. Figs. 3.4 and 5.3). These show the generic form of the pure aberrations obtained from (14.10) as the first term of the Lie series that we indicate by $\mathcal{G}_k(\alpha M_{k_+,k_0,k_-}) := \mathit{1} + \alpha\{M_{k_+,k_0,k_-}, \circ\}$ [cf. (13.59)]. They are:

$$\begin{aligned}
&\mathcal{G}_2(\alpha\, M_{2,0,0}) : \begin{pmatrix}\mathbf{p}\\ \mathbf{q}\end{pmatrix} = \begin{pmatrix}\mathbf{p}\\ \mathbf{q} - 4\alpha\,|\mathbf{p}|^2\mathbf{p}\end{pmatrix} && \text{spherical aberration,}\\
&\mathcal{G}_2(\alpha\, M_{1,1,0}) : \begin{pmatrix}\mathbf{p}\\ \mathbf{q}\end{pmatrix} = \begin{pmatrix}\mathbf{p} + \alpha\,|\mathbf{p}|^2\mathbf{p}\\ \mathbf{q} - \alpha(2\mathbf{p}\cdot\mathbf{q}\,\mathbf{p} + |\mathbf{p}|^2\mathbf{q})\end{pmatrix} && \text{coma,}\\
&\mathcal{G}_2(\alpha\, M_{1,0,1}) : \begin{pmatrix}\mathbf{p}\\ \mathbf{q}\end{pmatrix} = \begin{pmatrix}\mathbf{p} + 2\alpha\,|\mathbf{p}|^2\mathbf{q}\\ \mathbf{q} - 2\alpha\,|\mathbf{q}|^2\mathbf{p}\end{pmatrix} && \text{curvature of field,}\\
&\mathcal{G}_2(\alpha\, M_{0,2,0}) : \begin{pmatrix}\mathbf{p}\\ \mathbf{q}\end{pmatrix} = \begin{pmatrix}\mathbf{p} + 2\alpha\,\mathbf{p}\cdot\mathbf{q}\,\mathbf{p}\\ \mathbf{q} - 2\alpha\,\mathbf{p}\cdot\mathbf{q}\,\mathbf{q}\end{pmatrix} && \text{astigmatism,}\\
&\mathcal{G}_2(\alpha\, M_{0,1,1}) : \begin{pmatrix}\mathbf{p}\\ \mathbf{q}\end{pmatrix} = \begin{pmatrix}\mathbf{p} + \alpha(|\mathbf{q}|^2\mathbf{p} + 2\mathbf{p}\cdot\mathbf{q}\,\mathbf{q})\\ \mathbf{q} - \alpha\,|\mathbf{q}|^2\mathbf{q}\end{pmatrix} && \text{distorsion,}\\
&\mathcal{G}_2(\alpha\, M_{0,0,2}) : \begin{pmatrix}\mathbf{p}\\ \mathbf{q}\end{pmatrix} = \begin{pmatrix}\mathbf{p} + 4\alpha\,|\mathbf{q}|^2\mathbf{q}\\ \mathbf{q}\end{pmatrix} && \text{pocus.}
\end{aligned} \tag{14.15}$$

Except for the last one, the list of asignments is the one proposed by Dragt in [41]. In Figs. 14.2 we plot the aberrated image $\mathbf{q}'(\mathbf{p}, \mathbf{q}; \alpha)$ of a square grid of ray pencils of positions $\mathbf{q}_{i,j}$, with directions that draw out concentric circles of radii $|\mathbf{p}| = p_j$ (see Fig. 3.4). The axial symmetry of the aberrations is manifest in that the *shape* of the spots is independent of their angular position on the screen, while their *size* only depends on the distance to the optical center. In particular, spherical aberration is *field-independent*, i.e., the size and shape of the spots are independent of their location on the screen; the ray directions (optical momenta) are left invariant. The Fourier transform (14.4) of spherical aberration is pocus, which leaves the image points invariant but spherically aberrates the momentum subspace. Coma and distorsion are also Fourier conjugates of each other, so the spots diagram of the first is the aberration of momentum space in the second, and *viceversa.* Astigmatism and curvature of field are self-reciprocal under Fourier transformation, so the aberrations of both position and momentum are the same. •

The composition of elements in the aberration groups $\mathsf{A}_K^{\mathsf{AS}}\mathsf{Sp}(2,\mathsf{R})$ follows as in (13.31)–(13.33), namely

$$\mathcal{G}_K(\mathbf{A}; \mathbf{M})\, \mathcal{G}_K(\mathbf{B}; \mathbf{N}) = \mathcal{G}_K(\mathbf{A} \sharp \mathcal{M}(\mathbf{M}) : \mathbf{B};\, \mathbf{MN}). \tag{14.16}$$

The *gato* composition '$\sharp$' between pure aberrations is obtained in the same way as the ordered product in (13.40) [with the replacement of ranks $\frac{3}{2} \mapsto 2$, $2 \mapsto 3$, $\frac{5}{2} \mapsto 4$, i.e. of aberration orders $2 \mapsto 3$, $3 \mapsto 5$, $4 \mapsto 7$], but performed now with the aberration polynomials (14.13) of three variables which contain the Cartesian, axis-symmetric aberration coefficients $\mathbf{A} = \{\mathbf{A}_k\}_{k=2}^{K} = \{A_{k_+,k_0,k_-}\}$, etc., of $\mathbf{C} = \mathbf{A} \sharp \mathbf{B}$. The adjoint action of the radical subgroup $\mathbf{M} \in \mathsf{Sp}(2,\mathsf{R})$ on the aberrations in (14.16) is again linear, block-diagonal in k, and of the form (13.36)–(13.37), *viz.*,

$$\mathcal{M}(\mathbf{M}) : \mathbf{B}_k = \bar{\mathbf{D}}^k(\mathbf{M}^{-1})^\top \mathbf{B}_k \tag{14.17}$$

However, the matrix elements of $\bar{\mathbf{D}}^k(\mathbf{M})$ are now found from the transformation of the *three* variables $|\mathbf{p}|^2$, $\mathbf{p}\cdot\mathbf{q}$, and $|\mathbf{q}|^2$, by the 3×3 matrix $\mathbf{D}^1(\mathbf{M})$

in (13.14) – instead of the 2×2 matrix $\mathbf{M}$ of the $D = 1$-dim case of the previous chapter – and are *not* the irreducible matrices given in (13.6). Here, the matrices $\bar{\mathbf{D}}^k(\mathbf{M})$ for $k = 2, 3, 4$ have dimensions 6×6, 10×10, 15×15, and are *further* reducible to a minimal block-diagonal form, as will be shown forthwith.

14.2 The harmonic basis of aberrations

The simplicity of the basis of Cartesian monomials (14.1) is offset by the large matrices which represent the paraxial part $\mathsf{Sp}(2,\mathsf{R})$ of the axis-symmetric aberration groups. Here we present a basis of Hamilton–Lie aberrations which is better adapted to the axial symmetry of the system; this we shall call the *symplectic harmonic* basis, for reasons that will become clear below. In addition to rank and weight, k, m, the symplectic harmonics will have a *symplectic spin* label j, that will distinguish between the $\mathsf{sp}(2,\mathsf{R})$ submultiplets present in each aberration order.

We first note that in rank 2 we have the (*Casimir*) *skewness* invariant

$$Y_{2,0,0}(\mathbf{p},\mathbf{q}) := (\mathbf{p} \times \mathbf{q})^2 = |\mathbf{p}|^2|\mathbf{q}|^2 - (\mathbf{p}\cdot\mathbf{q})^2 = M_{1,0,1} - M_{0,2,0}, \tag{14.18}$$

$$\mathcal{M}(\mathbf{M}) : Y_{2,0,0}(\mathbf{p},\mathbf{q}) = Y_{2,0,0}(\mathbf{p},\mathbf{q}), \qquad \mathbf{M} \in \mathsf{Sp}(2,\mathsf{R}). \tag{14.19}$$

[*Cf.* (7.2) and (7.5).] In the monomial basis this is a linear combination of curvature of field (1,0,1) and astigmatism (0,2,0), and a 1-dim subspace within the 6-dim space of aberrations of third order which is invariant under $\mathsf{Sp}(2,\mathsf{R})$.

Remark: The canonical map of skewness. We can exponentiate the Poisson operator of skewness (14.18) to a one-parameter group of canonical transformations of phase space in closed form. This operator acts through

$$\{Y_{2,0,0}, \circ\} \begin{pmatrix} \mathbf{p} \\ \mathbf{q} \end{pmatrix} = 2 \begin{pmatrix} -\mathbf{p}\cdot\mathbf{q} & |\mathbf{p}|^2 \\ -|\mathbf{q}|^2 & \mathbf{p}\cdot\mathbf{q} \end{pmatrix} \begin{pmatrix} \mathbf{p} \\ \mathbf{q} \end{pmatrix} \tag{14.20}$$

[cf. (14.10)], where the matrix has determinant $\Delta = 4(\mathbf{p} \times \mathbf{q})^2$ that, being a function of axis-symmetric coordinates, is itself invariant under $\{Y_{2,0,0}, \circ\}$. Using (12.13)–(12.14) and $\phi := 2\alpha|\mathbf{p} \times \mathbf{q}|$, we find

$$\begin{aligned} \begin{pmatrix} \mathbf{p}'(\alpha) \\ \mathbf{q}'(\alpha) \end{pmatrix} &= \exp(\alpha\{Y_{2,0,0}, \circ\}) \begin{pmatrix} \mathbf{p} \\ \mathbf{q} \end{pmatrix} \\ &= \left[\mathbf{1} \cos\phi + \begin{pmatrix} -\mathbf{p}\cdot\mathbf{q} & |\mathbf{p}|^2 \\ -|\mathbf{q}|^2 & \mathbf{p}\cdot\mathbf{q} \end{pmatrix} \begin{pmatrix} \mathbf{p} \\ \mathbf{q} \end{pmatrix} \frac{\sin\phi}{|\mathbf{p} \times \mathbf{q}|} \right] \begin{pmatrix} \mathbf{p} \\ \mathbf{q} \end{pmatrix}. \end{aligned} \tag{14.21}$$

To unravel this expression, we may particularize for $\mathbf{q} = \binom{q}{0}$ and $\mathbf{p} = \binom{p_x}{p_y}$; then it is easy to prove that

$$\begin{pmatrix} p'_x(\alpha) \\ p'_y(\alpha) \end{pmatrix} = \begin{pmatrix} \cos\phi & -\sin\phi \\ \sin\phi & \cos\phi \end{pmatrix} \begin{pmatrix} p_x \\ p_y \end{pmatrix}, \quad \begin{pmatrix} q'_x(\alpha) \\ q'_y(\alpha) \end{pmatrix} = q \begin{pmatrix} \cos\phi \\ \sin\phi \end{pmatrix}. \tag{14.22}$$

The spots of this transformation are thus 2ϕ-arcs of centered circles, as shown in Fig. 14.3. The truncated third-order approximations are parabolas that osculate these circles; the fifth-order ones are quartic lines, etc. •

The vector complement of the skewness invariant in the 6-dim space of third-order aberrations is a 5-dim space which transforms within itself under linear (paraxial) transformations. This $\mathsf{sp}(2,\mathsf{R})$-*quintuplet* can be generated out of the (unique) highest weight monomial, $M_{2,0,0} = (|\mathbf{p}|^2)^2$, by applying the lowering operator $\widehat{M}_{0,0,1} = \{|\mathbf{q}|^2, \circ\}$ through (14.7) successively for $m = 2, 1, 0, -1, -2$. The process of generating this multiplet is shown in (14.9) and yields the $m = 0$ member, giving us the linear combination of monomials proportional to

$$Y_{2,2,0}(\mathbf{p},\mathbf{q}) := \tfrac{1}{3}[|\mathbf{p}|^2|\mathbf{q}|^2 + 2(\mathbf{p}\cdot\mathbf{q})^2] = \tfrac{1}{3}M_{1,0,1}(\mathbf{p},\mathbf{q}) + \tfrac{2}{3}M_{0,2,0}(\mathbf{p},\mathbf{q}). \quad (14.23)$$

This method of finding the members of a multiplet is exactly the same that we used in the $D = 1$ case, in the form of (13.23); it is independent of its realization by monomials in (p, q) or in $(|\mathbf{p}|^2, \mathbf{p}\cdot\mathbf{q}, |\mathbf{q}|^2)$, because it is a Lie-algebraic construction. In Fig. 14.3 we show the spots diagram of the aberrations in the same form as those of Fig. 14.2, namely, to third aberration order. The formulas can be obtained from the linear combination of aberrations in (14.15).

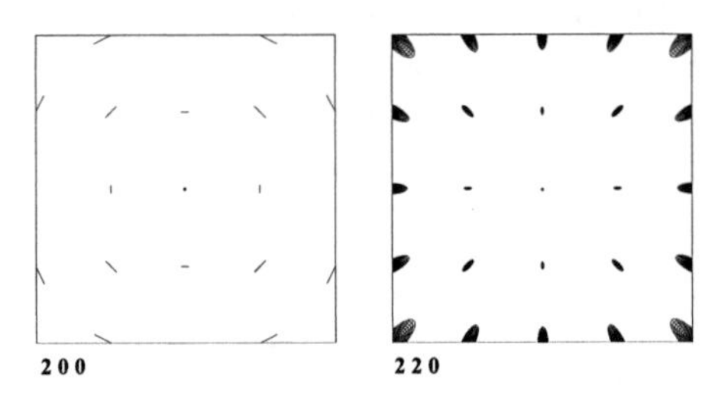

Fig. 14.3. Spots diagrams of the skewness $Y_{2,0,0}$, and the middle member of the quintuplet $Y_{2,2,0}$, both having rank 2 and weight 0. These spots and their field dependence should be compared with those of the Cartesian aberrations $M_{1,0,1}$ (curvature of field) and $M_{0,2,0}$ (astigmatism) in Fig. 13.2.

We now define the set of polynomials of reduced phase space, of rank k, *symplectic spin* j, and weight m, as linear combinations of the Cartesian monomials (14.1) with that rank and weight,

$$Y_{k,j,m}(\mathbf{p},\mathbf{q}) := (\mathbf{p}\times\mathbf{q})^{k-j} \sum_{\nu\in N(j,m)} C_{j,\nu,m}\, |\mathbf{p}|^{j+m-\nu}\, (\mathbf{p}\cdot\mathbf{q})^{\nu}\, |\mathbf{q}|^{j-m-\nu}, \quad (14.24)$$

where in axis-symmetric systems $k - j$ is even, $Y_{k,k,k} := (|\mathbf{p}|^2)^k$, and the algebraic property (13.23) holds, namely

$$\{|\mathbf{q}|^2, Y_{k,j,m}\} = 2(m+j)\, Y_{k,j,m-1}. \tag{14.25}$$

For rank k, the values of the symplectic spin, weight, and the index ν in the sum (14.24), are

$$j \in J(k) := \begin{cases} \{0,2,4,\ldots,k\}, & k \text{ even}, \\ \{1,3,5,\ldots,k\}, & k \text{ odd}, \end{cases} \qquad m|_{-j}^{j}, \tag{14.26}$$

$$\nu \in N(j,m) := \begin{cases} \{0,2,4,\ldots,j-|m|\}, & j-|m| \text{ even}, \\ \{1,3,5,\ldots,j-|m|\}, & j-|m| \text{ odd}. \end{cases} \tag{14.27}$$

The powers of the coordinates of reduced phase space, ν and $\frac{1}{2}(j \pm m - \nu)$, are nonnegative integers. The polynomials $Y_{k,j,m}(\mathbf{p},\mathbf{q})$ have homogeneous degree $2k$ in the coordinates $(\mathbf{p},\mathbf{q})$. To find the $2j+1$ aberrations which are partners in a symplectic multiplet of spin j under $\mathsf{sp}(2,\mathsf{R})$, we start from the member of highest weight $m=j$, and lower this weight iteratively through (14.25), producing thus a recursion relation for the coefficients $C_{j,\nu,m}$.

Remark: Solution of the recursion relation. Familiarity with recursions arising in Lie algebras suggests replacing (14.24) in (14.25) and thus, using (14.7) and collecting monomials, find the recursion relation between the coefficients:

$$(j+m-\nu+1)\, C_{j,\nu-1,m} + (\nu+1)\, C_{j,\nu+1,m} = (j+m)\, C_{j,\nu,m-1}, \tag{14.28}$$

and $C_{j,0,j} = 1$. Common *ansätze* lead us through the following succesive replacements

$$C_{j,\nu,m} =: 2^{\nu}\, D_{j,m} / [\tfrac{1}{2}(j+m-\nu)]!\, \nu!\, [\tfrac{1}{2}(j-m-\nu)]!, \tag{14.29}$$

$$\Rightarrow (j-m+1)\, D_{j,m} = (j+m)\, D_{j,m-1}, \tag{14.30}$$

$$\Rightarrow D_{j,m} = (j+m)!\,(j-m)!\, D_{j,j}/(2j)!, \tag{14.31}$$

$$C_{j,0,j} = 1 \Rightarrow D_{j,j} = j!; \qquad (2j)!/j! = 2^j (2j-1)!!, \tag{14.32}$$

where $(2j-1)!! = 1 \cdot 3 \cdot 5 \cdots (2j-1)$. •

Collecting the previous results, we write the expression for the rank-k polynomials belonging to irreducible spin-j multiplets, of weight m, as

$$Y_{k,j,m}(\mathbf{p},\mathbf{q}) = (\mathbf{p} \times \mathbf{q})^{k-j}\, Y_{j,j,m}(\mathbf{p},\mathbf{q}), \tag{14.33}$$

where $k-j$ is even in axis-symmetric systems, and

$$Y_{j,j,m}(\mathbf{p},\mathbf{q}) = \frac{(j+m)!\,(j-m)!}{2^j\,(2j-1)!!} \sum_{\nu \in N(j,m)} 2^{\nu} \frac{|\mathbf{p}|^{j+m-\nu}}{[\frac{1}{2}(j+m-\nu)]!} \frac{(\mathbf{p}\cdot\mathbf{q})^{\nu}}{\nu!} \frac{|\mathbf{q}|^{j-m-\nu}}{[\frac{1}{2}(j-m-\nu)]!}. \tag{14.34}$$

These polynomials form the *symplectic harmonic* basis for analytic functions of phase space, whose rank-k component in (14.13) is now written as

$$A_k(\mathbf{p},\mathbf{q}) := \sum_{j\in J(k)} \sum_{m=-j}^{j} A_{k,j,m}\, Y_{k,j,m}(\mathbf{p},\mathbf{q}), \tag{14.35}$$

where the coefficients $\{A_{k,j,m}\}$ for $2 \le k \le K$ are the *harmonic* Hamilton–Lie aberration coefficients. Under the paraxial part $\mathbf{M} \in \mathsf{Sp}(2,\mathsf{R})$ of an axis-symmetric optical system $\mathcal{G}_K(\mathbf{A};\mathbf{M})$, in particular its adjoint action for the product in (14.16), the column vectors $\mathbf{A}_{k,j} = (A_{k,j,m})$ formed with the harmonic aberration coefficients will transform through

$$\mathcal{M}(\mathbf{M}) : \mathbf{A}_{k,j} = \mathbf{D}^j(\mathbf{M}^{-1})^\top \mathbf{A}_{k,j} \tag{14.36}$$

[cf. (14.17)], with the $\mathsf{Sp}(2,\mathsf{R})$ irreducible representation matrices $\mathbf{D}^j(\mathbf{M})$ given in (13.6). The quintuplet of rank-2 aberrations will transform through the 5×5 matrix given in (13.16).

Remark: Solid spherical harmonics. We called the basis of polynomials $Y_{k,j,m}(\mathbf{p},\mathbf{q})$ in (14.33)–(14.34) *harmonic*, because they are closely related with the *solid spherical harmonics*, $\mathcal{Y}_{j,m}(\vec{x})$, $\vec{x} \in \mathsf{R}^3$. In fact [21, Sect. 3.10],

$$\begin{aligned}\mathcal{Y}_{j,m}(\vec{x}) &:= \sqrt{\frac{(2j+1)\,(j+m)!\,(j-m)!}{2^m\, 4\pi}} \\ &\quad\times \sum_n \frac{1}{2^n} \frac{x_+^{m+n}}{(m+n)!} \frac{x_0^{j-m-2n}}{(j-m-2n)!} \frac{x_-^n}{n!} \\ &= \sqrt{\frac{(2j-1)!!}{4\pi(2j+1)\,(j+m)!\,(j-m)!}}\, Y_{j,j,m}(\mathbf{p},\mathbf{q}),\end{aligned} \tag{14.37}$$

$$\text{with } x_+ = \tfrac{1}{\sqrt{2}}|\mathbf{p}|^2, \quad x_0 = \mathbf{p}\cdot\mathbf{q}, \quad x_- = \tfrac{1}{\sqrt{2}}|\mathbf{q}|^2. \tag{14.38}$$

The solid spherical harmonics are solutions to the Laplace equation, $\nabla^2 \mathcal{Y}_{j,m}(\vec{x}) = 0$ of homogeneous degree j; they can be written as $|\vec{x}\,|^j \times Y_{j,m}(\theta,\phi)$, with the ordinary spherical harmonics $Y_{j,m}(\theta,\phi) \sim P_j^{|m|}(\cos\theta)e^{im\phi}$, where P_j^m is the associated Legendre polynomial. When we pass to R^3-Cartesian coordinates, $x_0 = x_3$, $x_\pm = \mp\frac{1}{\sqrt{2}}(x_1 \pm i x_2)$, we see the invariant sphere

$$|\vec{x}\,|^2 = x_1^2 + x_2^2 + x_3^2 = x_0^2 - 2x_+x_- = (\mathbf{p}\cdot\mathbf{q})^2 - |\mathbf{p}|^2|\mathbf{q}|^2 = -(\mathbf{p}\times\mathbf{q})^2. \tag{14.39}$$

The Weyl trick (see Sect. 11.3) relates $\mathsf{SO}(3)$-rotations of the sphere and $\mathsf{Sp}(2,\mathsf{R})$-canonical transformations of phase space, with the pure imaginary $x_2 = i\frac{1}{2}(|\mathbf{p}|^2 + |\mathbf{q}|^2)$ and the negative square radius of constant ray skewness. •

Remark: Inverse expansion of monomials in harmonics. The expansion inverse to (14.34), namely of a Cartesian-basis monomial M_{k_+,k_0,k_-} as sum of symplectic harmonics $Y_{k,j,m}$ over $j \in J'(k,m) := \{k,\, k-2,\, \ldots\, |m|\}$, is [151]

$$M_{k_+,k_0,k_-} = \sum_{j\in J'(k,m)} L^{k,j,m}_{k_+,k_0,k_-}\, Y_{k,j,m}, \quad \begin{cases} k = k_+ + k_0 + k_-, \\ m = k_+ - k_-. \end{cases} \tag{14.40}$$

The coefficients are found from the three-term recurrence relation on ν,

$$\begin{aligned}(k-2\nu-m)(k-2\nu-m-1)\, L^{k,j,m}_{k_\sigma(\nu+1)} \\ + [2(k-2\nu-m)(2\nu+m+1) + 2\nu - (j-m)(j+m+1)]\, L^{k,j,m}_{k_\sigma(\nu)} \\ + 4\nu(\nu+m)\, L^{k,j,m}_{k_\sigma(\nu-1)} = 0.\end{aligned} \tag{14.41}$$

where $k_+(\nu) = \nu + m$, $k_0(\nu) = k - 2\nu - m$, $k_-(\nu) = \nu$; the recursion starts with

$$L^{k,j,m}_{m,2k-m,0} = \frac{(-\frac{1}{2})^{\frac{1}{2}(k-j)}}{[\frac{1}{2}(k-j)]!}\, \frac{(k-m)!}{(j-m)!}\, \frac{(2j+1)!!}{(k+j+1)!!}. \tag{14.42}$$

•

Remark: Fourier transform of harmonic aberrations. The Fourier transform $\mathcal{F}$ acts on the Cartesian basis of monomials through (14.4). In the explicit expression for the harmonic basis (14.33)–(14.34), the Fourier transform exchanges $|\mathbf{p}|$ and $|\mathbf{q}|$, inverts the sign of $\mathbf{p}\cdot\mathbf{q}$ (all of whose powers have parity $j-m$), and leaves the skewness $\mathbf{p}\times\mathbf{q}$ invariant. Thus it holds that

$$\mathcal{F} : Y_{k,j,m}(\mathbf{p},\mathbf{q}) = Y_{k,j,m}(\mathbf{q},-\mathbf{p}) = (-1)^{j-m} Y_{k,j,-m}(\mathbf{p},\mathbf{q}). \tag{14.43}$$

The result can be also proven abstractly from the action of $\mathcal{F} = \mathcal{M}\begin{pmatrix} 0 & 1 \\ -1 & 0 \end{pmatrix}$ on the symplectic spin multiplets through (13.8). •

Remark: Reflection of harmonic aberrations. The reflection $\mathcal{K}$ of an aberrating system by a mirror coincident with the standard screen involves the reflection of its aberrations. Reflection is represented by the paraxial matrix $\mathbf{K} = \begin{pmatrix} -1 & 0 \\ 0 & 1 \end{pmatrix}$; on the basis of symplectic harmonics it inverts the sign of each factor of $\mathbf{p}\cdot\mathbf{q}$ (the powers of $\mathbf{p}\times\mathbf{q}$ are even), so

$$\mathcal{K} : Y_{k,j,m}(\mathbf{p},\mathbf{q}) = Y_{k,j,m}(-\mathbf{p},\mathbf{q}) = (-1)^{j-m}\, Y_{k,j,m}(\mathbf{p},\mathbf{q}), \tag{14.44}$$

as follows from their form (14.24) and the range (14.27). •

Below we display the symplectic harmonic basis of up-to-seventh order aberrations, i.e., for ranks $k = 1, 2, 3, 4$, classified into multiplets of symplectic spin j, and weight m. We need only write explicitly the highest $j = k$ spin because all lower spins are repetitions times powers of $\mathbf{p}\times\mathbf{q}$. The negative-$m$ harmonics are given by their positive-m counterparts only exchanging $\mathbf{p}$'s and $\mathbf{q}$'s: $Y_{k,j,-m}(\mathbf{p},\mathbf{q}) = Y_{k,j,m}(\mathbf{q},\mathbf{p})$.

Rank $k = 1$: the $\mathsf{sp}(2,\mathsf{R})$ 3-vector of reduced phase space monomials is

$$Y_{1,1,1} = |\mathbf{p}|^2, \quad Y_{1,1,0} = \mathbf{p}\cdot\mathbf{q}, \quad Y_{1,1,-1} = |\mathbf{q}|^2. \tag{14.45}$$

Rank $k = 2$: the quintuplet and singlet polynomials which generate third-order aberrations, are

$$\begin{aligned} Y_{2,2,2} &= (|\mathbf{p}|^2)^2, \\ Y_{2,2,1} &= |\mathbf{p}|^2 \mathbf{p}\cdot\mathbf{q}, \\ Y_{2,2,0} &= \tfrac{1}{3}|\mathbf{p}|^2|\mathbf{q}|^2 + \tfrac{2}{3}(\mathbf{p}\cdot\mathbf{q})^2; \qquad Y_{2,0,0} = (\mathbf{p}\times\mathbf{q})^2. \end{aligned} \tag{14.46}$$

Rank $k = 3$: the septuplet and triplet of generators of fifth-order aberrations, are

$$\begin{aligned} Y_{3,3,3} &= (|\mathbf{p}|^2)^3, \\ Y_{3,3,2} &= (|\mathbf{p}|^2)^2 \mathbf{p}\cdot\mathbf{q}, \\ Y_{3,3,1} &= \tfrac{1}{5}(|\mathbf{p}|^2)^2|\mathbf{q}|^2 + \tfrac{4}{5}|\mathbf{p}|^2(\mathbf{p}\cdot\mathbf{q})^2, \qquad Y_{3,1,1} = (\mathbf{p}\times\mathbf{q})^2\, Y_{1,1,1}, \\ Y_{3,3,0} &= \tfrac{3}{5}|\mathbf{p}|^2\mathbf{p}\cdot\mathbf{q}|\mathbf{q}|^2 + \tfrac{2}{5}(\mathbf{p}\cdot\mathbf{q})^3; \qquad Y_{3,1,0} = (\mathbf{p}\times\mathbf{q})^2\, Y_{1,1,0}. \end{aligned} \tag{14.47}$$

Rank $k = 4$: the nonuplet, quintuplet and singlet of seventh-order aberrations, are

$$\begin{aligned} Y_{4,4,4} &= (|\mathbf{p}|^2)^4, \qquad Y_{4,2,2} = (\mathbf{p}\times\mathbf{q})^2\, Y_{2,2,2}, \\ Y_{4,4,3} &= (|\mathbf{p}|^2)^3 \mathbf{p}\cdot\mathbf{q}, \qquad Y_{4,2,1} = (\mathbf{p}\times\mathbf{q})^2\, Y_{2,2,1}, \\ Y_{4,4,2} &= \tfrac{1}{7}(|\mathbf{p}|^2)^3|\mathbf{q}|^2 + \tfrac{6}{7}(|\mathbf{p}|^2)^2(\mathbf{p}\cdot\mathbf{q})^2, \qquad Y_{4,2,0} = (\mathbf{p}\times\mathbf{q})^2\, Y_{2,2,0}; \\ Y_{4,4,1} &= \tfrac{3}{7}(|\mathbf{p}|^2)^2\mathbf{p}\cdot\mathbf{q}|\mathbf{q}|^2 + \tfrac{4}{7}|\mathbf{p}|^2(\mathbf{p}\cdot\mathbf{q})^3, \\ Y_{4,4,0} &= \tfrac{3}{35}(|\mathbf{p}|^2)^2(|\mathbf{q}|^2)^2 + \tfrac{24}{35}|\mathbf{p}|^2(\mathbf{p}\cdot\mathbf{q})^2|\mathbf{q}|^2 + \tfrac{8}{35}(\mathbf{p}\cdot\mathbf{q})^4; \; Y_{4,0,0} = (\mathbf{p}\times\mathbf{q})^4. \end{aligned} \tag{14.48}$$

Remark: On computational efficiency. The solid spherical harmonics $\mathcal{Y}_{j,m}$ in (14.37) are an orthonormal set of functions on the unit sphere, and this introduces square-root coefficients which are uncomfortable for hand or machine computation. We have chosen the highest-weight symplectic harmonics to have coefficient 1, and this leads to polynomials the sum of whose monomial coefficients is always unity (check above!), because every lowering step (14.7) that duplicates the number of monomials, does so with coefficients which sum to $4k_+ + 2k_0 = 2(m+j)$, and this is the coefficient in front of the next lower harmonic in (13.23) and (14.25). This feature provides a good check for computation. And also, when we concatenate optical subsystems, the adjoint action of the paraxial part on the harmonic aberration basis (14.36), is block-diagonal. This means that for aberration orders $k = 2, 3, 4$, the monomial basis requires products with $6^2 = 36$, $10^2 = 100$, $15^2 = 225$ matrix elements ($\sim \frac{1}{4}k^4$ for large k), while in the harmonic basis the number of active elements is $5^2 + 1^2 = 26$, $7^2 + 3^2 = 58$, $9^2 + 5^2 + 1^2 = 107$ ($\sim \frac{2}{3}k^3$) – 2:1 in favor of group theory for aberration order seven. Also under the *gato* operation of aberration composition, the harmonic basis provides shorter formulas than the monomial basis, as we shall see below in Sect. 14.5. •

14.3 Harmonic aberration family features

The classification of axis-symmetric Hamilton–Lie aberrations by families of rank, symplectic spin, and weight, leads to a corresponding organization of their common features. Although actual optical setups cannot produce isolated, pure harmonic (or monomial) aberrations, knowing their distinctive signatures unravels the nonlinear maps of rays in phase space into components having specific symmetries and covariances. In Fig. 14.4 we place the (k, j, m)-aberrations in a pattern which is the same as the familiar arrangement of the states of a 3-dim isotropic quantum harmonic oscillator with energy $(k + \frac{1}{2})\hbar$, orbital angular momentum j and its projection m. In this section we shall examine the pure aberrations as we did in (14.10) and (14.15), namely

$$\begin{aligned} \begin{pmatrix} \mathbf{p}'(\mathbf{p}, \mathbf{q}; \alpha) \\ \mathbf{q}'(\mathbf{p}, \mathbf{q}; \alpha) \end{pmatrix} &= \mathcal{G}_k(\alpha Y_{k,j,m}) : \begin{pmatrix} \mathbf{p} \\ \mathbf{q} \end{pmatrix} \approx (1 + \alpha\{Y_{k,j,m}, \circ\}) \begin{pmatrix} \mathbf{p} \\ \mathbf{q} \end{pmatrix} \\ &= \begin{pmatrix} \mathbf{p} \\ \mathbf{q} \end{pmatrix} + \alpha \begin{pmatrix} \partial/\partial\mathbf{q} \\ -\partial/\partial\mathbf{p} \end{pmatrix} Y_{k,j,m}(\mathbf{p}, \mathbf{q}), \end{aligned} \tag{14.49}$$

will yield the spot diagram (cf. the spots in Figs. 14.2 and 14.3).

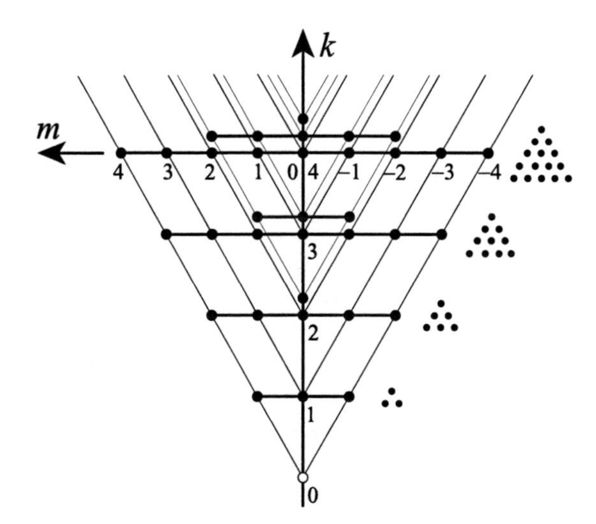

Fig. 14.4. The pattern of axis-symmetric harmonic aberrations (k, j, m) coincides with the 3-dim quantum oscillator states classified by energy (rank k), and angular momentum projection (m); thick horizontal lines indicate multiplets of symplectic spin $j = k, k-2, \ldots, 1$ or 0. The Fourier transform reflects the pattern across the vertical axis (with signs). Full V-lines join the families of primary spherical aberrations (k, k, k) and comas $(k, k, k-1)$, and their Fourier conjugate distorsions $(k, k, -k+1)$ and pocuses $(k, k, -k)$; secondary repeaters $(k, k-2, m)$ are also marked by V-lines. On the right, the corresponding Cartesian pattern of aberrations.

First we note in Fig. 14.4 that there is a *primary* aberration set $(k, j{=}k, m)$ with repetitions that occur as we displace this set along k each time by two units, $(k,\ j{=}k{-}2,\ m)$, $(k,\ j{=}k{-}4,\ m)$, etc. These *repeater* aberrations are related by (14.49) to the primary aberrations of phase space through

$$\{Y_{k,j,m}, \circ\}\begin{pmatrix}\mathbf{p}\\ \mathbf{q}\end{pmatrix} = (k-j)\,Y_{k-1,j,m}(\mathbf{p},\mathbf{q})\begin{pmatrix}\left(\begin{smallmatrix}0 & -1\\ 1 & 0\end{smallmatrix}\right)\mathbf{p}\\ \left(\begin{smallmatrix}0 & -1\\ 1 & 0\end{smallmatrix}\right)\mathbf{q}\end{pmatrix} + (\mathbf{p}\times\mathbf{q})^{k-j}\begin{pmatrix}\partial/\partial\mathbf{q}\\ -\partial/\partial\mathbf{p}\end{pmatrix} Y_{j,j,m}(\mathbf{p},\mathbf{q}). \tag{14.50}$$

We find the action of the primary aberrations $Y_{j,m} := Y_{j,j,m}$ on phase space from their explicit expression (14.34),

$$\{Y_{j,m}, \circ\}\begin{pmatrix}\mathbf{p}\\ \mathbf{q}\end{pmatrix} = \frac{1}{2j-1}\begin{pmatrix}(j{+}m)(j{-}m)Y_{j-1,m}(\mathbf{p},\mathbf{q}) & (j{-}m)(j{-}m{-}1)Y_{j-1,m+1}(\mathbf{p},\mathbf{q})\\ -(j{+}m)(j{+}m{-}1)Y_{j-1,m-1}(\mathbf{p},\mathbf{q}) & -(j{+}m)(j{-}m)Y_{j-1,m}(\mathbf{p},\mathbf{q})\end{pmatrix}\begin{pmatrix}\mathbf{p}\\ \mathbf{q}\end{pmatrix}. \tag{14.51}$$

Next, we note that aberrations have *parity.* When at a fixed position $\mathbf{q}$ on the screen we invert the direction of the ray, $\mathbf{p} \leftrightarrow -\mathbf{p}$, so $\angle(\mathbf{p},\mathbf{q}) =: \theta \leftrightarrow \pi + \theta$, then the corresponding symplectic harmonics and aberration spots undergo

$$Y_{k,j,m}(-\mathbf{p},\mathbf{q}) = (-1)^{j+m}\,Y_{k,j,m}(\mathbf{p},\mathbf{q}) \quad \Rightarrow \tag{14.52}$$

$$\mathbf{q}'(-\mathbf{p},\mathbf{q};\alpha) - \mathbf{q} = (-1)^{j+m+1}(\mathbf{q}'(\mathbf{p},\mathbf{q};\alpha) - \mathbf{q}), \tag{14.53}$$

as can be ascertained from (14.33)–(14.34), (14.49), and (14.51). This means that when $j+m$ is even (such as spherical aberrations $m=j$), the two rays $\pm\mathbf{p}$ will fall on points that are diametrically opposite on the spot. In the case when $j+m$ is odd (such as comas $m=j-1$), the two rays will fall on the *same* point of the screen, so the θ-circle of ray directions maps *twice* on the closed curve of the spot parametrized by θ. (The same equations tell us that the opposite parity is observed by $\mathbf{p}'(\mathbf{p},\mathbf{q};\alpha) - \mathbf{p}$; no degeneracy occurs in R^4 phase space.) Thus we divide all aberrations into *even* and *odd* according to the parity of $j-m$; these lie on alternate V-lines in Fig. 14.4.

The spots of axis-symmetric aberrations are invariant under rotations around the optical center, so we can place the object source of rays on the x-axis of the figures, and let the ray directions roam over a polar grid of circles and radii in momentum space, as in Fig. 3.4, with coordinates

$$\mathbf{q} = q\begin{pmatrix}1\\ 0\end{pmatrix}, \quad \mathbf{p} = p\begin{pmatrix}\cos\theta\\ \sin\theta\end{pmatrix} \quad\Rightarrow\quad \begin{cases}|\mathbf{p}|^2 = p^2, & \mathbf{p}\cdot\mathbf{q} = pq\cos\theta,\\ |\mathbf{q}|^2 = q^2, & \mathbf{p}\times\mathbf{q} = -pq\sin\theta,\end{cases} \tag{14.54}$$

$$\Rightarrow Y_{k,j,m}(\mathbf{p},\mathbf{q}) \sim p^{k+m}\,q^{k-m}\,(\sin\theta)^{k-j}\,P_j^{|m|}(\cos\theta), \tag{14.55}$$

where $k-j \geq 0$ is even, and the associate Legendre polynomial has $j-|m|$ zeros for $\theta \in (0,\pi)$. Below, in Figs. 14.5–14.11, we plot the spots associated to the axis-symmetric aberrations up to rank $k = 4$ (7th aberration order). The aberration spot in (14.49) is a function of the distance $q \geq 0$ of the position of the ray on the screen to the optical center, of the initial ray direction $p \geq 0$, and of the relative angle $\theta \in \mathcal{S}^1$ between the vectors $\mathbf{q}, \mathbf{p} \in \mathsf{R}^2$. Also, it is directly proportional to the subgroup parameter α, so we can choose $\alpha = 1$ and plot the *spot function*

$$\mathbf{s}_{k,j,m}(q,p,\theta) := \mathcal{G}_k(Y_{k,j,m}) : \mathbf{q} - \mathbf{q} \;=\; \{Y_{k,j,m}, \mathbf{q}\}. \tag{14.56}$$

Finally, to draw the figures uniformly we have chosen $q = 1$ for the center or apex of the spot, $p = 1$ for the (unphysical) angle of the outermost cone of rays, and divided by a factor of $2k$ to bring the spots to comparable sizes.

Spherical aberrations (k, k, k)**.** Members of the spherical aberration family have appeared very often in this text [see (3.36), (3.49), (5.9), (13.58), and (13.64)]. They are trivially exponentiated because the Lie series is truncated after the first term in the group parameter. The spot diagrams of the primary spherical aberrations are

$$\mathbf{s}_{k,k,k}(p,q,\theta) = \{(|\mathbf{p}|^2)^k, \mathbf{q}\} = -2k(|\mathbf{p}|^2)^{k-1}\mathbf{p} = -2k\,p^{2k-1}\begin{pmatrix}\cos\theta\\ \sin\theta\end{pmatrix}. \tag{14.57}$$

and are shown in Fig. 14.5 for aberration orders 3,5,7 (ranks $k = 2, 3, 4$). For small angles (say, $p \approx 0.1$) the spots become very small ($|\mathbf{s}| \approx 2k \times 10^{-k}$); so they are plotted with the considerations that end the previous paragraph.

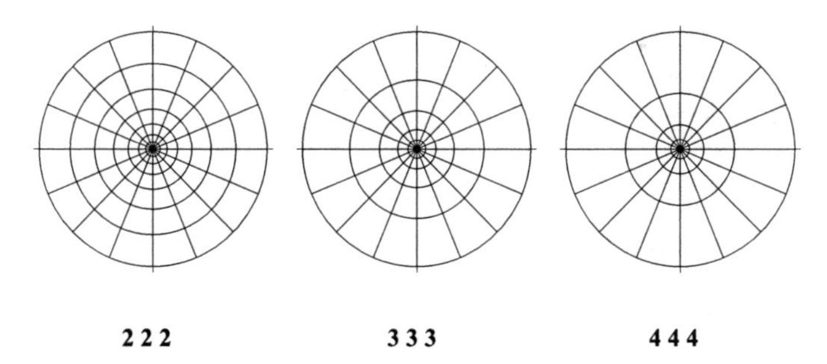

Fig. 14.5. Primary spherical aberration spots, for aberration orders 3,5,7 (ranks $k = 2, 3, 4$). The spots display concentric circles of radii $2kp_j^{2k-1}$, for $p_j \in \{0, 0.2, \ldots, 1.0\}$. We have divided by $2k$ to have uniformity of size.

Comas $(k, k, k-1)$**.** The comatic family of aberrations has been prominent in several topics involving distorsions of ray directions [see (3.23), (3.59), (5.36)–(5.37), (5.55) and (5.61), (5.64)–(5.67), and (13.58)]. The spots of the primary rank-k comas result from

$$\begin{aligned} \mathbf{s}_{k,k,k-1}(p,q,\theta) &= \{(|\mathbf{p}|^2)^{k-1}\mathbf{p}\cdot\mathbf{q},\mathbf{q}\} \\ &= -2(k-1)(|\mathbf{p}|^2)^{k-2}\mathbf{p}\cdot\mathbf{q}\,\mathbf{p} - (|\mathbf{p}|^2)^{k-1}\mathbf{q}) \\ &= -p^{2k-2}q\left[k\begin{pmatrix}1\\0\end{pmatrix} + (k-1)\begin{pmatrix}\cos 2\theta\\ \sin 2\theta\end{pmatrix}\right], \end{aligned} \tag{14.58}$$

and appear in Fig. 14.6 for $k = 2, 3, 4$. Comas are odd aberrations with linear field dependence, and a 2:1 map of circles that fit into a sector bounded by caustics on the screen whose opening angle is $2\arcsin(1 - 1/k)$ towards the center. Third-, fifth-, and seventh-order primary comas thus open at angles of 60°, 83.6208°, and 97.1808°. Finally, we note that the exponential series of the primary coma aberration, $\exp\{\alpha Y_{k,k,k-1}, \circ\}$, can be summed in closed form, with the result that the edges of the sectors will bend inwards [154].

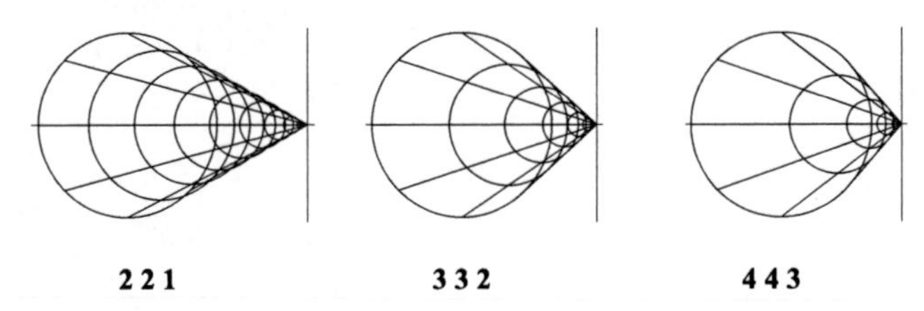

Fig. 14.6. Primary comatic aberration spots, for aberration orders 3,5,7 (ranks $k = 2, 3, 4$). The circles are the images of rays belonging to nested cones (in the paraxial model), opening with 'angles' $p_j \in \{0, 0.2, \ldots, 1.0\}$. We divide by $2k$ as in the previous figure to show the relative sizes.

Distorsions $(k, k, 1-k)$ **and pocus** $(k, k, -k)$**.** Distorsions are aberrations Fourier conjugate to comas. Their spots are

$$\mathbf{s}_{k,k,1-k}(p,q,\theta) = \{\mathbf{p}\cdot\mathbf{q}(|\mathbf{q}|^2)^{k-1},\mathbf{q}\} = -(k-1)\,q^{2k-2}\mathbf{q}, \tag{14.59}$$

i.e., field-dependent point-to-point maps of the positions of rays on the screen, which magnify the image in the $\mathbf{q}$-plane, with characteristic field dependences $\sim q^{2k-1}$ (linear for $k = 1$, cubic for third order, etc.). The Fourier conjugate of spherical aberration is pocus, with $\mathbf{s}_{k,k,-k}(p,q,\theta) = \mathbf{0}$; there is no change in the ray positions. In both cases the exponential Lie series can be summed in closed form.

The degenerate family $(k, k, k-2) \,/\, (k, k-2, k-2)$**.** Curvature of field and astigmatism in third aberration order both fall on the apex of the third V-line in the scheme of Fig. 14.4. The left half of this V-line is $m = k - 2$, a family consisting of degenerate pairs that we now investigate. Each pair consists of one primary $j = k$ aberration, and the first repeater $j = k - 2$ of spherical aberration. We find their spot diagrams from the explicit form of the symplectic harmonics in (14.33) and (14.34), as we did for the previous aberrations. For the primary aberration $m = j - 2$ which is member of the maximal spin-$j = k$ multiplet, the symplectic harmonic polynomial and spot function are:

$$Y_{k,k,k-2}(\mathbf{p},\mathbf{q}) = \frac{1}{2k-1}(|\mathbf{p}|^2)^{k-1}|\mathbf{q}|^2 + \frac{2k-2}{2k-1}(|\mathbf{p}|^2)^{k-2}(\mathbf{p}\cdot\mathbf{q})^2, \quad (14.60)$$

$$\begin{aligned} \mathbf{s}_{k,k,k-2}(p,q,\theta) &= -\frac{2k-2}{2k-1}p^{2k-3}q^2 \\ &\times \left[\begin{pmatrix} 3\cos\theta \\ \sin\theta \end{pmatrix} + 2(k-2)\cos^2\theta \begin{pmatrix} \cos\theta \\ \sin\theta \end{pmatrix}\right]. \end{aligned} \quad (14.61)$$

In Fig. 14.7 we show these aberrations at ranks $k = 2, 3, 4$. The first summand of the spot (14.61) is a 'horizontal' ellipse of axis ratio 3 : 1, and is the only term in the third-order aberration; in higher orders, one adds to this basic ellipse a summand with the shape of a dumbbell: the θ-circle expanded by a $\cos^2\theta$ factor.

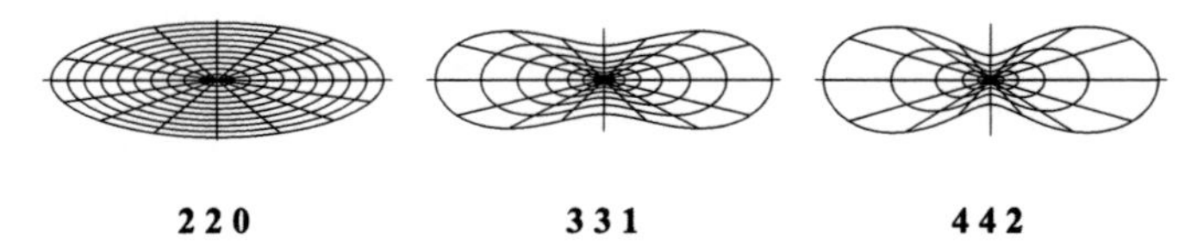

Fig. 14.7. Spots of the primary harmonic aberration family $(k, k, k-2)$, for ranks $k = 2, 3, 4$ (aberration orders 3,5,7).

The first repeater family of spherical aberrations $m = j$, $(\mathbf{p}\times\mathbf{q})^2(|\mathbf{p}|^2)^{k-2}$, belongs to the multiplet with spin $j = k-2$, and has

$$Y_{k,k-2,k-2}(\mathbf{p},\mathbf{q}) = (|\mathbf{p}|^2)^{k-1}|\mathbf{q}|^2 - (|\mathbf{p}|^2)^{k-2}(\mathbf{p}\cdot\mathbf{q})^2, \quad (14.62)$$

$$\begin{aligned} \mathbf{s}_{k,k-2,k-2}(p,q,\theta) &= -2\,p^{2k-3}q^2 \\ &\times \left[\begin{pmatrix} (k-2)\cos\theta \\ (k-1)\sin\theta \end{pmatrix} - (k-2)\cos^2\theta \begin{pmatrix} \cos\theta \\ \sin\theta \end{pmatrix}\right]. \end{aligned} \quad (14.63)$$

It is shown in Fig. 14.8 for ranks $k = 2, 3, 4$. Now there is a basic 'vertical' ellipse of k-dependent axis ratio $(k-2) : (k-1)$, which again is the only term in third order ($k = 2$), where it degenerates into a vertical line (cf. Fig. 14.3). In higher orders, the second summand is the same dumbbell as in (14.61), but now the x-coordinate of the two summands cancels for $\theta = 0, \pi$, so the spot is pinched at its waist and has a figure-8 shape.

The degenerate family $(k, k, 2-k) / (k, k-2, 2-k)$**.** Curvature of field and astigmatism in third aberration order *also* fall on the line $m = 2-k$ of the scheme in Fig. 14.4; on the *right half* of the third V-line. The pairs now consist of one primary aberration, and the first repeater of pocus. They are the Fourier transform $m \leftrightarrow -m$ of the degenerate family $(k, k, k-2) / (k, k-2, k-2)$ that we visited above, with even parity [recall (14.43)]. Thus we find

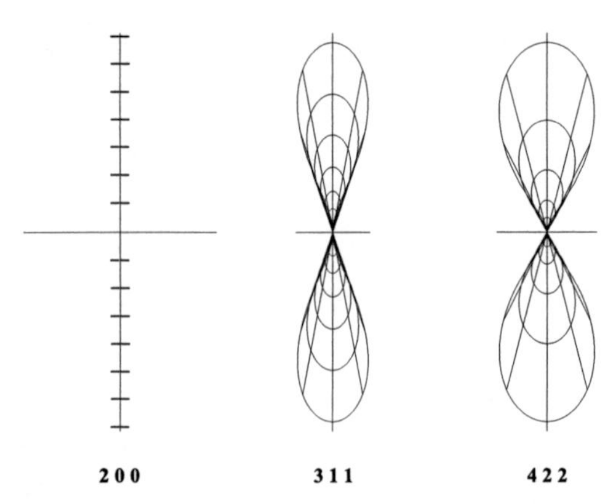

Fig. 14.8. Spots of the repeater spherical aberrations $(k, k-2, k-2)$, for ranks $k = 2, 3, 4$.

the aberration Y-polynomials from (14.60) and (14.62) simply exchanging $\mathbf{p}$ and $\mathbf{q}$; their spots can be calculated as in (14.61) and (14.63); they turn out to be

$$\mathbf{s}_{k,k,2-k}(p, q, \theta) = -2pq^{2k-1} \begin{pmatrix} (2k-1)\cos\theta \\ \sin\theta \end{pmatrix}, \tag{14.64}$$

$$\mathbf{s}_{k,k-2,2-k}(p, q, \theta) = -2pq^{2k-1} \begin{pmatrix} 0 \\ \sin\theta \end{pmatrix}. \tag{14.65}$$

The spots of the primary aberrations are 'horizontal' ellipses of axis ratios $(2k-1) : 1$, while the repeaters of pocus are lines in the y-direction, shown in Fig. 14.9. When the full Lie exponential series of the latter is considered, these lines are the first-term approximation to circle arcs which bend around the optical center, as in Fig. 14.3. Comparing with the previous degenerate-pair family, the spots in Fig. 14.9 represent the aberrations of momentum space which are Fourier-conjugate to those of Fig. 14.8, and *viceversa* with a minus sign.

The fourth V-line of aberrations in Fig. 14.4 includes primary aberrations $(k, k, \pm(k-3))$ and repeaters $(k, k-2, \pm(k-3))$; they are shown (for $k = 3, 4$) in Fig. 14.10, and have the generic shapes of comas. In the primaries, the closed curves resemble ellipses with displacing centers, while in the repeaters the curves are tangent to a common vertical caustic line. The apex of the fifth V-line – the last one shown in the pattern of Fig. 14.4 – is threefold degenerate with $(4, j, 0)$, $j = 4, 2, 0$; their spot diagrams are shown in Figures 14.11. We have thus displayed all up-to-seventh-order aberration spots, except for the 3 distorsions and 3 pocuses which have trivial spot diagrams, along family lines with common geometric features.

We recall from Sect. 3.5 that in every case there will be two invariants in phase space $(\mathbf{p}, \mathbf{q}) \in \mathsf{R}^4$. Because the system is axis-symmetric, all rays

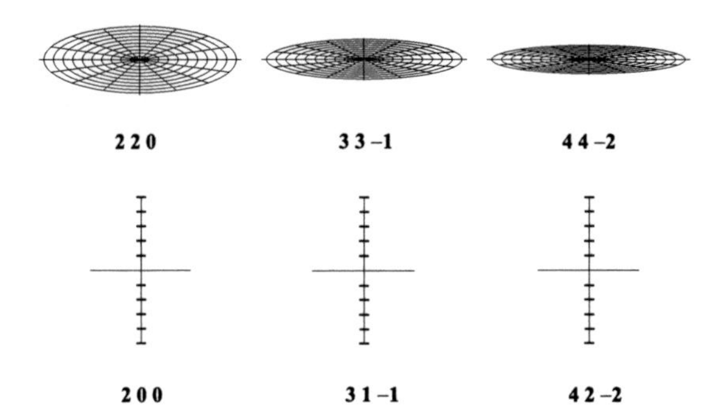

Fig. 14.9. *Top*: Spots of the primary harmonic aberrations $(k, k, 2-k)$, for ranks $k = 2, 3, 4$. *Bottom*: Linear spot (which is the first approximation to a circle) for all repeater aberrations of pocus $(k, k-2, 2-k)$.

of equal skewness $\mathbf{p} \times \mathbf{q}$ = constant form surfaces whose points are mapped amongst themselves [and this reduces phase space by one ignorable coordinate to (p, q, θ) – see Chap. 7]; and, because Hamilton–Lie aberrations have specific generator polynomials, $Y_{k,j,m}(p, q, \theta)$ = constant defines conserved Hamiltonian lines of flow on those surfaces, with the explicit form in (14.55). Further, we know that under the paraxial linear transformations, spots will also enter in linear combination (through vector addition) only with those of the same rank k, and within this set, only with its $2j + 1$ partners in the same symplectic spin-j multiplet.

As we proceed with V-lines of Hamilton–Lie aberrations with higher degeneracy, the geometric analysis of their spots becomes more arduous. As we saw above, their generic shapes are first organized into even and odd aberrations; the closed curves in even ones are centered, while those in odd ones form comet-shaped spots bordered by caustics. Let us characterize them with *ad hoc* symbols. Among the even aberrations there are circles ⊙ in the line of spherical aberrations $m = k$, and dots • in their Fourier-conjugates $m = -k$; then there are degenerate dumbbell-shapes ⋈ and figure-eights 8 in the lines $m = k-2$, $k-4$, and degenerate ellipses ⬭ and vertical lines | for $m = 2-k$, $4-k$. The odd aberrations on the other hand display circular comas ⊳ for $m = k-1$, and displaced dots in a distorted field ⏜ for their Fourier conjugates $m = 1-k$; and then generic comet shapes ⬭⊳, with ellipses of moving center, and ▷ with a common point of tangency. They can be tabulated as follows:

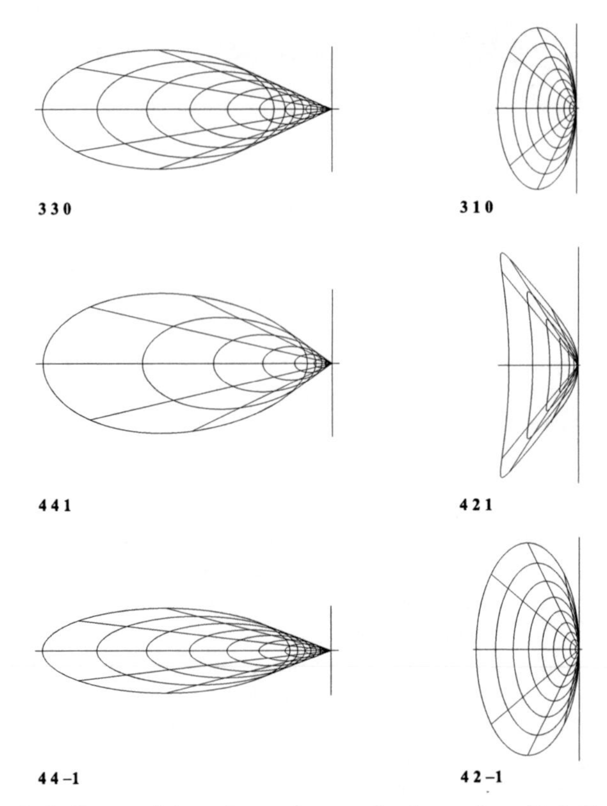

Fig. 14.10. *Left*: Spots of the primary harmonic aberrations $(3, 3, 0)$, $(4, 4, 1)$ and $(4, 4, -1)$. *Right*: Spots of the repeater aberrations $(3, 1, 0)$, $(4, 2, 1)$ and $(4, 2, -1)$.

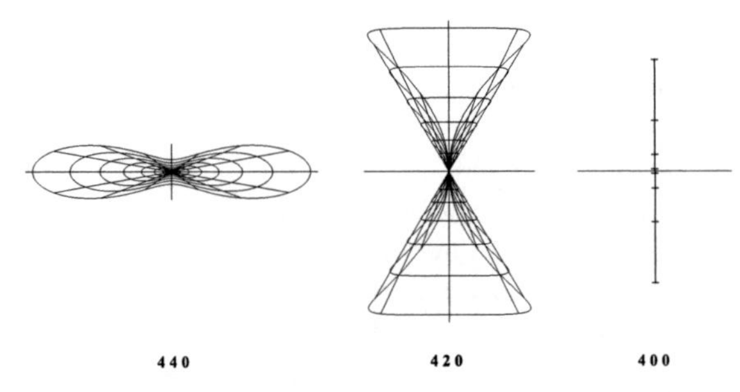

Fig. 14.11. Spots of the three degenerate aberrations $(4, 4, 0)$, $(4, 2, 0)$ and $(4, 0, 0)$.

RANK:	$k=2$		$k=3$		$k=4$		
SPIN:	$j=2,$	0	$j=3,$	1	$j=4,$	$2,$	0
$m=4$					⊙		
3			⊙		⊳		
2	⊙		⊳		⋈	8	
1	⊳		⋈	8	⬭▷	▷	
0	⬭	|	⬭▷	▷	⋈	8	|
−1	⌂		⬭	|	⬭▷	▷	
−2	•		⌂		⬭	|	
−3			•		⌂		
−4					•		
ORDER:	THIRD		FIFTH		SEVENTH		

14.4 Concatenation of aberrating systems

The elements of the rank-K axis-symmetric aberration group $\mathsf{A}_K^{\mathsf{AS}}\mathsf{Sp}(2,\mathsf{R})$ in the factored-product parameterization are a close generalization of the 1-dim case (13.27)–(13.30). There are again 3 coefficients $\{A_{1,1,m}\}_{m=-1}^{1}$ that determine the paraxial part $\mathsf{Sp}(2,\mathsf{R})$, and a classification of harmonic aberration coefficients $\{A_{k,j,m}\}$ expanded by symplectic spin $j \in \{k, k-2, \ldots 1 \text{ or } 0\}$, which determine the aberration polynomials $A_k(\mathbf{p},\mathbf{q})$ of ranks $2 \le k \le K$. As in (14.35), we have

$$\mathcal{G}_K(\mathbf{A};\mathbf{M}) := \mathcal{G}(\mathbf{A}_K,\ldots,\mathbf{A}_4,\mathbf{A}_3,\mathbf{A}_2;\mathbf{1})\,\mathcal{G}(\mathbf{0};\mathbf{M}(\mathbf{A}_1)), \tag{14.66}$$

$$\mathcal{G}(\mathbf{A};\mathbf{1}) := \exp\widehat{A}_K \times\cdots\times \exp\widehat{A}_4 \times \exp\widehat{A}_3 \times \exp\widehat{A}_2, \tag{14.67}$$

$$\widehat{A}_k := \textstyle\sum_{j\in J(k)}\sum_{m=-k}^{k} A_{k,j,m}\,\widehat{Y}_{k,j,m}, \tag{14.68}$$

$$\mathcal{G}(\mathbf{0},\mathbf{M}) := \exp\widehat{A}_1 = \mathcal{M}\left(\exp\begin{pmatrix} -A_{1,1,0} & -2A_{1,1,-1} \\ 2A_{1,1,1} & A_{1,1,0}\end{pmatrix}\right). \tag{14.69}$$

The concatenation of two or more axis-symmetric aberrating systems follows the product of their group elements, which is given by (13.31)–(13.33) and (14.16), *viz.*,

$$\mathcal{G}_K(\mathbf{A};\mathbf{M})\,\mathcal{G}_K(\mathbf{B};\mathbf{N}) = \mathcal{G}_K(\mathbf{A}\,\sharp\,\mathcal{M}(\mathbf{M}):\mathbf{B};\,\mathbf{MN}), \tag{14.70}$$

and the adjoint action of $\mathsf{Sp}(2,\mathsf{R})$ on the aberration coefficients is block-diagonal in j, as given by (14.36) [instead of (14.17)]. There only remains the specification of the *gato* product $\sharp$ between the harmonic aberration coefficients of $\mathbf{C} = \mathbf{A}\,\sharp\,\mathbf{B}$. This is found from (13.41)–(13.43) after replacing $\frac{3}{2} \mapsto 2$, $2 \mapsto 3$, $\frac{5}{2} \mapsto 4$ (i.e. aberration orders $2 \mapsto 3$, $3 \mapsto 5$, $4 \mapsto 7$). Machine calculation with the Poisson brackets using the monomials (14.8) yields the following results:

Aberration order 3:

$$C_{2,j,m} = A_{2,j,m} + B_{2,j,m}, \qquad j = 2, 0, \quad m = j, j-1, \ldots, -j. \qquad (14.71)$$

Aberration order 5:

$$\begin{aligned}
C_{3,3,3} &= 2A_{2,2,1}B_{2,2,2} - 2A_{2,2,2}B_{2,2,1} + A_{3,3,3} + B_{3,3,3},\\
C_{3,3,2} &= 4A_{2,2,0}B_{2,2,2} - 4A_{2,2,2}B_{2,2,0} + A_{3,3,2} + B_{3,3,2},\\
C_{3,3,1} &= 6A_{2,2,-1}B_{2,2,2} + 2A_{2,2,0}B_{2,2,1} - 2A_{2,2,1}B_{2,2,0} - 6A_{2,2,2}B_{2,2,-1} + A_{3,3,1} + B_{3,3,1},\\
C_{3,3,0} &= 8A_{2,2,-2}B_{2,2,2} + 4A_{2,2,-1}B_{2,2,1} - 4A_{2,2,1}B_{2,2,-1} - 8A_{2,2,2}B_{2,2,-2} + A_{3,3,0} +\\
&\quad B_{3,3,0},\\
C_{3,3,-1} &= 6A_{2,2,-2}B_{2,2,1} + 2A_{2,2,-1}B_{2,2,0} - 2A_{2,2,0}B_{2,2,-1} - 6A_{2,2,1}B_{2,2,-2} + A_{3,3,-1} +\\
&\quad B_{3,3,-1},\\
C_{3,3,-2} &= 4A_{2,2,-2}B_{2,2,0} - 4A_{2,2,0}B_{2,2,-2} + A_{3,3,-2} + B_{3,3,-2},\\
C_{3,3,-3} &= 2A_{2,2,-2}B_{2,2,-1} - 2A_{2,2,-1}B_{2,2,-2} + A_{3,3,-3} + B_{3,3,-3};
\end{aligned} \qquad (14.72)$$

$$\begin{aligned}
C_{3,1,1} &= 4/5\,A_{2,2,-1}B_{2,2,2} - 2/5\,A_{2,2,0}B_{2,2,1} + 2/5\,A_{2,2,1}B_{2,2,0} - 4/5\,A_{2,2,2}B_{2,2,-1} +\\
&\quad A_{3,1,1} + B_{3,1,1},\\
C_{3,1,0} &= 16/5\,A_{2,2,-2}B_{2,2,2} - 2/5\,A_{2,2,-1}B_{2,2,1} + 2/5\,A_{2,2,1}B_{2,2,-1} - 16/5\,A_{2,2,2}B_{2,2,-2} +\\
&\quad A_{3,1,0} + B_{3,1,0},\\
C_{3,1,-1} &= 4/5\,A_{2,2,-2}B_{2,2,1} - 2/5\,A_{2,2,-1}B_{2,2,0} + 2/5\,A_{2,2,0}B_{2,2,-1} - 4/5\,A_{2,2,1}B_{2,2,-2} +\\
&\quad A_{3,1,-1} + B_{3,1,-1}.
\end{aligned} \qquad (14.73)$$

Aberration order 7:

$$\begin{aligned}
C_{4,4,4} &= 8A_{2,2,1}^2B_{2,2,2} + 32/3\,A_{2,2,2}^2B_{2,2,0} - 16/3\,A_{2,2,0}B_{2,2,2}^2 - 32/3\,A_{2,2,0}A_{2,2,2}B_{2,2,2} -\\
&\quad 8A_{2,2,1}A_{2,2,2}B_{2,2,1} + 4A_{2,2,1}B_{2,2,1}B_{2,2,2} + 6A_{2,2,1}B_{3,3,3} - 4A_{2,2,2}B_{2,2,1}^2 + 16/3\,A_{2,2,2}\times\\
&\quad B_{2,2,0}B_{2,2,2} - 4A_{2,2,2}B_{3,3,2} + A_{4,4,4} + B_{4,4,4},\\
C_{4,4,3} &= 32A_{2,2,2}^2B_{2,2,-1} - 16A_{2,2,-1}B_{2,2,2}^2 - 32A_{2,2,-1}A_{2,2,2}B_{2,2,2} + 64/3\,A_{2,2,0}A_{2,2,1}\times\\
&\quad B_{2,2,2} - 80/3\,A_{2,2,0}A_{2,2,2}B_{2,2,1} - 8/3\,A_{2,2,0}B_{2,2,1}B_{2,2,2} + 12A_{2,2,0}B_{3,3,3} + 16/3\,A_{2,2,1}\times\\
&\quad A_{2,2,2}B_{2,2,0} + 40/3\,A_{2,2,1}B_{2,2,0}B_{2,2,2} + 2A_{2,2,1}B_{3,3,2} + 16A_{2,2,2}B_{2,2,-1}B_{2,2,2} - 32/3\,\times\\
&\quad A_{2,2,2}B_{2,2,0}B_{2,2,1} - 8A_{2,2,2}B_{3,3,1} + A_{4,4,3} + B_{4,4,3},\\
C_{4,4,2} &= 64/3\,A_{2,2,0}^2B_{2,2,2} + 8/3\,A_{2,2,1}^2B_{2,2,0} + 64A_{2,2,2}^2B_{2,2,-2} - 32A_{2,2,-2}B_{2,2,2}^2 - 64\times\\
&\quad A_{2,2,-2}A_{2,2,2}B_{2,2,2} + 16A_{2,2,-1}A_{2,2,1}B_{2,2,2} - 56A_{2,2,-1}A_{2,2,2}B_{2,2,1} - 20A_{2,2,-1}\times\\
&\quad B_{2,2,1}B_{2,2,2} + 18A_{2,2,-1}B_{3,3,3} - 4/3A_{2,2,0}B_{2,2,1}^2 - 8/3\,A_{2,2,0}A_{2,2,1}B_{2,2,1} - 64/3\,A_{2,2,0}\times\\
&\quad A_{2,2,2}B_{2,2,0} + 32/3\,A_{2,2,0}B_{2,2,0}B_{2,2,2} + 8A_{2,2,0}B_{3,3,2} + 40A_{2,2,1}A_{2,2,2}B_{2,2,-1} + 28\times\\
&\quad A_{2,2,1}B_{2,2,-1}B_{2,2,2} + 4/3\,A_{2,2,1}B_{2,2,0}B_{2,2,1} - 2A_{2,2,1}B_{3,3,1} - 32/3\,A_{2,2,2}B_{2,2,0}^2 + 32\times\\
&\quad A_{2,2,2}B_{2,2,-2}B_{2,2,2} - 8A_{2,2,2}B_{2,2,-1}B_{2,2,1} - 12A_{2,2,2}B_{3,3,0} + A_{4,4,2} + B_{4,4,2},\\
C_{4,4,1} &= 16/3\,A_{2,2,0}^2B_{2,2,1} + 16A_{2,2,1}^2B_{2,2,-1} - 96A_{2,2,-2}A_{2,2,2}B_{2,2,1} - 48A_{2,2,-2}B_{2,2,1}\times\\
&\quad B_{2,2,2} + 24A_{2,2,-2}B_{3,3,3} - 8A_{2,2,-1}B_{2,2,1}^2 + 160/3\,A_{2,2,-1}A_{2,2,0}B_{2,2,2} - 16A_{2,2,-1}\times\\
&\quad A_{2,2,1}B_{2,2,1} - 176/3\,A_{2,2,-1}A_{2,2,2}B_{2,2,0} - 8/3\,A_{2,2,-1}B_{2,2,0}B_{2,2,2} + 14A_{2,2,-1}B_{3,3,2} -\\
&\quad 16/3\,A_{2,2,0}A_{2,2,1}B_{2,2,0} + 16/3\,A_{2,2,0}A_{2,2,2}B_{2,2,-1} + 88/3\,A_{2,2,0}B_{2,2,-1}B_{2,2,2} + 8/3\,\times\\
&\quad A_{2,2,0}B_{2,2,0}B_{2,2,1} + 4A_{2,2,0}B_{3,3,1} - 8/3\,A_{2,2,1}B_{2,2,0}^2 + 96A_{2,2,1}A_{2,2,2}B_{2,2,-2} + 48\times\\
&\quad A_{2,2,1}B_{2,2,-2}B_{2,2,2} + 8A_{2,2,1}B_{2,2,-1}B_{2,2,1} - 6A_{2,2,1}B_{3,3,0} - 80/3\,A_{2,2,2}B_{2,2,-1}B_{2,2,0} -\\
&\quad 16A_{2,2,2}B_{3,3,-1} + A_{4,4,1} + B_{4,4,1},\\
C_{4,4,0} &= 40A_{2,2,-1}^2B_{2,2,2} + 40A_{2,2,1}^2B_{2,2,-2} - 20A_{2,2,-2}B_{2,2,1}^2 + 160/3\,A_{2,2,-2}A_{2,2,0}\times\\
&\quad B_{2,2,2} - 40A_{2,2,-2}A_{2,2,1}B_{2,2,1} - 320/3\,A_{2,2,-2}A_{2,2,2}B_{2,2,0} - 80/3\,A_{2,2,-2}B_{2,2,0}B_{2,2,2} +\\
&\quad 20A_{2,2,-2}B_{3,3,2} + 40/3\,A_{2,2,-1}A_{2,2,0}B_{2,2,1} - 80/3\,A_{2,2,-1}A_{2,2,1}B_{2,2,0} - 40A_{2,2,-1}\times\\
&\quad A_{2,2,2}B_{2,2,-1} + 20A_{2,2,-1}B_{2,2,-1}B_{2,2,2} - 20/3\,A_{2,2,-1}B_{2,2,0}B_{2,2,1} + 10A_{2,2,-1}B_{3,3,1} +
\end{aligned}$$

$$40/3\,A_{2,2,0}A_{2,2,1}B_{2,2,-1} + 160/3\,A_{2,2,0}A_{2,2,2}B_{2,2,-2} + 160/3\,A_{2,2,0}B_{2,2,-2}B_{2,2,2} + 40/3\,A_{2,2,0}B_{2,2,-1}B_{2,2,1} + 20A_{2,2,1}B_{2,2,-2}B_{2,2,1} - 20/3\,A_{2,2,1}B_{2,2,-1}B_{2,2,0} - 10 \times A_{2,2,1}B_{3,3,-1} - 20A_{2,2,2}B_{2,2,-1}^2 - 80/3\,A_{2,2,2}B_{2,2,-2}B_{2,2,0} - 20A_{2,2,2}B_{3,3,-2} + A_{4,4,0} + B_{4,4,0},$$

$$C_{4,4,-1} = 16A_{2,2,-1}^2B_{2,2,1} + 16/3\,A_{2,2,0}^2B_{2,2,-1} + 96A_{2,2,-2}A_{2,2,-1}B_{2,2,2} + 16/3\,A_{2,2,-2} \times A_{2,2,0}B_{2,2,1} - 176/3\,A_{2,2,-2}A_{2,2,1}B_{2,2,0} - 96A_{2,2,-2}A_{2,2,2}B_{2,2,-1} - 80/3\,A_{2,2,-2} \times B_{2,2,0}B_{2,2,1} + 16A_{2,2,-2}B_{3,3,1} - 8/3\,A_{2,2,-1}B_{2,2,0}^2 - 16/3\,A_{2,2,-1}A_{2,2,0}B_{2,2,0} - 16 \times A_{2,2,-1}A_{2,2,1}B_{2,2,-1} + 48A_{2,2,-1}B_{2,2,-2}B_{2,2,2} + 8A_{2,2,-1}B_{2,2,-1}B_{2,2,1} + 6A_{2,2,-1} \times B_{3,3,0} + 160/3\,A_{2,2,0}A_{2,2,1}B_{2,2,-2} + 88/3\,A_{2,2,0}B_{2,2,-2}B_{2,2,1} + 8/3\,A_{2,2,0}B_{2,2,-1} \times B_{2,2,0} - 4A_{2,2,0}B_{3,3,-1} - 8A_{2,2,1}B_{2,2,-1}^2 - 8/3\,A_{2,2,1}B_{2,2,-2}B_{2,2,0} - 14A_{2,2,1}B_{3,3,-2} - 48A_{2,2,2}B_{2,2,-2}B_{2,2,-1} - 24A_{2,2,2}B_{3,3,-3} + A_{4,4,-1} + B_{4,4,-1},$$

$$C_{4,4,-2} = 64A_{2,2,-2}^2B_{2,2,2} + 8/3\,A_{2,2,-1}^2B_{2,2,0} + 64/3\,A_{2,2,0}^2B_{2,2,-2} - 32/3\,A_{2,2,-2}B_{2,2,0}^2 + 40A_{2,2,-2}A_{2,2,-1}B_{2,2,1} - 64/3\,A_{2,2,-2}A_{2,2,0}B_{2,2,0} - 56A_{2,2,-2}A_{2,2,1}B_{2,2,-1} - 64 \times A_{2,2,-2}A_{2,2,2}B_{2,2,-2} + 32A_{2,2,-2}B_{2,2,-2}B_{2,2,2} - 8A_{2,2,-2}B_{2,2,-1}B_{2,2,1} + 12A_{2,2,-2} \times B_{3,3,0} - 8/3\,A_{2,2,-1}A_{2,2,0}B_{2,2,-1} + 16A_{2,2,-1}A_{2,2,1}B_{2,2,-2} + 28A_{2,2,-1}B_{2,2,-2}B_{2,2,1} + 4/3\,A_{2,2,-1}B_{2,2,-1}B_{2,2,0} + 2A_{2,2,-1}B_{3,3,-1} - 4/3\,A_{2,2,0}B_{2,2,-1}^2 + 32/3\,A_{2,2,0}B_{2,2,-2} \times B_{2,2,0} - 8A_{2,2,0}B_{3,3,-2} - 20A_{2,2,1}B_{2,2,-2}B_{2,2,-1} - 18A_{2,2,1}B_{3,3,-3} - 32A_{2,2,2}B_{2,2,-2}^2 + A_{4,4,-2} + B_{4,4,-2},$$

$$C_{4,4,-3} = 32A_{2,2,-2}^2B_{2,2,1} + 16/3\,A_{2,2,-2}A_{2,2,-1}B_{2,2,0} - 80/3\,A_{2,2,-2}A_{2,2,0}B_{2,2,-1} - 32A_{2,2,-2}A_{2,2,1}B_{2,2,-2} + 16A_{2,2,-2}B_{2,2,-2}B_{2,2,1} - 32/3\,A_{2,2,-2}B_{2,2,-1}B_{2,2,0} + 8 \times A_{2,2,-2}B_{3,3,-1} + 64/3\,A_{2,2,-1}A_{2,2,0}B_{2,2,-2} + 40/3\,A_{2,2,-1}B_{2,2,-2}B_{2,2,0} - 2A_{2,2,-1} \times B_{3,3,-2} - 8/3\,A_{2,2,0}B_{2,2,-2}B_{2,2,-1} - 12A_{2,2,0}B_{3,3,-3} - 16A_{2,2,1}B_{2,2,-2}^2 + A_{4,4,-3} + B_{4,4,-3},$$

$$C_{4,4,-4} = 32/3\,A_{2,2,-2}^2B_{2,2,0} + 8A_{2,2,-1}^2B_{2,2,-2} - 4A_{2,2,-2}B_{2,2,-1}^2 - 8A_{2,2,-2}A_{2,2,-1} \times B_{2,2,-1} - 32/3\,A_{2,2,-2}A_{2,2,0}B_{2,2,-2} + 16/3\,A_{2,2,-2}B_{2,2,-2}B_{2,2,0} + 4A_{2,2,-2}B_{3,3,-2} + 4A_{2,2,-1}B_{2,2,-2}B_{2,2,-1} - 6A_{2,2,-1}B_{3,3,-3} - 16/3\,A_{2,2,0}B_{2,2,-2}^2 + A_{4,4,-4} + B_{4,4,-4}; \tag{14.74}$$

$$C_{4,2,2} = -64/21\,A_{2,2,0}^2B_{2,2,2} - 8/21\,A_{2,2,1}^2B_{2,2,0} + 256/21\,A_{2,2,2}^2B_{2,2,-2} - 128/21 \times A_{2,2,-2}B_{2,2,2}^2 - 256/21\,A_{2,2,-2}A_{2,2,2}B_{2,2,2} + 176/21\,A_{2,2,-1}A_{2,2,1}B_{2,2,2} - 16/3 \times A_{2,2,-1}A_{2,2,2}B_{2,2,1} + 32/21\,A_{2,2,-1}B_{2,2,1}B_{2,2,2} + 24/7\,A_{2,2,-1}B_{3,3,3} + 4/21\,A_{2,2,0} \times B_{2,2,1}^2 + 8/21\,A_{2,2,0}A_{2,2,1}B_{2,2,1} + 64/21\,A_{2,2,0}A_{2,2,2}B_{2,2,0} - 32/21\,A_{2,2,0}B_{2,2,0}B_{2,2,2} - 8/7\,A_{2,2,0}B_{3,3,2} - 64/21\,A_{2,2,1}A_{2,2,2}B_{2,2,-1} + 8/3\,A_{2,2,1}B_{2,2,-1}B_{2,2,2} - 4/21\,A_{2,2,1} \times B_{2,2,0}B_{2,2,1} + 2A_{2,2,1}B_{3,1,1} + 24/35\,A_{2,2,1}B_{3,3,1} + 32/21\,A_{2,2,2}B_{2,2,0}^2 + 128/21 \times A_{2,2,2}B_{2,2,-2}B_{2,2,2} - 88/21\,A_{2,2,2}B_{2,2,-1}B_{2,2,1} - 4A_{2,2,2}B_{3,1,0} - 24/35\,A_{2,2,2}B_{3,3,0} + A_{4,2,2} + B_{4,2,2},$$

$$C_{4,2,1} = -16/7\,A_{2,2,0}^2B_{2,2,1} - 88/21\,A_{2,2,1}^2B_{2,2,-1} + 64/3\,A_{2,2,-2}A_{2,2,1}B_{2,2,2} - 704/21 \times A_{2,2,-2}A_{2,2,2}B_{2,2,1} - 128/21\,A_{2,2,-2}B_{2,2,1}B_{2,2,2} + 96/7\,A_{2,2,-2}B_{3,3,3} + 44/21\,A_{2,2,-1} \times B_{2,2,1}^2 - 32/21\,A_{2,2,-1}A_{2,2,0}B_{2,2,2} + 88/21\,A_{2,2,-1}A_{2,2,1}B_{2,2,1} - 32/21\,A_{2,2,-1}A_{2,2,2} \times B_{2,2,0} - 32/21\,A_{2,2,-1}B_{2,2,0}B_{2,2,2} + 16/7\,A_{2,2,0}A_{2,2,1}B_{2,2,0} + 64/21\,A_{2,2,0}A_{2,2,2} \times B_{2,2,-1} + 16/21\,A_{2,2,0}B_{2,2,-1}B_{2,2,2} - 8/7\,A_{2,2,0}B_{2,2,0}B_{2,2,1} + 4A_{2,2,0}B_{3,1,1} - 32/35 \times A_{2,2,0}B_{3,3,1} + 8/7\,A_{2,2,1}B_{2,2,0}^2 + 256/21\,A_{2,2,1}A_{2,2,2}B_{2,2,-2} + 352/21\,A_{2,2,1}B_{2,2,-2} \times B_{2,2,2} - 44/21\,A_{2,2,1}B_{2,2,-1}B_{2,2,1} - 2A_{2,2,1}B_{3,1,0} + 48/35\,A_{2,2,1}B_{3,3,0} - 32/3 \times A_{2,2,2}B_{2,2,-2}B_{2,2,1} + 16/21\,A_{2,2,2}B_{2,2,-1}B_{2,2,0} - 8A_{2,2,2}B_{3,1,-1} - 96/35\,A_{2,2,2} \times B_{3,3,-1} + A_{4,2,1} + B_{4,2,1},$$

$$C_{4,2,0} = -16/7\,A_{2,2,-1}^2B_{2,2,2} - 16/7\,A_{2,2,1}^2B_{2,2,-2} + 8/7\,A_{2,2,-2}B_{2,2,1}^2 + 128/7\,A_{2,2,-2} \times A_{2,2,0}B_{2,2,2} + 16/7\,A_{2,2,-2}A_{2,2,1}B_{2,2,1} - 256/7\,A_{2,2,-2}A_{2,2,2}B_{2,2,0} - 64/7\,A_{2,2,-2} \times B_{2,2,0}B_{2,2,2} + 48/7\,A_{2,2,-2}B_{3,3,2} - 24/7\,A_{2,2,-1}A_{2,2,0}B_{2,2,1} + 48/7\,A_{2,2,-1}A_{2,2,1} \times B_{2,2,0} + 16/7\,A_{2,2,-1}A_{2,2,2}B_{2,2,-1} - 8/7\,A_{2,2,-1}B_{2,2,-1}B_{2,2,2} + 12/7\,A_{2,2,-1}B_{2,2,0} \times$$

$$
\begin{aligned}
&B_{2,2,1} + 6A_{2,2,-1}B_{3,1,1} - 48/35\, A_{2,2,-1}B_{3,3,1} - 24/7\, A_{2,2,0}A_{2,2,1}B_{2,2,-1} + 128/7 \times \\
&A_{2,2,0}A_{2,2,2}B_{2,2,-2} + 128/7\, A_{2,2,0}B_{2,2,-2}B_{2,2,2} - 24/7\, A_{2,2,0}B_{2,2,-1}B_{2,2,1} - 8/7 \times \\
&A_{2,2,1}B_{2,2,-2}B_{2,2,1} + 12/7 A_{2,2,1}B_{2,2,-1}B_{2,2,0} - 6A_{2,2,1}B_{3,1,-1} + 48/35\, A_{2,2,1}B_{3,3,-1} + \\
&8/7\, A_{2,2,2}B^2_{2,2,-1} - 64/7\, A_{2,2,2}B_{2,2,-2}B_{2,2,0} - 48/7\, A_{2,2,2}B_{3,3,-2} + A_{4,2,0} + B_{4,2,0},
\end{aligned}
$$

$$
\begin{aligned}
C_{4,2,-1} = {}& -88/21\, A^2_{2,2,-1}B_{2,2,1} - 16/7\, A^2_{2,2,0}B_{2,2,-1} + 256/21\, A_{2,2,-2}A_{2,2,-1}B_{2,2,2} + \\
&64/21\, A_{2,2,-2}A_{2,2,0}B_{2,2,1} - 32/21\, A_{2,2,-2}A_{2,2,1}B_{2,2,0} - 704/21\, A_{2,2,-2}A_{2,2,2}B_{2,2,-1} - \\
&32/3 A_{2,2,-2}B_{2,2,-1}B_{2,2,2} + 16/21\, A_{2,2,-2}B_{2,2,0}B_{2,2,1} + 8A_{2,2,-2}B_{3,1,1} + 96/35 \times \\
&A_{2,2,-2}B_{3,3,1} + 8/7\, A_{2,2,-1}B^2_{2,2,0} + 16/7\, A_{2,2,-1}A_{2,2,0}B_{2,2,0} + 88/21\, A_{2,2,-1}A_{2,2,1} \times \\
&B_{2,2,-1} + 64/3\, A_{2,2,-1}A_{2,2,2}B_{2,2,-2} + 352/21\, A_{2,2,-1}B_{2,2,-2}B_{2,2,2} - 44/21\, A_{2,2,-1} \times \\
&B_{2,2,-1}B_{2,2,1} + 2A_{2,2,-1}B_{3,1,0} - 48/35\, A_{2,2,-1}B_{3,3,0} - 32/21\, A_{2,2,0}A_{2,2,1}B_{2,2,-2} + \\
&16/21\, A_{2,2,0}B_{2,2,-2}B_{2,2,1} - 8/7\, A_{2,2,0}B_{2,2,-1}B_{2,2,0} - 4A_{2,2,0}B_{3,1,-1} + 32/35\, A_{2,2,0} \times \\
&B_{3,3,-1} + 44/21\, A_{2,2,1}B^2_{2,2,-1} - 32/21\, A_{2,2,1}B_{2,2,-2}B_{2,2,0} - 128/21\, A_{2,2,2}B_{2,2,-2} \times \\
&B_{2,2,-1} - 96/7\, A_{2,2,2}B_{3,3,-3} + A_{4,2,-1} + B_{4,2,-1},
\end{aligned}
$$

$$
\begin{aligned}
C_{4,2,-2} = {}& 256/21\, A^2_{2,2,-2}B_{2,2,2} - 8/21\, A^2_{2,2,-1}B_{2,2,0} - 64/21\, A^2_{2,2,0}B_{2,2,-2} + 32/21 \times \\
&A_{2,2,-2}B^2_{2,2,0} - 64/21\, A_{2,2,-2}A_{2,2,-1}B_{2,2,1} + 64/21\, A_{2,2,-2}A_{2,2,0}B_{2,2,0} - 16/3\, A_{2,2,-2} \times \\
&A_{2,2,1}B_{2,2,-1} - 256/21\, A_{2,2,-2}A_{2,2,2}B_{2,2,-2} + 128/21 A_{2,2,-2}B_{2,2,-2}B_{2,2,2} - 88/21 \times \\
&A_{2,2,-2}B_{2,2,-1}B_{2,2,1} + 4A_{2,2,-2}B_{3,1,0} + 24/35\, A_{2,2,-2}B_{3,3,0} + 8/21\, A_{2,2,-1}A_{2,2,0} \times \\
&B_{2,2,-1} + 176/21\, A_{2,2,-1}A_{2,2,1}B_{2,2,-2} + 8/3\, A_{2,2,-1}B_{2,2,-2}B_{2,2,1} - 4/21\, A_{2,2,-1} \times \\
&B_{2,2,-1}B_{2,2,0} - 2A_{2,2,-1}B_{3,1,-1} - 24/35\, A_{2,2,-1}B_{3,3,-1} + 4/21\, A_{2,2,0}B^2_{2,2,-1} - 32/21 \times \\
&A_{2,2,0}B_{2,2,-2}B_{2,2,0} + 8/7\, A_{2,2,0}B_{3,3,-2} + 32/21\, A_{2,2,1}B_{2,2,-2}B_{2,2,-1} - 24/7\, A_{2,2,1} \times \\
&B_{3,3,-3} - 128/21\, A_{2,2,2}B^2_{2,2,-2} + A_{4,2,-2} + B_{4,2,-2};
\end{aligned} \qquad (14.75)
$$

$$
C_{4,0,0} = A_{4,0,0} + B_{4,0,0}. \qquad (14.76)
$$

These product tables were computed in [106], using `REDUCE` symbolic algorithms, to ninth aberration order. The results for rank 5 run for 14 pages in small type, so we omit them here.

The *inversion* of elements in the axis-symmetric aberration group was given for $D = 1$-dim aberrations in (13.47)–(13.51),

$$
\mathcal{G}(\mathbf{A};\mathbf{M})^{-1} = \mathcal{G}(\mathbf{0};\mathbf{M}^{-1})\,\mathcal{G}(\mathbf{A};\mathbf{1})^{-1}, \quad \mathcal{G}(\mathbf{A}^{\mathrm{I}};\mathbf{1}) := \mathcal{G}(\mathbf{A};\mathbf{1})^{-1}, \qquad (14.77)
$$

and written for the aberration polynomials A^{I}_k, $k = 2, 3, 4$. From the multiplication table for the $D = 2$-dim case, (14.71)–(14.76), we find the following explicit formulas for the harmonic aberration coefficients:

Aberration orders 3 and 5:

$$
A^{\mathrm{I}}_{2,j,m} = -A_{2,j,m}, \quad j = 2,\, 0, \quad m = j,\, j-1, \ldots, -j, \qquad (14.78)
$$

$$
A^{\mathrm{I}}_{3,j,m} = -A_{3,j,m}, \quad j = 3,\, 1, \quad m = j,\, j-1, \ldots, -j. \qquad (14.79)
$$

Aberration order 7:

$$
\begin{aligned}
A^{\mathrm{I}}_{4,4,4} &= 6A_{2,2,1}A_{3,3,3} - 4A_{2,2,2}A_{3,3,2} - A_{4,4,4}, \\
A^{\mathrm{I}}_{4,4,3} &= 12A_{2,2,0}A_{3,3,3} + 2A_{2,2,1}A_{3,3,2} - 8A_{2,2,2}A_{3,3,1} - A_{4,4,3}, \\
A^{\mathrm{I}}_{4,4,2} &= 18A_{2,2,-1}A_{3,3,3} + 8A_{2,2,0}A_{3,3,2} - 2A_{2,2,1}A_{3,3,1} - 12A_{2,2,2}A_{3,3,0} - A_{4,4,2}, \\
A^{\mathrm{I}}_{4,4,1} &= 24A_{2,2,-2}A_{3,3,3} + 14A_{2,2,-1}A_{3,3,2} + 4A_{2,2,0}A_{3,3,1} - 6A_{2,2,1}A_{3,3,0} - 16A_{2,2,2} \times \\
&\quad A_{3,3,-1} - A_{4,4,1},
\end{aligned}
$$

$$A^{\rm I}_{4,4,0} = 20A_{2,2,-2}A_{3,3,2} + 10A_{2,2,-1}A_{3,3,1} - 10A_{2,2,1}A_{3,3,-1} - 20A_{2,2,2}A_{3,3,-2} - A_{4,4,0},$$
$$A^{\rm I}_{4,4,-1} = 16A_{2,2,-2}A_{3,3,1} + 6A_{2,2,-1}A_{3,3,0} - 4A_{2,2,0}A_{3,3,-1} - 14A_{2,2,1}A_{3,3,-2} - 24 \times A_{2,2,2}A_{3,3,-3} - A_{4,4,-1},$$
$$A^{\rm I}_{4,4,-2} = 12A_{2,2,-2}A_{3,3,0}+2A_{2,2,-1}A_{3,3,-1}-8A_{2,2,0}A_{3,3,-2}-18A_{2,2,1}A_{3,3,-3}-A_{4,4,-2},$$
$$A^{\rm I}_{4,4,-3} = 8A_{2,2,-2}A_{3,3,-1} - 2A_{2,2,-1}A_{3,3,-2} - 12A_{2,2,0}A_{3,3,-3} - A_{4,4,-3},$$
$$A^{\rm I}_{4,4,-4} = 4A_{2,2,-2}A_{3,3,-2} - 6A_{2,2,-1}A_{3,3,-3} - A_{4,4,-4};$$
$$A^{\rm I}_{4,2,2} = 24/7\, A_{2,2,-1}A_{3,3,3} - 8/7\, A_{2,2,0}A_{3,3,2} + 2A_{2,2,1}A_{3,1,1} + 24/35\; A_{2,2,1}A_{3,3,1} - 4 \times A_{2,2,2}A_{3,1,0} - 24/35\; A_{2,2,2}A_{3,3,0} - A_{4,2,2},$$
$$A^{\rm I}_{4,2,1} = 96/7\, A_{2,2,-2}A_{3,3,3}+4A_{2,2,0}A_{3,1,1}-32/35\; A_{2,2,0}A_{3,3,1}-2A_{2,2,1}A_{3,1,0}+48/35 \times A_{2,2,1}A_{3,3,0} - 8A_{2,2,2}A_{3,1,-1} - 96/35\; A_{2,2,2}A_{3,3,-1} - A_{4,2,1},$$
$$A^{\rm I}_{4,2,0} = 48/7\, A_{2,2,-2}A_{3,3,2} + 6A_{2,2,-1}A_{3,1,1} - 48/35\; A_{2,2,-1}A_{3,3,1} - 6A_{2,2,1}A_{3,1,-1} + 48/35\; A_{2,2,1}A_{3,3,-1} - 48/7\, A_{2,2,2}A_{3,3,-2} - A_{4,2,0},$$
$$A^{\rm I}_{4,2,-1} = 8A_{2,2,-2}A_{3,1,1}+96/35\; A_{2,2,-2}A_{3,3,1}+2A_{2,2,-1}A_{3,1,0}-48/35\; A_{2,2,-1}A_{3,3,0}- 4A_{2,2,0}A_{3,1,-1} + 32/35\; A_{2,2,0}A_{3,3,-1} - 96/7\, A_{2,2,2}A_{3,3,-3} - A_{4,2,-1},$$
$$A^{\rm I}_{4,2,-2} = 4A_{2,2,-2}A_{3,1,0} + 24/35\; A_{2,2,-2}A_{3,3,0} - 2A_{2,2,-1}A_{3,1,-1} - 24/35\; A_{2,2,-1} \times A_{3,3,-1} + 8/7\, A_{2,2,0}A_{3,3,-2} - 24/7\, A_{2,2,1}A_{3,3,-3} - A_{4,2,-2};$$
$$A^{\rm I}_{4,0,0} = -A_{4,0,0}. \qquad (14.80)$$

Remark: On computational efficiency. As we argued above, the harmonic basis of aberration polynomials and coefficients provides computational efficiency to calculate the action of the paraxial part in products and inversion of optical subsystems, because the $\mathbf{D}(\mathbf{M})$ matrices are made block-diagonal with the extra symplectic spin label j. But the harmonic basis *also* provides shorter product and inversion tables because, as in the coupling of angular momenta j_1, j_2, there is the extra selection rule $|j_1-j_2| \le j \le j_1+j_2$, which becomes important for higher aberration orders. As a form of counting terms, for rank $k=2$ the product (14.71) performs 6 sums, for $k=3$ (14.73) computes 48, and for $k=4$ (14.76) has 308. By comparison, the group product written in terms of monomial aberration coefficients leads to 6, 44 and 407 terms respectively [107, 158]. •

14.5 Aberration coefficients for optical elements

In this section we display the Hamilton–Lie harmonic aberration coefficients of the basic axis-symmetric optical elements. Foremost, we have free-space displacement in a medium of refractive index n by a distance z along the optical axis $\mathcal{D}_{z/n}$, and thin lenses $\mathcal{L}_{\rm G}$. Their maps of phase space are given in (9.3) and (9.4)–(9.5), and they are realized by the linear canonical operators in (9.18). Beside homogeneous media, we also give results for the elliptic index-profile guides (2.15), and interfaces between two such media.

Free propagation by a distance z in a **homogeneous medium** n, is indicated by $\mathcal{D}_{n,z} =: \mathcal{G}(\mathbf{D}^{(n,z)}; \mathbf{M}_{\rm D}(n,z))$. The paraxial factor is given by (9.18),

$$\mathbf{M}_{\rm D}(n,z) = \begin{pmatrix} 1 & 0 \\ -z/n & 1 \end{pmatrix}. \tag{14.81}$$

We use the expansion (2.11) for the homogeneous medium Hamiltonian to see that all but the primary spherical aberration coefficients vanish. To aberration order 7 (rank 4) they are:

$$D^{(n,z)}_{2,2,2} = -z/(8n^3), \quad D^{(n,z)}_{3,3,3} = -z/(16n^5), \quad D^{(n,z)}_{4,4,4} = -5z/(128n^7). \tag{14.82}$$

Free propagation by length z in an **elliptic index-profile** waveguide has refractive index and Hamiltonian

$$n_{\rm e}(\mathbf{q}) = \sqrt{n^2 - \nu^2|\mathbf{q}|^2}, \qquad h_{\rm e}(\mathbf{p},\mathbf{q}) = -\sqrt{n^2 - (|\mathbf{p}|^2 + \nu^2|\mathbf{q}|^2)} \tag{14.83}$$

[see (2.15) and (2.16)]. The one-parameter subgroup of maps generated by this Hamiltonian is given in (2.18)–(2.19), and we indicate their aberration group operator by $\mathcal{E}_{n,\nu;z} =: \mathcal{G}(\mathbf{E}^{(n,\nu,z)}; \mathbf{M}_{\rm e}(n,\nu;z)) = \exp(-z\{h_{\rm e}, \circ\})$. Its paraxial factor is a fractional Fourier transformer (10.29), i.e.,

$$\mathbf{M}_{\rm e}(n,\nu;z) = \begin{pmatrix} \cos\frac{\nu}{n}z & \nu\sin\frac{\nu}{n}z \\ -\frac{1}{\nu}\sin\frac{\nu}{n}z & \cos\frac{\nu}{n}z \end{pmatrix}, \tag{14.84}$$

and the aberration polynomials for ranks $k = 2,3,4$ are

$$\begin{aligned} E_2(\mathbf{p},\mathbf{q}) &= -z(|\mathbf{p}|^2 + \nu^2|\mathbf{q}|^2)^2/8n^3, \\ E_3(\mathbf{p},\mathbf{q}) &= -z(|\mathbf{p}|^2 + \nu^2|\mathbf{q}|^2)^3/18n^5, \\ E_4(\mathbf{p},\mathbf{q}) &= -5z(|\mathbf{p}|^2 + \nu^2|\mathbf{q}|^2)^4/128n^7. \end{aligned} \tag{14.85}$$

The monomial aberration coefficients can be found by means of the binomial formula, and the symplectic coefficients from there. There is a selection rule though: there are *no* $\mathbf{p}\cdot\mathbf{q}$ terms in the aberration polynomials. Odd-parity aberrations will thus be absent (no coma nor distorsion among others). The nonzero primary and repeater aberrations will relate through factors of $|\mathbf{p}|^2|\mathbf{q}|^2 = Y_{2,2,0} + \frac{2}{3}Y_{2,0,0}$. It should be realized that these aberrations act on phase space coordinates that rotate in the (p_x, q_x) and (p_y, q_y) planes according to the paraxial part of the transformation. Since $\mathbf{p}\cdot\mathbf{q} = 0$ implies $|\mathbf{p}\times\mathbf{q}| = |\mathbf{p}||\mathbf{q}|$, and the largest binomial coefficient occurs in (14.85) for powers of $|\mathbf{p}|^2|\mathbf{q}|^2 = (\mathbf{p}\times\mathbf{q})^2$ the largest (and most deleterious) aberrations are those due to the singlets $Y_{k,0,0}$, whose spots turn into circle arcs that are concentric with the optical axis of the guide [see Fig. 14.3 and (14.49)].

The aberration coefficients of the root transformation have to be 'discovered' by extracting them from the phase space map that is available to us from Chap. 4, (4.8)–(4.9). The process to extract these coefficients was performed in Chap. 13 for the case of $D = 1$ dimension, between (13.65) and (13.73). In the present $D = 2$-dim axis-symmetric case, the previous results describe now only *meridional* rays, i.e., those which lie in a plane with the optical axis, so $\theta = 0, \pi$, $\mathbf{p}\cdot\mathbf{q} = \pm|\mathbf{p}||\mathbf{q}|$ and $\mathbf{p}\times\mathbf{q} = 0$; this eliminates all but the primary aberrations $j = k$ in (14.50). To follow the same process for all rays $(\mathbf{p},\mathbf{q}) \in \mathsf{R}^4$, we should find explicitly the generic phase space map produced by the generic aberration group element.

In the factored-product parametrization (14.66)–(14.69), the action of the right paraxial factor is linear and well known, so there essentially remains the task to apply the aberration part [cf. (13.55)],

$$\begin{aligned} \mathbf{q}_{\mathrm{A}}(\mathbf{p},\mathbf{q};\mathbf{A}) := \mathcal{G}_K^{\mathrm{AS}}(\mathbf{A};\mathbf{1})\,\mathbf{q} &= \exp\widehat{A}_4 \times \exp\widehat{A}_3 \times \exp\widehat{A}_2\,\mathbf{q} \\ &= \mathbf{q} + P(|\mathbf{p}|^2, \mathbf{p}\cdot\mathbf{q}, |\mathbf{q}|^2; \mathbf{A})\,\mathbf{p} + Q(|\mathbf{p}|^2, \mathbf{p}\cdot\mathbf{q}, |\mathbf{q}|^2; \mathbf{A})\,\mathbf{q}, \end{aligned} \tag{14.86}$$

where

$$P = (2A_{2,0,0} + 2/3\,A_{2,2,0})\,|\mathbf{p}|^2 + 2A_{2,2,-1}\,\mathbf{p}\cdot\mathbf{q} + 4A_{2,2,-2}\,|\mathbf{q}|^2 \qquad (3^{\mathrm{rd}}\text{ order})$$

$$+(2A_{2,0,0}A_{2,2,1} - 4A_{2,2,-1}A_{2,2,2} + 2/3\,A_{2,2,0}A_{2,2,1} + 2A_{3,1,1} + 2/5\,A_{3,3,1})\,(|\mathbf{p}|^2)^2$$

$$+(4A_{2,0,0}A_{2,2,0} - 16A_{2,2,-2}A_{2,2,2} - 2A_{2,2,-1}A_{2,2,1} + 4/3\,A_{2,2,0}^2 + 2A_{3,1,0} + 6/5 \times A_{3,3,0})\,|\mathbf{p}|^2\mathbf{p}\cdot\mathbf{q}$$

$$+(2A_{2,0,0}A_{2,2,-1} - 4A_{2,2,-2}A_{2,2,1} + 2/3\,A_{2,2,-1}A_{2,2,0} + 4A_{3,1,-1} + 4/5\,A_{3,3,-1})\,|\mathbf{p}|^2|\mathbf{q}|^2$$

$$+(4A_{2,0,0}A_{2,2,-1} - 8A_{2,2,-2}A_{2,2,1} + 4/3\,A_{2,2,-1}A_{2,2,0} - 2A_{3,1,-1} + 8/5\,A_{3,3,-1})\,(\mathbf{p}\cdot\mathbf{q})^2$$

$$+(8A_{2,0,0}A_{2,2,-2} - 16/3\,A_{2,2,-2}A_{2,2,0} + 2A_{2,2,-1}^2 + 4A_{3,3,-2})\,\mathbf{p}\cdot\mathbf{q}|\mathbf{q}|^2$$

$$+(6A_{3,3,-3})\,(|\mathbf{q}|^2)^2 \qquad (5^{\mathrm{th}}\text{ order})$$

$$+(-8/3\,A_{2,2,0}^2A_{2,2,2} + 3A_{2,0,0}A_{2,2,1}^2 - 8A_{2,0,0}A_{2,2,0}A_{2,2,2} + 2A_{2,0,0}A_{3,3,2} + 64/3 \times A_{2,2,-2}A_{2,2,2}^2 - 4/3\,A_{2,2,-1}A_{2,2,1}A_{2,2,2} - 12A_{2,2,-1}A_{3,3,3} + A_{2,2,0}A_{2,2,1}^2 + 2/3\,A_{2,2,0} \times A_{3,3,2} + 2A_{2,2,1}A_{3,1,1} + 2/5\,A_{2,2,1}A_{3,3,1})\,(|\mathbf{p}|^2)^3$$

$$+(4/3\,A_{2,2,0}^2A_{2,2,1} - 24A_{2,0,0}A_{2,2,-1}A_{2,2,2} + 4A_{2,0,0}A_{2,2,0}A_{2,2,1} + 4A_{2,0,0}A_{3,3,1} + 112/3\,A_{2,2,-2}A_{2,2,1}A_{2,2,2} - 48A_{2,2,-2}A_{3,3,3} + 5/3\,A_{2,2,-1}A_{2,2,1}^2 - 56/3\,A_{2,2,-1}A_{2,2,0} \times A_{2,2,2} - 10A_{2,2,-1}A_{3,3,2} + 4A_{2,2,0}A_{3,1,1} + 32/15A_{2,2,0}A_{3,3,1} + 2A_{2,2,1}A_{3,1,0} + 6/5 \times A_{2,2,1}A_{3,3,0})\,(|\mathbf{p}|^2)^2\mathbf{p}\cdot\mathbf{q}$$

$$+(-4/3\,A_{2,0,0}^2A_{2,2,0} - 20/3\,A_{2,2,-1}^2A_{2,2,2} - 4/9\,A_{2,0,0}A_{2,2,0}^2 - 16A_{2,0,0}A_{2,2,-2}A_{2,2,2} + 2A_{2,0,0}A_{2,2,-1}A_{2,2,1} + 2A_{2,0,0}A_{3,1,0} + 6/5\,A_{2,0,0}A_{3,3,0} + 10/3\,A_{2,2,-2}A_{2,2,1}^2 - 12 \times A_{2,2,-2}A_{3,3,2} + 2/3\,A_{2,2,-1}A_{2,2,0}A_{2,2,1} - 2A_{2,2,-1}A_{3,1,1} - 2/5\,A_{2,2,-1}A_{3,3,1} + 2/3 \times A_{2,2,0}A_{3,1,0} + 2/5\,A_{2,2,0}A_{3,3,0} + 4A_{2,2,1}A_{3,1,-1} + 4/5\,A_{2,2,1}A_{3,3,-1} - 4/3\,A_{2,0,0}^3 - 4/81 \times A_{2,2,0}^3)\,(|\mathbf{p}|^2)^2|\mathbf{q}|^2$$

$$+(4/3\,A_{2,0,0}^2A_{2,2,0} - 64/3\,A_{2,2,-1}^2A_{2,2,2} + 40/9\,A_{2,0,0}A_{2,2,0}^2 - 32A_{2,0,0}A_{2,2,-2}A_{2,2,2} - 8A_{2,0,0}A_{2,2,-1}A_{2,2,1} - 2A_{2,0,0}A_{3,1,0} + 24/5\,A_{2,0,0}A_{3,3,0} + 56/3\,A_{2,2,-2}A_{2,2,1}^2 - 32 \times A_{2,2,-2}A_{3,3,2} - 8/3\,A_{2,2,-1}A_{2,2,0}A_{2,2,1} + 8A_{2,2,-1}A_{3,1,1} - 32/5\,A_{2,2,-1}A_{3,3,1} + 10/3 \times A_{2,2,0}A_{3,1,0} + 4A_{2,2,0}A_{3,3,0} - 2A_{2,2,1}A_{3,1,-1} + 8/5\,A_{2,2,1}A_{3,3,-1} + 4/3\,A_{2,0,0}^3 + 112/81 \times A_{2,2,0}^3)\,|\mathbf{p}|^2(\mathbf{p}\cdot\mathbf{q})^2$$

$$+(-4A_{2,0,0}^2A_{2,2,-1} - 14/3\,A_{2,2,-1}^2A_{2,2,1} - 8A_{2,0,0}A_{2,2,-2}A_{2,2,1} + 4/3\,A_{2,0,0}A_{2,2,-1} \times A_{2,2,0} + 4A_{2,0,0}A_{3,1,-1} + 24/5\,A_{2,0,0}A_{3,3,-1} - 160/3\,A_{2,2,-2}A_{2,2,-1}A_{2,2,2} + 16A_{2,2,-2} \times A_{2,2,0}A_{2,2,1} - 8A_{2,2,-2}A_{3,1,1} - 128/5\,A_{2,2,-2}A_{3,3,1} + 8/9\,A_{2,2,-1}A_{2,2,0}^2 + 28/3\,A_{2,2,0} \times A_{3,1,-1} + 16/5\,A_{2,2,0}A_{3,3,-1} + 4A_{2,2,1}A_{3,3,-2})\,|\mathbf{p}|^2\mathbf{p}\cdot\mathbf{q}|\mathbf{q}|^2$$

$$+(-8A_{2,0,0}^2A_{2,2,-2} - 32A_{2,2,-2}^2A_{2,2,2} - A_{2,2,-1}^2A_{2,2,0}/3 - A_{2,0,0}A_{2,2,-1}^2 + 8/3\,A_{2,0,0} \times A_{2,2,-2}A_{2,2,0} + 2A_{2,0,0}A_{3,3,-2} + 16/9\,A_{2,2,-2}A_{2,2,0}^2 - 12A_{2,2,-2}A_{3,1,0} - 36/5\,A_{2,2,-2} \times A_{3,3,0} + 8A_{2,2,-1}A_{3,1,-1} + 8/5\,A_{2,2,-1}A_{3,3,-1} + 2/3\,A_{2,2,0}A_{3,3,-2} + 6A_{2,2,1}A_{3,3,-3}) \times |\mathbf{p}|^2(|\mathbf{q}|^2)^2$$

$$+(4A_{2,0,0}^2A_{2,2,-1} - 20/3\,A_{2,2,-1}^2A_{2,2,1} - 16A_{2,0,0}A_{2,2,-2}A_{2,2,1} + 8/3\,A_{2,0,0}A_{2,2,-1} \times A_{2,2,0} - 4A_{2,0,0}A_{3,1,-1} + 16/5\,A_{2,0,0}A_{3,3,-1} - 64/3\,A_{2,2,-2}A_{2,2,-1}A_{2,2,2} + 32/3 \times A_{2,2,-2}A_{2,2,0}A_{2,2,1} + 16A_{2,2,-2}A_{3,1,1} - 64/5\,A_{2,2,-2}A_{3,3,1} + 16/9\,A_{2,2,-1}A_{2,2,0}^2 + 6 \times$$

$$
\begin{aligned}
& A_{2,2,-1}A_{3,1,0} - 12/5\, A_{2,2,-1}A_{3,3,0} - 16/3\, A_{2,2,0}A_{3,1,-1} + 64/15 A_{2,2,0}A_{3,3,-1})\, (\mathbf{p}\cdot\mathbf{q})^3 \\
& +(8A^2_{2,0,0}A_{2,2,-2} - 128/3\, A^2_{2,2,-2}A_{2,2,2} - 4/3\, A^2_{2,2,-1}A_{2,2,0} + 4A_{2,0,0}A^2_{2,2,-1} - 32/3 \times \\
& A_{2,0,0}A_{2,2,-2}A_{2,2,0} + 8A_{2,0,0}A_{3,3,-2} + 128/9\, A_{2,2,-2}A^2_{2,2,0} - 64/3\, A_{2,2,-2}A_{2,2,-1} \times \\
& A_{2,2,1} + 20A_{2,2,-2}A_{3,1,0} - 24A_{2,2,-2}A_{3,3,0} - 2A_{2,2,-1}A_{3,1,-1} + 8/5\, A_{2,2,-1}A_{3,3,-1} + \\
& 32/3\, A_{2,2,0}A_{3,3,-2})\, (\mathbf{p}\cdot\mathbf{q})^2 |\mathbf{q}|^2 \\
& +(-32A^2_{2,2,-2}A_{2,2,1} + 12A_{2,0,0}A_{3,3,-3} + 8A_{2,2,-2}A_{2,2,-1}A_{2,2,0} + 8A_{2,2,-2}A_{3,1,-1} - \\
& 112/5\, A_{2,2,-2}A_{3,3,-1} + 10A_{2,2,-1}A_{3,3,-2} + 16A_{2,2,0}A_{3,3,-3} - A^3_{2,2,-1})\, \mathbf{p}\cdot\mathbf{q}(|\mathbf{q}|^2)^2 \\
& +(-16/3\, A^2_{2,2,-2}A_{2,2,0} + 2A_{2,2,-2}A^2_{2,2,-1} - 12A_{2,2,-2}A_{3,3,-2} + 18A_{2,2,-1}A_{3,3,-3}) \times \\
& (|\mathbf{q}|^2)^3, \qquad\qquad (7^{\text{th}} \text{ order})
\end{aligned}
$$

(14.87)

$$
\begin{aligned}
Q = {} & A_{2,2,1}\, |\mathbf{p}|^2 + (-2A_{2,0,0} + 4/3\, A_{2,2,0})\, \mathbf{p}\cdot\mathbf{q} + A_{2,2,-1}\, |\mathbf{q}|^2 \qquad\qquad (3^{\text{rd}} \text{ order}) \\
& +(-4A_{2,2,0}A_{2,2,2} + 3/2 A^2_{2,2,1} + A_{3,3,2})\, |\mathbf{p}|^2|\mathbf{q}|^2 \\
& +(-2A_{2,0,0}A_{2,2,1} - 8A_{2,2,-1}A_{2,2,2} + 4/3\, A_{2,2,0}A_{2,2,1} - 2A_{3,1,1} + 8/5\, A_{3,3,1})\, |\mathbf{p}|^2\mathbf{p}\cdot\mathbf{q} \\
& +(-4/3\, A_{2,0,0}A_{2,2,0} - 8A_{2,2,-2}A_{2,2,2} + A_{2,2,-1}A_{2,2,1} - 2A^2_{2,0,0} - 2/9\, A^2_{2,2,0} + A_{3,1,0} + \\
& 3/5\, A_{3,3,0})\, |\mathbf{p}|^2|\mathbf{q}|^2 \\
& +(-8/3\, A_{2,0,0}A_{2,2,0} - 2A_{2,2,-1}A_{2,2,1} + 2A^2_{2,0,0} + 8/9\, A^2_{2,2,0} - 3A_{3,1,0} + 6/5\, A_{3,3,0})\, (\mathbf{p}\cdot\mathbf{q})^2 \\
& +(-6A_{2,0,0}A_{2,2,-1} - 2A_{3,1,-1} + 8/5\, A_{3,3,-1})\, \mathbf{p}\cdot\mathbf{q}|\mathbf{q}|^2 \\
& +(-8A_{2,0,0}A_{2,2,-2} + 4/3\, A_{2,2,-2}A_{2,2,0} - A^2_{2,2,-1}/2 + A_{3,3,-2})\, (|\mathbf{q}|^2)^2 \qquad (5^{\text{th}} \text{ order}) \\
& +(16A_{2,2,-1}A^2_{2,2,2} - 28/3\, A_{2,2,0}A_{2,2,1}A_{2,2,2} - 12A_{2,2,0}A_{3,3,3} + 3A_{2,2,1}A_{3,3,2} + 5/2 \times \\
& A^3_{2,2,1})\, (|\mathbf{p}|^2)^3 \\
& +(-16A^2_{2,2,0}A_{2,2,2} - 3A_{2,0,0}A^2_{2,2,1} + 8A_{2,0,0}A_{2,2,0}A_{2,2,2} - 2A_{2,0,0}A_{3,3,2} + 128/3 \times \\
& A_{2,2,-2}A^2_{2,2,2} + 16/3\, A_{2,2,-1}A_{2,2,1}A_{2,2,2} - 24A_{2,2,-1}A_{3,3,3} + 14/3\, A_{2,2,0}A^2_{2,2,1} - 20/3 \times \\
& A_{2,2,0}A_{3,3,2} - 2A_{2,2,1}A_{3,1,1} + 28/5\, A_{2,2,1}A_{3,3,1})\, (|\mathbf{p}|^2)^2\mathbf{p}\cdot\mathbf{q} \\
& +(-2A^2_{2,0,0}A_{2,2,1} - 2/9\, A^2_{2,2,0}A_{2,2,1} + 8A_{2,0,0}A_{2,2,-1}A_{2,2,2} - 4/3\, A_{2,0,0}A_{2,2,0}A_{2,2,1} - \\
& 4A_{2,0,0}A_{3,1,1} - 4/5\, A_{2,0,0}A_{3,3,1} + 8/3 A_{2,2,-2}A_{2,2,1}A_{2,2,2} - 24A_{2,2,-2}A_{3,3,3} + 17/6 \times \\
& A_{2,2,-1}A^2_{2,2,1} - 20/3\, A_{2,2,-1}A_{2,2,0}A_{2,2,2} - A_{2,2,-1}A_{3,3,2} - 16/3\, A_{2,2,0}A_{3,1,1} - 16/15 \times \\
& A_{2,2,0}A_{3,3,1} + 3A_{2,2,1}A_{3,1,0} + 9/5\, A_{2,2,1}A_{3,3,0})\, (|\mathbf{p}|^2)^2|\mathbf{q}|^2 \\
& +(2A^2_{2,0,0}A_{2,2,1} + 8/9\, A^2_{2,2,0}A_{2,2,1} + 16A_{2,0,0}A_{2,2,-1}A_{2,2,2} - 8/3\, A_{2,0,0}A_{2,2,0}A_{2,2,1} + \\
& 4A_{2,0,0}A_{3,1,1} - 16/5\, A_{2,0,0}A_{3,3,1} + 32A_{2,2,-2}A_{2,2,1}A_{2,2,2} + 6A_{2,2,-1}A^2_{2,2,1} - 80/3 \times \\
& A_{2,2,-1}A_{2,2,0}A_{2,2,2} - 16A_{2,2,-1}A_{3,3,2} + 4/3\, A_{2,2,0}A_{3,1,1} - 16/15 A_{2,2,0}A_{3,3,1} - 5A_{2,2,1} \times \\
& A_{3,1,0} + 6A_{2,2,1}A_{3,3,0})\, |\mathbf{p}|^2(\mathbf{p}\cdot\mathbf{q})^2 \\
& +(-8/3\, A^2_{2,0,0}A_{2,2,0} - 40/3\, A^2_{2,2,-1}A_{2,2,2} - 20/9\, A_{2,0,0}A^2_{2,2,0} + 48A_{2,0,0}A_{2,2,-2}A_{2,2,2} + \\
& 2A_{2,0,0}A_{2,2,-1}A_{2,2,1} - 6A_{2,0,0}A_{3,1,0} - 18/5\, A_{2,0,0}A_{3,3,0} + 32/3\, A_{2,2,-2}A^2_{2,2,1} - 64/3 \times \\
& A_{2,2,-2}A_{2,2,0}A_{2,2,2} - 16A_{2,2,-2}A_{3,3,2} + 8/3\, A_{2,2,-1}A_{2,2,0}A_{2,2,1} - 14A_{2,2,-1}A_{3,1,1} - \\
& 24/5\, A_{2,2,-1}A_{3,3,1} + 2A_{2,2,1}A_{3,1,-1} + 32/5\, A_{2,2,1}A_{3,3,-1} + 4/3\, A^3_{2,0,0} - 32/81 A^3_{2,2,0}) \times \\
& |\mathbf{p}|^2\mathbf{p}\cdot\mathbf{q}|\mathbf{q}|^2 \\
& +(-2A^2_{2,0,0}A_{2,2,-1} + 5/6 A^2_{2,2,-1}A_{2,2,1} + 8A_{2,0,0}A_{2,2,-2}A_{2,2,1} - 4/3\, A_{2,0,0}A_{2,2,-1}A_{2,2,0} - \\
& 8A_{2,0,0}A_{3,1,-1} - 8/5\, A_{2,0,0}A_{3,3,-1} - 40/3\, A_{2,2,-2}A_{2,2,-1}A_{2,2,2} + 4/3\, A_{2,2,-2}A_{2,2,0} \times \\
& A_{2,2,1} - 16A_{2,2,-2}A_{3,1,1} - 16/5\, A_{2,2,-2}A_{3,3,1} - 2/9\, A_{2,2,-1}A^2_{2,2,0} - A_{2,2,-1}A_{3,1,0} - \\
& 3/5\, A_{2,2,-1}A_{3,3,0} + 4/3\, A_{2,2,0}A_{3,1,-1} + 4/15 A_{2,2,0}A_{3,3,-1} + 3A_{2,2,1}A_{3,3,-2})\, |\mathbf{p}|^2(|\mathbf{q}|^2)^2 \\
& +(8/3\, A^2_{2,0,0}A_{2,2,0} - 32/3\, A^2_{2,2,-1}A_{2,2,2} - 16/9\, A_{2,0,0}A^2_{2,2,0} + 4A_{2,0,0}A_{2,2,-1}A_{2,2,1} + 6 \times \\
& A_{2,0,0}A_{3,1,0} - 12/5\, A_{2,0,0}A_{3,3,0} + 16/3\, A_{2,2,-2}A^2_{2,2,1} + 8A_{2,2,-1}A_{3,1,1} - 32/5\, A_{2,2,-1} \times \\
& A_{3,3,1} - 4A_{2,2,0}A_{3,1,0} + 8/5\, A_{2,2,0}A_{3,3,0} - 4A_{2,2,1}A_{3,1,-1} + 16/5\, A_{2,2,1}A_{3,3,-1} - 4/3 \times \\
& A^3_{2,0,0} + 32/81 A^3_{2,2,0})\, (\mathbf{p}\cdot\mathbf{q})^3 \\
& +(2A^2_{2,0,0}A_{2,2,-1} - 2A^2_{2,2,-1}A_{2,2,1} + 16A_{2,0,0}A_{2,2,-2}A_{2,2,1} - 8/3\, A_{2,0,0}A_{2,2,-1}A_{2,2,0} + \\
& 8A_{2,0,0}A_{3,1,-1} - 32/5\, A_{2,0,0}A_{3,3,-1} - 32A_{2,2,-2}A_{2,2,-1}A_{2,2,2} + 16/3\, A_{2,2,-2}A_{2,2,0} \times \\
& A_{2,2,1} + 8A_{2,2,-2}A_{3,1,1} - 32/5\, A_{2,2,-2}A_{3,3,1} + 8/9\, A_{2,2,-1}A^2_{2,2,0} - 5A_{2,2,-1}A_{3,1,0} - 6 \times \\
& A_{2,2,-1}A_{3,3,0} - 16/3\, A_{2,2,0}A_{3,1,-1} + 64/15 A_{2,2,0}A_{3,3,-1} + 8A_{2,2,1}A_{3,3,-2})\, (\mathbf{p}\cdot\mathbf{q})^2|\mathbf{q}|^2
\end{aligned}
$$

$+(-64/3\, A_{2,2,-2}^2 A_{2,2,2} + 2/3\, A_{2,2,-1}^2 A_{2,2,0} - 3A_{2,0,0}A_{2,2,-1}^2 + 8A_{2,0,0}A_{2,2,-2}A_{2,2,0} - 10A_{2,0,0}A_{3,3,-2} - 8/3\, A_{2,2,-2}A_{2,2,-1}A_{2,2,1} - 8A_{2,2,-2}A_{3,1,0} - 24/5\, A_{2,2,-2}A_{3,3,0} - 6 \times A_{2,2,-1}A_{3,1,-1} - 16/5\, A_{2,2,-1}A_{3,3,-1} + 20/3\, A_{2,2,0}A_{3,3,-2} + 12A_{2,2,1}A_{3,3,-3})\, \mathbf{p}\cdot\mathbf{q}(|\mathbf{q}|^2)^2$
$+(-12A_{2,0,0}A_{3,3,-3} - 4/3\, A_{2,2,-2}A_{2,2,-1}A_{2,2,0} - 8A_{2,2,-2}A_{3,1,-1} - 8/5\, A_{2,2,-2} \times A_{3,3,-1} - A_{2,2,-1}A_{3,3,-2} + 8A_{2,2,0}A_{3,3,-3} + A_{2,2,-1}^3/2)\, (|\mathbf{q}|^2)^3.$ (7th order)

(14.88)

A corresponding expansion for the aberration of momentum is obtained through Fourier transformation of (14.86). Using (14.43), we find

$$\begin{aligned}\mathbf{p}_{\mathrm{A}}(\mathbf{p},\mathbf{q};\mathbf{A}) = -\mathcal{F} : \mathbf{q}_{\mathrm{A}}(\mathbf{p},\mathbf{q};\mathbf{A}) \;=\; \mathcal{F}\,\mathcal{G}_K^{\mathrm{AS}}(\mathbf{A};1)\,\mathcal{F}^{-1}\mathbf{p} \;=:\; \mathcal{G}_K^{\mathrm{AS}}(\widetilde{\mathbf{A}};1)\,\mathbf{p}\\ = \mathbf{p} + R(|\mathbf{p}|^2, \mathbf{p}\cdot\mathbf{q}, |\mathbf{q}|^2; \mathbf{A})\,\mathbf{p} + S(|\mathbf{p}|^2, \mathbf{p}\cdot\mathbf{q}, |\mathbf{q}|^2; \mathbf{A})\,\mathbf{q},\end{aligned} \tag{14.89}$$

where $\widetilde{A}_{k,j,m} = (-1)^{j-m} A_{k,j,-m}$, and

$$\begin{aligned}R(|\mathbf{p}|^2, \mathbf{p}\cdot\mathbf{q}, |\mathbf{q}|^2; \mathbf{A}) &= Q(|\mathbf{p}|^2, \mathbf{p}\cdot\mathbf{q}, |\mathbf{q}|^2; \widetilde{\mathbf{A}}),\\ S(|\mathbf{p}|^2, \mathbf{p}\cdot\mathbf{q}, |\mathbf{q}|^2; \mathbf{A}) &= -P(|\mathbf{p}|^2, \mathbf{p}\cdot\mathbf{q}, |\mathbf{q}|^2; \widetilde{\mathbf{A}}).\end{aligned} \tag{14.90}$$

We can now follow the process (13.65)—(13.73) to extract the paraxial part and, recursively, the coefficients of increasing aberration order. We display these coefficients in the form of (13.72) and (13.73), namely

$$\begin{aligned}\mathbf{A} &= \{\mathbf{A}_4, \mathbf{A}_3, \mathbf{A}_2\},\\ \mathbf{A}_k &= \{\mathbf{A}_{k,k}, \mathbf{A}_{k,k-2}, \ldots, \mathbf{A}_{k,1 \text{ or } 0}\},\\ \mathbf{A}_{k,j} &= \{A_{k,j,j}, A_{k,j,j-1}, \ldots, A_{k,j,-j}\}.\end{aligned} \tag{14.91}$$

The root transformation (4.8)–(4.9), denoted $\mathcal{R}_{n;\zeta} =: \mathcal{G}(\mathbf{R};\mathbf{M}_{\mathrm{R}})$, in a medium n, by the eight-order polynomial surface of revolution

$$\zeta(|\mathbf{q}|^2) = \zeta_2|\mathbf{q}|^2 + \zeta_4(|\mathbf{q}|^2)^2 + \zeta_6(|\mathbf{q}|^2)^3 + \zeta_8(|\mathbf{q}|^2)^4, \tag{14.92}$$

has the paraxial part [see (9.7) and (9.18)]

$$\mathbf{M}_{\mathrm{R}} = \begin{pmatrix} 1 & -2n\zeta_2 \\ 0 & 1 \end{pmatrix}, \tag{14.93}$$

and the harmonic Hamilton–Lie aberration coefficients:[1]

$\mathbf{R}_2 = \{\{0,\ 0,\ -\zeta_2/2n,\ 0,\ n\zeta_4\},$
$\{-\zeta_2/3n\}\};$

(14.94)

$\mathbf{R}_3 = \{\{0,\ 0,\ -\zeta_2/8n^3,\ -\zeta_2^2/2n^2,\ -\zeta_4/2n,\ 2\zeta_2\zeta_4,\ n\zeta_6\},$
$\{-\zeta_2/10n^3,\ -\zeta_2^2/5\,n^2,\ -2/5\,\zeta_4/n\}\};$

(14.95)

[1] Recall that the output of `mexLIE` [161] writes numerical fractions to the left and symbolic quantities to the right, so that when two '/'s appear, as in '$3/56\,\zeta_2/n^5$,' we should read $\frac{3}{56}\zeta_2/n^5$.

$$\mathbf{R}_4 = \{\{0,\ 0,\ -\zeta_2/16n^5,\ -\zeta_2^2/4n^4,\ -\zeta_4/8n^3 - 5/6\zeta_2^3/n^3,\ -2\zeta_2\zeta_4/n^2,\ -\zeta_6/2n + 16/3 \times \zeta_2^2\zeta_4/n,\ 6\zeta_2\zeta_6,\ 8/3\,n\zeta_2\zeta_4^2 + n\zeta_8\},\ \{-3/56\,\zeta_2/n^5,\ -\zeta_2^2/7\,n^4,\ -\zeta_4/7\,n^3 - 2/7\,\zeta_2^3/n^3,\ -8/7\,\zeta_2\zeta_4/n^2,\ -3/7\,\zeta_6/n - 16/21 \times \zeta_2^2\zeta_4/n\},\ \{-\zeta_4/15n^3\}\}. \tag{14.96}$$

Refracting surface transformation $\mathcal{S}_{n,m;\zeta} = \mathcal{R}_{n;\zeta}\mathcal{R}^{-1}_{m;\zeta} =: \mathcal{G}(\mathbf{S};\mathbf{M}_\mathrm{s})$ by the surface (14.92), from medium n to medium m. The paraxial part is

$$\mathbf{M}_\mathrm{s} = \begin{pmatrix} 1 & 2(m-n)\zeta_2 \\ 0 & 1 \end{pmatrix} \tag{14.97}$$

[see (9.8) and (9.18)], while the Lie-Hamilton aberration coefficients are:

$$\mathbf{S}_2 = \{\{0,\ 0,\ \zeta_2/2m - \zeta_2/2n,\ 2n\zeta_2^2/m - 2\zeta_2^2,\ -m\zeta_4 + 2m\zeta_2^3 + n\zeta_4 - 4n\zeta_2^3 + 2n^2 \times \zeta_2^3/m\},\ \{\zeta_2/3\,m - \zeta_2/3\,n\}\}; \tag{14.98}$$

$$\mathbf{S}_3 = \{\{0,\ 0,\ \zeta_2/8m^3 - \zeta_2/8n^3,\ n\zeta_2^2/m^3 - \zeta_2^2/2m^2 - \zeta_2^2/2n^2,\ 3n^2\zeta_2^3/m^3 - 3n\zeta_2^3/m^2 + \zeta_4/2m + 2\zeta_2^3/m - \zeta_4/2n - 2\zeta_2^3/n,\ -2m\zeta_2\zeta_4/n + 4m\zeta_2^4/n - 2\zeta_2\zeta_4 + 4n^3\zeta_2^4/m^3 - 6n^2\zeta_2^4/m^2 + 4n\zeta_2\zeta_4/m + 4n\zeta_2^4/m - 6\zeta_2^4,\ -m\zeta_6 + 6m\zeta_2^2\zeta_4 - 2m\zeta_2^5 + n\zeta_6 - 12n \times \zeta_2^2\zeta_4 + 4n\zeta_2^5 + 2n^4\zeta_2^5/m^3 - 4n^3\zeta_2^5/m^2 + 6n^2\zeta_2^2\zeta_4/m\},\ \{\zeta_2/10m^3 - \zeta_2/10n^3,\ 2/5\,n \times \zeta_2^2/m^3 - \zeta_2^2/5\,m^2 - \zeta_2^2/5\,n^2,\ 2/5\,n^2\zeta_2^3/m^3 - 2/5\,n\zeta_2^3/m^2 + 2/5\zeta_4/m - 2/5\,\zeta_2^3/m - 2/5 \times \zeta_4/n + 2/5\zeta_2^3/n\}\}; \tag{14.99}$$

$$\mathbf{S}_4 = \{\{0,\ 0,\ \zeta_2/16m^5 - \zeta_2/16n^5,\ 3/4n\zeta_2^2/m^5 - \zeta_2^2/4m^4 - \zeta_2^2/4m^3n - \zeta_2^2/4n^4,\ 15/4n^2 \times \zeta_2^3/m^5 - 5/2n\zeta_2^3/m^4 + \zeta_4/8m^3 - 5/12\zeta_2^3/m^3 - \zeta_4/8n^3 - 5/6\zeta_2^3/n^3,\ 10n^3\zeta_2^4/m^5 - 10n^2 \times \zeta_2^4/m^4 + 3n\zeta_2\zeta_4/m^3 + 8/3\,n\zeta_2^4/m^3 - 2\zeta_2\zeta_4/m^2 - 4\zeta_2^4/m^2 + \zeta_2\zeta_4/mn + 4\zeta_2^4/mn - 2 \times \zeta_2\zeta_4/n^2 - 8/3\,\zeta_2^4/n^2,\ -16/3\,m\zeta_2^2\zeta_4/n^2 + 32/3\,m\zeta_2^5/n^2 + 15n^4\zeta_2^5/m^5 - 20n^3\zeta_2^5/m^4 + 15n^2\zeta_2^2\zeta_4/m^3 + 6n^2\zeta_2^5/m^3 - 62/3\,n\zeta_2^2\zeta_4/m^2 - 20/3\,n\zeta_2^5/m^2 + \zeta_6/2m + 19\zeta_2^2\zeta_4/m + 11 \times \zeta_2^5/m - \zeta_6/2n - 8\zeta_2^2\zeta_4/n - 16\zeta_2^5/n,\ -6m\zeta_2\zeta_6/n + 116/3\,m\zeta_2^3\zeta_4/n - 52/3\,m\zeta_2^6/n + 4 \times \zeta_2\zeta_6 + 12n^5\zeta_2^6/m^5 - 20n^4\zeta_2^6/m^4 + 28n^3\zeta_2^3\zeta_4/m^3 - 4/3\,n^3\zeta_2^6/m^3 - 176/3\,n^2\zeta_2^3\zeta_4/m^2 + 64/3\,n^2\zeta_2^6/m^2 + 2n\zeta_2\zeta_6/m + 72n\zeta_2^3\zeta_4/m - 36n\zeta_2^6/m + 4n\zeta_4^2/m - 80\zeta_2^3\zeta_4 + 124/3 \times \zeta_2^6 - 4\zeta_4^2,\ 8m\zeta_2\zeta_4^2 - m\zeta_8 + 2m\zeta_2^2\zeta_6 + 86/3\,m\zeta_2^4\zeta_4 - 124/3\,m\zeta_2^7 - 24n\zeta_2\zeta_4^2 + n\zeta_8 - 4n\zeta_2^2 \times \zeta_6 - 40n\zeta_2^4\zeta_4 + 200/3\,n\zeta_2^7 + 4n^6\zeta_2^7/m^5 - 8n^5\zeta_2^7/m^4 + 18n^4\zeta_2^4\zeta_4/m^3 - 20/3\,n^4\zeta_2^7/m^3 - 152/3n^3\zeta_2^4\zeta_4/m^2 + 112/3\,n^3\zeta_2^7/m^2 + 40/3\,n^2\zeta_2\zeta_4^2/m + 2n^2\zeta_2^2\zeta_6/m + 164/3\,n^2\zeta_2^4\zeta_4/m - 188/3\,n^2\zeta_2^7/m + 8/3\,m^2\zeta_2\zeta_4^2/n - 32/3m^2\zeta_2^4\zeta_4/n + 32/3\,m^2\zeta_2^7/n\},\ \{3/56\,\zeta_2/m^5 - 3/56\,\zeta_2/n^5,\ 3/7\,n\zeta_2^2/m^5 - \zeta_2^2/7\,m^4 - \zeta_2^2/7\,m^3n - \zeta_2^2/7\,n^4,\ 9/7\,n^2\zeta_2^3/m^5 - 6/7\,n \times \zeta_2^3/m^4 + \zeta_4/7\,m^3 - \zeta_2^3/7\,m^3 - \zeta_4/7n^3 - 2/7\,\zeta_2^3/n^3,\ 12/7\,n^3\zeta_2^4/m^5 - 12/7n^2\zeta_2^4/m^4 + 12/7\,n\zeta_2\zeta_4/m^3 - 8/7\,n\zeta_2^4/m^3 - 8/7\,\zeta_2\zeta_4/m^2 + 12/7\zeta_2^4/m^2 + 4/7\,\zeta_2\zeta_4/mn - 12/7 \times \zeta_2^4/mn - 8/7\,\zeta_2\zeta_4/n^2 + 8/7\zeta_2^4/n^2,\ 16/21\,m\zeta_2^2\zeta_4/n^2 - 32/21\,m\zeta_2^5/n^2 + 6/7\,n^4\zeta_2^5/m^5 - 8/7\,n^3\zeta_2^5/m^4 + 20/7\,n^2\zeta_2^2\zeta_4/m^3 - 20/7n^2\zeta_2^5/m^3 - 64/21\,n\zeta_2^2\zeta_4/m^2 + 104/21\,n \times \zeta_2^5/m^2 + 3/7\,\zeta_6/m - 12/7\,\zeta_2^2\zeta_4/m - 18/7\,\zeta_2^5/m - 3/7\,\zeta_6/n + 8/7\zeta_2^2\zeta_4/n + 16/7 \times \zeta_2^5/n\},\ \{\zeta_4/15m^3 - \zeta_4/15n^3\}\}. \tag{14.100}$$

Again we draw attention to the selection rule which, for surfaces that are tangent to the screen at the optical center, sets to zero the first two coefficients of each multiplet, i.e., spherical aberration $S_{k,k,k}$ and coma $S_{k,k,k-1}$. Also, if the surface is flat up to some homogeneous degree $2k_f$ [i.e., when $\zeta(|\mathbf{q}|^2)$ in (14.92) has polynomial coefficients $\zeta_2 = \cdots = \zeta_{2k_f} = 0$ and $\zeta_{2(k_f+1)} \neq 0$] then all aberrations are zero up to rank k_f, except for $S_{k_f,k_f,-k_f} = n\zeta_{2k_f}$.

Root transformation in elliptic index-profile guides. Lastly, we display the harmonic aberration coefficients for the root transformation between the standard screen and a tangent, axis-symmetric surface $\zeta(|\mathbf{q}|^2)$ of polynomial form (14.92), in an elliptic index-profile medium (14.83). We denote it by $\mathcal{R}^{\mathrm{e}}_{n,\nu;\zeta} =: \mathcal{G}(\mathbf{R}^{\mathrm{e}};\mathbf{M}_{\mathrm{R}})$. This will be useful in analyzing the performance of axis-symmetric graded-index lenses for fractional Fourier transformers in the next chapter. The paraxial matrix $\mathbf{M}_{\mathrm{R}}$ is the *same* as for the root transformation in a homogeneous medium given in (14.93). The aberration part, organized in the same form as in (14.94)–(14.96), is characterized by the following harmonic coefficients [153]:

$$\mathbf{R}^{\mathrm{e}}_2 = \{\{0,\,0,\,-\zeta_2/2n,\,0,\,n\zeta_4-\zeta_2\nu^2/2n\},\,\{-\zeta_2/3n\}\}; \tag{14.101}$$

$$\mathbf{R}^{\mathrm{e}}_3 = \{\{0,\,0,\,-\zeta_2/8n^3,\,-\zeta_2^2/2n^2,\,-\zeta_2\nu^2/4n^3-\zeta_4/2n,\,2\zeta_2\zeta_4-\zeta_2^2\nu^2/2n^2,\,n\zeta_6-\zeta_2\times \nu^4/8n^3-\zeta_4\nu^2/2n\},\,\{-\zeta_2/10n^3,\,-\zeta_2^2/5\,n^2,\,-\zeta_2\nu^2/5\,n^3-2/5\,\zeta_4/n\}\}; \tag{14.102}$$

$$\mathbf{R}^{\mathrm{e}}_4 = \{\{0,\,0,\,-\zeta_2/16n^5,\,-\zeta_2^2/4\,n^4,\,-3/16\,\zeta_2\nu^2/n^5-\zeta_4/8n^3-5/6\,\zeta_2^3/n^3,\,-\zeta_2^2\nu^2/2n^4-2\zeta_2\zeta_4/n^2,\,-3/16\,\zeta_2\nu^4/n^5-\zeta_4\nu^2/4\,n^3-\zeta_2^3\nu^2/3\,n^3-\zeta_6/2n+16/3\,\zeta_2^2\zeta_4/n,\,6\zeta_2\zeta_6-\zeta_2^2\nu^4/4\,n^4-2\zeta_2\zeta_4\nu^2/n^2,\,8/3\,n\zeta_2\zeta_4^2+n\zeta_8-\zeta_2\nu^6/16n^5-\zeta_4\nu^4/8n^3+\zeta_2^3\nu^4/2n^3-8/3\,\zeta_2^2\zeta_4\nu^2/n-\nu^2\zeta_6/2n\},\,\{-3/56\,\zeta_2/n^5,\,-\zeta_2^2/7\,n^4,\,-3/14\zeta_2\nu^2/n^5-\zeta_4/7\,n^3-2/7\times \zeta_2^3/n^3,\,-2/7\,\zeta_2^2\nu^2/n^4-8/7\,\zeta_2\zeta_4/n^2,\,-9/56\,\zeta_2\nu^4/n^5-3/14\zeta_4\nu^2/n^3+8/21\zeta_2^3\nu^2/n^3-3/7\,\zeta_6/n-16/21\zeta_2^2\zeta_4/n\},\,\{-\zeta_2\nu^2/10n^5-\zeta_4/15n^3\}\}. \tag{14.103}$$

The tables of this chapter cover the aberrations of axis-symmetric systems, whose screens are $D=2$-dimensional [phase space $(\mathbf{p},\mathbf{q}) \in \mathsf{R}^4$], and whose elements are axis-symmetric, i.e., invariant under a common group $\mathsf{O}(2)$ of rotations around the optical axis, and reflections across planes containing this axis. The theory of higher-order perturbations in classical mechanics of D-dim systems [phase space $(\vec{p},\vec{q}) \in \mathsf{R}^{2D}$] with $\mathsf{O}(D)$ invariance ('spherical' symmetry), has a reduced phase space which is again 3-dim, namely $(|\vec{p}|^2,\,\vec{p}\cdot\vec{q},\,|\vec{q}|^2)$. The aberration multiplets will correspond to the states degenerate in energy of a $(D+1)$-dim quantum harmonic oscillator, and can be classified accordingly.

15 Parametric correction of fractional Fourier transformers

Fractional Fourier transformations $\mathcal{F}^\alpha$ are linear, canonical operators which rotate the phase space of rays. In the $D = 1$-dim paraxial régime,[1]

$$\mathcal{F}^\alpha := \mathcal{M}(\mathbf{F}^\alpha), \qquad \mathbf{F}^\alpha = \begin{pmatrix} \cos\frac{\pi}{2}\alpha & -\sin\frac{\pi}{2}\alpha \\ \sin\frac{\pi}{2}\alpha & \cos\frac{\pi}{2}\alpha \end{pmatrix}, \tag{15.1}$$

$$\mathcal{F}^\alpha : \begin{pmatrix} p \\ q \end{pmatrix} = \mathbf{F}^{-\alpha}\begin{pmatrix} p \\ q \end{pmatrix} = \begin{pmatrix} \cos\frac{\pi}{2}\alpha & \sin\frac{\pi}{2}\alpha \\ -\sin\frac{\pi}{2}\alpha & \cos\frac{\pi}{2}\alpha \end{pmatrix}\begin{pmatrix} p \\ q \end{pmatrix} =: \begin{pmatrix} p_\alpha \\ q_\alpha \end{pmatrix}. \tag{15.2}$$

The set of fractional Fourier transforms constitutes a Lie group $\mathsf{U}(1)$.

The Fourier transform $\mathcal{F}$ has appeared often; it is the foundation of the Hamiltonian formulation of geometric optics (and of classical and quantum mechanics) on phase space. In attempting to build an optical Fourier transformer to perform the map (15.2), one problem we have argued before [see pages 31 and 90], is that a true *global* Fourier transform of optical phase space $\wp$ (see Chap. 2) *cannot* be linear, except in the first-order, paraxial approximation. A second problem is that actual optical systems composed of refracting surfaces and/or graded-index guides *cannot* reproduce exactly any desired canonical transformation.[2]

In the following three sections we analyze again three structurally different optical setups which served as *paraxial* fractional Fourier transformers, but now in the metaxial régime. They are: a 1-lens DLD arrangement [see (9.14), (10.8)–(10.10) and Fig. 10.2], a length of harmonic optical guide [see Sect. 2.3, and (3.38) and (14.83)–(14.84)], and a system *cum* reflection in a warped mirror (see Sect. 4.6 and 10.4). These arrangements are shown schematically in Fig. 15.1. They are important non-imaging devices which can be characterized by Hamilton–Lie aberration coefficients that depend on the structural parameters of nonspherical lenses and mirrors. In every case, the factored-product parametrization (13.26)–(13.30) separates the desired paraxial part from the unwanted aberrations; and our purpose is to find strategies to minimize the latter, so that the system will perform closer to the ideal linear fractional Fourier transformer (15.2). We work in $D = 1$ dimensions because formulas are simpler, the monomial and symplectic harmonic

[1] Recall our careful discussion on page 161: the matrix which characterizes the canonical transformation acts through its *inverse* on the phase space vector.

[2] This statement still awaits a proper proof ...

bases are the same (labelled by rank k and weight m, $|m| \leq k$), and phase space deformations can be plotted in plane figures; but also because planar optical chips may be more realistic devices for finite signal analysis. Some thoughts in this direction will be placed in the concluding section.

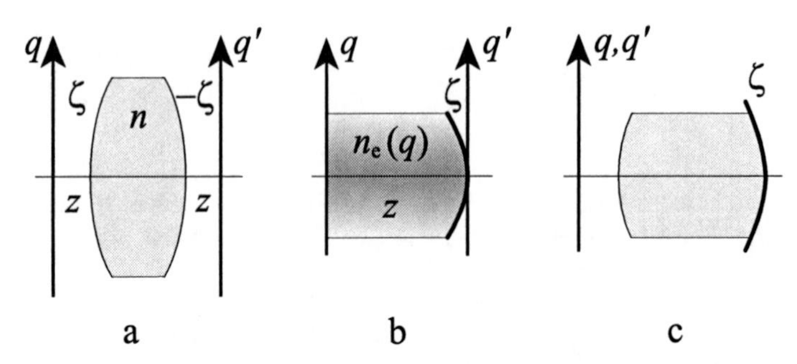

Fig. 15.1. Three fractional Fourier transformer devices from the paraxial régime (with the extra metaxial parameters): (**a**) Symmetric arrangement of one thick lens with polynomial surfaces; (**b**) Elliptic index-profile guide with a polynomial exit face; (**c**) 'cat-eye' system of one spherical refracting surface, *cum* reflection in an adjustable warped mirror of polynomial shape.

15.1 The lens arrangement

Fractional Fourier transformers can be approximated by symmetric optical systems composed of a lens that is equidistant between the object and image screen-lines, as shown in Fig. 15.1a. We concatenate the five elements of this arrangement: two free spaces of length $z \geq 0$ in the medium n_1, and a lens of thickness $\tau \geq 0$ and refractive index n_2, bounded by the two symmetric refracting interfaces $\pm z = \zeta := \zeta_2 q^2 + \zeta_4 q^4 + \cdots$ as in (14.92), namely

$$\mathcal{F}^\alpha \approx \mathcal{G}^{\rm F}(\mathbf{A}^{\rm F}; \mathbf{F}^\alpha) = \mathcal{D}(z/n_1)\, \mathcal{S}(n_1, n_2; \zeta)\, \mathcal{D}(\tau/n_2)\, \mathcal{S}(n_2, n_1; -\zeta)\, \mathcal{D}(z/n_1). \tag{15.3}$$

Since ζ is centered, only integer-rank aberrations will be present in $\mathbf{A}^{\rm F}$. For simplicity, we shall set $n_1 = 1$ and remain with generic $n := n_2$; the optical thickness of the lens we denote by $t := \tau/n$.

In the paraxial approximation, the two faces of the lens have Gaussian powers $g := 2(n-1)\zeta_2 = 2(1-n)(-\zeta_2)$, and the 2×2 matrices of this régime multiply to produce the fractional Fourier transform (see Sect. 10.3.1),

$$\begin{aligned} \mathbf{F}^\alpha &= \begin{pmatrix} 1 & 0 \\ -z & 1 \end{pmatrix} \begin{pmatrix} 1 & g \\ 0 & 1 \end{pmatrix} \begin{pmatrix} 1 & 0 \\ -t & 1 \end{pmatrix} \begin{pmatrix} 1 & g \\ 0 & 1 \end{pmatrix} \begin{pmatrix} 1 & 0 \\ -z & 1 \end{pmatrix} \\ &= \begin{pmatrix} 1 - gt - 2gz + g^2 zt & g(2 - gt) \\ -2z - t + 2z^2 g + 2gzt - g^2 z^2 t & 1 - gt - 2gz + g^2 zt \end{pmatrix}. \end{aligned} \tag{15.4}$$

When we specify the desired Fourier power $0 \leq \alpha < \pi$ and the available Gaussian power of the lens faces g, then the two length parameters are determined:

$$z = \frac{1}{g} + \cot\frac{\alpha}{2} \geq 0, \quad t = \frac{\tau}{n} = \frac{\sin\alpha + 2g}{g^2} \geq 0. \tag{15.5}$$

The aberration part of the arrangement contains the coefficients $\mathbf{A}^{\rm F}$ of the ordered product (15.3), which is computed with the semidirect multiplication of (13.31)–(13.33), where the matrices $\mathbf{D}^k(\mathbf{M})$ represent the action of the paraxial part of the first factor on the aberrations of the second [see (13.34)–(13.37)], and the *gato* composition '$\sharp$' of the aberrations. [See (13.38), detailed for the coefficients of aberration orders 3, 5, and 7 in (14.71), (14.72), and (14.74) respectively, for the maximal symplectic spin $j = k$ applicable in the $D = 1$ case.] Since associativity holds [see (13.44) and (13.52)], we calculate first the aberration coefficients of the lens in (15.3),

$$\mathcal{G}(\mathbf{L};\mathbf{M}_{\rm L}) := \mathcal{S}(1,n;\zeta)\,\mathcal{D}(t)\,\mathcal{S}(n,1;-\zeta), \tag{15.6}$$

$$\mathbf{L} = \mathbf{S}^{(1,n;\zeta)} \sharp \mathcal{M}(\mathbf{M}_{\rm S}^{(1,n;\zeta)}) : [\mathbf{D}^{(t)} \sharp \mathcal{M}(\mathbf{M}_{\rm D}^{(t)}) : \mathbf{S}^{(n,1;-\zeta)}], \tag{15.7}$$

with the paraxial matrices and aberration coefficients of the free displacement [$\mathbf{M}_{\rm D}^{(t)}$ in (14.81) and $\mathbf{D}_k^{(t)}$ in (14.82)], of the refracting surfaces [$\mathbf{M}_{\rm S}^{(n_1,n_2;\pm\zeta)}$ in (14.97) and $\mathbf{S}_k^{(n_1,n_2;\pm\zeta)}$ in (14.98)–(14.100)], and the lens thickness τ determined by (15.5). And finally, we place this lens between the two free displacements by z with similar operations. Clearly, this is a task best done by machine. To third aberration order, the $\sharp$ operation is sum between 5-vectors acted by triangular 5×5 matrices. Generally, in rank k only the first k coefficients ζ_{2k} of the surface appear, and ζ_2 is fixed by the paraxial g. Thus, only ζ_4 is present as an adjustable new parameter [see (14.94)–(14.96)].

The one-lens optical fractional Fourier transformer acts on phase space by rotating it linearly with the paraxial part, and then aberrating it with terms of third homogeneous degree:

$$\begin{aligned}\begin{pmatrix} p^{[3]}(p,q;\alpha) \\ q^{[3]}(p,q;\alpha)\end{pmatrix} &= \mathcal{G}_2^{\rm F}(\mathbf{A}^{\rm F};\mathbf{F}^\alpha)\begin{pmatrix} p \\ q\end{pmatrix} = \mathcal{G}_2^{\rm F}(\mathbf{A}^{\rm F};\mathbf{1})\begin{pmatrix} p_\alpha \\ q_\alpha\end{pmatrix} \\ &= \begin{pmatrix} p_\alpha + A^{\rm F}_{2,1}p_\alpha^3 + 2A^{\rm F}_{2,0}p_\alpha^2 q_\alpha + 3A^{\rm F}_{2,-1}p_\alpha q_\alpha^2 + 4A^{\rm F}_{2,-2}q_\alpha^3 \\ q_\alpha - 4A^{\rm F}_{2,2}p_\alpha^3 - 3A^{\rm F}_{2,1}p_\alpha^2 q_\alpha - 2A^{\rm F}_{2,0}p_\alpha q_\alpha^2 - A^{\rm F}_{2,-2}q_\alpha^3 \end{pmatrix},\end{aligned} \tag{15.8}$$

where $\begin{pmatrix} p_\alpha \\ q_\alpha\end{pmatrix}$ is defined in (15.2). Now, how should we 'minimize' the aberrations over the phase space plane? There may be more than one criterion.

Example: The uncorrected one-lens Fourier transformer. As a basis for comparison, we analyze first an 'uncorrected' Fourier transformer of power $\alpha = 1$ built with *circular* refracting lines $z = \pm\zeta_r(q)$ of radius r (the $D = 1$ counterpart of spherical-face lenses), where each surface

$$\pm z = \zeta_r(q) := r - \sqrt{r^2 - q^2} \approx \frac{q^2}{2r} + \frac{q^4}{8r^3} + \frac{q^6}{16r^5} + \frac{5q^8}{128r^7}, \tag{15.9}$$

has Gaussian power $g = (n-1)/r$. We choose length units by setting $r = 1$, so that $\zeta_2 = \frac{1}{2}$, $\zeta_4 = \frac{1}{8}$, $\zeta_6 = \frac{1}{16}$, etc., and set the value $n = 1.5$ for the refractive index of the lens, so the Gaussian power of each of its lines is fixed to $g = \frac{1}{2}$; the remaining parameters result from (15.5). In Figs. 15.2 we show the deformation of phase space given by (15.8) with respect to the rotated axes (p_1, q_1), and computed to aberration orders 3, 5, and 7. There we can see the patch of phase space where the aberration expansion cogently converges, and where it may diverge. •

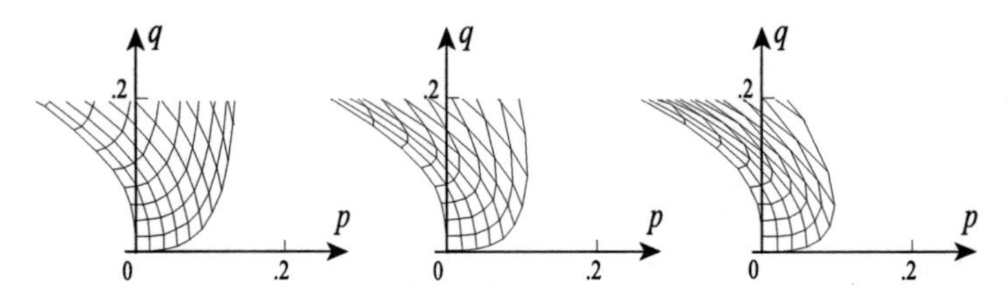

Fig. 15.2. Phase space maps of an (*uncorrected*) Fourier transformer of power $\alpha = 1$, built with one thick lens of circular refracting faces, in aberration orders 3, 5, and 7, *left to right* respectively. The axes have been rotated clockwise by $\frac{1}{2}\pi$ to $p_1 = q$ and $q_1 = -p$.

Seeing Figs. 15.2 it is natural to ask *what* should and can be corrected with the parameters ζ_4, ζ_6, etc., so that the image of the rotated Cartesian coordinate grid of phase space be as undeformed as possible. The vectors tangent to the aberrated grid (15.8) are

$$\mathbf{u}_\alpha(p,q) := \frac{\partial}{\partial p}\begin{pmatrix} p^{[3]}(p,q;\alpha) \\ q^{[3]}(p,q;\alpha) \end{pmatrix}, \quad \mathbf{v}_\alpha(p,q) := \frac{\partial}{\partial q}\begin{pmatrix} p^{[3]}(p,q;\alpha) \\ q^{[3]}(p,q;\alpha) \end{pmatrix}, \quad (15.10)$$

and we want them orthogonal. To third aberration order, their scalar product $\mathbf{u}_\alpha(p,q)\cdot\mathbf{v}_\alpha(p,q)$ at each point $(p,q)\in\mathsf{R}^2$ is

$$2(A^{\mathrm{F}}_{2,0}-6A^{\mathrm{F}}_{2,2})\,p^2 + 6(A^{\mathrm{F}}_{2,-1}-A^{\mathrm{F}}_{2,1})\,pq + 2(6A^{\mathrm{F}}_{2,-2}-A^{\mathrm{F}}_{2,0})\,q^2. \quad (15.11)$$

This expression generally cannot be made zero by some value of ζ_4, so the aberrated grid cannot be straightened out everywhere. But we can follow the tactic of asking for grid orthogonality on a *line* in phase phase space – say, on the quadrant bisectors $q = p =: r$. Then, (15.11) reduces to

$$\mathbf{u}_\alpha(r,r)\cdot\mathbf{v}_\alpha(r,r) = [12(A^{\mathrm{F}}_{2,-2} - A^{\mathrm{F}}_{2,2}) + 6(A^{\mathrm{F}}_{2,-1} - A^{\mathrm{F}}_{2,1})]\,r^2 + \text{terms in } r^4, \quad (15.12)$$

where now the term in r^2 can be made zero (but not together with the term in r^4). Or we can resort to a second tactic: asking for orthogonality of the vectors tangent to the edges of the aberrated quadrants, $(r,0)$ and $(0,r)$. The relevant scalar product is then

$$\mathbf{u}_\alpha(r,0)\cdot\mathbf{v}_\alpha(0,r) = 12(A^{\rm F}_{2,-2} - A^{\rm F}_{2,2})\,r^2 + 36(A^{\rm F}_{2,1}A^{\rm F}_{2,-2} + A^{\rm F}_{2,-1}A^{\rm F}_{2,2})\,r^4. \tag{15.13}$$

Again, the first term can be made zero with the single parameter ζ_4, but the second cannot. In deciding how to best fix ζ_4, we may follow a third tactic: setting

$$A^{\rm F}_{2,2} = A^{\rm F}_{2,-2}, \tag{15.14}$$

because spherical aberration and pocus are Fourier transform aberrations of each other, and the first is usually the largest. By choosing the criterion (15.14) we may hope that the non-orthogonality of the grid at the phase space bisectors (15.12) *and* at the edges (15.13) will be reduced. Equation (15.14) is linear in the parameter ζ_4 of the polynomial refracting interfaces, so its value can be obtained easily.

Once the system has been 'corrected' to third aberration order with the parameter ζ_4, we subject the fifth-order aberrations to a similar treatment. There are now seven independent coefficients $\{A^{\rm F}_{3,m}\}^3_{m=-3}$, and the previous deformed cubic-line Cartesian coordinate grid will be further mapped on quintic curves that osculate the previous cubics. The departure from orthogonality can be analyzed as above [see (15.10)–(15.12)] and is again dominated by the fifth-order spherical aberration and pocus coefficients, $A^{\rm F}_{3,3}$ and $A^{\rm F}_{3,-3}$, where now ζ_6 is again present linearly. The same compromise tactic to correct as far as possible leads to the condition $A^{\rm F}_{3,3} = A^{\rm F}_{3,-3}$ which determines the value of ζ_6. This is then introduced to the seventh-order correction which fixes ζ_8. Hand algebra is impractical, but the task can be handled conveniently with the symbolic computation algorithms of `mexLIE` [161].

Figures 15.3 show the phase space aberrations that thick-lens arrangements suffer, compared with the ideal linear fractional Fourier transform, to seventh aberration order. The aberrations of the 'uncorrected' arrangements built with lenses of circular faces (15.9) are in **a**, to be compared with the 'corrected' polynomial-face lenses in **b**. The criterion for correction was the equality (15.14) for ranks $k = 2, 3, 4$ to determine the values of ζ_4, ζ_6, ζ_8 that appear below, and all other parameters set to the numerical values specified in (15.9), namely: $g = \frac{1}{2}$ for each surface and the fractional Fourier rotation angle of phase space, $\phi = \frac{1}{2}\pi\alpha$, at intervals of 15°. The refractive index of the lens is $n = \frac{3}{2}$, so $\zeta_2 = \frac{1}{2}$, and the lengths z and τ of the arrangements are determined from (15.5).

Fourier angle ϕ	15°	30°	45°	60°	75°	90°
ζ_4	−0.0159	−0.0303	−0.0350	−0.0310	−0.0183	0.0000
ζ_6	0.0213	0.0179	0.0087	−0.0010	−0.0044	0.0000
ζ_8	−0.0211	−0.0152	−0.0022	−0.0025	−0.0070	−0.0062

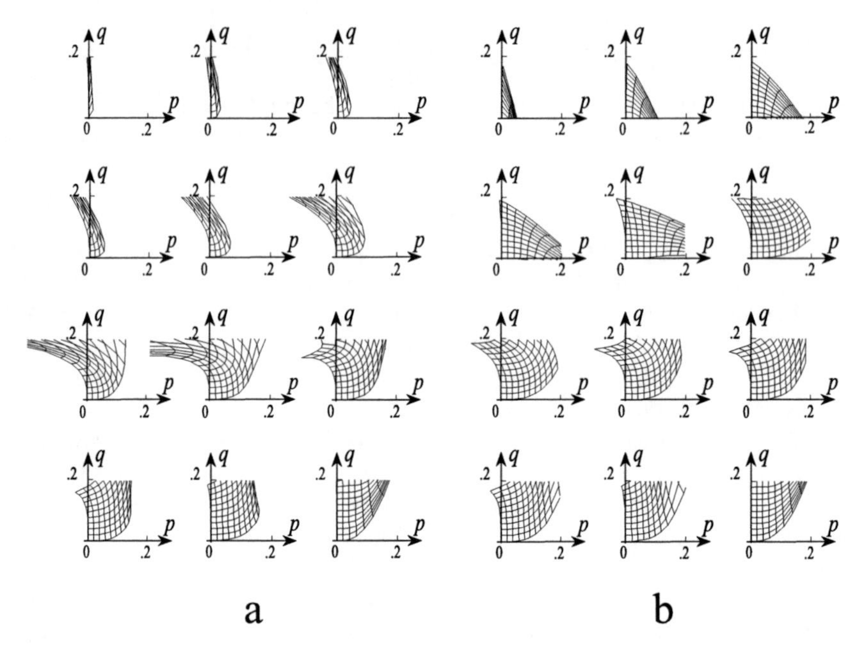

Fig. 15.3. Phase space maps of the aberrations of one-lens arrangements for fractional Fourier transformation. (**a**) 'uncorrected' lens with circular refracting face lines, and (**b**) 'corrected' lens with polynomial face lines as described in the text, for Fourier angles $\phi = \frac{1}{2}\pi\alpha = 15°, 30°, 45°, \ldots, 180°$.

Fourier angle ϕ	105°	120°	135°	150°	165°	180°
ζ_4	0.0194	0.0428	0.0782	0.1264	0.1786	0.0816
ζ_6	0.0115	0.0153	0.0515	−0.3021	−0.6600	0.0250
ζ_8	0.0063	0.0701	0.2703	1.3753	5.3782	0.0098

By comparison, the polynomial coefficients for circular refracting lines are $\zeta_2 := \frac{1}{2}$, $\zeta_4 = 1/8 = 0.125$, $\zeta_6 = 1/16 = 0.0625$, and $\zeta_8 = 5/128 = 0.039062$.

The figures suggest that the polynomial correction to the lens arrangement of fractional Fourier transformers is appropriate for Fourier angles ϕ greater than 60°, including the Fourier transform of unit power at $\phi = 90°$, up to the inverted imager $\phi = 180°$, and in a sensibly larger patch of phase space. For smaller transform angles ϕ, the free flight distance z becomes very large [*cf.* (15.5)] and the aberration expansion itself appears to be unreliable. From the numerical values of the polynomial coefficients ζ_{2k} tabulated above, we note that they vary smoothly with ϕ, and peak at $\phi = 165°$. Particularly, we point out that for the Fourier transformer $\alpha = 1$ ($\phi = 90°$), the 'best' profile for the lens is essentially a parabola.

15.2 The warped-face guide arrangement

We have often encountered the harmonic guide whose refractive index $n_{\rm e}(q)$ has the elliptic profile (14.83), which rotates phase space (see Sect. 2.3). As the standard screen (line) advances along the guide, it registers a paraxial inverse fractional Fourier transformations (see Sect. 10.3.1 and Sect. C.3–4). But as Fig. 2.5 on page 23 shows, the circles farther away from the optical center rotate with increasing angular frequency, producing the aberrations whose polynomials are given in (14.85). With these formulas we find the 'uncorrected' deformed grid of Fig. 15.4, corresponding to the operator $\mathcal{E}_{n,\nu;z} = \exp(-z\{h_{\rm e}, \circ\})$ in (14.84)–(14.85) for $D = 1$-dim systems, and calculated for increasing aberration orders 3, 5, 7 as before in Fig. 15.2; these are polynomial approximants to the exact formula used for Fig. 2.5.

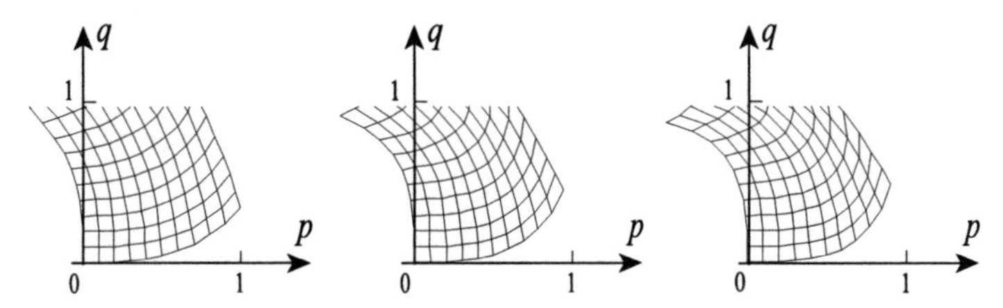

Fig. 15.4. Phase space maps of an inverse Fourier transformer of power $\alpha = 1$, built with a length of guide of elliptic index profile and straight faces, in aberration orders 3, 5, and 7, *left to right* respectively. As in Fig. 13.2, The axes have been rotated clockwise by $\frac{1}{2}\pi$ to show the deformation of the first quadrant.

The leftward shear in the deformed grid in Fig. 15.4 can be counteracted by the rightward shear of phase space produced by a lens (cf. Fig. 9.2); so if we replace the straight ends of the guide by a warped output face or screen-line, we may produce a better inverse-Fourier transformer. We thus design a polynomial line, $z = \zeta := \zeta_2 q^2 + \zeta_4 q^4 + \cdots$ as in Fig. 15.1b and (14.92), where we wrap the photographic paper or place light sensors to detect ray positions or/and directions. (The 'straightening out' of this phase space to one referred back to the flat standard screen is then performed by *non-optical* means.) The change of reference surface for phase space is carried by the *root* transformation in the guide, $\mathcal{R}^{\rm e}_{n,\nu;\zeta}$, characterized by its aberration coefficients, (14.101)–(14.103). [Recall Chap. 4 for the geometric meaning of the root transformation between any two surfaces (lines) in an inhomogeneous medium.]

We have annotated the explicit aberration coefficients of the root transformation in a guide in (14.101)–(14.103). Firstly, we note that $\zeta_2 = 0$, because there should not be any parxial shear that would deform the circular fractional Fourier trajectories into ellipses. So ζ_4, ζ_6, $\zeta_8, \ldots$ are the only effective correction parameters for aberration orders 3, 5, 7, ... The aberrations in this root transformation, $\mathcal{R}^{\rm e}_{n,\nu;\zeta} = \mathcal{G}(\mathbf{R}^{\rm e};\mathbf{1})$, are quite sparse:

$$\begin{aligned}&\mathbf{R}^{\rm e}_2 = \{0,0,0,0,n\zeta_4\}, \qquad \mathbf{R}^{\rm e}_3 = \{0,0,0,0,-\zeta_4/2n,\, n\zeta_6-\zeta_4\nu^2/2n\},\\ &\mathbf{R}^{\rm e}_4 = \{0,0,0,0,-\zeta_4/8n^3,0,-\zeta_6/2n-\zeta_4\nu^2/4n^3,0,n\zeta_8-\zeta_6\nu^2/2n-\zeta_4\nu^4/8n^3\}.\end{aligned} \tag{15.15}$$

The guide of length z with the refractive index-profile $n_{\rm e}(n,\nu)$ in (14.83), and with the symmetric quartic-and-higher polynomial exit face ζ in (14.92), approximates the fractional Fourier transform $\mathcal{F}^\alpha$, through

$$\mathcal{F}^\alpha \approx \mathcal{E}_{n,\nu;z}\,\mathcal{R}^{\rm e}_{n,\nu;\zeta} = \mathcal{G}(\mathbf{E}^{(n,\nu,z)};\mathbf{M}_{\rm e}(n,\nu;z))\,\mathcal{G}(\mathbf{R}^{(n,\nu,\zeta)};\mathbf{1}) \tag{15.16}$$

$$= \mathcal{G}(\mathbf{A}^{\rm F}(\alpha);\,\mathbf{F}^\alpha) := \mathcal{G}(\mathbf{E}\sharp\mathbf{D}(\mathbf{F}^{-\alpha})^\top\mathbf{R};\,\mathbf{F}^\alpha), \tag{15.17}$$

with $\mathbf{M}_{\rm e}(n,\nu;z) = \mathbf{F}^\alpha$ and the negative Fourier rotation angle $\phi = \frac{1}{2}\pi\alpha = -\nu\, z/n < 0$.

To aberration order 3 (rank 2), the vectors in (15.17) are sparse enough to allow the explicit computation of $\mathbf{A}^{\rm F}_2(\alpha)$,

$$\begin{aligned}A^{\rm F}_{2,2} &= -(\phi/8n^2) + n\zeta_4\sin^4\phi, & A^{\rm F}_{2,1} &= 0,\\ A^{\rm F}_{2,0} &= -(\phi/4n^2) + 6n\zeta_4\sin^2\phi\cos^2\phi, & A^{\rm F}_{2,-1} &= 0,\\ A^{\rm F}_{2,-2} &= -(\phi/8n^2) + n\zeta_4\cos^4\phi.\end{aligned} \tag{15.18}$$

Again, as in the previous section, we are faced with the question of *what* to correct with ζ_4. [If we were to demand the condition (15.14) then, except for $\phi = \frac{1}{4}\pi$, we would get $\zeta_4 = 0$.] For reference, in Figs. 15.5 we plot the deformation of a phase space grid in the 'uncorrected' guide with straight faces, to seventh aberration order, together with its correction according to the following criterion: set spherical aberration coefficients successively to zero, $A^{\rm F}_{k,k} = 0$. This will glue the corrected grid of phase space to the rotated-q axis of the figures, which is vertical. The rationale of this tactic lies on imagining that we want to analyze the ray direction (momentum) content of a narrow-waisted beam (a thin strip in phase space along the horizontal p-axis) through the (inverse) Fourier transform, $\mathcal{F}^{-1}$. The paraxial part will rotate this to a thin vertical strip, which will project on the q-axis with minimal unfocusing. This criterion determines ζ_4, ζ_6, ζ_8 to the values given below, which were used to produce the corrected grids in Figs. 15.5; their values are negative, so the guide will have a convex exit face. On the basis of these figures, we conclude that waveguide arrangements to produce (inverse) fractional Fourier transforms are adequate for Fourier angles up to about 90°.

Fourier angle ϕ	15°	30°	45°	60°	75°	90°
$-\zeta_4$	0.0097	0.0194	0.0291	0.0388	0.0485	0.0582
$-\zeta_6$	0.0043	0.0086	0.0129	0.0172	0.0215	0.0259
$-\zeta_8$	0.0018	0.0038	0.0060	0.0086	0.0117	0.0155

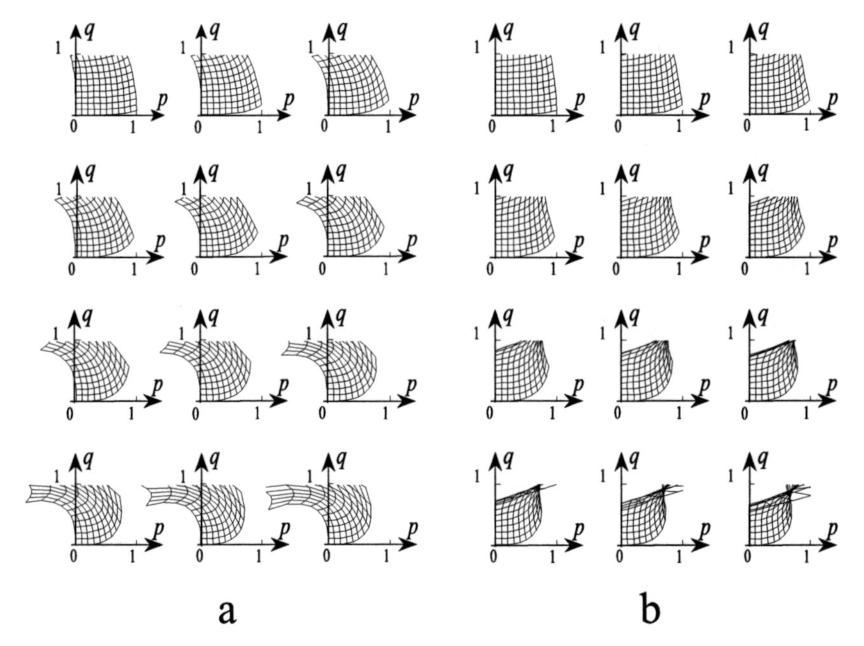

Fig. 15.5. Phase space maps of the aberrations of a guide of elliptic index-profile with a warped exit face, to produce inverse fractional Fourier transformation. (**a**) 'uncorrected' guide with straight faces and (**b**) 'corrected' guide with a warped face and no spherical aberration, as described in the text, for Fourier angles $\phi = \frac{1}{2}\pi\alpha = 15°, 30°, 45°, \ldots, 180°$.

Fourier angle ϕ	105°	120°	135°	150°	165°	180°
$-\zeta_4$	0.0679	0.0776	0.0873	0.0970	0.1067	0.1164
$-\zeta_6$	0.0302	0.0345	0.0388	0.0431	0.0474	0.0517
$-\zeta_8$	0.0201	0.0256	0.0321	0.0398	0.0489	0.0594

Remark: Guides in the $D = 2$ case. Cylindrical guides with $D = 2$-dim faces and any axis-symmetric refractive index profile, suffer from the ailment that, according to their skewness, rays will spiral around the guide [see (14.18)–(14.22) and Fig. 14.3a], with no possible correction. •

15.3 The cat's eye arrangement

We dedicated Sect. 10.4 to study systems *cum* reflection in the paraxial régime. These are optical arrangements such as that shown in Fig. 15.1c, where light, after traversing a system $\mathcal{G}$, is reflected back to the object screen by a mirror, and goes through the same system that now acts as $\overline{\mathcal{G}}$; given thus $\mathcal{G}$, the system *cum* reflection is $\mathcal{G}^{\text{II}} := \mathcal{G}\,\overline{\mathcal{G}}$. There we showed that *any* paraxial optical system (except imagers), *cum* reflection in an appropriately warped mirror, is a pure fractional Fourier transformer. We exemplified such a device by the cat's eye arrangement given by (10.46)–(10.47), where the reflecting surface is –paraxially – an ajustable parabola.

In this section we examine the aberrations of the cat's eye in the metaxial régime. We consider plane optical systems that are symmetric across the optical axis, and whose aberrations are thus of integer rank; we shall work with $k = 1$ (paraxial), $k = 2$ and 3 (aberration orders 3 and 5) only. In the factored product parametrization defined in (13.27)–(13.30), the symmetric rank-3 aberration group elements

$$\mathcal{G}_3(\mathbf{A};\mathbf{M}) := \exp\{A_3, \circ\} \times \exp\{A_2, \circ\} \times \mathcal{G}(\mathbf{0};\mathbf{M}), \tag{15.19}$$

realize optical systems whose paraxial part is $\mathbf{M} \in \mathsf{Sp}(2,\mathsf{R})$, and whose aberrations $\mathbf{A}$ are given by polynomials $A_2(p,q)$, $A_3(p,q)$, of homogeneous degrees 4 and 6 in the phase space coordinates.

The basic properties of reflection were given for the global régime in Sect. 4.6 (*e.g.* the inversion of the order of factors); in Sect. 10.4, where we studied the paraxial part, we determined the action of reflection on the aberration monomial basis in (13.10) [and in (14.44) for the $D = 2$-dim axis-symmetric case]. It follows that the reflection of the generic metaxial system (15.19) is

$$\begin{aligned}\overline{\mathcal{G}_3(\mathbf{A};\mathbf{M})} &= \overline{\mathcal{G}(\mathbf{0};\mathbf{M})} \times \exp\{\overline{A}_2, \circ\} \times \exp\{\overline{A}_3, \circ\} \\ &= \mathcal{G}_3\Big(\mathbf{D}^2(\overline{\mathbf{M}}^{-1})^{\top}\,\overline{\mathbf{A}}_2;\, \mathbf{1}\Big)\, \mathcal{G}_3\Big(\mathbf{D}^3(\overline{\mathbf{M}}^{-1})^{\top}\,\overline{\mathbf{A}}_3;\, \mathbf{1}\Big)\, \mathcal{G}(\mathbf{0};\overline{\mathbf{M}}),\end{aligned} \tag{15.20}$$

where $\mathbf{D}^k(\mathbf{M})$ are the $(2k+1) \times (2k+1)$ representation matrices for $\mathbf{M} \in \mathsf{Sp}(2,\mathsf{R})$ [see (13.6) and (13.13)–(13.16)]. The reflection of the paraxial part was given in (10.36) through the nonsymplectic reflection matrix $\mathbf{K} := \begin{pmatrix} -1 & 0 \\ 0 & 1 \end{pmatrix} = \mathbf{K}^{-1}$, as $\overline{\mathbf{M}} = \mathbf{K}\,\mathbf{M}^{-1}\,\mathbf{K}$. The reflection of the monomial basis of the aberration polynomials $\overline{A}_2(p,q)$ and $\overline{A}_3(p,q)$ follow from (13.10), *viz.*,

$$\overline{\begin{pmatrix} a & b \\ c & d \end{pmatrix}} = \begin{pmatrix} d & b \\ c & a \end{pmatrix}, \qquad \begin{aligned}\overline{A}_{k,m} &= (-1)^{k+m}\, A_{k,m}, \\ \overline{\mathbf{A}}_k &= \mathbf{D}^k(\mathbf{K})\, \mathbf{A}_k.\end{aligned} \tag{15.21}$$

And because the rank-2 and -3 aberrations commute within the rank-3 symmetric aberration Lie algebra [recall (13.24)], the two leftmost factors in (15.20) also commute. To rank 3, thus

$$\overline{\mathcal{G}_3(\mathbf{A};\mathbf{M})} = \mathcal{G}_3\Big(\mathbf{D}(\overline{\mathbf{M}}^{-1})^{\top}\,\overline{\mathbf{A}};\, \overline{\mathbf{M}}\Big). \tag{15.22}$$

The system $\mathcal{G}_3(\mathbf{A};\mathbf{M})$ *cum* its reflection is now given by

$$\begin{aligned}\mathcal{G}^{\text{II}}(\mathbf{A};\mathbf{M}) &:= \mathcal{G}_3(\mathbf{A};\mathbf{M})\,\overline{\mathcal{G}_3(\mathbf{A};\mathbf{M})} = \mathcal{G}_3(\mathbf{A}^{\text{II}};\mathbf{M}^{\text{II}}),\\ \mathbf{M}^{\text{II}} = \mathbf{M}\,\overline{\mathbf{M}}, &\quad \mathbf{A}^{\text{II}} = \mathbf{A}\,\sharp\,\mathbf{D}([\mathbf{M}^{\text{II}}]^{-1})^{\top}\overline{\mathbf{A}} = \mathbf{A}\,\sharp\,\mathbf{D}(\mathbf{M}^{\text{II}}\mathbf{K})^{\top}\,\mathbf{A},\end{aligned} \tag{15.23}$$

where we used the property of $\mathbf{M}^{\text{II}}$ to be self-reflecting, $\mathbf{M}^{\text{II}} = \mathbf{K}(\mathbf{M}^{\text{II}})^{-1}\mathbf{K}^{-1}$ so $(\mathbf{M}^{\text{II}}\mathbf{K})^2 = \mathbf{1}$, the fact that $\mathbf{D}^k(\mathbf{K})$ is symmetric, and the *gato* composition $\sharp$ of pure aberrations. Within rank 3 (aberration order 5), this operation [as we recall from (13.41) and (13.42), replacing $\frac{3}{2} \mapsto 2$ and $2 \mapsto 3$] initiates with a sum of the highest-rank aberrations, so that

$$\mathbf{A}^{\text{II}}_k = \left(\mathbf{1} + \mathbf{D}^k(\mathbf{M}^{\text{II}}\mathbf{K})^{\top}\right)\mathbf{A}_k + \begin{cases} n\text{-linear terms in } A_{k',m}, \\ k'_1 + \cdots + k'_n - (n-1) = k. \end{cases} \tag{15.24}$$

Example: Aberrations of an imager *cum* reflection. When the system $\mathcal{G}_3(\mathbf{A};\mathbf{M})$ is paraxially a (generally *impure*) imaging device of magnification μ,

$$\mathbf{M} = \begin{pmatrix} 1/\mu & g \\ 0 & \mu \end{pmatrix} \quad \Rightarrow \quad \mathbf{M}^{\text{II}} = \begin{pmatrix} 1 & \beta \\ 0 & 1 \end{pmatrix}, \quad \beta := 2g/\mu, \tag{15.25}$$

cum reflection it becomes paraxially a unit, upright imager. This system corresponds to the point $\chi = 3$ of Fig. 10.8 in Sect. 10.4. Its five third-order aberrations can be found by inspection of (15.23),

$$\begin{pmatrix} A^{\text{II}}_{2,2} \\ A^{\text{II}}_{2,1} \\ A^{\text{II}}_{2,0} \\ A^{\text{II}}_{2,-1} \\ A^{\text{II}}_{2,-2} \end{pmatrix} = \begin{pmatrix} A_{2,2} \\ A_{2,1} \\ A_{2,0} \\ A_{2,-1} \\ A_{2,-2} \end{pmatrix} + \begin{pmatrix} 1 & 0 & 0 & 0 & 0 \\ -4\beta & 1 & 0 & 0 & 0 \\ 6\beta^2 & -3\beta & 1 & 0 & 0 \\ -4\beta^3 & 3\beta^2 & -2\beta & 1 & 0 \\ \beta^4 & -\beta^3 & \beta^2 & -\beta & 1 \end{pmatrix} \begin{pmatrix} A_{2,2} \\ -A_{2,1} \\ A_{2,0} \\ -A_{2,-1} \\ A_{2,-2} \end{pmatrix}. \tag{15.26}$$

In particular, when the original system is a perfect paraxial magnifier $g = 0$, the odd aberrations of the system *cum* reflection will vanish. Magnifiers *cum* reflection have therefore no coma nor distorsion in aberration orders 3 and 5; the even aberrations double on the other hand, as

$$\begin{aligned} A^{\text{II}}_{2,\pm 2} &= 2A_{2,\pm 2}, \quad A^{\text{II}}_{3,\pm 3} = 2A_{3,\pm 3} + 4A_{2,\pm 2}A_{2,\pm 1}, \\ A^{\text{II}}_{2,0} &= 2A_{2,0}\;, \quad A^{\text{II}}_{3,\pm 1} = 2A_{3,\pm 1} + 12A_{2,\pm 2}A_{2,\mp 1} - 4A_{2,\pm 1}A_{2,0}. \end{aligned} \tag{15.27}$$

In higher orders, the left-to-right reshuffling between (15.20) and (15.22) contributes with multilinear terms of lower-order aberration coefficients; this is an artifact in the parametrization of the factored-product expansion. •

Example: Fourier transformation by a cat's eye. A Fourier transformer can be built in the cat's eye arrangement of Fig. 15.1c, with a mirror ζ appropriately warped in the paraxial régime to produce $\mathbf{M}^{\text{II}} = \mathbf{F}$ [see (15.23), having fixed the coefficient ζ_2 in (14.92)]. As in the previous example, this system will be free of some of its aberrations. When the mirror ζ is warped to a higher-order polynomial shape with parameters, ζ_4, $\zeta_6, \ldots$, the aberration coefficients of the one-pass system $\mathcal{G}(\mathbf{A};\mathbf{M})$ will depend on these parameters, rank by rank, as in the previous two sections; and so will the system *cum* reflection $\mathcal{G}(\mathbf{A}^{\text{II}};\mathbf{M}^{\text{II}})$. Since $\mathbf{D}^k(\mathbf{F}^{\pm 1})$ is antidiagonal [see (13.8)], the third-order aberrations of the Fourier cat's eye can be found as in (15.26), resulting now in

$$A^{\text{II}}_{k,m} = A_{k,m} + A_{k,-m} = A^{\text{II}}_{k,-m}, \qquad k = 2, 3. \tag{15.28}$$

As before, the warp correction parameter ζ_4 of the mirror is present linearly in the third-order spherical aberration $A_{2,2}$, while $A_{3,3}$ is linear in ζ_6 and quadratic in ζ_4. A value of ζ_4 can thus be found which sets $A^{\text{II}}_{2,2} = 0 = A^{\text{II}}_{2,-2}$; and then a ζ_6 for $A^{\text{II}}_{3,3} = 0 = A^{\text{II}}_{3,-3}$. Or alternatively, ζ_4 can be found to set $A^{\text{II}}_{3,2} = A^{\text{II}}_{3,-2}$ to zero, and ζ_6 as before, or a number of other possibilities. In the Fourier basis for aberrations [see (13.9)] where $\mathcal{F}$ is diagonal, the zeros can be placed as in the previous example of imaging systems *cum* reflection. •

Restrictions among aberrations $A^{\text{II}}_{k,m}$ are present in any system *cum* reflection, because the matrix $\mathbf{E}^k(\mathbf{M}^{\text{II}}) := \mathbf{1} + \mathbf{D}^k(\mathbf{M}^{\text{II}}\mathbf{K})^\top$ in (15.24) is an idempotent *projector*, as we now show. Indeed, from the representation property of the $\mathbf{D}$-matrices in (13.7), their determinant in (13.17), and the identity $(\mathbf{M}^{\text{II}}\mathbf{K})^2 = \mathbf{1}$ which is valid for systems *cum* reflection, it follows that

$$\Big(\mathbf{E}^k(\mathbf{M}^{\text{II}})\Big)^2 = \mathbf{1} + 2\mathbf{D}^k(\mathbf{M}^{\text{II}}\mathbf{K})^\top + \mathbf{D}^k\Big((\mathbf{M}^{\text{II}}\mathbf{K})^2\Big)^\top = 2\mathbf{E}^k(\mathbf{M}^{\text{II}}), \tag{15.29}$$

Now, if $\det \mathbf{E}^k \neq 0$, then $\mathbf{E}^k = 2\mathbf{1} \Rightarrow \mathbf{M}^{\text{II}} = \mathbf{K}$, which contradicts $\mathbf{M}^{\text{II}} \in \mathsf{Sp}(2,\mathsf{R})$; hence $\det \mathbf{E}^k = 0$, and thus the matrix $\mathbf{E}^k(\mathbf{M}^{\text{II}})$ in (15.24) projects a nonzero subspace of aberrations of the system *cum* reflection, to zero.

The ideal fractional Fourier transformers $\mathcal{F}^\alpha = \mathcal{G}(\mathbf{0};\mathbf{F}^\alpha)$ in (15.1)–(15.2) are self-reflecting systems,

$$\overline{\mathbf{F}^\alpha} = \mathbf{K}\,\mathbf{F}^{-\alpha}\,\mathbf{K} = \mathbf{F}^\alpha, \qquad (\mathbf{F}^\alpha)^{\text{II}} = \mathbf{F}^{2\alpha}. \tag{15.30}$$

Built with an optical arrangement $\mathcal{G}(\mathbf{A}^{\text{II}};\mathbf{F}^\alpha) \approx \mathcal{F}^\alpha$, the aberrations $\mathbf{A}^{\text{II}}$ of the apparatus will be projected by $\mathbf{E}^k(\mathbf{F}^\alpha)$ to a lower-dimensional manifold: in rank k, there are k restrictions on the $2k+1$ aberrations. This can be seen cleanly for aberration order 3 (rank $k = 2$), where (15.24) has no multilinear terms in lower-rank coefficients and 2 out of 5 aberrations are zero. For fractional Fourier transformers and higher-order aberrations similar results follow.

Remark: Fourier basis of aberrations of $\mathbf{F}^\alpha$-transformers. In the first example of this section, the imager *cum* reflection in (15.25) referred to $\mathcal{F}^4 = \mathcal{F}^0 = \mathit{1}$ where, as we saw, $\mathbf{E}^k(\mathbf{1}) = \mathbf{1} + \mathbf{D}^k(\mathbf{K})$ maps to zero the k aberrations $A^{\text{II}}_{k,m}$ with $k - m$ odd, such as coma and distorsion, and duplicates the remaining $k{+}1$ even aberrations. In the Fourier transformer *cum* reflection of the second example of this section, $\mathbf{E}^k(\mathbf{F})$ was the sum of $\mathbf{1}$ plus the antidiagonal matrix $\mathbf{D}^k(\mathbf{F}\,\mathbf{K})$, with $\mathbf{F}\,\mathbf{K} = -\left(\begin{smallmatrix} 0 & 1 \\ 1 & 0 \end{smallmatrix}\right)$, whose result is in (15.28): the k coefficient combinations $A^{\text{II}}_{k,m} - A^{\text{II}}_{k,-m}$ are zero. To extend this result to systems *cum* reflection that are fractional Fourier transformers, it is best to work in the Fourier aberration basis (13.9), where this transform is represented by the diagonal matrix $\widetilde{\mathbf{F}}^\alpha := \mathbf{B}\,\mathbf{F}^\alpha\,\mathbf{B}^{-1}$, obtained through similarity with the complex symplectic Bargmann–Segal transform matrix of (C.44), i.e. $\mathbf{B} := \frac{1}{\sqrt{2}}\left(\begin{smallmatrix} 1 & -i \\ -i & 1 \end{smallmatrix}\right)$. In that basis, $\widetilde{\mathbf{F}}^\alpha\widetilde{\mathbf{K}}$ is an antidiagonal matrix, and so is $\mathbf{D}^k(\widetilde{\mathbf{F}}^\alpha\widetilde{\mathbf{K}})$. •

15.4 Afterword

It would not serve the purpose of this book to go into finer detail on fractional Fourier transform arrangements, or indeed their use in optical signal analysis, which has been the title and field of one recent and excellent text [112]. Our concern has been to understand the symmetries and aberrations of Fourier transformers in geometric optics because, as detailed in appendix C, there is a precise correspondence in the paraxial régime with the Fourier *integral* transform for wave optics. Geometric and wave theories also mantain correspondence under the nonlinear Hamilton–Lie transformations of spherical aberration and coma, generated by functions $f(p)$ and $q\,g(p)$, and by $f(q)$ and $p\,g(q)$ for pocus (Gaussian phase) and distorsion, – but under the composition of these infinite-parameter algebras, the correspondence breaks down into quantization-rule and operator-ordering problems.

For 'arbitrary' aberrations, the relation between geometric and wave theories complicates, and is not 1:1. One result which brings some order into the problem of 'wavizing' aberrations, is that under the *linear* transformations $\mathbf{D}^k(\mathbf{M})$ that participate in the product of two aberration group elements (13.31)–(13.33), the classical Poisson operators mantain their correspondence *only* with the *Weyl*–ordered operators[3] [54, 148]. The composition of two pure rank-K aberrations in geometric optics through the *gato* operation $\sharp$ with Poisson brackets (13.41)–(13.43), *differs* from the commutator Lie brackets of the corresponding operators, by terms with powers of the wavelength λ^k, $2 \le k \le K$. One should in principle resort to experiment to settle any ordering-rule ambiguity. But symmetries *will* hold fast in the leap between geometric and wave models. So we should expect that geometric Fourier transformers, corrected by Lie methods, will provide better Fourier integral transformers.

---

[3] The Weyl order of the noncommuting operator factors in a monomial is the sum of all their permutations as individual objects, divided by the factorial of their number.

References

1. I.D. Ado, The representation of Lie algebras by matrices, *Usphekhi Mat. Nauk* **3**, 159–173 (1947); English translation: *Amer. Math. Soc. Transl.* No. 2, *ibid.* (1949).
2. S.T. Ali, N.M. Atakishiyev, S.M. Chumakov, and K.B. Wolf, The Wigner function for general Lie groups and the wavelet transform, *Ann. H. Poincaré* **1**, 685–714 (2000).
3. M.A. Alonso, Radiometry and wide-angle fields. I. Coherent fields in two dimensions, *J. Opt. Soc. Am. A* **18**, 902–909 (2001); II. Coherent fields in three dimensions, *ibid.*, 910–918; III. Partial coherence, *ibid.*, 2501–2511.
4. Arvind, B. Dutta, N. Mukunda, and R. Simon, The real symplectic groups in quantum mechanics and optics, *Pramana* **45**, 471–513 (1995); `arXiv:quant-ph/9509002v3`.
5. N.M. Atakishiyev, W. Lassner, and K.B. Wolf, The relativistic coma aberration. I. Geometrical optics, *J. Math. Phys.* **30**, 2457–2462 (1989).
6. N.M. Atakishiyev W. Lassner and K.B. Wolf, The relativistic coma aberration. II. Helmholtz wave optics, *J. Math. Phys.* **30**, 2463–2468 (1989).
7. N.M. Atakishiyev, S.M. Chumakov, and K.B. Wolf, Wigner distribution function for finite systems, *J. Math. Phys.* **39**, 6247–6261 (1998).
8. E.J. Atzema, *The Structure of Systems of Lines in 19th Century Geometrical Optics: Malus' Theorem and the Description of the Infinitely Thin Pencil*, Ph.D. Thesis, Rijksuniversiteit te Utrecht, 1993.
9. E.J. Atzema, G. Krötzsch and K.B. Wolf, Canonical transformations to warped surfaces: correction of aberrated optical images, *J. Phys. A* **30**, 5793–5803 (1997).
10. H. Bacry, The de Sitter group $L_{4,1}$ and the bound states of the hydrogen atom, *N. Cimento* **41A**, 222–234 (1966).
11. H. Bacry and M. Cadilhac, Metaplectic group and Fourier optics, *Phys. Rev. A* **23**, 2533–2536 (1981).
12. A. Baker, *Matrix groups. An Introduction to Lie Group Theory*, Springer Undergraduate Texts in Mathematics Series (Springer-Verlag, London, 2002).
13. M. Bander and C. Itzykson, Group theory and the hydrogen atom (I), *Rev. Mod. Phys.* **38**, 330–345 (1966); (II) *ibid.*, 346–358.
14. V. Bargmann, Zur Theorie des Wasserstoffatoms (Bemerkungen zur gleichnamigen Arbeit von V. Fock), *Z. Phys.* **99**, 576–582 (1936).
15. V. Bargmann, Irreducible unitary representations of the Lorentz group, *Ann. Math.* **48**, 568–642 (1947).
16. V. Bargmann, On a Hilbert space of analytic functions and an associated integral transform, Part I, *Commun. Pure Appl. Math.* **14**, 187–214 (1961);

(II) A family of related function spaces and application to distribution theory, *ibid.*, 1–101.

17. V. Bargmann, Group representation in Hilbert spaces of analytic functions. In: *Analytical Methods in Mathematical Physics*, Ed. by P. Gilbert and R.G. Newton (Gordon & Breach, New York, 1970), pp. 27–63.
18. H.O. Bartelt, K.-H. Brenner and H. Lohmann, The Wigner distribution function and its optical production, *Opt. Comm.* **32**, 32–38 (1980); H. Bartelt and K.-H. Brenner, The Wigner distribution function: an alternate signal representation in optics, *Israel J. Techn.* **18**, 260–262 (1980); K.-H. Brenner and H. Lohmann, Wigner distribution function display of complex 1D signals, *Opt. Comm.* **42**, 310–314 (1982).
19. A.O. Barut, *Dynamical Groups and Generalized Symmetries in Quantum Theory* (Univ. of Canterbury, Christchurch, N.Z., 1972).
20. D. Basu and K.B. Wolf, The unitary irreducible representations of SL(2,R) in all subgroup reductions, *J. Math. Phys.* **23**, 189–205 (1982).
21. L.C. Biedenharn and J.D. Louck, *Angular Momentum in Quantum Physics, Theory and Application*, Encyclopedia of Mathematics and Its Applications, Vol. 8, Ed. by G.-C. Rota (Addison-Wesley, Reading, MA, 1981).
22. G. Birkhoff, Lie groups isomorphic with no linear group, *Bull. Amer. Math. Soc.* **42**, 882–888 (1936).
23. *Dynamical Groups and Spectrum Generating Algebras*, Vols. 1 & 2, Ed. by A. Bohm, Y. Ne'eman, and A.O. Barut (World Scientific, Singapore, 1988).
24. C.P. Boyer and K.B. Wolf, Deformations of inhomogeneous classical Lie algebras to the algebras of the linear groups, *J. Math. Phys.* **14**, 1853–1859 (1973); *ib.*, The algebra and group deformations $I^m[SO(n) \otimes SO(m)] \Rightarrow SO(n,m)$, $I^m[U(n) \otimes U(m)] \Rightarrow U(n,m)$, and $I^m[Sp(n) \otimes Sp(m)] \Rightarrow Sp(n,m)$, for $1 \leq m \leq n$, *J. Math. Phys.* **15**, 2096–2100 (1974).
25. H.A. Buchdahl, *Optical Aberration Coefficients* (Dover, New York, 1968).
26. H.A. Buchdahl, *An Introduction to Hamiltonian Optics* (Cambridge Univ. Press, Cambridge UK, 1970).
27. H.A. Buchdahl, Kepler problem and Maxwell fish-eye, *Am. J. Phys.* **46**, 840–843 (1978).
28. J.F. Cariñena and J. Nasarre, On symplectic structures arising from geometric optics, *Fortschr. Phys.* **44**, 181–198 (1996).
29. P. Carruthers and M.M. Nieto, Phase and angle variables in quantum mechanics, *Rev. Mod. Phys.* **40**, 411–440 (1968).
30. É. Cartan, *Sur la Structure des Groupes Finis et Continus*, Ph.D. Thesis, Université de Paris, 1894; see: *É. Cartan, Œuvres complètes* (Gauthier-Villars, Paris, 1952).
31. O. Castaños, E. López-Moreno and K.B. Wolf, Canonical transforms for paraxial wave optics. In: *Lie Methods in Optics*, Lecture Notes in Physics, Vol. 250, Ed. by J. Sánchez-Mondragón and K.B. Wolf (Springer-Verlag, Heidelberg, 1986), pp. 159–182.
32. C.C. Chevalley, *Theory of Lie Groups* (Princeton Univ. Press, Princeton, N.J., 1946).
33. S.M. Chumakov, V.G. Kadyshevsky, and K.B. Wolf, Conformal optical systems. In: *Proceedings of the VII International Conference on Symmetry Methods in Physics*, Vol. 1, Ed. by A.N. Sissakian and G.S. Pogosyan (Joint Institute for Nuclear Research, Dubna, 1996), pp. 95–110.

34. E.U. Condon, Immersion of the Fourier transform in a continuous group of functional transformations, *Proc. Nat. Acad. Sci.* **23**, 158–164 (1937).
35. B. Cordani, *The Kepler problem. Group Theoretical Aspects, Regularization and Quantization, with Application to the Study of Perturbation.* Progress in Mathematical Physics Book Series, Vol. 29 (Birkhäuser, Basel, 2002).
36. M.L. Curtis, *Matrix Groups* (Springer-Verlag, New York, 2nd Ed., 1984).
37. A. Deprit, Canonical transformations depending on a small parameter, *Celestial Mech.* **1**, 12–30 (1969); A. Deprit, The main problem of artificial satellite theory for small and moderate excentricities, *Celestial Mech.* **2**, 166–206 (1970).
38. P.A.M. Dirac, *The Principles of Quantum Mechanics* (Oxford Univ. Press, London, 1st ed. 1930; 4th ed. 1958).
39. P.A.M. Dirac, A remarkable representation of the 3+2 de Sitter group, *J. Math. Phys.* **4**, 901–909 (1963).
40. A.J. Dragt and J. Finn, Lie series and invariant functions for analytic symplectic maps, *J. Math. Phys.* **17**, 2215–2227 (1976).
41. A.J. Dragt, Lie algebraic theory of geometrical optics and optical aberrations, *J. Opt. Soc. Am.* **72**, 372–379 (1982).
42. A.J. Dragt, *Lectures on Nonlinear Orbit Dynamics*, AIP Conference Proceedings N° 87 (American Institute of Physics, New York, 1982).
43. A.J. Dragt, E. Forest, and K.B. Wolf, Foundations of a Lie algebraic theory of geometrical optics. In: *Lie Methods in Optics*, Lecture Notes in Physics, Vol. 250, Ed. by J. Sánchez-Mondragón and K.B. Wolf (Springer-Verlag, Heidelberg, 1986).
44. A.J. Dragt, Elementary and advanced Lie algebraic methods with applications to accelerator design, electron microscopes, and light optics, *Nucl. Instr. Meth. Phys. Res. A* **258**, 339–354 (1987).
45. A.J. Dragt, *Lie Methods for Nonlinear Dynamics with Applications to Accellerator Physics*, (University of Maryland) `http://www.physics.umd.edu/dsat/dsatliemethods.html`
46. C. Eckart, The application of group theory to the quantum dynamics of monoatomic systems, *Rev. Mod. Phys.* **2**, 305–380 (1930).
47. M.J. Englefield, *Group Theory and the Coulomb Problem* (Wiley-Interscience, New York, 1972).
48. A. Erdélyi, W. Magnus, F. Oberhettinger, and F.G. Tricomi, *Higher Transcendental Functions*, 3 Vols. (McGraw-Hill, New York, 1953–1955).
49. H. Fassbender, *Symplectic Methods for the Symplectic Eigen-problem* (Kluwer Academic, New York, 2000).
50. V. Fock, Zur Theorie des Wasserstoffatoms, *Z. Phys.* **98**, 145–154 (1935); V.A. Fock, Wasserstoffatom und nicht-euklidische Geometrie, *Izv. Akad. Nauk SSSR* **2**, 169–188 (1935), abstract in German.
51. A. Frank, F. Leyvraz, and K.B. Wolf, Hidden symmetry and potential group of the Maxwell fish-eye, *J. Math. Phys.* **31**, 2757–2768 (1990).
52. C. Fronsdal, Elementary particles in a curved space, *Rev. Mod. Phys.* **37**, 201–223 (1965).
53. W. Fulton and J. Harris, *Representation Theory. A First Course*, Graduate Texts in Mathematics, Vol. 129 (Springer-Verlag, London, 1997).
54. M. García-Bullé, W. Lassner, and K.B. Wolf, The metaplectic group within the Heisenberg-Weyl ring, *J. Math. Phys.* **27**, 29–36 (1986).

55. I.M. Gel'fand, *Lectures on Linear Algebra* (Interscience, New York, 1961).
56. R. Gilmore, *Lie Groups, Lie Algebras, and Some of their Applications* (Wiley Interscience, New York, 1978).
57. H. Goldstein, Prehistory of the Runge-Lenz vector, *Amer. J. Phys.* **43**, 737–738 (1975).
58. H. Goldstein, *Classical Mechanics* (Addison Wesley, Reading, MA, 2nd ed. 1980).
59. G. Golub and C.F. van Loan, *Matrix Computations* (John Hopkins Univ. Press, Baltimore, MD, 1983).
60. P. González-Casanova and K.B. Wolf, Interpolation of solutions to the Helmholtz equation, *Numerical Meth. Part. Diff. Eq.* **11**, 77–91 (1995).
61. I.S. Gradshteyn and I.M. Ryzhik, *Tablitsy Integralov, Summ, Ryadov i Proizvedeniĭ* (Nauka, Moscow, 1971); *ib.*, *Tables of Integrals, Series and Products* (Academic Press, San Diego CA, 6th ed. 2000).
62. V. Guillemin and S. Sternberg, *Symplectic Techniques in Physics* (Cambridge University Press, 1984).
63. P. Günter, *Sophus Lie.* In: *100 Jahre Mathematisches Seminar der Karl-Marx-Universität Leipzig*, Ed. by H. Beckert and H. Schumann (VEB Deutscher Verlag der Wissenschaften, Berlin, 1981), pp. 111–133.
64. F. Gürsey, Color quarks and octonions. In: *Current Problems in High Energy Particle Theory*, Ed. by G. Domokos and S. Kövesi-Domokos (John Hopkins Univ., Baltimore, MD, 1974), pp. 15–42.
65. B.C. Hall, *Lie Groups, Lie Algebras, and Representations*, Graduate Texts in Mathematics, Vol. 222 (Springer-Verlag, London, 2003.
66. W.R. Hamilton, On the application of the method of quaternions to some dynamical questions, *Proc. Roy. Irish Acad.* **3**, 441–448 (1847).
67. M. Hávliček and W. Lassner, Canonical realizations of the Lie algebra sp$(2n, R)$, *Int. J. Theor. Phys.* **15**, 867–890 (1976).
68. R. Herman, *Vector Bundles in Mathematical Physics* (Benjamin, New York, 1970).
69. G. Hori, Theory of general perturbations with unspecified canonical variables, *Publ. Astron. Soc. Japan* **18**, 287–296 (1966); *ib.*, Theory of general perturbations, *Recent Advances in Dynamical Astronomy*, Ed. by B.B. Tapley and V. Szebehely (Reidel, Dordrecht, 1973), pp. 231–249.
70. D.D. Holm and K.B. Wolf, Lie-Poisson description of Hamiltonian ray optics, *Physica D* **51**, 189–199 (1991).
71. T. Husain, *Introduction to Topological Groups* (Saunders, Philadelphia, 1966).
72. E. Inönü and E.P. Wigner, On the contraction of groups and their representations, *Proc. Natl. Acad. Sci. U.S.* **39**, 510–524 (1953).
73. K. Iwasawa, On the representation of Lie algebras, *Japan J. Math.* **19**, 513–523 (1948); *ib.*, On some types of topological groups, *Ann. Math.* **50**, 507–558 (1948).
74. E.G. Kalnins, *Separation of Variables for Riemannian Spaces of Constant Curvature* (Longman Scientific & Technical, Longman House, UK, 1986).
75. E.G. Kalnins, W. Miller Jr., and G.S. Pogosyan, Completeness of multiseparable superintegrability on the complex 2-sphere, *J. Phys. A* **33**, 6791–6806 (2000); *ib.*, Complete sets of invariants for dynamical systems that admit a separation of variables, *J. Phys. A* **43**, 3592–3609 (2002).
76. A. Katz, *Classical Mechanics, Quantum Mechanics, Field Theory* (Academic Press, New York, 1965).

77. M. Kauderer, *Symplectic Matrices, First Order Systems and Special Relativity* (World Scientific, Singapore, 1994).
78. S.A. Khan and K.B. Wolf, Hamiltonian orbit structure of the set of paraxial optical systems, *J. Opt. Soc. Am. A* **19**, 2436–2444 (2002).
79. Y.S. Kim and M.E. Noz, *Theory and Applications of the Poincaré Group* (Reidel, Dordrecht, 1986).
80. J.R. Klauder and E.C.G. Sudarshan, *Fundamentals of Quantum Optics* (Benjamin, New York, 1968).
81. H. Kogelnik, On the propagation of Gaussian beams through lenslike media including those with a loss or gain variation, *Appl. Opt.* **4**, 1562–1569 (1965).
82. P. Kramer, M. Moshinsky, and T.S. Seligman, Complex extensions of canonical transformations in quantum mechanics. In *Group Theory and its Applications*, Vol. III, Ed. by E.M. Loebl (Academic Press, New York, 1975).
83. Yu.A. Kravtsov and Yu.I. Orlov, *Geometrical Optics of Inhomogeneous Media* (Springer-Verlag, Berlin, 1990).
84. V. Lakshminarayanan and A.K. Ghatak, *Lagrangian Optics* (Kluwer, Boston, 2001).
85. A.J. Laub and K. Meyer, Canonical forms for symplectic and Hamiltonian matrices, *Celestial Mech.* **9**, 213–238 (1974).
86. R. Lenz, *Group Theoretical Methods in Image Processing*, Lecture Notes in Computer Science, Vol. 413 (Springer-Verlag, Heidelberg, 1987).
87. J.-M. Lévy-Leblond, Galilei Group and Galilean Invariance. In: *Group Theory and Its Applications*, Vol. II, Ed. by E.M. Loebl (Academic Press, New York, 1971), pp. 221–299.
88. S. Lie, Die infinitesimalen Berührungstransformationen der Optik, *Berichte über die Verhandlungen der Königlich Sächsischen Akademie del Wissenschaften zu Leipzig. Mathematisch-Physische Klasse* **47**, 131–133 (1896).
89. E. López Moreno and K.B. Wolf, De la ley de Snell-Descartes a las ecuaciones de Hamilton en el espacio fase de la óptica geométrica, *Rev. Mex. Fís.* **35**, 291–300 (1989).
90. R.K. Luneburg, *Mathematical Theory of Optics* (Univ. of California, Berkeley CA, 1964).
91. B. Macukow and H.H. Arsenault, Matrix decompositions for nonsymmetrical optical systems, *J. Opt. Soc. Am.* **73**, 1360–1366 (1983).
92. V.I. Man'ko and K.B. Wolf, The mapping between Heisenberg-Weyl and Euclidean optics is comatic. In *Lie Methods in Optics, II Workshop*, Lecture Notes in Physics, Vol. 352, Ed. by K.B. Wolf (Springer-Verlag, Heidelberg, 1989), pp. 163–197.
93. J.E. Marsden, E. Weinstein, and P. Holmes, *Geometry, Mechanics & Dynamics* (Springer-Verlag, New York, 2002).
94. J.C. Maxwell, Solution of a problem, *Cambridge and Dublin Math. J.* **8**, 188-193 (1854).
95. H.V. McIntosh, Symmetry and Degeneracy. In: *Group Theory and Its Applications*, Vol. II, Ed. by E.M. Loebl (Academic Press, New York, 1971), pp. 75–144.
96. B. Mielnik and J.F. Plebański, Combinatorial approach to Baker-Campbell-Hausdorff exponents, *Ann. Inst. H. Poincaré* **12**, 215–254 (1970).
97. W. Miller Jr., *Symmetry and Separation of Variables*, Encyclopedia of Mathematics and Its Applications, Vol. 4, Ed. by G.-C. Rota (Addison-Wesley, Reading, MA, 1977).

98. C.B. Moler and C.F. van Loan, Nineteen dubious ways to compute the exponential to a matrix, *SIAM Review* **20**, 801–836 (1979).
99. P.M. Morse and H. Feshbach, *Methods of Theoretical Physics*, 2 Vols. (McGraw-Hill, New York, 1953).
100. M. Moshinsky, *Group Theory and the Many Body Problem* (Gordon & Breach, New York, 1967); See also: M. Moshinsky and Yu.F. Smirnov, *The Harmonic Oscillator in Modern Physics* (Hardwood Academic Publ., The Netherlands, 1996).
101. M. Moshinsky and C. Quesne, Oscillator Systems. In: *Proceedings of the 15th Solvay Conference in Physics (1970)* (Gordon & Breach, New York, 1974); *ib.*, Linear canonical transformations and their unitary representation, *J. Math. Phys* **12**, 1772–1780 (1971); C. Quesne and M. Moshinsky, Canonical transformations and matrix elements, *J. Math. Phys* **12**, 1780–1783 (1971); M. Moshinsky, Canonical transformations and quantum mechanics, *SIAM J. Appl. Math.* **25**, 193–203 (1973).
102. M. Moshinsky, T.H. Seligman and K.B. Wolf, Canonical transformations and the radial oscillator and Coulomb problems, *J. Math. Phys.* **13**, 901–907 (1972).
103. M. Moshinsky, J. Patera, R.T. Sharp, and P. Winternitz, Everything you ever wanted to know about $SU(3) \supset O(3)$, *Ann. Phys.* **95**, 139–160 (1975).
104. M. Moshinsky and P. Winternitz, Quadratic Hamiltonians in phase space and their eigenstates, *J. Math. Phys.* **21**, 1667–1682 (1980).
105. V. Namias, The fractional order Fourier transform and its application to quantum mechanics, *J. Inst. Maths. Applics.* **25**, 241–265 (1980); *ib.*, Fractionalization of the Hankel transform, *J. Inst. Maths. Applics.* **26**, 187–197 (1981).
106. M. Navarro Saad, *Cálculo de aberraciones en sistemas ópticos con teoría de grupos.* B.Sc. Thesis, Universidad Nacional Autónoma de México, 1984.
107. M. Navarro Saad and K.B. Wolf, Applications of a factorization theorem for ninth–order aberration optics, *J. Symbolic Comp.* **1**, 235–239 (1985).
108. M. Navarro Saad and K.B. Wolf, Factorization of the phase-space transformation produced by an arbitrary refracting surface, *J. Opt. Soc. Am. A* **3**, 340–346 (1986).
109. M. Nazarathy and J. Shamir, First order systems — a canonical operator representation: lossless systems, *J. Opt. Soc. Am. A* **72**, 356–364 (1982).
110. Y. Ne'eman, *Algebraic Theory of Particle Physics* (Benjamin, New York, 1967).
111. O.T. O'Meara, *Symplectic Groups* (American Mathematical Society, Providence, R.I., 1982).
112. H.M. Ozaktas, Z. Zalevsky, and M. Alper Kutay, *The Fractional Fourier Transform, with Applications in Optics and Signal Processing* (Wiley, Chichester, 2001).
113. J. Patera, R.T. Sharp, P. Winternitz, and H. Zassenhaus, Continuous subgroups of the fundamental groups of physics. III. The de Sitter groups, *J. Math. Phys.* **18**, 2259–2288 (1977).
114. J. Patera, P. Winternitz, and H. Zassenhaus, Maximal abelian subalgebras of real and complex symplectic Lie algebras, *J. Math. Phys.* **24**, 1973–1985 (1983).
115. T.O. Philips y E.P. Wigner, De Sitter space and positive energy. In: *Group Theory and its Applications*, Ed. by E.M. Loebl (Academic Press, New York, 1968), pp. 631–676.

116. G.S. Pogosyan and P. Winternitz, Separation of variables and subgroup bases on the N-dimensional hyperboloid, *J. Math. Phys.* **43**, 3387–3410 (2002).
117. M.A. Preston, *Physics of the Nucleus* (Addison-Wesley, Reading, MA, 1962).
118. R.K. Srinivasa Rao, *The Rotation and Lorentz Groups, and their Representations for Physicists* (Wiley, New Delhi, 1988).
119. R. Rashed, A pioneer in anaclastics —Ibn Sahl on burning mirrors and lenses, *ISIS* **81**, 464–491 (1990); R. Rashed, *Géometrie et dioptrique au Xe siècle: Ibn Sahl, al-Qūhī, et Ibn al-Hayatham* (Collection Sciences et Philosophie Arabes, Textes et Études, Les Belles Lettres, Paris, 1993).
120. A.L. Rivera, S.M. Chumakov, and K.B. Wolf, Hamiltonian foundation of geometrical anisotropic optics, *J. Opt. Soc. Am. A* **12**, 1380–1389 (1995); *ib.*, Optica geométrica *vs.* física en medios anisotrópicos, *Rev. Mex. Fís.* **43**, 1027–1043 (1997).
121. B.A. Rosenfel'd and I.D. Sergeeva, *Stereograficheskaya Proektsiya* (Nauka, Moscow, 1973).
122. C. Runge, *Vektoranalysis* (English translation: Dutton, New York, 1919).
123. W.P. Schleich, *Quantum Optics in Phase Space* (Wiley, London, 2001).
124. I.E. Segal, Distributions in Hilbert spaces and canonical systems of operators, *Trans. Amer. Math. Soc.* **88**, 12–42 (1958).
125. L. Seidel, Zur Dioptrik, *Astronomische Nachrichten* **871**, 105–120 (1853).
126. L.S. Shively, *An Introduction to Modern Geometry* (Wiley, New York, 1939).
127. C.L. Siegel, Symplectic geometry, *Amer. J. Math.* **65**, 1–86 (1943).
128. R. Simon, E.C.G. Sudarshan, and N. Mukunda, Anisotropic Gaussian Schell-model beams: Passage through optical systems and associated invariants, *Phys. Rev. A* **31**, 2419–2434 (1985).
129. R. Simon and N. Mukunda, Iwasawa decomposition in first-order optics: universal treatment of shape-invariant propagation for coherent and partially coherent beams, *J. Opt. Soc. Am. A* **15**, 2146–2155 (1998).
130. R. Simon and K.B. Wolf, Structure of the set of paraxial optical systems, *J. Opt. Soc. Am. A* **17**, 342–355 (2000).
131. R. Simon and K.B. Wolf, Fractional Fourier transforms in two dimensions, *J. Opt. Soc. Am. A* **17**, 2368–2381 (2000).
132. J.-M. Souriau, *Structure des Systèmes Dynamiques* (Dunod, Paris, 1970); *ib.* Géometrie symplectique et physique mathématique, *Gazette des Mathématiciens* **10**, 90–130 (1978); *ib. Structure of Dynamical Systems; a Symplectic View of Physics* (Birkhäuser, Boston, MA, 1997).
133. O. Stavroudis, *The Optics of Rays, Wavefronts, and Caustics* (Academic Press, New York, 1978).
134. S. Steinberg and K.B. Wolf, Invariant inner products on spaces of solutions of the Klein–Gordon and Helmholtz equations, *J. Math. Phys.* **22**, 1660–1663 (1981).
135. S. Steinberg, Lie series, Lie transformations and their applications. In *Lie Methods in Optics*, Lecture Notes in Physics, Vol. 250, Ed. by J. Sánchez-Mondragón and K.B. Wolf (Springer-Verlag, Heidelberg, 1986), pp. 45–103.
136. E.C.G. Sudarshan, N. Mukunda and R. Simon, Realization of first order optical systems using thin lenses, *Optica Acta* **32**, 855–872 (1985).
137. G.F. Torres del Castillo and F. Aceves de la Cruz, Connection between the Kepler problem and the Maxwell fish-eye, *Rev. Mex. Fís.* **44**, 546–549 (1998).
138. B.L. van der Waerden, The classification of simple Lie groups, *Math. Z.* **37**, 446–462 (1933).

139. A. Weinstein, *Lectures on Symplectic Manifolds* (University of California / American Mathematical Society, Providence, R.I., 1977).
140. H. Weyl, Gruppentheorie und Quantenmechanik, *Z. Phys.* **46**, 1–46 (1928).
141. H. Weyl, *Gruppentheorie und Quantenmechanik* (Hirzel, Leipzig, 1st ed., 1928; 2nd ed., 1931). Translated by H.P. Robertson as *The Theory of Groups and Quantum Mechanics* (Methuen, London, 1931; Dover, New York, 1949).
142. H. Weyl, *The Classical Groups: Their Invariants and Representations* (Princeton Univ. Press, Princeton, NJ, 2nd ed. 1946).
143. E.P. Wigner, *Group Theory and its Applications to the Quantum Mechanics of Atomic Spectra* (Academic Press, New York, 1959), translation by J.J. Griffin of the 1931 German edition.
144. E.P. Wigner, On the quantum correction for thermodynamic equilibrium, *Phys. Rev.* **40**, 749-759 (1932); M. Hillery, R.F. O'Connell, M.O. Scully, and E.P. Wigner, Distribution functions in physics: fundamentals, *Phys. Rep.* **259**, 121–127 (1984); H.-W. Lee, Theory and applications of the quantum phase-space distribution functions, *Phys. Rep.* **259**, 147–211 (1995).
145. K.B. Wolf, Dynamical groups for the point rotor and the hydrogen atom, *Suppl. N. Cimento* **5**, 1041–1050 (1967).
146. K.B. Wolf, A recursive method for the calculation of the SO_n, $SO_{n,1}$, and ISO_n representation matrices, *J. Math. Phys.* **12**, 197–206 (1971).
147. K.B. Wolf, Canonical transforms. I. Complex linear transforms, *J. Math. Phys.* **15**, 1295–1301 (1974); II. Complex radial transforms, *ibid.* **15**, 2101–2111 (1974).
148. K.B. Wolf, The Heisenberg-Weyl Ring in Quantum Mechanics. In *Group Theory and its Applications*, Vol. III, Ed. by E.M. Loebl (Academic Press, New York, 1975), pp. 190–247.
149. K.B. Wolf *Integral Transforms in Science and Engineering* (Plenum, New York, 1979).
150. K.B. Wolf, Canonical transforms. IV. Hyperbolic transforms: continuous series of SL(2,R) representations, *J. Math. Phys.* **21**, 680–688 (1980).
151. K.B. Wolf, Symmetry in Lie optics, *Ann. Phys.* **172**, 1–25 (1986).
152. K.B. Wolf, The group–theoretical treatment of aberrating systems. II. Axis–symmetric inhomogeneous systems and fiber optics in third aberration order, *J. Math. Phys.* **27**, 1458–1465 (1986).
153. K.B. Wolf, Symmetry-adapted classification of aberrations, *J. Opt. Soc. Am. A* **5**, 1226–1232 (1988).
154. K.B. Wolf, Nonlinearity in aberration optics. In: *Symmetries and Nonlinear Phenomena*, Proceedings of the International School of Applied Mathematics (Centro Internacional de Física, Paipa, Colombia), Ed. by D. Levi and P. Winternitz (World Scientific, Singapore, 1988), pp. 376–429.
155. K.B. Wolf, Elements of Euclidean optics. In: *Lie Methods in Optics*, Lecture Notes in Physics, Vol. 352, Ed. by K.B. Wolf (Springer-Verlag, Heidelberg, 1989), pp. 115–162.
156. K.B. Wolf, The Fourier transform in metaxial geometric optics, *J. Opt. Soc. Am. A* **8**, 1399–1403 (1991).
157. K.B. Wolf, Refracting surfaces between fibers, *J. Opt. Soc. Am. A* **8**, 1389–1398 (1991).
158. K.B. Wolf and G. Krötzsch, Group-classified polynomials of phase space in higher-order aberration expansions, *J. Symbolic Comp.* **12**, 673–695 (1991).

159. K.B. Wolf, The Euclidean root of Snell's law. I. Geometric polarization optics, *J. Math. Phys.* **33**, 2390–2408 (1992).
160. K.B. Wolf, Relativistic aberration of optical phase space, *J. Opt. Soc. Am. A* **10**, 1925–1934 (1993).
161. K.B. Wolf and G. Krötzsch, `mexLIE 2`, *A set of symbolic computation functions for geometric aberration optics* (in `muSIMP`), Manuales IIMAS-UNAM No. 10, 1995, 103 pp.
162. K.B. Wolf, M.A. Alonso, and G.W. Forbes, Wigner functions for Helmholtz wavefields, *J. Opt. Soc. Am. A* **16**, 2476–2487 (1999).
163. K.B. Wolf and G. Krötzsch, El problema de las tres lentes, *Rev. Mex. Fís.* **47**, 291–298 (2001).
164. K.B. Wolf, Porqué y cómo exponenciamos matrices hamiltonianas, *Rev. Mex. Fís.* **49**, 465–476 (2003).
165. C.E. Wulfman, Dynamical Groups in Atomic and Molecular Physics. In: *Group Theory and its Applications*, Vol. 2, Ed. by E.M. Loebl (Academic Press, New York, 1971).

Index

Printing: Mercedes-Druck, Berlin
Binding: Stein+Lehmann, Berlin

Texts and Monographs in Physics

Series Editors: R. Balian W. Beiglböck H. Grosse E. H. Lieb
N. Reshetikhin H. Spohn W. Thirring

Essential Relativity Special, General, and Cosmological Revised 2nd edition
By W. Rindler

The Elements of Mechanics
By G. Gallavotti

Generalized Coherent States and Their Applications
By A. Perelomov

Quantum Mechanics II
By A. Galindo and P. Pascual

Geometry of the Standard Model of Elementary Particles
By. A. Derdzinski

From Electrostatics to Optics
A Concise Electrodynamics Course
By G. Scharf

Finite Quantum Electrodynamics
The Causal Approach 2nd edition
By G. Scharf

Path Integral Approach to Quantum Physics An Introduction
2nd printing By G. Roepstorff

Supersymmetric Methods in Quantum and Statistical Physics By G. Junker

Relativistic Quantum Mechanics and Introduction to Field Theory
By F. J. Ynduráin

Local Quantum Physics Fields, Particles, Algebras 2nd revised and enlarged edition
By R. Haag

The Mechanics and Thermodynamics of Continuous Media By M. Šilhavý

Quantum Relativity A Synthesis of the Ideas of Einstein and Heisenberg
By D. R. Finkelstein

Scattering Theory of Classical and Quantum *N*-Particle Systems
By. J. Derezinski and C. Gérard

Effective Lagrangians for the Standard Model By A. Dobado, A. Gómez-Nicola, A. L. Maroto and J. R. Peláez

Quantum The Quantum Theory of Particles, Fields, and Cosmology By E. Elbaz

Quantum Groups and Their Representations
By A. Klimyk and K. Schmüdgen

Multi-Hamiltonian Theory of Dynamical Systems By M. Błaszak

Renormalization An Introduction
By M. Salmhofer

Fields, Symmetries, and Quarks
2nd, revised and enlarged edition By U. Mosel

Statistical Mechanics of Lattice Systems
Volume 1: Closed-Form and Exact Solutions
2nd, revised and enlarged edition
By D. A. Lavis and G. M. Bell

Statistical Mechanics of Lattice Systems
Volume 2: Exact, Series and Renormalization Group Methods
By D. A. Lavis and G. M. Bell

Conformal Invariance and Critical Phenomena By M. Henkel

The Theory of Quark and Gluon Interactions
3rd revised and enlarged edition
By F. J. Ynduráin

Quantum Field Theory in Condensed Matter Physics By N. Nagaosa

Quantum Field Theory in Strongly Correlated Electronic Systems
By N. Nagaosa

Information Theory and Quantum Physics
Physical Foundations for Understanding the Conscious Process By H.S. Green

Magnetism and Superconductivity
By L.-P. Lévy

The Nuclear Many-Body Problem
By P. Ring and P. Schuck

Perturbative Quantum Electrodynamics and Axiomatic Field Theory By O. Steinmann

Quantum Non-linear Sigma Models
From Quantum Field Theory to Supersymmetry, Conformal Field Theory, Black Holes and Strings By S. V. Ketov

>springeronline.com

Texts and Monographs in Physics

Series Editors: R. Balian W. Beiglböck H. Grosse E. H. Lieb N. Reshetikhin H. Spohn W. Thirring

The Statistical Mechanics of Financial Markets 2nd Edition By J. Voit

Statistical Mechanics A Short Treatise
By G. Gallavotti

Statistical Physics of Fluids
Basic Concepts and Applications
By V. I. Kalikmanov

Many-Body Problems and Quantum Field Theory An Introduction 2nd Edition
By Ph. A. Martin and F. Rothen

Foundations of Fluid Dynamics
By G. Gallavotti

High-Energy Particle Diffraction
By E. Barone and V. Predazzi

Physics of Neutrinos
and Applications to Astrophysics
By M. Fukugita and T. Yanagida

Relativistic Quantum Mechanics
By H. M. Pilkuhn

The Geometric Phase in Quantum Systems
Foundations, Mathematical Concepts, and Applications in Molecular and Condensed Matter Physics
By A. Bohm, A. Mostafazadeh, H. Koizumi, Q. Niu and J. Zwanziger

The Atomic Nucleus as a Relativistic System
By L.N. Savushkin and H. Toki

The Frenkel–Kontorova Model
Concepts, Methods, and Applications
By O. M. Braun and Y. S. Kivshar

Aspects of Ergodic, Qualitative and Statistical Theory of Motion
By G. Gallavotti, F. Bonetto and G. Gentile

Quantum Entropy and Its Use
By M. Ohya and D. Petz

General Relativity
With Applications to Astrophysics
By N. Straumann

Coherent Dynamics of Complex Quantum Systems By V. M. Akulin

Geometric Optics on Phase Space
By K. B. Wolf

>springeronline.com

Printed in the United States
137707LV00006B/16/P

9 783540 220398

This book was once
read by a
"Beautiful Soul"